Fundamentals of Brain Network Analysis

Fundamentals of Brain Network Analysis

Alex Fornito

Monash University, Australia

Andrew Zalesky

The University of Melbourne, Australia

Edward T Bullmore

University of Cambridge, United Kingdom

AMSTERDAM • BOSTON • HEIDELBERG • LONDON
NEW YORK • OXFORD • PARIS • SAN DIEGO
SAN FRANCISCO • SINGAPORE • SYDNEY • TOKYO

Academic Press is an imprint of Elsevier

Academic Press is an imprint of Elsevier
125 London Wall, London, EC2Y 5AS, UK
525 B Street, Suite 1800, San Diego, CA 92101-4495, USA
50 Hampshire Street, 5th Floor, Cambridge, MA 02139, USA
The Boulevard, Langford Lane, Kidlington, Oxford OX5 1GB, UK

Notices
Knowledge and best practice in this field are constantly changing. As new research and experience broaden
our understanding, changes in research methods, professional practices, or medical treatment may become
necessary.

Practitioners and researchers must always rely on their own experience and knowledge in evaluating
and using any information, methods, compounds, or experiments described herein. In using such
information or methods they should be mindful of their own safety and the safety of others, including
parties for whom they have a professional responsibility.

To the fullest extent of the law, neither the Publisher nor the authors, contributors, or editors, assume
any liability for any injury and/or damage to persons or property as a matter of products liability,
negligence or otherwise, or from any use or operation of any methods, products, instructions, or ideas
contained in the material herein.

Library of Congress Cataloging-in-Publication Data
A catalog record for this book is available from the Library of Congress

British Library Cataloguing in Publication Data
A catalogue record for this book is available from the British Library

ISBN: 978-0-12-407908-3

For information on all Academic Press publications
visit our website at www.elsevier.com

Printed in USA

Working together
to grow libraries in
developing countries

www.elsevier.com • www.bookaid.org

Fundamentals of Brain Network Analysis

Alex Fornito
Monash University, Australia

Andrew Zalesky
The University of Melbourne, Australia

Edward T Bullmore
University of Cambridge, United Kingdom

ELSEVIER

AMSTERDAM • BOSTON • HEIDELBERG • LONDON
NEW YORK • OXFORD • PARIS • SAN DIEGO
SAN FRANCISCO • SINGAPORE • SYDNEY • TOKYO
Academic Press is an imprint of Elsevier

Academic Press is an imprint of Elsevier
125 London Wall, London, EC2Y 5AS, UK
525 B Street, Suite 1800, San Diego, CA 92101-4495, USA
50 Hampshire Street, 5th Floor, Cambridge, MA 02139, USA
The Boulevard, Langford Lane, Kidlington, Oxford OX5 1GB, UK

Library of Congress Cataloging-in-Publication Data
A catalog record for this book is available from the Library of Congress

British Library Cataloguing in Publication Data
A catalogue record for this book is available from the British Library

ISBN: 978-0-12-407908-3

For information on all Academic Press publications
visit our website at www.elsevier.com

Printed in USA

Working together
to grow libraries in
developing countries

www.elsevier.com • www.bookaid.org

Contents

Author Biographies

Alex Fornito

Alex Fornito completed a PhD in the Departments of Psychology and Psychiatry at the University of Melbourne, Australia, followed by Post-Doctoral training at the University of Cambridge, UK. He is an associate professor, Australian Research Council Future Fellow, and Deputy Director of the Brain and Mental Health Laboratory in the Monash Institute of Cognitive and Clinical Neurosciences, Australia. Alex's research uses cognitive neuroscience, network science, and graph theory to understand brain network organization in health and disease. He has published over 100 scientific articles, much of which are focused on the development and application of new methods to understand how brain networks dynamically adapt to changing task demands, how they are disrupted by disease, and how they are shaped by genetic influences.

Andrew Zalesky

Andrew Zalesky completed his PhD in the Department of Electrical and Electronic Engineering at the University of Melbourne, Australia. He works with neuroscientists, utilizing his engineering expertise in networks to understand human brain organization in health and disease. He has developed widely used methods for modeling and performing statistical inference on brain imaging data. His methods are utilized to investigate brain connectivity abnormalities in disease. He identified some of the first evidence of connectome pathology in schizophrenia. Andrew currently holds a fellowship from the National Health and Medical Research Council of Australia. He is based at the University of Melbourne and holds a joint appointment between the Melbourne Neuropsychiatry Centre and the Melbourne School of Engineering. He leads the Systems Neuropsychiatry Group.

Edward Bullmore

Ed Bullmore trained in medicine at the University of Oxford and St Bartholomew's Hospital, London, and then in psychiatry at the Bethlem Royal and Maudsley Hospital, London. In 1993, he was a Wellcome Trust (Advanced) Research Fellow at the Institute of Psychiatry, King's College London, where he completed a PhD in the statistical analysis of MRI data, before moving to Cambridge as Professor of Psychiatry in 1999. Currently, he is co-Chair of Cambridge Neuroscience, Scientific Director of the Wolfson Brain Imaging Centre, and Head of the Department of Psychiatry in the University of Cambridge. He is also an honorary Consultant Psychiatrist, and Director of R&D in Cambridgeshire and Peterborough Foundation NHS Trust. Since 2005, he has worked half-time for GlaxoSmithKline, currently focusing on immuno-psychiatry. He has been elected as a Fellow of the Royal College of Physicians, the Royal College of Psychiatrists, and the Academy of Medical Sciences. He has published about 500 scientific papers, and his work has been highly cited. He has played an internationally-leading role in understanding brain connectivity and networks by graph theoretical analysis of neuroimaging and other neuroscientific datasets.

Foreword

For over a century, the notion that individual nerve cells or neurons are the basic structural and functional units of the nervous system has been the bedrock of neuroscience. Probably the most influential contribution came from the work of Ramón y Cajal on brain anatomy and histology, which revealed a staggering diversity of neuronal cell types and morphologies arranged into intricate circuitry. With astounding intuition, and preceding the discovery of the basic physiological mechanisms of neuronal communication, Ramón y Cajal not only sketched the anatomical arrangement of the cellular elements of neuronal circuits, he also devised functional "wiring diagrams" that specified the direction and flow of neural signals. Ramón y Cajal is now recognized as a major architect of the "neuron doctrine," which continues to be foundational for all of modern neuroscience.

A very different alternative framework was the reticular theory, whose main proponent was Cajal's antagonist Camillo Golgi. The theory was based on an alternative model of how neurons were anatomically and functionally related—instead of forming discrete elements, Golgi's view was that neurons formed a continuum, by being physically conjoined into a single "network" (or reticulum). Even as the evidence supporting the neuron doctrine accumulated, Golgi persevered in this view, as he could not fathom how discrete elements could support integrative function. As he stated in his 1906 Nobel lecture, "However opposed it may seem to the popular tendency to individualize the elements, I cannot abandon the idea of a unitary action of the nervous system."

As we now know, Golgi's conception of neurons as forming a continuous syncytium was wrong. And yet, in pointing to the gap between the organization of the brain into individual nerve cells on one side and its integrative activity on the other, Golgi articulated a fundamental problem that has animated theoretical neuroscience ever since. In a sense, Golgi's reticular theory was an early antecedent of models that attempted to account for neural integration and computation, as exemplified in the speculative ideas of early connectionists and later in mathematical theories of neural networks. While networks, often in the guise of "neural circuits," have been a long-standing intellectual current in theoretical neuroscience, until fairly recently it was quite difficult, if not impossible, to directly observe and measure the anatomical and functional networks of the brain. The necessary sophisticated tools for mapping extended anatomical networks and for recording functional brain activity across large neuronal populations, or even the whole brain, were slow to emerge.

All this has changed in recent years. New technologies for mapping and recording large-scale neuronal systems have finally arrived, creating an abundance of data on the anatomical layout and functional dynamics of neural systems. As a result, neuroscience is rapidly transitioning into an era of "big data." This transition creates a number of new opportunities. For example, the development of new methods for mapping anatomical connections in entire nervous systems now allows the construction of comprehensive structural networks or connectomes, across species and individuals. In parallel, the development of a broad range of neurophysiological, optical, and magnetic resonance imaging techniques now enables continuous recording of the dynamics of neural activity across hundreds of neurons, or indeed the whole brain. How can we make sense of these rich and comprehensive anatomical and physiological data sets? How can we extract principles of how brains are structurally and dynamically organized?

These are the challenges that this volume attempts to address. Its focus is on the application of the tools and methods of network science, which has already proven extraordinarily fruitful and has gathered significant momentum over the past decade. This volume is the first to offer a comprehensive overview of network modeling and analysis techniques to brain data, both structural and functional. The book fills the growing need for a systematic and didactic treatment of the major types of network measures and models that are relevant to neuroscience research.

The book's basic plan goes from a brief introduction of major concepts and terms to a more in-depth consideration of measures that capture nodal statistics (degree and strength), to other ways of expressing centrality and influence in a network (centrality), and characterizing important network elements (hubs). Then, the book turns to a discussion of how network elements cooperate as network cores or rich clubs, before introducing models that characterize the efficiency with which neural information is communicated in brain networks, according to different models based on shortest path routing, diffusion, and greedy navigation. It then examines how nodes and edges form stereotypic processing units (motifs), and how network organization is shaped by principles related to the conservation of time, space, and material; principles that were first described by Ramón y Cajal over a century ago. A major avenue in contemporary network analysis is the decomposition of large networks into smaller communities or modules, usually on the basis of their dense internal connectivity. Here, the book goes into considerable depth by providing an overview of some of the classic as well as the very latest approaches for identifying network modules. Finally, important chapters address issues related to null models (an emerging methodological focus in network science) and methods for comparing and classifying networks. The latter will be particularly appreciated by readers interested in applying network approaches to characterizing differences in network organization associated with developmental changes or clinical conditions.

The book does much more than survey the methodological underpinnings of the burgeoning field of connectomics. Importantly, it manages to translate the often highly technical jargon of networks into language that connects with the practice and terminology of empirical neuroscience. While comprehensive in its treatment of each specific topic, the material is presented in a style that makes network methods accessible to the average neuroscientist who is interested in tapping into the enormous potential of network science for unlocking principles of brain organization. The book not only makes network science approachable, it addresses the many domain-specific issues that surround brain networks such as the construction of networks from appropriately chosen nodal partitions and edge measures, or the construction of null models that are both rigorous and biologically plausible. These discussions are generally missing from other more abstract accounts of complex networks, and yet they are critically important for generating and analyzing neurobiologically meaningful network models. This focus on network neuroscience will be especially appreciated by students who are eager to learn the basics of this rapidly evolving new field.

As neuroscience data continue to grow in volume and complexity, this book fills an important gap by providing a set of theoretically grounded tools that can address at least some of the big data challenges facing the discipline today. I am certain that this much-needed primer on brain networks will become an indispensable addition to the bookshelves of all neuroscientists interested in the organization and function of nervous systems, from cellular to systems scales.

Olaf Sporns, PhD
Distinguished Professor, Robert H. Shaffer Chair
Indiana University

Acknowledgments

Alex Fornito is thankful to his coauthors for their hard work and perseverance in putting this book together; to the various mentors and collaborators over the course of his career who have had a profound influence on his understanding of the brain; and to his students and postdocs, in particular Ben Fulcher, Nigel Rogasch, Linden Parkes, Ari Pinar, and Stuart Oldham, who provided helpful feedback on preliminary drafts. He acknowledges the ongoing support of the Australian Research Council and National Health and Medical Research Council in funding his work. Finally, he expresses his eternal gratitude to his family, especially Olinda and little Nicholas, whose love, support, and patience have made this project a reality.

Andrew Zalesky is grateful for the wonderfully rich discussions with colleagues that have shaped his thinking about brain networks, much of which is reflected in these pages. He is thankful for the support of the staff and students at the Melbourne Neuropsychiatry Centre, especially those who helped him get started in a new faculty and a new field. He is also pleased to acknowledge the continuing support of the Australian Research Council, the National Health and Medical Research Council, the University of Melbourne, and the many sponsors and friends who have supported his research activities over the years.

Ed Bullmore thanks the many colleagues and students who have contributed to his own understanding of brain network analysis over the years; funding agencies including the Wellcome Trust, the Medical Research Council, the National Institute of Health Research, and the National Institute of Health Graduate Partnership Program, that have materially supported his work on connectomics; and all the inhabitants of our home on Clarendon Street, Cambridge, for their love and encouragement.

CHAPTER 1

An Introduction to Brain Networks

It is often said that the brain is the most complex network known to man. A human brain comprises about 100 billion (10^{11}) neurons connected by about 100 trillion (10^{14}) synapses, which are anatomically organized over multiple scales of space and functionally interactive over multiple scales of time. This vast system is the biological hardware from which all our thoughts, feelings, and behavior emerge. Clinical disorders of human brain networks, like dementia and schizophrenia, are among the most disabling and therapeutically intractable global health problems. It is therefore unsurprising that understanding brain network connectivity has long been a central goal of neuroscience, and has recently catalyzed an unprecedented era of large-scale initiatives and collaborative projects to map brain networks more comprehensively and in greater detail than ever before (Bohland et al., 2009; Kandel et al., 2013; Van Essen and Ugurbil, 2012). As we will see, one of the implications of modern brain network science is that the human brain may not, in fact, be a uniquely complex system. However, it is certainly timely, challenging and important to understand its organization more clearly.

Central to current thinking about brain networks is the concept of the **connectome**. This word was first coined in 2005 by Olaf Sporns, Giulio Tononi, and Rolf Kötter (2005) and independently in a PhD dissertation by Patric Hagmann (2005) to define a **matrix** representing all possible pairwise anatomical connections between neural elements of the brain (Figure 1.1). The term connectome, in the first and strictest sense of the word, thus stands for an ideal or canonical state of knowledge about the cellular wiring diagram of a brain. The truly exponential growth of research in this area in the last 10 years has led to investigations of a more general concept of the connectome that includes the matrix of anatomical connections between large-scale brain areas as well as between individual neurons; and the matrix of functional interactions that is revealed by the analysis of physiological processes unfolding as slowly as the fluctuations of cerebral blood oxygenation measured with **functional magnetic resonance imaging** (MRI; spanning frequencies below 0.1 Hz), or as fast as the high-frequency neuronal oscillations detectable with invasive and noninvasive electrophysiology (over 500 Hz; see also Chapter 2). A consistent conceptual focus on quantifying, visualizing, and understanding brain network organization

1

Fundamentals of Brain Network Analysis. http://dx.doi.org/10.1016/B978-0-12-407908-3.00001-7

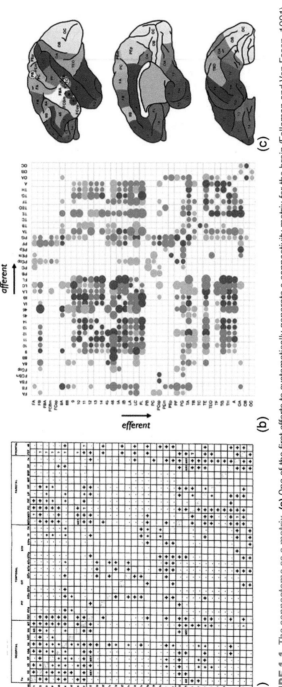

FIGURE 1.1 The connectome as a matrix. **(a)** One of the first efforts to systematically generate a connectivity matrix for the brain (Felleman and Van Essen, 1991). This matrix represents the connectivity of 32 neocortical areas involved in visual function in the macaque monkey, constructed by collating the results of a large number of published **tract-tracing** studies in this animal. In this matrix, a black cross indicates an outgoing projection from the region listed in the row to the region listed in the column. **(b)** An updated connectivity matrix of the macaque comprising 39 cortical areas, as reconstructed from an online database of tract-tracing studies. This matrix is organized such that colored elements represent a projection from the region listed in the column to the region listed in the row (see Chapter 3). The size of the dots in each matrix element is proportional to the projection distance and darker colors indicate stronger average reported connectivity strength. **(c)** The anatomical locations of the areas listed in the matrix in **(b)**. Darker colors identify regions with higher total connectivity to the rest of the network. *(a) Reproduced from Felleman and Van Essen (1991) and (b, c) from Scholtens et al. (2014) with permission.*

across multiple scales of space and time is a fundamental characteristic of the burgeoning field of **connectomics** (Bullmore and Sporns, 2009).

The relatively recent birth of connectomics should not be interpreted as evidence of a prior lack of neuroscientific interest in brain networks. In fact, many nineteenth and twentieth century neuroscientists—like Ramón y Cajal, Golgi, Meynert, Wernicke, Flechsig, and Brodmann—were well aware of the importance of connectivity and networks in understanding nervous systems. These and other foundational neuroscientists made seminal discoveries and wrote down enduring conceptual insights that have since underpinned the way that we think about nervous systems.

So, what's new? Why do we need new words to label a neuroscientific program that is arguably as old as neuroscience itself? Why now for the connectome and connectomics? Is it just a passing fad, a fashionable blip in professional jargon? Or are there more fundamental factors that can explain why the connectome has exploded as a distinctive focus for neuroscience in the twenty-first century?

In our view, there are two convergent factors driving the scientific ascendancy of connectomics. First, recent years have witnessed rapid growth in the science of networks in general. Since the 1980s there have been major conceptual developments in the statistical physics of complex networks and ever-wider applications of network science to the analysis and modeling of big data. New ways have been found of quantifying the topological complexity of large systems of interacting agents, and striking commonalities have been observed in the organizational properties of a broad array of real-life networks, including, but not limited to, air transportation networks, microchip circuits, the internet, and brains.

The second factor driving the growth of connectomics is the technological evolution of methods to measure and visualize brain organization, across multiple scales of resolution. Since the 1990s there has been significant progress in human neuroimaging science, especially using MRI to map whole brain anatomical and functional networks at macroscopic scale ($\sim$1-10 mm^3, order of 10^{-2} m) in healthy volunteers and patients with neurological and psychiatric disorders (Bullmore and Sporns, 2009; Fornito et al., 2015). In the last 10 years, there have also been spectacular methodological developments in **tract tracing**, optical microscopy, optogenetics, multielectrode recording, histological gene expression, and many other neuroscience techniques that can now be used to map brain systems at mesoscopic ($\sim$$10^{-4}$ m) and microscopic scales ($\sim$$10^{-6}$ m), under more controlled experimental conditions, and in a wider range of species (Kennedy et al., 2013; Oh et al., 2014; Chung and Deisseroth, 2013; Fenno et al., 2011).

The convergence of these two powerful trends—(1) the mathematical and conceptual developments in complex network science; and (2) the evolution of technologies for measuring nervous systems—is the crux of what motivates and is distinctively characteristic of the new field of connectomics. This book is about how we can apply the science of complex networks to understand brain connectivity. In particular, we focus on the use of **graph theory** to model, estimate and simulate the **topology** and dynamics of brain networks. Graph theory is a branch of mathematics concerned with understanding systems of interacting elements. A graph is used to model such systems simply as a set of **nodes** linked by **edges**. This representation is remarkably flexible and, despite its formal simplicity, can be used to investigate many aspects of brain organization in diverse kinds of data.

In this introductory chapter, we provide a motivation for why graph theory is useful for understanding brain networks, and offer a brief historical overview of how brain graphs have become a key tool in systems neuroscience. This background provides context for the subsequent chapters, which concentrate in more technical detail on specific aspects of graph theory and their application to connectomic analysis of neuroscientific data.

1.1 GRAPHS AS MODELS FOR COMPLEX SYSTEMS

Complex systems have properties that are neither completely random nor completely regular, instead showing nontrivial characteristics that are indicative of a more elaborate, or complex, organization. Such systems are all around us, and range from societies, economies, and ecosystems, to infrastructural systems, information processing networks, and molecular interactions occurring within biological organisms (Barabási, 2002). These are all big systems—often comprising millions of agents interacting with each other—and they are represented by very diverse kinds of data. It is only in the last 20 years or so that it has become mathematically tractable and scientifically interesting to quantify this daunting complexity.

As methods have been developed to deal with such data, and as these methods have been applied more widely in different domains or fields, it has become clear that superficially different systems—such as friendship networks, metabolic interaction pathways, and very large-scale integrated computer circuits—can express remarkably general properties in terms of their network organization (Albert and Barabási, 2002; Newman, 2003a). From these developments, an interdisciplinary field of network science has formed around the use of general analytic methods to model complex networks, and to explore the scope of common or near-universal principles of network organization, function, growth, and evolution. Principal among these general methods is graph theory.

1.1.1   A Brief History of Graph Theory

The first use of a graph to understand a real-world system is widely credited to the Swiss mathematician Leonhard Euler (1707-1783). In 1735, Euler lived in the Prussian town of Königsberg (now the Russian city of Kaliningrad), which was built around seven bridges across the river Pregel, linking the two main riverbanks and two islands in the middle of the river (Figure 1.2a). An unresolved problem at that time was whether it was possible to walk around the town via a route that crossed each bridge once and only once. Euler solved this problem by representing the four land masses divided by the river as nodes, and the seven bridges as interconnecting edges (Figure 1.2b). From this proto- typical graph, he was able to show that no more than two nodes (the start and end points of the walk) should have an odd number of edges connecting them to the rest of the graph for such a walk to be possible. In fact, all four nodes in the Königsberg graph had an odd number of edges, meaning that it was impos- sible to find any route around the city that crossed each and every bridge only once. In this way, Euler proved once and for all that the system of bridges and islands that comprised the city was organized such that the "Königsberg walk" was topologically prohibited.

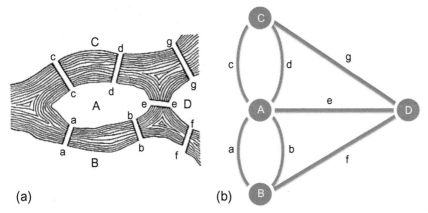

(a)   (b)

FIGURE 1.2 The origins of graph theory: the first topological analysis by Euler. **(a)** Simplified geographic map of the Prussian town of Königsberg, which comprised four landmasses (marked A–D) linked by seven bridges (marked a–f). The specific problem that Euler solved was whether it was possible to find a **path** that crossed each bridge once and only once. The geographical map lavishes detail on features like the shape of the islands, and the currents in the river, that are completely irrelevant to solving this problem. **(b)** Graphical representation of the problem. Euler's topological analysis successfully simplified the system as a **binary graph**—with nodes for landmasses and edges for bridges—and focused on the **degree**, or number of edges, that connected each node to the graph. **(a)** *Reproduced from Kraitchik (1942) with permission.*

The importance of Euler's analysis is *not* in the details of the geography of eighteenth century Königsberg; rather it is important precisely because it successfully ignored so many of those details and focused attention on what later became known as the topology of the problem. The topology of a graph defines how the **links** between system elements are organized. Indeed, this is exactly what Euler's graph focuses on; the network of bridges connecting islands and riverbanks. It is not informed by any other physical aspect of Königsberg, such as the physical distance (length) of the bridges, the distances between them, and so on. More generally, the results of this and any other topological analysis will be invariant under any continuous spatial transformation of the system. To see this, imagine taking the physical map of Königsberg depicted in Figure 1.2a and increasing or decreasing its physical scale, or rotating it, or reflecting it, or stretching it. None of these (or any other) continuous spatial transformations will have any effect on the number of bridges connecting any particular island to the rest of the town.

Topology developed strongly as a field of mathematics from the late nineteenth century, preceding important developments in the statistical analysis of graphs in the twentieth century. In the 1950s, this work was spearheaded by Paul Erdös and Alfred Rényi, who introduced an influential statistical model for generating random graphs and for predicting some of their topological properties (Erdös and Rényi, 1959; see Bollobás, 1998 for overview). In an **Erdös-Rényi graph**, there are N nodes and a uniform probability p of each possible edge between them. If p is close to one, the graph is densely connected and if p is close to zero, the graph is sparsely connected. Erdös and Rényi showed that many important properties of these graphs, such as the mean number of connections attached to any single node (also called the mean degree of the graph), and whether the graph is a single **connected component** or contains isolated nodes (which are not connected to other nodes), could be predicted analytically from their **generative model** (Chapters 4, 6, and 10).

Both the Königsberg graph invented by Euler and the random graphs generated by the Erdös-Rényi model are examples of the simplest class of graphs: binary undirected graphs. They are binary graphs because the edges are either absent or present or, equivalently, the edge weight is either zero or one. They are **undirected graphs** because the edges connect nodes symmetrically; no distinction is made between the source and target of a connection. The principles of topological analysis have since been extended to more sophisticated graphs that include both weighted and directed connectivity (Chapter 3). As we will see in later chapters, these extensions are particularly important for characterizing certain kinds of brain network data.

A critical step from the mathematics of random graph theory to the physics of complex networks was taken by Duncan Watts and Steven Strogatz (Watts and Strogatz, 1998; Figure 1.3). Like Erdös and Rényi, they defined a generative model for graphs; but they began their analysis with a simple regular **lattice** of N nodes, each connected directly to an arbitrary number of other nodes. The **Watts-Strogatz model** then randomly selects an edge connecting nodes i and j in the lattice and incrementally rewires the graph so that this edge now connects node i to another randomly selected node, h, such that $h \neq j$. This generative process of random mutation of connectivity can be applied to each edge with an arbitrary probability of rewiring p_{WS}, so that when $p_{WS} = 1$, all the edges have been randomly rewired and the lattice has been topologically transformed to an Erdös-Rényi random graph (further details of these models are considered in Chapter 10).

Watts and Strogatz (1998) focused on two key properties of their network model: the **clustering coefficient** and the **characteristic path length**. The clustering coefficient provides an index of the cliquishness or clustering of connectivity in a graph, and is the probability that two nodes each directly connected to a third node will also be directly linked to each other (Chapter 8). The characteristic path length is commonly used to index the integrative capacity of a network and is a measure of the topological distance between nodes, computed as the minimum number of edges required to link any two nodes in the network, on average (Chapter 7). The intuition is that a shorter average **path length** results in more rapid and efficient integration across the network (Latora and Marchiori, 2001). Random graphs have a short characteristic path length and low clustering. On the other hand, the regular lattice analyzed by Watts and Strogatz has high clustering and long characteristic path length (Figure 1.3).

The first critical discovery revealed by computer simulations of the Watts-Strogatz model was that the rate of change in path length was much faster than the rate of change in clustering, as the probability of rewiring an edge in the lattice was progressively increased from zero towards one. Specifically, changing just a few edges in the lattice dramatically decreased the characteristic path length of the graph, but did not greatly reduce the high average clustering that characterized the lattice. In other words, there was a range of rewiring probabilities that generated graphs with a hybrid combination of topological properties: short path length, like a random graph, and high clustering, like a lattice. By analogy to the qualitatively similar properties of social networks, first explored by Milgram (1967), these nearly-regular and nearly-random graphs were called **small-world** networks. The second main discovery reported by Watts and Strogatz was based on empirical analysis. They measured the path length and clustering of graphs representing three real-life systems and found that the small-world combination of greater-than-random clustering with nearly-random path length was characteristic of all three: a social network (costarring movie actors), an infrastructural network (an electrical power grid), and the neuronal network of the nematode worm, *Caenorhabditis elegans*.

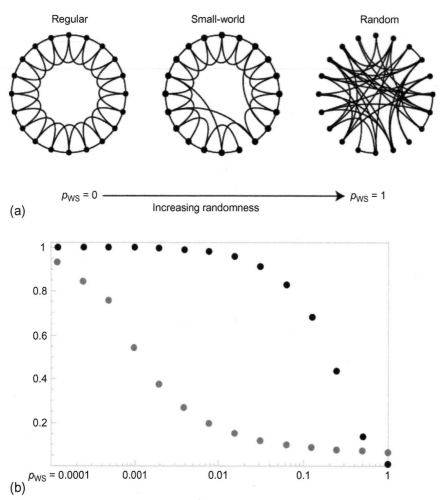

FIGURE 1.3 Small-world networks. **(a)** The seminal work of Watts and Strogatz (1998) identified a continuum of network topologies ranging from completely regular and lattice-like (left) to completely random (right). Interposed between these extremes is a class of networks with a so-called small-world topology, which can be generated by randomly rewiring (with probability, p_{WS}) an arbitrary proportion of edges in a regular network. **(b)** Watts and Strogatz found that rewiring just a few edges (small p_{WS}) led to a dramatic reduction in path length (red line), whereas high clustering was more resilient to random rewiring (black line), resulting in a regime of rewiring probability in which the network showed high clustering, like a lattice, and low path length, like a random network. The red and black lines correspond to the average clustering coefficient and average path length, respectively, of the graph at a given p_{WS}, divided by the corresponding value computed in a comparable lattice. The combination of high clustering and short path length is the defining characteristic of small-world networks and has since been described in many real-world systems. *Images reproduced from Watts and Strogatz (1998) with permission.*

At about the same time, Barabási and Albert (1999) introduced another generative model that built a complex graph by adding nodes incrementally (Chapter 10). In this model, as each new node i is added to the graph, the probability that it will form a connection or edge to any other node, j, is proportional to the number of connections, or degree, of node j. In other words, new nodes connect preferentially to existing nodes that already have a large number of connections and thus represent putative network **hubs**. By this generative process of **preferential attachment**, the "rich get richer," or nodes that have high degree initially tend to have even higher degree as the graph grows by iterative addition of new nodes. As a result, the distribution of degree across network nodes is not the unimodal Poisson-like distribution that is characteristic of the Erdös-Rényi model; instead it is characteristically fat-tailed, conforming to what is called a **scale-free** or power-law distribution. A scale-free **degree distribution** means that the probability of finding a very high degree hub node in the graph is greater than would be expected if the degree distribution was unimodal, like a Poisson or Gaussian function. More simply, it is likely that a scale-free network will contain at least a few highly connected hub nodes (Chapters 4 and 5). Barabási and Albert also found evidence of power-law degree distributions in several empirically observed complex systems.

Like the apparent ubiquity of small-worldness, the fact that so many substantively different systems share the property of scale-free degree distributions suggested that some key topological principles might be nearly universal for a large class of complex networks (Barabási and Albert, 1999). However, it is important to appreciate that the empirical ubiquity of small-worldness or scale-free degree distributions does not by itself mean that random edge rewiring or preferential attachment is the generative mechanism that built all these systems in real life. Systems with complex topological properties, like small-worlds and hubs, can be generated by many different growth rules.

A third major development in the application of graph theory to real-world systems has been the discovery that such networks are modular—they can be nearly decomposed into subsets of nodes that are more densely connected to each other than to nodes in any other **modules** (Simon, 1962). Following work by Mark Newman, Michelle Girvan, others (Girvan and Newman, 2002; Newman and Girvan, 2004; Newman, 2004c; 2006; for a detailed review, see Fortunato, 2010), the quantification of network **modularity** has become a large and rapidly developing area of complex network science (Chapter 9). For example, the global airline network, where nodes are airports and edges are direct flights between them, has a hierarchical modular structure, in which modules can be further decomposed into submodules and so on (Guimerà et al., 2005; Figure 1.4). Each topological module of the airline network corresponds approximately to a geographical continent or political territory, like the

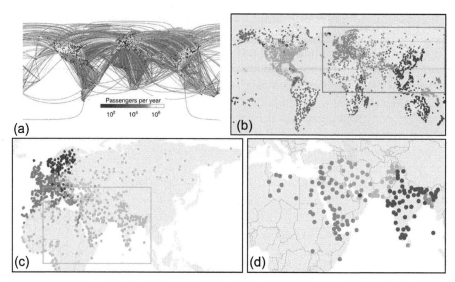

FIGURE 1.4 Hierarchical modular organization of the global air transportation network. **(a)** A geographical representation of the worldwide airline network. Black dots represent airports (nodes), and colored lines represent passenger traffic between airports (edges). **(b)** This network has a hierarchical modular structure, such that modules can be further divided into submodules and so on. Shown here is the highest level of the module hierarchy, plotted on a geographic map of the world. Colors correspond to different modules. The modular organization is strongly dominated by geographic location, such that airports in the same continent are in the same module. This is consistent with the fact that most flights link airports in the same country or continent. **(c)** The next level in the hierarchy, focusing on the orange Eurasian module (box) in **(b)**. This module now splits into different submodules, such as Scandinavia, central Europe, Western Russia, the Middle East, and North Africa. **(d)** The next level in the hierarchy, focusing on the Middle-Eastern submodule in **(c)**. Again, we see a tendency for airports to segregate into sub-submodules (like India) according to their geographic locations and political affiliations. *(a) Reproduced from Grady et al. (2012) with permission. (b-d) reproduced from Sales-Pardo et al. (2007), Copyright (2007) National Academy of Sciences, U.S.A., with permission.*

United States or the European Union. This represents the familiar experience that most flights from a US or EU airport are to other airports in the same territory or continental land mass; only a few big airports, corresponding to high degree hubs, such as New York JFK and London Heathrow, have many intercontinental flights. In a network with high topological modularity, the density of intramodular connections is much greater than the density of intermodular connections. Typically most of the intermodular communications are mediated by a few so-called connector hubs that link different modules (Guimerà et al., 2005). It turns out that many real-life systems share this topological property of modularity, again suggesting that it represents a near universal characteristic of complex networks (Simon, 1962).

These three key concepts of graph topology—small-worldness, degree distribution, and modularity—are the tip of an iceberg of complex network science. Additionally, there has been growing interest in the dyadic subdivision of a network into a relatively small core or **rich club** of highly interconnected high degree hubs and a larger periphery of sparsely interconnected lower degree nodes (Colizza et al., 2006; Chapter 6). There has also been important work to identify the topological **motifs** of a network: basic building blocks of connection profiles between small sets of three or four nodes that recur in a network with a frequency that is greater than expected by chance (Chapter 8). Graphs further provide a powerful approach for simulating the effects of damage to a network by studying how global topological properties, such as network connectedness or characteristic path length, are affected as the nodes or edges of a graph are computationally deleted (Chapter 6). Most complex networks are fairly resilient to random attack on their nodes, but are much more vulnerable to a targeted attack that prioritizes the highest degree hub nodes (Albert et al., 2000). For example, if the global airline network was attacked one airport at a time, but the choice of which node to attack next was made at random, a very large number of airports would need to be disabled before intercontinental traffic was affected. This is because only a small number of airports service long-haul flights. However, if the attacks were targeted on those few major hub airports, like JFK or London Heathrow, this would be equivalent to removing most of the intermodular flights between the US and EU modules. The result would be a dramatic increase in the number of flights required to link two cities on different continents (i.e., increased path length) and potentially a **fragmentation** of the network into two or more completely isolated modules. In this way, hubs can increase the vulnerability of many complex networks to targeted attack (Chapter 6).

1.1.2 Space, Time, and Topology

As is hopefully becoming clear, topology is an important aspect of how many networks are organized; but other dimensions must also be considered in the analysis of most, if not all, types of networks. Some complex networks, like the World Wide Web or the semantic web of Shakespeare's sonnets (Motter et al., 2002), are quite purely topological: they don't really exist in space (the web), or space and time (sonnets). There are, however, many other complex networks, particularly biological networks such as the brain, that are embedded in spatial dimensions and are dynamically active over time. For brain network analysis, the familiar three-plus-one dimensions of space and time must thus be incorporated with the more novel fifth dimension of topology.

For spatial networks generally there will be inevitable constraints on how the topological plan can actually be built in the three dimensions of the world (Barthélemy, 2011). For example, to build a high-performance computer chip, each processing node or logic gate must be physically wired to a number of other nodes according to a topologically complex design for high performance. It is also important that the total amount of wiring used should be as small as possible, because wiring is expensive and generates thermal noise. Furthermore, it is usually mandated that the topology must be embedded in only two dimensions, on the surface of a silicon chip. Empirical analysis and generative modeling of computer circuits and other spatial networks has indicated that conservation of wiring **cost** is an important factor in network formation that often drives the physical location and connectivity of nodes, such that spatially proximal nodes have a higher probability of connectivity than spatially distant nodes (Christie and Stroobandt, 2000). However, minimization of wiring cost is clearly not the only selection pressure, otherwise all spatial networks would be low-dimensional lattices with topologically clustered nodes embedded as close to each other as possible in physical space. Accordingly, generative models that posit a trade-off or competition between cost minimization and some other topological factor that provides functional benefits, have been more successful in simulating the organization of spatial networks (Vértes et al., 2012). Such economical principles of competition between physical cost and topological value may be generally influential in the formation of networks embedded in space (Latora and Marchiori, 2001; Achard and Bullmore, 2007; Chapter 8).

Most networks will also be active over time with dynamics that are related to the functional performance of the system. Perhaps unsurprisingly, the structural topology of a network has an important influence on the functional dynamics that emerge from interactions between nodes over time. Networks with complex topology have complex dynamics, broadly speaking. For example, networks displaying high dynamical complexity—that is, dynamics which are neither fully segregated nor fully integrated—show a complex, small-world topology (Sporns et al., 2000; see Chapter 8).

It has also been shown that small-world and scale-free network topologies are associated with the emergence of so-called critical dynamics (Chialvo, 2010). Self-organized critical dynamics are often inferred when functional interactions between nodes exist at all scales of space and time encompassed by the system and are statistically distributed as power laws (Chapter 4). Topologically complex networks have been linked to the emergence of such scale-invariant network dynamics, which are consistent with the system being in a self-organized critical state that is interposed between completely ordered and disordered dynamics (Bak et al., 1987). Critical dynamics have been shown computationally to have advantages for information processing and memory, and experimentally they have been inferred from power-law scaling of dynamics in many nervous systems, from cultured cellular networks to whole brain electrophysiological and haemodynamic recordings (Linkenkaer-Hansen et al., 2001; Beggs and Plenz, 2003; Kitzbichler et al., 2009). In general, network topology plays an important role in constraining system dynamics; and, reciprocally, system dynamics can often drive the evolution or development of network topology.

1.2 GRAPH THEORY AND THE BRAIN

As we have seen, graph theory has played an integral role in recent efforts to understand the structure and function of complex systems. Nervous systems are undoubtedly complex, so it is natural to assume that graph theory may also prove useful for neuroscience. Importantly, graph-based representations of brain networks—brain graphs—can easily be constructed from neural connectivity matrices, such as the ones depicted in Figure 1.1. Each row or column representing a different brain region in the matrix is drawn as a node in the graph, and the values in each matrix element are drawn as edges. In fact, as we will see throughout this book, matrix and graph representations of a network are formally equivalent, and much of the mathematics of graph theory is applied through the analysis of matrices. In this section, we consider how graph theory has been applied to understand brain networks and how it has emerged as a powerful analytic tool for connectomics.

1.2.1 The Neuron Theory and Connectivity at the Microscale

When did the ideas of graph theory and network science first begin to permeate neuroscience? Formally, the first applications of graph theory to neuroscientific data were not published until the end of the twentieth century (Felleman and Van Essen, 1991; Young, 1992; Watts and Strogatz, 1998). However, the seminal neuron theory, established by Ramón y Cajal's brilliant microscopic studies and theoretical thinking in the late nineteenth and early twentieth century (Ramón y Cajal, 1995), set the scene for graph theory to later make sense as a model of nervous systems (Figure 1.5). Ramón y Cajal and some others, using the then-revolutionary technique of silver impregnation to visualize the complex branching processes of individual neurons, claimed that neurons were discrete cells that contacted each other very closely by synaptic junctions. This model contradicted the principal alternative paradigm, the reticular theory advocated by Camille Golgi, who invented the neuronal staining method used by Ramón y Cajal. According to Golgi, there was a continuous syncytial connection between the cell bodies of the nervous system. Both men shared the Nobel Prize in 1906, but it was not until the 1950s when **electron microscopy** finally resolved the theoretical question in favor of Ramón y Cajal (DeRobertis and Bennett, 1955).

It is now accepted beyond doubt that synaptic junctions are generally points of close contiguity, but not continuity, between connected neurons. Ramón y Cajal's model of discrete neurons interconnected by synapses is naturally suited

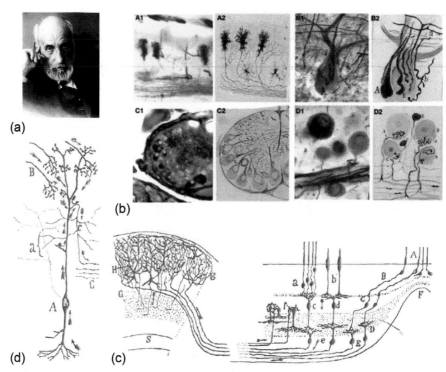

FIGURE 1.5 The pioneering work of Ramón y Cajal (1852-1934). **(a)** Santiago Ramón y Cajal made countless microscopic slide preparations of nervous tissue (colored slides in **(b)**), and recorded the data for publication by pen and ink drawings (monochrome panels in **(b)**). More than a century after the publication of his work, it is easy to recognize many examples of close correspondence between stained cells and drawn cells. **(c)** From these observations, Ramón y Cajal evolved neuron theory. For example, he showed how activity in retinal cone cells (marked A in the drawing) passed via synaptic junctions to foveal bipolar and ganglion cells (C) and was then projected to axonal arborizations (g) terminating in close proximity to neurons in the superior colliculus (H). Ramón y Cajal also formulated laws of conservation of space, time, and material to explain the morphological adaptations of individual neurons. **(d)** The so-called shepherd's crook cell of the reptilian optic lobe, for example, had the unusual characteristic that the axon (marked C on the drawing) did not emerge close to the cell body (A). Ramón y Cajal argued that this apparently odd axonal location was mandated by the conservation of material or cytoplasm (in modern parlance, minimization of wiring cost). Assuming correctly that electrical activity must flow from all other parts of the neuron towards the axon and its collateral ramifications (marked by arrows), he reasoned that all other possible locations of the axonal hillock would be associated with "waste of material" or "unnecessary lengthening" of the axonal projection. **(b)** Reproduced from García-Lopez et al. (2010) and **(c, d)** from Ramón y Cajal (1995) with permission.

to a graph theoretic representation, whereby neurons are represented by nodes and axonal projections or synaptic junctions are represented by edges. In this way, one might argue, Ramón y Cajal was the giant on whose shoulders graph theoretic analysis of neural systems formally emerged some 100 years later.

One other theoretical contribution by Ramón y Cajal that remains influential in connectomics is his proposal of a few apparently simple general laws to govern most, if not all, aspects of nervous system anatomy. He summarized these rules, also called Cajal's conservation laws, in the following words:

> Doubt for us is unacceptable, and all of the morphological features displayed by neurons appear to obey precise rules that are accompanied by useful consequences. What are these rules and consequences? We have searched in vain for them over the course of many years… Finally however we realized that all of the various conformations of the neuron… are simply morphological adaptations governed by laws of conservation for time, space and material… which must be considered the final cause of all variations in the shape of neurons, [and] should in our view be immediately obvious to anyone thinking about or trying to verify them.
>
> **Ramón y Cajal (1995), p. 116, Volume I.**

In more modern language, Ramón y Cajal anticipated that many aspects of brain network organization would be driven both by minimization of axonal wiring cost, which conserves cellular material and space; and by minimization of conduction delay in the transmission of information between neurons, which conserves time (Figure 1.5; see also Chapter 8).

It turns out that many aspects of brain network organization do indeed seem to have been selected to minimize wiring cost and/or to minimize metabolic expenditure (Niven and Laughlin, 2008). Topological features like modules and clusters are often anatomically colocalized, which conserves material. Other aspects of the connectome that promote the efficient integration of information across the network, such as short characteristic path length, may have been selected to minimize conduction delay, thus increasing the speed at which information can be exchanged between neurons or conserving time. Connectomics has thus begun to restate and refine Ramón y Cajal's conservation laws in terms of a competition between minimization of wiring cost and maximization of integrative topology (Bullmore and Sporns, 2012; Budd and Kisvárday, 2012).

1.2.2 Clinicopathological Correlations and Connectivity at the Macroscale

The first attempts to understand macroscopic networks of interconnected cortical areas roughly paralleled Ramón y Cajal's seminal work on microscopic neuronal connectivity. Network diagrams were drawn by clinical pioneers like

Theodor Meynert, Carl Wernicke, and Ludwig Lichtheim to summarize white matter connections between cortical areas, and to explain how the symptoms of brain disorder could be related to pathological lesions. The Wernicke-Lichtheim model of language remains the most successful of these early models of macroscale brain network organization, linking a language production area in the frontal cortex to a language comprehension area in the temporal cortex (as well as a more vaguely located association area). Some aspects of this model were able to account convincingly for the generation of specific symptoms: in particular, a lesion of the arcuate fasciculus linking frontal and temporal language areas was predicted and shown to cause an inability to repeat heard words despite otherwise apparently normal language, so-called conduction aphasia (Lichtheim, 1885; Figure 1.6a). Wernicke generalized these ideas as an associative theory of brain function, in which higher-order cognitive abilities (and their disorders, such as psychosis) were thought to emerge from the integration (or pathological disintegration) of anatomically distributed yet connected cortical areas (Wernicke, 1906). The nineteenth century diagrams of large-scale brain network organization drawn by these pioneers, comprising a few spatially circumscribed areas (nodes) interconnected by white matter tracts (edges), set the scene for graph theoretical analysis of nervous systems at the macroscopic scale, just as the neuron theory laid the foundation for graph-based models at microscopic (cellular) scales.

One important difference between the macroscopic diagram makers, like Wernicke and Lichtheim, and the microscopic anatomists, especially Ramón y Cajal, was the quality of the data available to them. Benefitting from contemporary technical developments in optics and tissue staining, Ramón y Cajal, Golgi, and others were able to produce very detailed, high-quality images of neurons and microscopic circuits (Figure 1.5). In contrast, the macroscopic diagram makers worked with poorer quality data. For example, Wernicke's work was based on clinicopathological correlation, linking the pattern of symptoms and signs expressed by a few patients in the clinic with the postmortem appearance of their brains (Wernicke, 1970). Even at the time, the unreliability of this method and the diagnostic formulations that followed were sharply criticized by contemporaries, including Sigmund Freud (1891; Figure 1.6b).

Partly as a result of the methodological weakness of postmortem clinicopathological correlations, large-scale network concepts of neurological and psychiatric disorders were somewhat eclipsed for the first half of the twentieth century by more locally anatomical or purely psychological models of cognitive function and brain disorders (Shallice, 1988). The localizationist ambition predominantly focused on how specific psychological processes arose from the function of discrete brain regions; or how specific facets of cognition, emotion, and behavior were anatomically localized and segregated in the brain. Although this tradition has conceptual roots in Gall's discredited phrenology

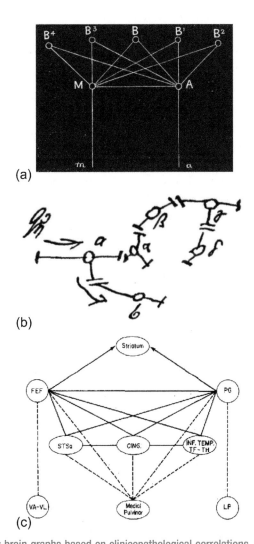

(a)

(b)

(c)

FIGURE 1.6 Early brain graphs based on clinicopathological correlations. **(a)** Lichtheim's sketch of a large-scale human brain network for language. Broca's and Wernicke's areas, and the arcuate fasciculus which directly connected them, were anatomically localized in the left inferior frontal and superior temporal cortex. However, the number of association areas that should be included in the network was not known, nor was their anatomical location. In this map, Lichtheim speculated that there may be many association areas and that Broca's and Wernicke's areas were the most densely connected nodes in the network. A represents an auditory area (i.e., Wernicke's area), M a motor area (i.e., Broca's area), and B other association areas. **(b)** A network diagram drawn by Freud, who was a critic of contemporary efforts to map clinical disorders onto large-scale brain networks. Nonetheless, he originally framed his nascent theory of psychoanalysis in terms of a neuronal network. The libido, designated by an enigmatic Q-like glyph, flows across weighted synaptic junctions between charged cells, engendering primary (id) or secondary (ego) process thinking according to the anatomical path of cathexis (or charge of libidinal energy) in the network. Psychoanalysis subsequently moved away from such an explicitly neuronal model, although libido retained a biological meaning in Freud's work until about 1910. **(c)** Mesulam's later and more anatomically accurate "hub and spoke" model of spatial attention, which comprises multiple cortical and subcortical areas interconnected by white matter tracts; the frontal eye fields (FEF) and posterior parietal cortex (PG) are the most densely connected hubs of the network. *Panels (a), (b), and (c) reproduced from Lichtheim (1885), Freud (1891, 1895), and Mesulam (1990), respectively, with permission.*

(Fodor, 1983), it received considerable support from famous clinical case studies, such as Broca's Leborgne (Broca, 1861), Harlow's Phineas Gage (Harlow, 1848), and Scoville and Milner's HM (Scoville and Milner, 1957), each of which demonstrated how highly selective cognitive and behavioral deficits could arise from focal brain damage. Further support came from Hubel and Wiesel's (1959) seminal single cell recordings of visual neurons in the cat, and Penfield and Jasper's (1954) intraoperative cortical stimulation studies of human epilepsy patients, both of which demonstrated an extraordinary degree of functional specialization at the level of individual neurons (in the cat) and small patches of cortex (in the human).

More integrative, network-based models of the brain were reinvigorated by the work on disconnexion or dysconnectivity syndromes conducted by Norman Geschwind, Marsel Mesulam and colleagues in the 1960s and subsequent years (Geschwind, 1965a,b; Mesulam, 1990). This work showed how many psychological deficits observed clinically could be explained in terms of the network anatomy of the brain. It also demonstrated how normal functions were often not localized to a single, specialized cortical area, but were instead anatomically represented by a large-scale network; for example, spatial attention was linked to a distributed ensemble of frontal and parietal cortical areas and subcortical nuclei that were interconnected by white matter tracts (Figure 1.6c). From this perspective, lesions or other disease processes attacking any component of these networks could be linked to symptomatic disturbances in patients. With better quality data available to link cortical anatomy to psychological functions and clinical symptoms, the importance of large-scale networks for understanding brain function and brain disorders, first advocated in the nineteenth century, was securely reaffirmed around 100 years later.

1.2.3   The Dawn of Connectomics

In the perfect light of hindsight, we can see that the conceptual precedents for connectomics go back a long way in the history of neuroscience. This makes it difficult to say when exactly the first connectome was drawn. The Wernicke-Lichtheim model, for example, is a directed, binary graph; but it was not described or analyzed as such. Mesulam's spatial attention model explicitly included network hubs, but they were not quantitatively defined in terms of degree or any other measure of topological **centrality** (Chapters 4 and 5). We might therefore characterize these maps as proto-connectomes that preceded the conscious and deliberate application of graph theory to neuroscience data in the 1990s.

The first brain graphs, representing a large number of cortical or subcortical nodes interconnected by axonal edges, were based on tract-tracing data in the cat and the macaque monkey. Tract-tracing measures the axonal propagation of a tracer or signal from the site of its injection to or from all other brain regions which are directly, anatomically connected to the injection site. It thus offers an excellent technique for exploring the wiring diagram of mammalian cortex, although each experiment will only generate data on the connectivity of a small number of injection sites and many experiments must be combined to estimate the connectivity of large nervous systems. To get around this problem, the first connectomes were generated by collating findings across a large number of published tract-tracing studies. With this approach, David van Essen and Dan Felleman represented the wiring diagram of the macaque visual cortex as a hierarchical system of areal nodes interconnected by directed, lamina-specific edges (Felleman and van Essen, 1991; Figures 1.1a and 1.7a and b). Another topological analysis of the large-scale macaque brain network disclosed a non-trivial organization, with some clusters of highly interconnected areas but over-all sparse connectivity (Young, 1992; Figure 1.7c), akin to the modular organization that was later found to be characteristic of many complex networks (Newman, 2003a). These studies also spurred more systematic attempts to collate and organize the findings from large numbers of tract-tracing studies in the macaque, resulting in the first, freely available connectomic data repository—the CoCoMac database (Stephan et al., 2001).

Despite these efforts, the next key step in the formation of connectomics was based not on mammalian tract-tracing data but on the neuronal network of *C. elegans*. The nervous system of this animal was mapped completely at the level of 302 individual neurons, with approximately 5000 chemical synaptic connections between them (as well as 2000 neuromuscular junctions and 600 gap junctions) reconstructed by expert visual inspection of serial electron micrographs (Figure 1.8). The experimental work by John White, Sydney Brenner, and colleagues took more than a decade to complete, and was published as a single 340-page volume of the *Philosophical Transactions of the Royal Society* (White et al., 1986). This canonical dataset on anatomical connectivity at a cellular scale in a relatively small nervous system has since informed many studies examining its spatial and topological organization as a network (Chen et al., 2006; Varshney et al., 2011; Nicosia et al., 2013; Towlson et al., 2013).

In 1998, Watts and Strogatz modeled the *C. elegans* connectome as a binary graph, where the neurons were represented as nodes and the synapses as edges. They showed that this network had a short characteristic path length and high clustering; in other words, its global organization conformed to a small-world topology. In assessing the clustering and path length of the *C. elegans* nervous system, these measures were compared to their values in a **null model**, which is an important step in many network analyses. We return to the concept of using null models to normalize graph theoretical measures on connectivity matrices in Chapter 10.

In addition to their technical novelty, the Watts and Strogatz results on *C. elegans* represented the first point of contact between modern network science and

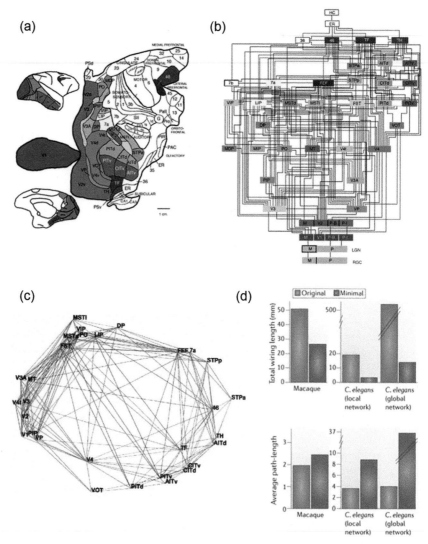

FIGURE 1.7 From brain anatomy to network topology. **(a)** Felleman and Van Essen (1991) compiled tract-tracing data on 32 regions comprising the macaque visual system (shown anatomically in flat map representation) and **(b)** used information on the cortical lamina of tract projection and termination to infer a hierarchical wiring diagram. **(c)** Young (1992) was the first to represent brain network data (in this case, macaque tract-tracing data) in a topological configuration that locates directly connected nodes in close proximity, regardless of the anatomical distance between them. **(d)** Kaiser and Hilgetag (2006) showed that the macaque and *C. elegans* connectomes could be computationally rewired to minimize their wiring costs (top barchart); but the cost-minimized graphs (red bars in both charts) had longer path length than the naturally selected connectomes (bottom barchart), pointing to a trade-off between wiring cost and integrative topology in brain networks. *(a, b) Reproduced from Felleman and Van Essen (1991), (c) from Young (1992), and (d) from Bullmore and Sporns (2012) with permission.*

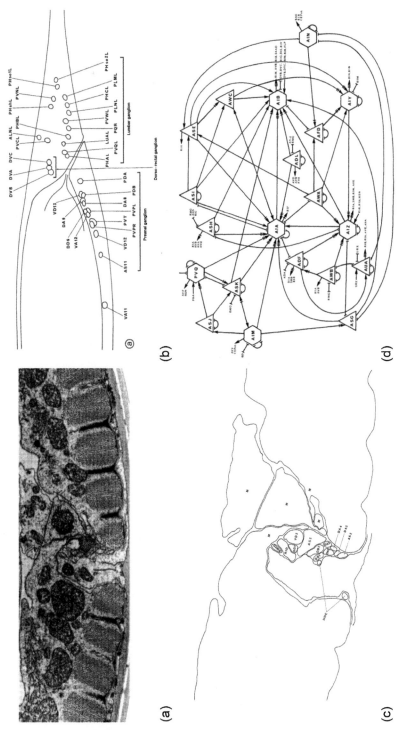

FIGURE 1.8 Mapping the connectome of the nematode worm *C. elegans*. White and colleagues used serial electron micrographs **(a)** to visually dissect and define synaptic terminals **(b)** between specific, named cells **(c)** allowing them to construct circuit maps or wiring diagrams of the anatomical network of synaptic connections between neurons **(d)**. In this representation, sensory neurons are represented as triangles, motor neurons as circles, and interneurons as hexagons. Arrows indicate the directionality of chemical synapses and T-shaped edges represent gap junctions. *Images reproduced from White et al. (1986) with permission.*

neuroscience, the first quantification of the complex topology of a nervous system, and the first hint that the organization of brain networks might share properties in common with other complex networks (Watts and Strogatz, 1998).

In the first years of the twenty-first century, Olaf Sporns, Claus Hilgetag, Rolf Kötter, and others translated the quantitative methods of Watts and Strogatz from the neuronal connectome of *C. elegans* to the large-scale interareal tract-tracing data then available for the cat and the monkey. It was demonstrated that these mammalian networks also had the characteristically small-world property of short path length and high clustering; and it was argued that this architecture might serve a dual function. The short path length of brain networks might favor integrated processing of information over the network as a whole, whereas high clustering might favor segregated processing within functionally specialized cliques of nodes (Sporns et al., 2004). In other words, the small-world organization of the brain offered a topological substrate for both functional segregation and integration, thus resolving the apparent paradox of how these two seemingly opposing tendencies could be supported by a single architecture. Moreover, by computationally rewiring the edges of the *C. elegans* and the macaque brain networks, it was shown that path length was inversely related to wiring cost (Kaiser and Hilgetag, 2006). In other words, the wiring cost of brain networks, approximated by the Euclidean distance between connected nodes, could be reduced in silico by rewiring the connections between nodes specifically to minimize their total physical distance; but in doing so the characteristic path length was increased. This was the first rigorous example of a trade-off between topological and spatial properties of the connectome (Figure 1.7d).

These and other early graph theoretical studies of brain networks demonstrated proof of concept—they showed that the mathematical tools of graph theory were applicable to suitably simplified nervous systems. However, these analyses were limited to historical data on *C. elegans*, the cat or the macaque. Around 2005, the nascent field of connectomics took a decisive step closer to mainstream systems neuroscience with the development of processing pipelines that allowed the application of graph theoretic techniques to human neuroimaging data.

1.2.4 Neuroimaging and Human Connectomics

Since the 1980s, there had been growing interest in the idea of looking at the interactions between pairs of neurophysiological time series recorded simultaneously in two distinct anatomical locations. The simplest version of the analysis is to estimate the correlation coefficient between two time series: brain regions that demonstrate correlated patterns of signal change over time are then said to be functionally connected. More formally, **functional connectivity** is defined as a statistical dependence between the time series of measured neurophysiological signals. This concept was originally developed for the analysis of spike trains recorded from single units (Gerstein and Perkel, 1969; Aertsen et al. 1989) and was then translated to the analysis of human functional neuroimaging data by Karl Friston, Barry Horwitz, Randy McIntosh, and others (Friston, 1994; McIntosh et al., 1996; Horwitz, 2003). The basic concept—that two locations can be said to be functionally connected if they have coherent or synchronized dynamics—generalizes to many different modalities of neurophysiological imaging, including functional MRI, **electroencephalography** (EEG), **magnetoencephalography** (MEG), **multielectrode array** (MEA) recording of **local field potentials** (either in vivo or in vitro), and **positron emission tomography** (PET; some of these methods are considered in more detail in Chapter 2). For example, analysis of functional connectivity by **independent component analysis** of resting-state functional MRI data has been used to identify a small set of around ten spatially distributed, large-scale neural systems showing coherent, low-frequency oscillations: so-called resting-state networks (Damoiseaux et al., 2006; Fox and Raichle, 2007; Smith et al., 2009; Fornito and Bullmore, 2010).

The first graph theoretic analyses of human brain functional networks were based on functional connectivity matrices estimated from functional MRI and M/EEG data (Stam, 2004; Eguíluz et al., 2005; Salvador et al., 2005; Achard et al., 2006). In this work, correlation or coherence between time series recorded at different brain locations (nodes) was estimated for every possible pair of nodes and pair-wise correlations were arbitrarily thresholded to define binary edges constituting a graph of the large-scale functional network (Figure 1.9a; Chapters 2 and 3). This work has shown that human brain functional connectivity networks show similar organizational properties to those that had been discovered in the anatomical networks of the macaque, cat, and *C. elegans*, as well as many other naturally complex systems. For example, functional MRI networks are small-world, contain hubs, have a hierarchical modular structure, and appear to be constrained by a pressure to minimize wiring costs (approximated by the Euclidean distance of edges; Vértes et al., 2012; Figure 1.9b). Indeed, trade-offs between spatial constraints (minimization of wiring cost) and topology are a prominent and heritable feature of human brain networks (Bassett et al., 2010; Fornito et al., 2011b; Betzel et al., 2016).

The first graph theoretic analyses of human brain anatomical networks were based on tractographic analysis of **diffusion MRI** data (Hagmann et al., 2008; Figure 1.9c–e) and **structural covariance** analysis of conventional MRI data (He et al., 2007; Alexander-Bloch et al., 2013a). Regardless of the method of construction, such human brain anatomical networks have consistently been found to show the same nontrivial organizational properties discovered in other types of brain networks.

While this convergence between brain graphs constructed from tract tracing, structural MRI, diffusion MRI, and functional MRI is encouraging, it is important to note that functional connectivity and **structural connectivity** are quite

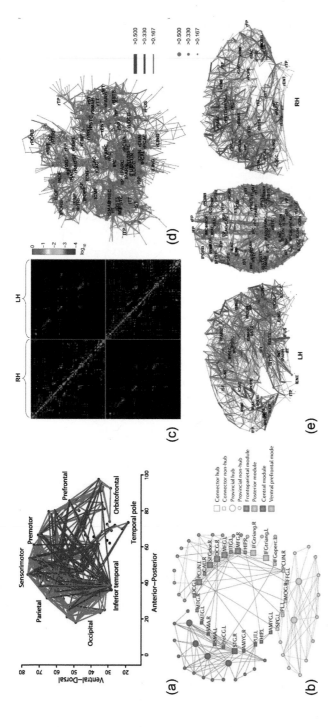

FIGURE 1.9 Brain graphs from human magnetic resonance imaging (MRI). **(a)** Brain graphs can be constructed to represent functional connectivity networks. This is an unweighted, undirected graph representing thresholded correlations (edges) between pairs of resting-state functional MRI time series recorded simultaneously from 90 cortical and subcortical areas (nodes). The nodes were defined by an anatomical parcellation and are located in anatomical space; red edges represent shorter distance connections and blue edges represent longer distance connections. **(b)** The modular organization of a functional MRI graph is shown in a topological configuration. Major modules are differently colored and the connector hubs, which mediate the majority of intermodular links, are highlighted as square nodes. **(c)** Diffusion MRI **tractography** can be used to construct a weighted, undirected connectivity matrix, where each element quantifies the density of axonal projections between a specific pair of regions. This network can be displayed either as a brain graph projected using an algorithmically defined layout to emphasize topological aspects of network organization **(d)**, or with nodes positioned according to their anatomical location **(e)**. In these graphs, edge thickness is proportional to connectivity weight and node size is proportional to the degree of each node. **(a)** Reproduced from Achard et al. (2006), **(b)** from Bullmore and Sporns (2012), and **(c–e)** from Hagmann et al. (2008) with permission.

different concepts in many ways (Chapter 2). Anatomical connectivity can be defined most fundamentally as an axonal projection from one cell to another, or at coarser spatial scales by a tract of axons projecting from one area to another. Such axonal projections are expected to change only slowly over time. In contrast, functional connectivity is a statistical measure of synchronized activity that does not necessarily imply an underlying anatomical connection, and which can change rapidly over time.

So far, most graph theoretic studies of functional connectivity networks have summarized coherent neurophysiological activity over an extended period with a single scalar metric, such as a correlation coefficient, effectively "averaging out" any dynamic changes in the network configuration over time. Functional connectivity when analyzed in this way is correlated with, although not identical to, anatomical connectivity (Skudlarski et al., 2008; Honey et al., 2009). In a computational model, the convergence between functional connectivity and the known underlying connectivity between nodes increased as a function of the period of time over which functional connectivity was averaged (Honey et al., 2007). Thus, as functional connectivity is averaged over longer time periods, it may converge onto structural connectivity, although it is important to remember that structural and functional connectivity are different measures and may thus yield connectomes with different values of some topological parameters (Zalesky et al., 2012b). Moreover, such a time-averaged analysis of functional connectivity neglects the physiological reality that the brain's information processing systems must be rapidly reconfigurable in response to changing environmental conditions or experimental task demands (Palva et al., 2010; Kitzbichler et al., 2011; Bassett et al., 2011; Fornito et al., 2012a). Even in the so-called resting state it is clear that functional connectivity is not stationary (Chang and Glover, 2010) but spontaneously transitions through a series of different network configurations (Zalesky et al., 2014). The time-resolved analysis of functional network topology is likely to be an important focus for the future development of connectomics.

Despite the obvious methodological limitations of MRI, such as its limited resolution of network organization in space and time, and the uncertain neurobiological interpretation of MRI estimators of anatomical and functional connectivity, it nonetheless offers the best available tool for characterizing brain connectivity in living humans. MRI-based connectomics will likely continue to offer unique insights into the cognitive and clinical significance of certain organizational properties of brain networks.

The relationship between network topology and cognitive function is an important question that has so far been relatively under-explored, although initial work has shown that interindividual variations in the topological properties of structural and functional networks, including path length, cost-efficiency, and modular organization, correlate with variability in cognitive performance (Li et al., 2009; van den Heuvel et al., 2009; Bassett et al., 2009, 2011; Zalesky et al., 2011; Fornito et al., 2012a; Dwyer et al., 2014). Indeed, the modular topology of the human connectome might be expected to correspond somehow to the neophrenological concept of modules as fast, specialized, automatic, faculties for algorithmic information processing (Fodor, 1983). In support of this prediction, meta-analysis of more than 1000 primary functional MRI studies of brain activation showed that the modules of the interregional coactivation network were functionally specialized for specific cognitive processes (Crossley et al., 2013). For instance, perception was represented by a module comprising areas of visual cortex localized mainly in the occipital lobe; whereas action was represented by a module comprising areas of motor cortex. Conversely, it may be that other topological features, such as connector hubs and intermodular connections, support more conscious, effortful, domain-general cognitive processes that depend on the costly, integrated activity of a global neuronal workspace (Dehaene et al., 1998). For example, in the same meta-analysis of task-related functional MRI data (Crossley et al., 2013), the most topologically central subset of regions, comprising a rich club in frontal and parietal association cortex, was associated with the greatest diversity of experimental tasks, especially "higher order" executive tasks demanding both cognition and action (Figures 1.10 and 1.11). Using ultra-high field functional

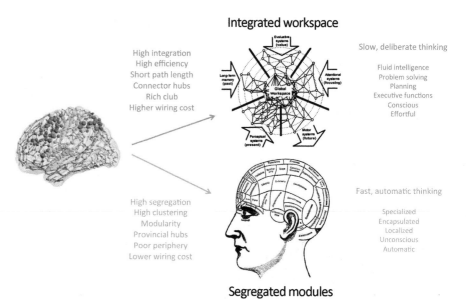

FIGURE 1.10 Schematic linking brain network topology and psychological function. Complex brain network topology can be subdivided into (top stream) high-cost components supporting a so-called global workspace architecture that favors integrated information processing, deliberate thinking, and flexible intelligence; and (bottom stream) low-cost components supporting a segregated architecture that favors fast, specialized, and automated information-processing. *Brain graph reproduced from Crossley et al. (2013) and workspace image from Dehaene et al. (1998), Copyright (2007) National Academy of Sciences, U.S.A., with permission.*

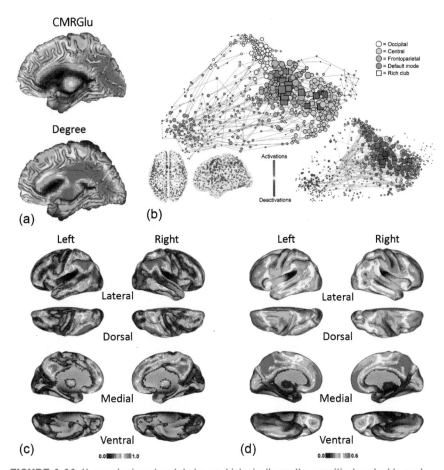

CMRGlu

Degree

○ = Occipital
○ = Central
○ = Frontoparietal
● = Default mode
□ = Rich club

Activations

Deactivations

(a)

(b)

Left   Right

Lateral

Dorsal

Medial

Ventral

(c)   0.0 ■■■ ■■1.0

Left   Right

Lateral

Dorsal

Medial

Ventral

(d)   0.0 ■■■ ■■0.6

FIGURE 1.11 Human brain network hubs are biologically costly, cognitively valuable, and vulnerable to pathology. **(a)** Hubs are metabolically expensive. Shown here is the anatomical overlap of brain regions with high cerebral metabolic rates of glucose metabolism (CMRglu; top) measured using PET, and brain regions with high degree in functional MRI networks (bottom). **(b)** Hubs are activated by "higher order" cognitive tasks. Meta-analysis of coactivation patterns across more than 1000 task-based functional MRI experiments found that a rich club of hubs (marked as square nodes in a topological representation) was coactivated by a diverse range of experimental conditions, especially executive tasks involving both action and cognition. **(c)** and **(d)** Hubs are commonly implicated in clinical brain disorders. Shown here is the anatomical overlap between regions with high functional connectivity **(c)** and high levels of amyloid deposition measured by PET in patients with Alzheimer's disease **(d)**.
(a) Reproduced from Tomasi et al. (2013), *(b)* from Crossley et al. (2013), and *(c, d)* from Buckner et al. (2009) with permission.

MRI to optimize the time resolution of functional network changes in response to briefly presented visual stimuli, it was likewise shown that conscious perception was distinctively associated with increased global integration and decreased modularity of network topology (Godwin et al., 2015).

MRI connectomics also has important clinical implications (Fornito et al., 2015; Box 1.1). One key observation that has already been made with remarkable consistency concerns the clinical significance of brain network hubs. In Alzheimer's disease, it was shown that high degree nodes in functional MRI graphs have greater local deposition of amyloid protein (measured using PET) than less topologically central brain regions (Buckner et al., 2009). Across a range of neurodegenerative disorders, node degree and other measures of topological centrality in functional connectivity networks have been positively correlated with local gray matter atrophy measured using MRI (Zhou et al., 2012). More generally, a meta-analysis of case-control MRI studies of 26 different brain disorders demonstrated that the probability of pathological loss of gray matter signal in a given brain region was significantly increased for regions

BOX 1.1 CLINICAL APPLICATIONS OF BRAIN NETWORK ANALYSIS

Like the ripples caused by a stone falling into a pond, pathological perturbations or lesions that are initially localized to specific parts of a nervous system often propagate along axonal fibers to affect the functioning of otherwise intact areas. The ancient Roman physician Galen (c. 129 CE to c. 200 CE) was among the first to note this fact, proposing that animal spirits could flow through nervous pathways to affect distant areas (Galen, 1976). More than 1800 years later, von Monakow (Finger et al., 2004; von Monakow, 1969) coined the term **diaschisis** (Greek for "shocked throughout") to describe the interruption of function in remote, intact brain regions that were connected to an injured site. The importance of neuronal connectivity in disease was also acknowledged by nineteenth and twentieth century proponents of clinicopathological correlation as a means for uncovering the network architecture of the brain (Section 1.2.2), culminating in Geschwind's (1965a,b) characterization of a new class of neurological "disconnexion" syndromes.

Connectomics and graph theory offer a powerful framework for mapping, tracking, and predicting patterns of disease spread in brain disorders (Fornito et al., 2015; Stam, 2014; Figure 1.1.1; see also Box 11.2). From this perspective, brain changes in disease can be characterized at the level of network connectivity or topology, and the development of statistical techniques that allow for valid inference on group differences is an active area of research (Chapter 11). The

clinical application of connectomics is now confirming the early insights of Galen, von Monakow, and others by demonstrating that the propagation of pathology in the brain is indeed constrained by its network architecture. For example, neurodegeneration occurs in functionally and structurally connected networks (Raj et al., 2012; Seeley et al., 2009), pathology accumulates in highly connected hub areas of the brain (Buckner et al., 2009; Crossley et al., 2014), and the cognitive sequelae of brain injury or disease are closely related to the connection topology of the affected region (Warren et al., 2014).

Computational models of large-scale brain network dynamics allow neural activity to be simulated on structural network architectures (Deco and Kringelbach, 2014; Deco et al., 2008), and have shown that the functional effects of virtual "lesions" induced in silico are determined by the connection topology of the lesioned nodes (Alstott et al., 2009; Cabral et al., 2012; Honey and Sporns, 2008). Moreover, the therapeutic efficacy of invasive and noninvasive brain stimulation therapies across a range of disorders is critically related to the connectivity of the site targeted for stimulation (Fox et al., 2014; Riva-Posse et al., 2014). These findings suggest that an understanding of network topology may allow us to predict expected levels of impairment and prospects for recovery following insult (Fornito et al., 2015), and to select individually tailored interventions with maximum chances of therapeutic success.

BOX 1.1 CLINICAL APPLICATIONS OF BRAIN NETWORK ANALYSIS—CONT'D

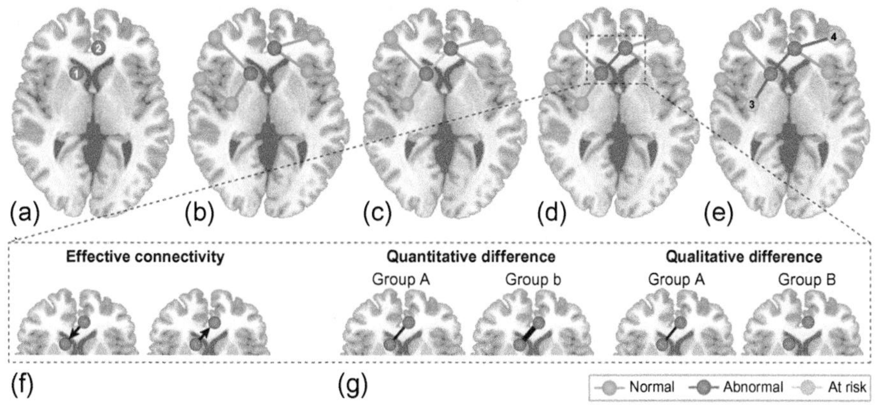

FIGURE 1.1.1 Using brain graphs to map, track, and predict patterns of disease propagation across the connectome. **(a)** Historically, most studies of clinical disorders have focused on characterizing pathology in discrete brain regions. For example, in a typical human MRI experiment, a patient and control group are compared on some measure of tissue structure or function at many different locations in the brain. The result is a map identifying areas of localized group differences. In this example, two regions, 1 and 2, show an abnormality in the patient group. This map localizes the abnormalities, but offers no further information about whether these changes are related in some way or are independent. **(b)** Mapping the connectivity of the brain reveals the broader network context of the local changes. In this case, we see that regions 1 and 2 are connected to other areas, but are not connected to each other, suggesting that they are subject to independent pathophysiological processes. **(c)** Regions 1 and 2 are connected, but the connection itself is not abnormal. In this case, the pathology may have originated in one area and spread to the other along the intact pathway. Identifying the primary abnormality would require longitudinal analysis, or a model of the causal interactions between regions (i.e., **effective connectivity**, see Chapter 2). The connection between regions 1 and 2 is considered "at-risk" of subsequent deterioration because it links two dysfunctional areas. **(d)** Regions 1 and 2 and their connection are abnormal. In this case, primary pathology in one area may have resulted in secondary deterioration of its axons and, subsequently, the other area. Alternatively, a primary pathology of the connecting fiber bundle may have caused dysfunction in both regions. **(e)** Dysfunction in regions 1 or 2 can alter their connectivity with other areas, and place those other areas (regions 3 and 4) at risk of deterioration. **(f)** Resolving directions of information flow and building a model of effective connectivity allows a more precise characterization of directions of disease spread in brain networks. For example, if region 2 projects to 1, but 1 does not project to 2 (left), it is more likely that pathology propagated from region 1 to 2 than vice versa. **(g)** Connectivity differences between patients and controls can be either quantitative (left) or qualitative (right). A quantitative difference occurs when there is a difference in connection strength, but patients and controls share the same underlying network architecture. A qualitative difference occurs when a connection is present in one group but not in the other. *Reproduced from Fornito et al. (2015) with permission.*

with high degree in the healthy connectome (Crossley et al., 2014). These and other findings strongly suggest that topologically integrative, but biologically costly network hubs, may be points of special vulnerability for the expression of pathogenetically diverse brain disorders (Figure 1.11).

At a more mechanistic level, computational modeling suggests that the spatial pattern of neurodegeneration in Alzheimer's and frontotemporal dementia can be predicted by a relatively simple process of disease diffusion simulated on connectomes reconstructed from diffusion MRI (Raj et al., 2012). Other work indicates that neurodegeneration occurs within structurally and functionally connected networks (Seeley et al., 2009). Collectively, these findings suggest that the topology of large-scale brain networks constrains the way in which neurodegenerative disease spreads throughout the brain. To paraphrase Hebb, it seems that brain areas that are wired together tend to die together. It has also been shown that local reductions of cortical thickness in patients with childhood-onset schizophrenia, a neurodevelopmental disorder, were concentrated within a single module of brain regions that shared a common developmental trajectory. In other words, the normal adolescent processes of brain network development may constrain the expression of brain abnormalities associated with schizophrenia and other psychiatric disorders (Alexander-Bloch et al., 2014). MRI measures of topological centrality and modular organization could thus be important biomarkers for early diagnosis and prediction of clinical outcomes in neurology and psychiatry (Bullmore and Sporns, 2012; Stam, 2014; Fornito et al., 2015).

1.2.5   Back to Basics: From Macro to Meso and Micro Connectomics

By this account of recent history, graph theory impacted decisively on neuroscience at the microscopic scale of *C. elegans* (1998); then jumped to the mesoscopic scale of tract-tracing data on the cat and the macaque (circa 2001); and then translated to the macroscopic scale of human neuroimaging (circa 2005). In the decade since then, graph theoretical studies based on MRI, MEG, and EEG have rapidly proliferated and much of the methodological and conceptual development of connectomics has rested on analysis of human neuroimaging data representing macroscale networks ($\sim 10^2$ m). An exciting recent development, and one that promises to be game-changing for the next decade of connectomics, is the increasing availability of high quality data on brain networks at mesoscales ($\sim 10^{-4}$ m) and microscales ($\sim 10^{-6}$ m).

At the mesoscopic scale of tract-tracing, advances in the automation and standardization of experimental procedures, and improvements in how tracer propagation is imaged and quantified, have driven rapid changes in the scale and richness of data available on the mouse and the macaque monkey (Kennedy et al., 2013; Oh et al., 2014; Zingg et al., 2014). These datasets are sufficiently comprehensive, comprising tens or hundreds of individual tracer injection experiments, to support estimation of the anatomical connectivity between all possible pairs of brain regions defined according to specified anatomical criteria and standard experimental procedures (Figure 1.12; see also Chapter 2).

Pioneering studies using such techniques in the macaque and the mouse have demonstrated that the weight of anatomical connectivity varies over five orders of magnitude; and that mesoscale connectomes may be more densely connected (with a **connection density** over 60%, in some cases) than had been assumed, due to the presence of many weak and previously unrecognized connections (Markov et al., 2014; Oh et al., 2014). The connectivity matrices generated by these methods can be used to construct **directed graphs**, because tracer propagation is either anterograde or retrograde from the injection site (Figure 1.12). The resulting graphs can also be weighted according to the number or fraction of neurons in the connected region that have been labeled by the tracer, thus providing an estimate of connectivity strength. As will be evident throughout this book, many graph theoretic techniques are readily applicable to such weighted and directed connectomes and often enable a richer characterization of network topology.

At a microscopic scale, there are many major efforts underway to map cellular networks in model organisms ranging from *Drosophila* (Chiang et al., 2011) to zebra fish (Ahrens et al., 2013), and these are supported by significant advances in technologies for reconstructing single cells and their synaptic connections (Lichtman et al., 2008). In the near future, there will certainly be a wave of data available to support graph theoretical analysis of anatomical networks at

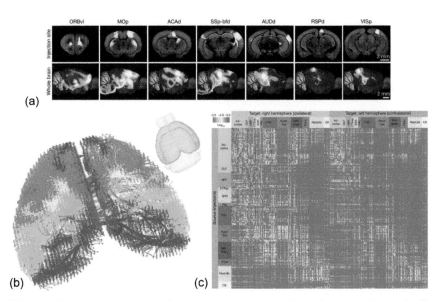

FIGURE 1.12 Large-scale standardized mesoscale connectomics in the mammalian brain. The automation of tissue processing, dissection, and imaging pipelines has enabled the construction of mesoscale connectomes in model species with viral tract tracing. Shown here is one specific pipeline used to map neuronal connectivity in the mouse brain. Separate injections to different brain regions in different animals can be used to map regional connection profiles **(a)**, which can then be tracked and collated to map connectivity between multiple brain regions **(b)**. These data can be used to generate an interregional connectivity matrix **(c)**. *Figure reproduced from Oh et al. (2014) with permission.*

cellular scale, and to link cellular characteristics to mesoscale and macroscale properties of inter-regional connectivity (Figure 1.13a). It is also likely that there will be greater interest in the graph theoretical analysis of functional networks recorded from neuronal cultures at cellular scale by MEA. In these datasets, each electrode measures a local field potential and the functional connectivity between each pair of electrodes can be estimated to construct a network where nodes represent electrodes (each recording from a few local neurons), and edges represent highly synchronized electrical activity. Preliminary results have again demonstrated the apparent ubiquity of complex topological features: MEA graphs have hubs, cores, modules, and small-world organization, and develop in vitro according to a rich-get-richer rule (Downes et al., 2012; Schroeter et al., 2015; Figure 1.13b). This experimental paradigm clearly offers some attractive opportunities for future research to manipulate

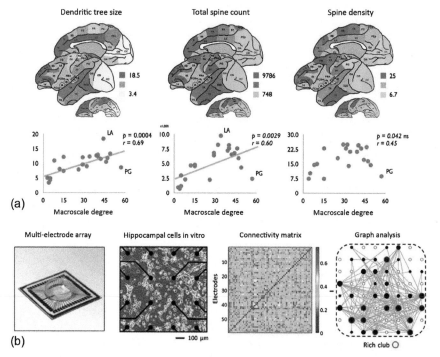

FIGURE 1.13 New approaches to connectomics. **(a)** Integrating databases that describe regional variations in microscale neuronal properties and macroscale connectivity patterns allows analysis of the relationship between regional cytoarchitecture and interregional connection topology. For example, high degree nodes in the macaque cortical network have higher dendritic tree size and higher spine count. **(b)** Multi-electrode array recording is another technology with potential to support greater understanding of networks at the cellular scale. Neuronal cultures are grown on an electrode array and the functional connectivity between local field potentials recorded from each pair of electrodes can be summarized as an association matrix and thresholded to generate binary undirected graphs with hubs and rich clubs. *(a) Reproduced from Scholtens et al. (2014) and (b) from Schroeter et al. (2015), with permission.*

the system pharmacologically, or optogenetically, and so elucidate the molecular and cellular mechanisms driving preferential attachment, the emergence of hubs, and other topological features of structural and functional cellular networks.

1.3 ARE GRAPH THEORY AND CONNECTOMICS USEFUL?

This might seem like an odd question to ask at the end of the first chapter of a book about analyzing brain networks. But any new method demands critical evaluation. In our view, there are two factors arguing in favor of graph theory for neuroscience: its simplicity and its generalizability.

Graphs are fairly simple models. In fact, they are often ruthless abstractions of minutely detailed natural systems, by which the myriad biological intricacies of the brain are reduced to a collection of (often homogenous) nodes and edges. Simplicity is always attractive theoretically, especially in the age of "big data" when there is a real risk of being overwhelmed by terabytes of biological measurements. Moreover, the mathematics of graphs are rigorous while also being accessible to neuroscientists, physiologists, and biologists who may not have a formal training in quantitative sciences. It certainly helps that many of the key concepts of connectomics can be represented graphically and illustrated by analogy to well-known complex systems, such as the global airline network. A brain graph thus has the benefit of providing a conceptually simple model implemented using an accessible mathematical language. However, does a brain graph entirely follow the instruction, sometimes credited to Einstein, that "everything should be made as simple as possible, but not simpler"? Or are brain graphs not so much attractively simple as unacceptably simplistic?

The answer to this question may be in the goals of the scientist. Certainly, a graph cannot be used to model every single detail of the brain. However, it offers a useful, simplified abstraction that allows us to formally address critical questions, such as how does brain network structure constrain function? What are the general organizational principles of brain networks? What developmental processes can give rise to networks that look and function like the brain? If those kinds of questions are of interest, then graph theory is useful. But the challenge to link topological metrics on abstract, simple graphs to biological mechanisms at cellular and molecular levels remains important and has provoked some interesting recent work. For example, the hubs of macroscopic brain structural networks derived from tract-tracing data show distinct microscopic properties related to neuronal morphology and density (Scholtens et al., 2014; Figure 1.13a); associations have been found between network topology and local gene expression profiles (Wolf et al., 2011; French and Pavlidis, 2011; Rubinov et al., 2015; Fulcher and Fornito, 2016); and the hubs of human functional MRI networks have been located in brain regions that have high rates of glucose metabolism measured by PET (Tomasi et al., 2013; Figure 1.11). These are all early examples of successful efforts to explore the local biological characteristics of nodal topological properties estimated by the systems-level mathematical analysis of graph theory. We anticipate that as more is known about the biological substrates of brain graphs, topological properties (like hubs and modules) will turn out to be meaningfully linked to underlying biological mechanisms, such as gene expression and developmentally determined patterns of variation in cytoarchitectonics and myeloarchitectonics. In this event, any current concern that brain graphs are unacceptably simplistic will naturally recede.

The generalizability of graph theory is its second major advantage. As we have seen, graphs have already been used to model a diverse array of complex systems. Graph theoretical methods can also be adapted to analyze gene expression arrays and other -omic biological datasets, which typically also have many of the same complex network properties as the connectome (Oldham et al., 2008). In neuroscience, the applicability of graph theory to all kinds of neuroimaging and neurophysiological data opens up the scientific opportunity to look at the same topological and spatial properties of brain networks across a wide range of species, scales, and modalities, thus providing a lingua franca for neuroscientists working in diverse domains. Multiscale measurements of brain networks could serve to translate insights about the topology, dynamics, development, and function of brain networks from one experimentally specialized field of neuroscience to another. Perhaps the ultimate scientific goal in this pursuit is the elucidation of very general selection pressures, or conservation laws, that can explain brain network organization from microscales to macroscales. Graph theory is well suited to address this and other strategic challenges in understanding the network organization of brains.

1.4 SUMMARY

We have seen in this chapter that the importance of understanding brain connectivity was apparent to pioneers, such as Ramón y Cajal and Golgi, from the moment they first peered through the microscope to observe the morphology of neurons and their processes. However, progress in subsequent years was slow, due to a lack of appropriate tools and conceptual frameworks for measuring, mapping, and simulating large-scale neuronal networks. The past few decades have witnessed unprecedented advances not only in our ability to map brain connectivity at micro-, meso- and macroscales, but also in our ability to quantify and generate the connection topology of such intricate systems. These developments have coincided with technical advances in methods for acquiring, curating, and disseminating large-scale data on brain connectivity, and with the birth of a more general science of complex networks. In particular, graph theory has emerged as a powerful tool for developing a coherent understanding of brain network organization that cuts across spatial and temporal scales, and which allows us to understand how the connectome relates to a much broader class of naturally occurring complex systems.

Our goal in this book is to introduce the fundamentals of graph theory and network science, as applied to the brain. This book is not intended as a general introduction to the mathematics of graph theory and/or the physics of complex networks, for which we refer interested readers to the excellent text book on networks by Newman (2010), as well as detailed review articles by Newman (2003a), Albert and Barabási (2002), and Boccaletti et al. (2006). A more mathematical treatment is provided by Bollobás (1998). More general introductions to connectomics and brain networks have been provided by Sporns (2011a, 2012; see also Buzsáki, 2006). In contrast to these texts, this book will focus in detail on the specific aspects of graph theory that are proving to be the most useful for understanding brain connectivity. We will examine measures that have been adapted from the study of other complex networks, as well as those that have been developed specifically in the neuroscientific context. We will also consider examples of how these measures and methods have been applied to understand brain network organization across a diverse range of species, resolution scales, and experimental techniques. In the next chapter, we begin our discussion of this work with a consideration of perhaps the most fundamental aspect of brain network analysis: how to define nodes and edges.

CHAPTER 2

Nodes and Edges

Nodes and **edges** are the elemental building blocks of networks, and their accurate definition is of paramount importance for any valid graph theoretical model of network organization (Butts, 2009). The nodes and edges of some networks are clearly defined: in friendship networks, nodes represent people and edges represent social relations; in the global air transportation network, nodes represent airports and edges represent direct flights between them; in the World Wide Web, nodes represent web pages and edges represent hyperlinks. What are the nodes and edges of a nervous system?

The answer to this question is complicated by the multiscale organization of the brain. In humans, neural anatomy is organized over six orders of magnitude, from the level of individual neurons and synapses (<1 μm), through populations of neurons such as cortical columns and cytoarchitectonic regions (~ 1 cm), to large-scale divisions such as lobes, systems, and hemispheres (~ 10 cm; Figure 2.1; Lichtman and Denk, 2011). The dynamics unfolding on these structures evolve over a similar range of temporal scales, spanning frequencies between 0.05 and 500 Hz (Figure 2.2; Buzsáki and Draguhn, 2004). This multiscale organization means that there is no single, privileged scale for the analysis of brain networks. Rather, each scale may provide important information that can be both scale-specific and scale-invariant.

Unfortunately, there is no single technology that can measure brain networks over all biologically relevant scales of space or time. This means that a consideration of **connectomics** across multiple scales must inevitably also consider multiple different measurement modalities. **Graph theory** provides a unifying language to understand brain network **topology**, regardless of scale and measurement method, because the network is abstractly rendered as a set of nodes and edges. However, it is important to appreciate the way in which we define nodes and edges at each scale, as our definition is critically dependent on the measurement technique that is chosen. For example, a graph constructed from a **multielectrode array** recording from a neuronal culture can, in principle, be analyzed in the same way as a graph constructed from human **functional MRI** data. However, as we will see throughout this chapter, the nodes and edges derived from such diverse data do not necessarily refer to the same neurobiological structures or processes. In fact, the scale and method of measurement

37

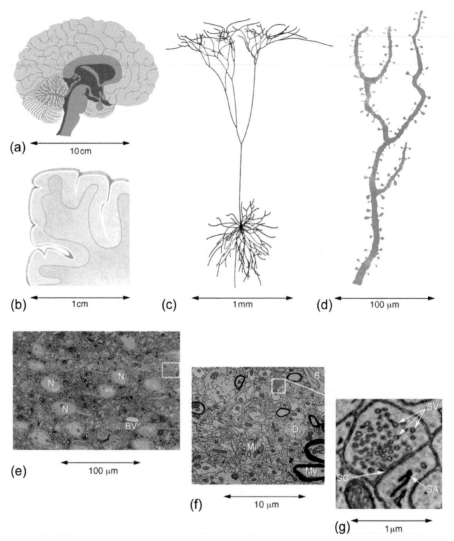

FIGURE 2.1 The multiscale spatial organization of brain anatomy. **(a–g)** The brain is organized across multiple scales of spatial resolution. **(a)** At the scale of centimeters, brain areas can be divided into broad anatomical divisions such as cortical lobes and cytoarchitectonic areas. **(b)** At sub-centimeter resolution, neurons aggregate into columns, layers, and other cell groups on the scale of a few millimeters. **(c)** At sub-millimeter resolution, it is possible to resolve neuronal processes such as dendritic trees (on a scale of around 100 µm) and axons that project over both short and long distances. Pyramidal neurons can have dendritic trees that span a cubic millimeter. **(d)** At the scale of around 10 µm, the structure of individual fibers and dendritic spines can be resolved. **(e)** Electron micrograph showing a set of structures that can be visualized within a section of tissue that spans 100 µm. Neurons (N), their processes, and blood vessels (BV) can be identified. **(f)** Zooming in on a small patch that spans 10 µm (white box in **(e)**) reveals further detail. The dark rings are cross-sections of myelinated axons (for example, My). Small dark objects are mitochondria (Mi). Neuronal and glial somata (S) and cross-sections of dendrites are also visible (D). Blue shows glial processes, red a presynaptic terminal and green shows dendritic spines. **(g)** Zooming in on the synapse shown in red in **(f)**, we see synaptic vesicles (SV), the synaptic cleft (SC) and a spine apparatus (SA) in the postsynaptic spine. *Figure reproduced from Lichtman and Denk (2011), with permission.*

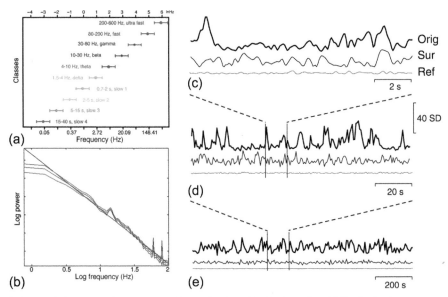

FIGURE 2.2 The multiscale temporal organization of brain dynamics. **(a)** The brain shows hierarchical organization over temporal scales ranging from 0.05 to 500 Hz. The mean frequencies of neuronal oscillations show a linear progression on a logarithmic scale with a constant ratio between neighboring frequencies, leading to the delineation of distinct frequency bands. The oscillations shown here were measured in rat cortex, but these frequency bands are ubiquitous in many different types of neuronal activity recordings. **(b)** Power spectrum of neuronal activity, measured using intracranial EEG of the right temporal lobe of a sleeping human participant. On the doubly logarithmic axes, there is a near-linear decrease in power as a function of frequency, f. This $1/f$-like scaling implies that low-frequency oscillations can drive high-frequency dynamics, and is consistent with a self-similar, so-called **fractal** organization of brain activity across temporal scales. **(c–e)** Example of self-similar temporal organization in a human MEG signal temporally filtered at 10 Hz. **(c)** The original signal (Orig), a surrogate signal (Sur) and a signal recorded from a reference electrode (Ref) sampled at 15 Hz. The amplitude scale is measured in units of standard deviation from the reference time course. **(d)** and **(e)** show the same signal sampled at 1.5 and 0.15 Hz, respectively. The original signal shows strong fluctuations across all time scales whereas the surrogate data only shows pronounced fluctuations at the temporal scale shown in **(c)**. Thus, the MEG recording shows hierarchical and statistically self-similar organization across multiple scales of time, consistent with a fractal organization. *Reproduced **(a,b)** from Buzsáki and Draguhn (2004) and **(c–e)** from Linkenkaer-Hansen et al. (2001), with permission.*

often constrains the types of analyses that are possible, and how the findings of these analyses should be interpreted.

In connectomics, a distinction is commonly drawn between three spatial scales: microscopic, mesoscopic, and macroscopic (Bohland et al., 2009). The *microscopic* scale refers to properties that are too small to resolve with the naked eye and thus require the use of microscopic techniques for visualization.

This scale has become synonymous with networks reconstructed at the level of individual neurons and synapses. At the other end of the spectrum, the *macroscopic* scale refers to properties that can be resolved without the aid of microscopic methods. In connectomics, this scale most commonly refers to analyses of structural and functional interactions between large-scale populations of neurons, such as those characterized with MRI, MEG or EEG. The *mesoscopic* scale bridges the microscopic and macroscopic. Analyses at this scale often combine microscopic and macroscopic techniques to understand neuronal connectivity with high precision across the entire brain (or a large part of it). The special techniques used at each scale constrain the way in which nodes and edges are defined.

In this chapter, we overview some of the key approaches to define nodes and edges at microscopic, mesoscopic, and macroscopic scales. We focus on methods at each scale that are currently being used to generate **connectomes** for large portions of nervous systems, and consider how these methods influence the way in which nodes and edges are defined. In particular, we consider how each scale can provide useful insights into two of the major types of brain connectivity that will be considered throughout this book: **structural connectivity** and **functional connectivity** (Box 2.1).

BOX 2.1 CLASSES OF BRAIN CONNECTIVITY

A distinction is often drawn between three classes of brain connectivity: structural connectivity, functional connectivity, and **effective connectivity** (Bullmore and Sporns, 2009; Fornito et al., 2013; Friston, 1994, 2011). Structural connectivity refers to the anatomical connections between neural elements, such as axons and synapses between neurons at the microscale, as well as large-scale fiber bundles or fasciculi linking cortical areas and subcortical nuclei at meso- and macroscales. Structural connectivity is measured using techniques such as **electron microscopy** (micro), axonal **tract-tracing** (meso), and **diffusion MRI** (macro).

Functional connectivity refers to a statistical dependence between physiological recordings that have been acquired from distinct neural elements. Many different measures can be used to quantify these dependencies in the time domain, the frequency (**Fourier**) domain, or the **wavelet**

domain (Box 2.2). Their interpretation depends on the type of recordings being analyzed. For example, a statistical dependence in the spiking output of two neurons can be unambiguously interpreted as having a neuronal origin, whereas correlations in hemodynamic signals measured with functional MRI offer only an indirect window into neuronal communication. In general, the different methods used to record brain activity show a trade-off between spatial resolution, temporal resolution, and the invasiveness and scalability of the technique (Figure 2.1.1). High-precision, invasive methods such as single unit recordings have high spatial and temporal resolution, but are not scalable to large-scale neural systems. Noninvasive methods, such as EEG and functional MRI, are scalable but have poor spatial resolution.

Functional connectivity is estimated at the level of measured neurophysiological signals. Effective connectivity is defined

BOX 2.1   CLASSES OF BRAIN CONNECTIVITY—CONT'D

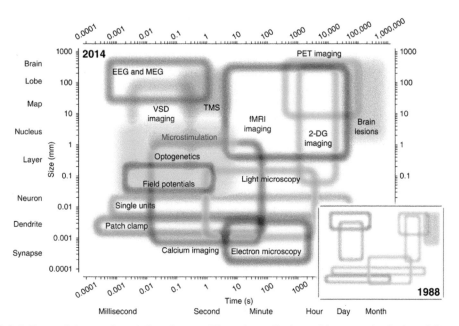

FIGURE 2.1.1 The spatiotemporal resolution of some of the major methods used for measuring brain activity, as available in 2014. Colored areas represent the range of spatial and temporal resolution available for each technique. Open areas are measurement techniques and filled areas are methods based on perturbation, such as lesion studies. The inset is a comparison of methods available in 1988. Note the large gaps for certain domains of spatiotemporal resolution that were subsequently filled by the development of new technologies. In general, higher resolution techniques are more invasive. EEG, **electroencephalography**; fMRI, functional MRI; MEG, **magnetoencephalography**; PET, **positron emission tomography**; TMS, **transcranial magnetic stimulation**; VSD, voltage-sensitive dye; 2-DG, 2-deoxyglucose. *Image reproduced from Sejnowski et al. (2014), with permission.*

at the *neuronal* level and explicitly aims to uncover the direct, causal influences that neural elements exert over each other's activity. Effective connectivity was first used to understand dependencies in the spiking activity of individual neurons, and was defined as the minimum neuronal circuit model that could reproduce the observed signal dependencies (Aertsen et al., 1989, 1994). By emphasizing interactions at the neuronal level, effective connectivity often requires a model of how neuronal dynamics generate the measured signal; for example, a mapping between neuronal activity and the hemodynamic signals recorded with functional MRI (Friston et al., 2003). This mapping allows us to test which of a set of competing models of effective connectivity, specified at the neuronal level, can best account for the observed fluctuations in the measured signals. Such analyses can be computationally intensive, but recent advances suggest that they can be scaled to the analysis of large-scale networks (Friston et al., 2011; Seghier and Friston, 2013).

Continued

BOX 2.1 CLASSES OF BRAIN CONNECTIVITY—CONT'D

Directionality is an important characteristic of structural, functional, and effective connectivity. Structural connectivity is inherently directed, since each axon has a source and a target. However, some measures of structural connectivity, such as diffusion MRI, cannot resolve this directionality and only tell us whether two regions are connected or not. Functional connectivity can be either directed or undirected, depending on the method used for measurement. For example, computing a correlation between two neurophysiological signals results in an undirected measure of functional connectivity, whereas Granger causality analysis, which quantifies the degree to which the activity of one signal can predict future variations of another, yields a directed measure of functional connectivity (Box 2.2).

Effective connectivity is always directed, as it is based on a model of causal interactions between neural systems. The primary distinction between effective connectivity and directed functional connectivity is that the former is specified at the neuronal level whereas the latter is quantified at the level of measured signals. Measures of directed functional connectivity may provide a good index of effective connectivity when the measured signals are a close proxy of actual neuronal dynamics, such as those acquired with invasive electrophysiology. When the measured signals are only indirect markers of neuronal activity, such as the hemodynamic fluctuations recorded with functional MRI, measures of directed functional connectivity can yield imprecise estimates of effective connectivity. The distinction between effective and functional connectivity is discussed in detail by Friston (2009) and Friston et al. (2013). Our focus in this book will be on studies of structural and functional connectivity, as there are relatively few analyses of effective connectivity in large neural systems.

2.1 MICROSCALE CONNECTOMICS

Ramón y Cajal's work firmly established the neuron as the fundamental element of a nervous system (López-Muñoz et al., 2006; Yuste, 2015). It therefore seems logical to conclude that the cellular scale is a natural scale—perhaps even *the* natural scale—at which to study brain connectivity. This logic underpinned the first attempt to generate a connectome—the reconstruction of the nervous system of the nematode worm *C. elegans* at the level of each and every neuron and synapse (White et al., 1986). The principal advantages of connectomic investigations at this scale include the capacity for precise and detailed delineation of nodes and edges. The principal disadvantages include computational burden, labor intensive data acquisition and processing, and poor scalability to large neural systems.

2.1.1 Structural Connectivity at the Microscale

Neural tissue is dense—rough estimates suggest that there are ~90,000-100,000 neurons, about a billion synapses, and kilometers of axons and dendrites in the depth of cortex beneath a 1 mm^2 patch of the human pial surface (Logothetis, 2008). Within any given block of tissue, cells and their processes are densely intertwined, making it very difficult to resolve their fine structure. In small model systems such as the mouse and fly, dendrites can have diameters of 30-50 nm; the resolution of any imaging technique must be around half this size if these structures are to be resolved accurately (Helmstaedter, 2013). These resolutions are smaller than the visible wavelength of light, meaning that such structures cannot be resolved with traditional **light microscopy**.

This resolution limit posed a problem for early investigators such as Ramón y Cajal and Golgi, for whom the light microscope was the only tool available. Camillo Golgi's seminal innovation was to show that light microscopy could resolve a single neuron and its processes if the neuron was stained by dye, and without staining other nearby neurons. He thus developed a method for randomly staining a sparse subset of neurons via the immersion of a tissue specimen in a solution of silver nitrate (Pannese, 1999). The dye labels spatially separated individual neurons and their branches in their entirety. Ramón y Cajal subsequently improved the technique, and used it as the basis for his foundational studies of neuronal morphology.

Since that time, a variety of other staining and fluorescence-based methods have been developed. These methods have become an essential part of the neuroanatomists' toolkit. However, the sparsity of the labeling that they provide means that they cannot be used for comprehensive mapping of neuronal connectivity. Another limitation is that light microscopy cannot easily resolve individual synapses, although super-resolution methods are available (Betzig et al., 2006; Klar et al., 2000). Ramón y Cajal circumvented this problem by inferring a connection between two neurons if they made contact, but he was unable to demonstrate a direct synaptic **link**. This ambiguity led to a major theoretical debate between Ramón y Cajal and Golgi: Ramón y Cajal proposed that neurons were spatially distinct entities that communicated via synaptic junctions, whereas Golgi argued that neurons formed a single, continuous network. It was only with the advent of electron microscopy in the 1950s that this debate was resolved in favor of Ramón y Cajal's model (Chapter 1; for a historical account, see López-Muñoz et al., 2006).

Electron microscopy images the passage of electrons through a tissue specimen that has been stained with heavy metal elements. Some electrons pass through the tissue while others are scattered, creating a contrast that can be used to resolve anatomical features (Agar et al., 1974). The shorter wavelength of electrons compared to photons allows an electron microscope to resolve much smaller structures than a typical light microscope. In fact, electron microscopy is able to resolve all cells and processes within a block of tissue, thereby enabling the dense reconstruction of neuronal connectivity. The method was a natural choice when White et al. (1986) set out to comprehensively reconstruct the nervous system of *C. elegans*.

Commonly used variants of electron microscopy are transmission electron microscopy (TEM) and scanning electron microscopy (SEM; Briggman and Bock, 2012). With TEM, thick sections of tissue are first extracted and embedded within a fixing solution. The thick sections are manually cut into thinner sections using an ultramicrotome with a diamond knife. They are then placed, one at a time, onto a support film that is transparent to electrons. The sections are stained and placed under the TEM column, which fires electrons through the specimen. The scattered electrons pass through magnifiers before being detected by a digital camera. Each pixel in the camera corresponds to a location in the section. The pixels are imaged in parallel, up to a section size of ~1 mm × 2 mm (Figure 2.3a).

SEM uses a finely focused electron beam across a small, rectangular area of the specimen, serially scanning one rectangle at a time in a process that is called *raster* scanning. Some variants of SEM allow for the automated sectioning and transfer of slices to the scanner, thus removing the need for time-consuming and error-prone manual intervention (Figure 2.3b; Hayworth et al., 2006). In serial blockface SEM, a diamond knife is used to section the tissue within a low-vacuum chamber of the microscope (Figure 2.3c and d). The image is acquired directly on the tissue blockface, thus avoiding the need for postscan slice alignment (Denk and Horstmann, 2004). For overviews of TEM and SEM, interested readers are referred to articles by Briggman and Denk (2006) and Briggman and Bock (2012). Other useful discussions are provided by Kleinfeld et al. (2011) and Lichtman et al. (2014).

After imaging, the sections of neural tissue are digitally stacked to form a three-dimensional volume. This volume is then segmented into different tissue compartments such as neurons, glia, blood vessels, and so on (Figure 2.3e–i). Current automated algorithms are prone to error and often require manual intervention (Figure 2.3e–h). Once reconstruction is complete, the structures need to be annotated so that the locations of individual synapses can be identified, thus forming the basis of a brain graph (Figure 2.3j).

Reconstructing connectomes in this way is computationally demanding, time-consuming, and labor intensive. In terms of computation, a single cubic millimeter of rat cortex imaged with a resolution in the order of a few nanometers will generate ~2 PB (2 million GB) of data, meaning that a complete atlas of rat cortex, corresponding to a volume of ~500 mm^3, would require around 1 EB (1000 PB) and that a complete human cortex would require about 1000 EB (Lichtman et al., 2014). In terms of time and labor, White et al.'s (1986) efforts to reconstruct the relatively small nervous system of *C. elegans* took over 10 years. State-of-the art acquisition methods require about 800 h to reconstruct 1 mm^3 of neural tissue, meaning that a single machine would require over a decade to scan the mouse brain in its entirety (Lichtman et al., 2014).

Once the images have been acquired, a major bottleneck is the substantial time required to accurately segment and annotate even modestly sized patches of neuropil. To get around this problem, one study created a dense reconstruction of connectivity between 950 neurons of the mouse retina by

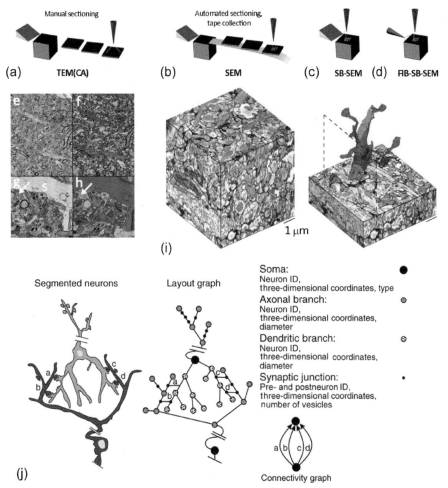

FIGURE 2.3 Mapping neuronal connectivity with electron microscopy. **(a–d)** Examples of different approaches to electron microscopy. **(a)** With TEM, thin sections are cut manually and placed under the electron beam of the microscope for imaging. TEM(CA) corresponds to TEM combined with fast camera arrays. **(b)** In SEM, the sectioning can be automated and the slides are placed on a conveyor-like belt, which passes through the microscope. **(c)** In serial block face SEM (SB-SEM), a diamond knife is used to cut sections within a chamber of the scanner, and imaging is performed on the block face. **(d)** In another variant of blockface SEM, a focused ion beam (FIB) is used to cut the sections. **(e)** An electron micrograph of a 30 nm-thick section of mouse cortex, showing the density of cells and processes within a 40 µm × 20 µm patch of tissue. **(f)** Example output of an automated segmentation algorithm that assigns a distinct color to different tissue classes. **(g)** A magnified section of the cortical patch shown in **(e)** and **(f)**. **(h)** The same magnified section segmented by a human operator. The white arrow points to a discrepancy between the human and computer segmentation. The human operator correctly identifies this structure as a single axon, whereas the computer assigns different tissue classes to the vesicles within the axon. This discrepancy demonstrates the complexity of the segmentation task, and the need for human intervention. **(i)** Segmentations must be consistent across all two-dimensional sections of the three-dimensional volume that has been imaged. Shown here is a volume of rat barrel cortex (left) sampled with in-plane resolution of 13.2 nm/pixel. The volume comprises 253 sections with 30 nm thickness. Right shows a manually traced dendritic spine (red). Once such structures have been constructed, individual neurons and synapses must be annotated to generate a connectome. **(j)** The segmented data can then be converted into a graph. Left image (segmented neurons) shows a schematic of synaptic contacts between two neurons. The dark gray neuron has four synaptic contacts with the spines of the light gray neuron (a–d). The middle image (layout graph) shows a simplified summary of connectivity that retains information about the location of every synaptic contact and dendritic branch. Circles of different sizes and shading correspond to different neuron identifiers, as shown in the legend. This graph can be further simplified as a graph of two nodes linked by four edges (bottom right). If we take the number of synapses between two neurons as an index of connectivity strength, we can further simplify the graph by drawing a single edge with a weight of four. *Reproduced **(a–d)** from Helmstaedter (2013), **(e–h,j)** from Lichtman et al. (2014), and **(i)** from Briggman and Denk (2006).*

using a crowd-sourcing approach in which the decisions of over 200 part-time undergraduates were combined into a consensus reconstruction (Helmstaedter et al., 2013). Even so, completion of this task took over 20,000 man hours of annotation time.

Despite its poor scalability, electron microscopy remains a powerful approach for mapping neuronal connectivity within spatially confined circuits. At first glance, the subcellular resolution afforded by the technique naturally provides a clear way of defining nodes (as neurons) and edges (as synapses). Important decisions are nonetheless required. In defining nodes, we must consider that neurons vary in their size, shape, and function, and that these variations can be used to distinguish between different classes of neurons. This heterogeneity can be vast. For example, 56 distinct cell types were identified from a subset of just 290 cells of the *D. melanogaster* medulla (Takemura et al., 2013), and rough estimates suggest that there may be as many as 5000 different types of neuron in the mammalian brain (Bota et al., 2003). Given this heterogeneity, should each neuron be treated the same? Should we only represent connectivity between specific classes of neurons? Should we represent edges differently depending on the types of neurons that they connect? Graph theory offers tools for representing such variations (Chapter 3), but the answer to these questions will depend on the specific research aims of the investigator.

Defining edges in densely reconstructed neuronal networks also requires careful consideration. Manual annotation of presynaptic terminals and postsynaptic densities is time-consuming, laborious, and prone to human error (Plaza et al., 2014). Moreover, in some cases, it may be necessary to image a large volume of tissue to accurately determine whether two neurons are connected or not. More specifically, in order to comprehensively rule out a synapse between two neurons, it is necessary to reconstruct either the entire extent of the output part (that is, the axon) of a presynaptic neuron, or the input part (that is, the soma and dendritic tree) of a postsynaptic neuron (Figure 2.4a). As a result, specific neural circuits will have a minimum volume that must be imaged in order for their connectivity to be mapped comprehensively (Figure 2.4b–e). The smallest dimension of this volume is called the minimal circuit dimension (Figure 2.4f; Helmstaedter, 2013). The axons of some neurons can extend over very long distances, such as across cerebral hemispheres or down the spinal column. Full reconstruction of the morphology of the neurons and neurites in these circuits thus requires a large proportion of the brain to be imaged (Lichtman and Denk, 2011).

Beyond simply determining the presence or absence of a synaptic contact between neurons, ambiguities also arise in estimating synaptic strength. In some organisms, synapses are of comparable size, and the number of synapses between two neurons is a suitable index of connection weight (Takemura et al., 2013).

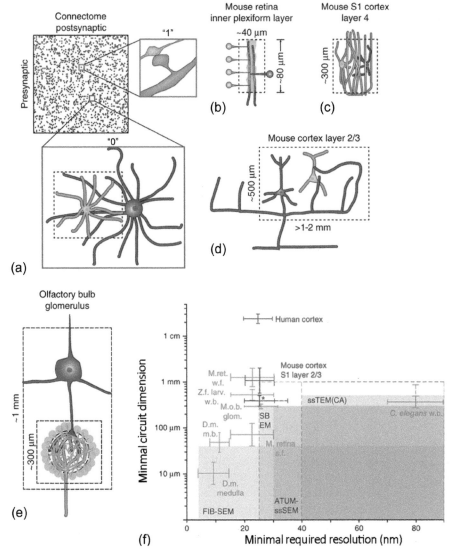

FIGURE 2.4 The minimal circuit dimension for accurate mapping of neuronal connectivity. **(a)** An example connectivity **matrix** of a neuron-level connectome. Blue elements correspond to the presence of a synapse ("1" in the matrix) and white elements correspond to the absence of a synapse ("0" in the matrix). As shown in the green box, determining the presence of a synapse between two neurons is a local decision on the order of a few micrometers. As shown in the red box, ruling out a synapse requires at least one of the neurons to be reconstructed in its entirety. In the case shown here, the full dendritic tree of the blue neuron has been imaged (dashed box), allowing us to comprehensively search for the presence or absence of a synapse. The minimal circuit volume is the volume of tissue required to make such a decision for a given circuit (in this case, the volume of the dashed box); the minimal circuit dimension is the smallest dimension of the imaging volume required to make such a decision (that is, the smallest dimension of the dashed box encasing the blue neuron). **(b–e)** Approximate minimal circuit dimensions (dashed boxes) for some example circuits. **(f)** Approximate minimal circuit dimension and minimal required resolution for accurate reconstruction of neuronal connectivity across different species commonly studied in connectomics. Gray shading corresponds to the resolutions attainable with current electron microscopic techniques and dashed lines indicate likely future improvements. ssTEM(CA), serial section TEM with fast camera arrays; SB-SEM, serial blockface SEM; ATUM-ssSEM, automated serial section tape-collection SEM; FIB-SEM, focused ion beam SEM; M.ret. w.f., mouse retina, wide field; Z.f. larv., zebrafish larva; D.m., *D. melanogaster*; w.b., whole brain; M.o.b. glom., m.b. = mouse olfactory bulb, 1 glomerulus, only intraglomerular circuitry; M. retina s.f. = mouse retina, small-field; D.m.m.b. = *Drosophila melanogaster* mushroom body. *Images reproduced from Helmstaedter (2013).*

In other species, the size of each synapse, or the number of vesicles within each synaptic terminal, may be a more appropriate measure (Jarrell et al., 2012; Plaza et al., 2014).

Electron microscopy also allows us to distinguish between different types of connections. For example, White et al.'s (1986) initial map of the *C. elegans* nervous system differentiated between chemical synapses and gap junctions. Other distinctions are possible, such as the separate identification of excitatory and inhibitory synapses based on cell morphology (Bock et al., 2011), or the classification of synapses based on their location, such as whether the synapse is axosomatic, axodendritic, axoaxonal, and so on. Neurons can also communicate without a direct synapse or gap junction, via mechanisms such as volume transmission, the spillover of neurotransmitters into nearby synapses, and diffuse hormonal and peptidergic signals (Diamond, 2002; Oláh et al., 2009). Moreover, dendrites can process information by interacting with other spines in the local dendritic branch, thus generating outputs independently of the soma (Branco and Häusser, 2010). These interactions can be missed by a purely morphological approach.

An important consideration in microscale connectomics is that the location and spatial distribution of synapses can change rapidly as a function of development, experience-dependent plasticity, and sleep-wake cycles (Tononi and Cirelli, 2014). For example, one study found that the formation and dissolution of synapses in macaque primary visual cortex over the course of 1 week affected around 7% of the total cell population, in the absence of overt changes in visual stimulation (Stettler et al., 2006). This variability suggests that the static picture obtained by a dense reconstruction may offer only a partial representation of the true complexity of synaptic connectivity. A discussion of this and other issues pertaining to microscale connectomics is provided by Morgan and Lichtman (2013).

2.1.2 Functional Connectivity at the Microscale

Structure constrains function in many areas of biology. The brain is no exception, and it follows that accurate measurement of structural connectivity is critical for mapping the anatomical backbone of brain function. Indeed, this logic underpins the central focus in connectomics on the generation of structural wiring diagrams for nervous systems (Sporns et al., 2005). However, brain function is not completely determined by structure. Rather, anatomical networks act as constraints on the possible repertoire of functional network states (Marder, 2015; Marder and Goaillard, 2006; Weimann and Marder, 1994). Structural connectivity is thus a foundation from which myriad functional circuits can be activated, depending on the prevailing context (Bargmann and Marder, 2013). Ultimately, it is the activity that unfolds on the network structure that drives behavior; a consideration of functional connectivity is thus a central component of connectomics.

The importance of functional interactions between neurons is embodied in the concept of a cell assembly, proposed by Hebb (1949) as a mechanism by which multiple, distinct features of a stimulus can be represented as a coherent entity through the joint activity of distributed, interconnected neurons. Neuronal inhibition is known to play a critical role in forming these assemblies, as it allows the precise alignment of the spiking activity of different cells within specific time windows (Singer, 2013). In particular, precisely timed inhibitory signals synchronize oscillations between populations of neurons at specific phases, thus coordinating fluctuations in neuronal excitability (Fries et al., 2001; Singer, 1999; Womelsdorf et al., 2014). This coordination is important for maximizing signal-to-noise and circuit specificity, since the probability that any individual neuron will fire depends on the integration of its afferent signals. Put another way, an action potential is more likely to be triggered in a target neuron if afferent signals arrive at the same time, and in periods when the target neuron is highly excitable. Coordinated oscillatory activity is thus thought to mediate neuronal communication—the so-called communication-through-coherence hypothesis (Bastos et al., 2015a; Fries, 2005).

The classic method for measuring synchronized neuronal activity at cellular resolution uses invasive electrophysiology. Electrodes inserted into distinct parts of the brain record the spiking activity of either individual or multiple cells. Measures of statistical dependence between the activity traces recorded from these cells are then computed (Aertsen et al., 1994; Gerstein and Perkel, 1969; Gerstein et al., 1978). For example, multiunit recordings in the primary visual cortex of the cat were used to show that adjacent neurons within a single orientation column show synchronous activity when exposed to a visual stimulus of optimal orientation, velocity, and preferred direction of movement (Gray and Singer, 1989). Subsequent work showed that responses were also synchronized across spatially segregated columns in response to global, rather than local stimulus properties (Gray et al., 1989), suggesting a potential mechanism through which functionally specialized cell assemblies are bound together to form coherent perceptions of the world (Singer, 1999; Singer and Gray, 1995).

Invasive electrophysiology can offer a detailed window into the activity of individual neurons, and can be adapted to investigate interactions between reasonably large neuronal networks. For instance, one study was able to demonstrate **small-world** properties in functional connectivity networks generated from recordings acquired in 63 individual neurons of the cat visual cortex (Yu et al., 2008). In this analysis, each node represented a specific neuron and edges were computed as correlations in spiking activity (see also Box 2.2).

BOX 2.2 MEASURING FUNCTIONAL CONNECTIVITY

Functional connectivity is quantified using measures of statistical dependence between neural time series. These time series can either be discrete (for example, spike trains) or continuous (for example, LFPs). This means that, in principle, functional connectivity can be quantified using any generic method for estimating statistical dependencies in time series data. Here, we focus on some key considerations when selecting an appropriate measure of functional connectivity. Interested readers are referred to an introductory text on the analysis of neural time series by Cohen (2014). Commonly used measures of signal dependence in neuroscience are also considered by Greenblatt et al. (2012) and Smith et al. (2011).

The Pearson correlation coefficient is perhaps the most widely used measure of functional connectivity. It estimates the linear correlation between pairs of simultaneously measured time series. A high correlation indicates in-phase coupling between the signals; a correlation close to zero indicates no linear relationship; and a negative correlation indicates an out-of-phase relationship. It is also possible to shift one of the time series with respect to the other to investigate lagged correlations. The correlation coefficient is both simple and versatile, having been adapted to quantify dependencies in high-resolution recordings of neuronal spike activity (Averbeck et al., 2006) and low-resolution recordings of hemodynamic signals (Biswal et al., 1995). To ensure valid statistical inference, prewhitening or wavelet methods can be used to correct for the **autocorrelation** of neural activity, which otherwise violates the assumption of independence between data points (Bullmore et al., 1996a, 2001; Patel and Bullmore, 2015).

Key properties of the correlation coefficient usefully highlight some important considerations when measuring functional connectivity. One such property is that the correlation coefficient only quantifies linear dependencies. Other measures that are sensitive to both linear and nonlinear interactions, such as mutual information (Salvador et al., 2007; Chapter 8), and synchronization likelihood (Stam and van Dijk, 2002), can offer a more complete characterization of functional connectivity if strong nonlinear interactions are expected.

A second property is that the correlation coefficient is sensitive to indirect, "third-party" effects. These effects arise when two network nodes show correlated activity in the absence of a direct anatomical connection because they are both connected to a third, intermediary node (Box 7.1; Chapter 10). Other shared inputs, such as the effect of a common driving stimulus (so-called coactivation) or fluctuations in physiological and neuromodulatory processes, can also cause correlated signal fluctuations (Ostojic et al., 2009; Scholvinck et al., 2011). In brain networks, this means that the correlation coefficient will be sensitive to polysynaptic and other indirect influences. It also means that the values of certain network measures, such as the **clustering coefficient**, can be artificially inflated (Zalesky et al., 2012b; Chapter 10). Partial correlations attempt to deal with this problem by estimating the correlation between two nodes after removing variance that is shared with other system elements (Marrelec et al., 2006). However, partial correlations can sometimes over-compensate and artificially reduce the clustering of large-scale networks (Zalesky et al., 2012b). Regularized measures of partial correlation have been suggested, which shrink the values of indirectly correlated regions to zero and can result in improved estimation of direct interactions (Varoquaux and Craddock, 2013; Yatsenko et al., 2015).

A third property of the correlation coefficient is that it does not distinguish between the two principal coupling modes hypothesized to underlie brain function: phase coupling and amplitude coupling (Section 2.2.2; Figure 2.12a and b). Phase coupling is measured within specific frequency bands, so one option is to compute correlations in the spectral or wavelet domain, after the signals have been filtered within a given frequency range. A popular method for quantifying covariance in the spectral domain is coherence (Gardner, 1992), which is very similar to the correlation of frequency-filtered signals (Guevara and Corsi-Cabrera, 1996). However, coherence can be sensitive to changes in signal amplitude and thus bias estimates of phase synchrony. More specific measures of phase coupling have been developed that are not sensitive to amplitude fluctuations (Lachaux et al., 1999). In general, these methods examine differences in the phase angles between two time courses at different lags; if two signals show a consistent phase relationship over time, they have higher phase coupling (Figure 2.12a).

Amplitude coupling can be estimated as the correlation of the power fluctuations (that is, the amplitude envelope) of the frequency-filtered signals, provided that the signals are sampled with sufficiently high temporal resolution

BOX 2.2 MEASURING FUNCTIONAL CONNECTIVITY—CONT'D

(Figure 2.12b). In functional MRI, the measured signals are relatively slow and functional connectivity in this context is often measured as the correlation in amplitude fluctuations of the full signal, rather than just the amplitude envelope.

A fourth property of the Pearson correlation is that it is a scalar summary—it describes activity recorded over an extended period of time with a single value. It thus assumes that connectivity is stationary and that it does not vary dynamically over time. This assumption contradicts both intuition and experimental evidence that neuronal networks rapidly synchronize and desynchronize their activity under different psychological states (Varela et al., 2001). A simple solution to this problem is to perform a **sliding window analysis**. For example, given two time series with 500 data points each, we could compute a correlation coefficient using a window that includes only the first 30 points. We then move the window by one point and compute a correlation between values 2 through 31, and continue sliding the window along one time point at a time, each time estimating a correlation. The result is a time series of correlation coefficients estimated from overlapping windows (nonoverlapping windows can also be used). This approach is useful for examining time-resolved networks measured with low temporal-resolution techniques such as functional MRI (Zalesky et al., 2014). It is also useful for mapping dynamic network changes using high-resolution signals, such as those measured with MEG, where the evolution of neural synchronization can be tracked in relation to changes in mental states at time scales in the order of 10 ms (Kitzbichler et al., 2011).

A fifth property of the correlation coefficient is that it is undirected—it tells us how two signals A and B are related, but it does not tell us whether signal A drives fluctuations in signal B or vice-versa. Alternative measures of directed functional connectivity are available. For example, Granger causality is one popular method that infers causal influences between signals based on temporal precedence: if fluctuations in signal A can predict subsequent fluctuations of signal B, then A is said to cause B (in a Granger sense; see Figure 2.12c). Inferences about directed functional connectivity are generally more accurate when analyzing signals that are a more direct marker of neuronal activity, such as those recorded with invasive electrophysiology. For low-resolution methods such as functional MRI, variations in the speed of the hemodynamic response across different areas, coupled with the complex mapping between the recorded signal and underlying neuronal activity, can distort inferences about actual causal interactions occurring at the neuronal level (Friston et al., 2013). Models of effective connectivity can prove useful in this regard (Box 2.1). We will see throughout this book that the distinction between directed and undirected measures of connectivity has important implications for how we construct and analyze a brain network.

Finally, values of the correlation coefficient are bounded between −1 and 1, and the polarity of the coefficient tells us something about the type of interaction between nodes: a positive value implies cooperation and integration whereas a negative value implies antagonism and segregation. Signed edge weights mandate the use of special methods in brain network analysis (see Chapters 4, 7, and 9). Other measures of functional connectivity, such as mutual information, only assume positive values. This may simplify analysis, but it prevents us from distinguishing between different types of functional interactions.

Multi-electrode arrays (MEAs) have also been used to record the activity of neurons cultured in vitro. Primary cultures of disaggregated mouse embryonic neurons form structural and functional networks over the course of a few weeks of growth. Simultaneous recording of **local field potentials** (LFPs) (see also Section 2.2.2) from an array of electrodes embedded in the culture surface provides a rich measure of activity with spatial resolution in the order of a few neurons per electrode. Bursts of high amplitude oscillation (activation) often cascade across all or some of the recording electrodes, with each spatially patterned cascade of activation comprising a so-called neuronal avalanche (Beggs and Plenz, 2003). Comparable cascading behavior has been identified using in vivo electrophysiology in the macaque monkey (Petermann et al., 2009) and by functional MRI and magnetoencephalography (MEG) studies in humans (Kitzbichler et al., 2009). This convergence indicates that some dynamical principles of functional microscale networks in culture may be consistently represented at larger spatial scales and in more intact nervous systems.

There is also evidence that the topological properties of functional networks derived from MEA recordings are similar to those observed in larger systems. For example, a functional connectivity matrix was estimated repeatedly over time from an array of 64 electrodes recording from a growing network of mouse hippocampal neurons (Schroeter et al., 2015). For graph analysis, each node was defined as an electrode, with each electrode expected to record electrical activity from about three neighboring neurons. Each edge represented the temporal covariance between electrophysiological signals recorded from distinct electrodes. The resulting graphs exhibited many of the topological properties observed in vivo, such as the presence of highly connected **hubs**, and dense interconnectivity between hubs, consistent with a **rich-club** organization (Chapter 6). However, these electrophysiological methods, whether applied in vivo or in vitro, are generally not adept at comprehensively mapping functional interactions between large numbers of cells or cell assemblies in an intact nervous system, particularly when neuronal populations are separated by long anatomical distances.

One method that has proven powerful in mapping neuronal interactions across large distances with cellular resolution is calcium imaging, which measures intracellular calcium levels by introducing specific molecules, called calcium indicators, to neural tissue either by injection or transgenic methods (Figure 2.5a). These indicator molecules change their fluorescence in response to the binding of Ca^{2+} ions, and the changes in fluorescence are detected using optical imaging techniques, such as confocal microscopy or two-photon microscopy (Figure 2.5b). Action potentials cause the cellular influx of calcium through voltage-dependent calcium channels; intracellular calcium can thus be used as a very accurate marker of spiking activity (Kerr et al., 2005). These activity-dependent calcium signals can be resolved at the level of presynaptic neurons, postsynaptic terminals, axons, dendrites, and local populations (Grienberger and Konnerth, 2012; Jennings and Stuber, 2014; Kerr et al., 2005). An overview of the principles and applications of calcium imaging is provided by Grienberger and Konnerth (2012).

Calcium imaging can be combined with dense electron microscopic reconstructions of structural connectivity to understand structure-function relationships within specific neural circuits. For example, Bock et al. (2011) used two-photon

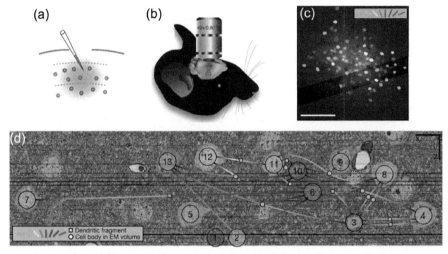

FIGURE 2.5 Calcium imaging of the mouse brain. **(a)** A schematic showing the injection of a calcium indicator dye into a local patch of cortex. Dotted lines represent different cortical layers. **(b)** An experimental preparation for calcium imaging of the mouse brain, in which a two-photon microscope is fixed to the head. **(c)** Orientation preferences of individual neurons in mouse primary visual cortex, as measured with calcium imaging. Each color identifies a neuron with a distinct orientation preference. **(d)** A subset of the functionally defined neurons in **(c)**, overlaid on the anatomical volume of cortex in which they are embedded, as reconstructed with electron microscopy. Colored lines represent synapses made by these neurons with either the soma (cyan squares) or dendrites (cyan circles) of local inhibitory interneurons. The connections from multiple functionally defined neurons converge on the interneurons. *Reproduced **(a)** from Grienberger and Konnerth (2012), **(b)** from Jennings and Stuber (2014), and **(c,d)** from Bock et al. (2011), with permission.*

calcium imaging to map the orientation preferences of a population of layer 2/3 neurons in mouse visual cortex (Figure 2.5c). They then used electron microscopy to reconstruct the synaptic links between a subset of 14 neurons within this larger, functionally defined population, and used morphological criteria to distinguish excitatory from inhibitory cells. The authors found that inhibitory interneurons often received convergent input from multiple excitatory cells, and that this convergent input was determined by spatial proximity rather than orientation preference (Figure 2.5d). In particular, the strongest predictor of convergent input to an interneuron was the number of spatially nearby synaptic terminals. This finding suggests that geometry rather than functional specialization is the major predictor of synaptic connectivity, at least with respect to excitatory input to interneurons within local cortical circuits. This result is consistent with Ramón y Cajal's laws of conservation of material and space, which imply that the minimization of axonal volume (representing the wiring **cost** of the network) is an important constraint on network organization (see also Chapter 8).

Calcium imaging can be used to record the activity of much larger sets of neurons. For instance, functional connectivity graphs have been constructed from calcium recordings of the activity of neuronal populations comprising about 1000 cells (Gururangan et al., 2014; Sadovsky and MacLean, 2013, 2014). In these networks, each neuron is a node and edges correspond to correlations in the spiking activity of each cell, as estimated from the recorded calcium signals. Analyses of these networks have revealed many of the same topological properties that have been identified in the analysis of macroscopic brain networks, such as the small-world combination of high clustering and low average **path length** (Chapter 8), the presence of highly connected hub nodes (Chapter 4), and hierarchical, modular organization (Chapter 9).

Despite this impressive capacity of calcium imaging to record the activity of relatively large numbers of cells, a major drawback of the technique for the study of mammalian brains is that it can only sample a restricted, superficial patch of cortex at any given time (although some developments are enabling the imaging of deeper tissue; Jung, 2004). This limitation can be overcome in smaller model organisms. The larval zebrafish has proven to be a powerful model in this regard, as it has a small and transparent brain comprising around 10^5 neurons. The fish can be paralyzed and placed in a virtual swimming environment so that neuronal dynamics can be monitored, across the entire brain, during specific behaviors (Figure 2.6a; Ahrens et al., 2012). In one study, high-speed light-sheet microscopy was used to record genetically encoded calcium signals from more than 80% of all neurons, at a sampling rate of 0.8 Hz (Figure 2.6b and c; Ahrens et al., 2013). In a preliminary investigation of functional connectivity, the time series of calcium signal fluctuations from individual neurons were correlated with signals recorded everywhere else in the brain to map distributed neural systems showing correlated fluctuations in spontaneous neural activity. These fluctuations evolved over slow time scales, with a period of 20-30 s. For a chosen reference neuron in the hindbrain, these slow fluctuations were correlated across populations of neurons that were symmetrically located about the midline: an ipsilateral population fluctuated in-phase with the reference neuron, whereas a contralateral population fluctuated in antiphase (Figure 2.6d). Similar combinations of anatomically distributed in-phase and antiphase covariations in activity are commonly observed in analyses of slowly fluctuating hemodynamic signals measured with human functional MRI (Fox et al., 2005).

Calcium imaging has also been used for large-scale, cellular-resolution recording in *C. elegans*, with one team being able to sample activity at 4-6 Hz in over 70% of neurons in the head ganglia (Schrödel et al., 2013). The integration of calcium imaging with genetic labeling of specific neuronal subtypes and optogenetics further allows for the targeted manipulation and monitoring of activity in precisely defined neural systems (Jennings and Stuber, 2014). Calcium imaging is

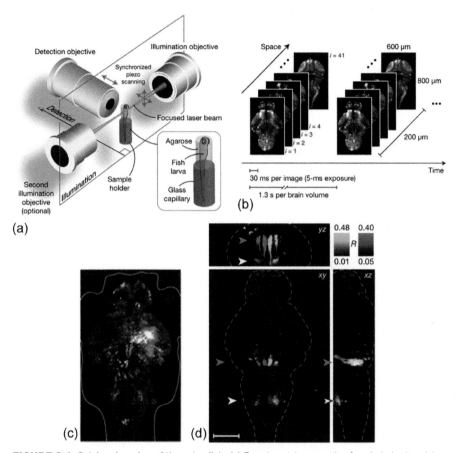

FIGURE 2.6 Calcium imaging of the zebrafish. **(a)** Experimental preparation for whole-brain calcium imaging of the larval zebrafish with light-sheet microscopy. A focused laser is passed through the fish, causing the calcium indicators to fluoresce. This fluorescence is detected by an appropriately positioned microscope. **(b)** High resolution images are acquired in steps of 5 μm every 30 ms, resulting in a volume of 800 μm × 600 μm × 200 μm (which contains the entire zebrafish brain) once every 1.3 s. **(c)** Dorsal projection of whole-brain activity in the larval zebrafish, recorded using a genetically encoded calcium indicator and at single-cell resolution. **(d)** Spatially distributed neurons in the anterior hindbrain (blue arrows) and inferior olive (white arrows) that show positively correlated (green) and negatively correlated (magenta) calcium signal fluctuations with a single reference neuron. Front view shown on top, top view on bottom, and side view on right. *Reproduced **(a,b)** from Ahrens et al. (2012) and **(c,d)** from Ahrens et al. (2013), with permission.*

thus a powerful and promising tool for large-scale functional connectomics at cellular resolution in vivo. It allows nodes to be defined at the level of individual neurons, and edges to be defined as statistical dependencies either in temporal fluctuations of the calcium signals (Ahrens et al., 2013), or of spiking activity estimated from these signals (Lütcke et al., 2013; Vogelstein et al., 2010). A limitation of both calcium imaging and invasive electrophysiology is that they are not applicable to the analysis of distributed brain dynamics in large animals. Alternative methods for measuring activity at single-cell resolution in large neural circuits are under development and are reviewed by Alivisatos et al. (2012).

2.2   MESOSCALE CONNECTOMICS

Microscale connectomics offers unparalleled precision for resolving the synaptic connectivity and spiking activity of individual neurons, but its techniques are not scalable to large-scale neural systems. The high plasticity of synaptic connectivity also makes it difficult to distinguish stable characteristics of neuronal networks from more transient features. A coarse-grained approach, such as that afforded by a mesoscale analysis of the whole-brain connectome, can smooth out some of this variability and offer a more robust means for characterizing time-invariant aspects of brain architecture (Bohland et al., 2009; Mitra, 2014).

Mesoscale connectomics typically uses microscopic methods with the aim of mapping connectivity between neuronal populations or cell assemblies, rather than individual neurons. It takes advantage of the fact that neurons aggregate into functionally related, spatially circumscribed populations, which can range in size from cortical columns to larger cytoarchitectonic areas and subcortical nuclei. Each of these larger areas, which may comprise thousands or millions of cells, is treated as a single node. Using this coarser approach to defining nodes inevitably results in the loss of information encapsulated by the interactions and connectivity patterns between individual neurons. For example, if we treat the primary visual area V1 as a single node, we will ignore variability in the response properties of individual neurons that are tuned to different stimulus properties. This loss of information is counterbalanced by an improved ability to map network structure over long distances. That is, by mapping connectivity of large populations of cells rather than individual neurons, we can study interactions between areas distributed throughout the entire brain.

2.2.1   Structural Connectivity at the Mesoscale

Invasive **tract tracing** is the primary method used to map structural connectivity in mesoscale connectomics. In a typical tract tracing experiment, a fluorescent dye or other tracer molecule is injected into a specific part of the brain. Cellular membranes are permeable to these tracers. Once inside the cell, the tracer is transferred between the soma and peripheral axon terminals via active axonal transport. After the tracer has had sufficient time to fill the entire extent of the labeled neurons, the animal is sacrificed, its brain dissected, and sites of tracer uptake are mapped.

Tracers are principally distinguished by their direction of transport. Antero-grade tracers are transported from the cell body to the axon terminal and are thus used to map the efferent projection sites of an injected area. Retrograde tracers are transported from the cell periphery to the soma and are used to iden-tify the upstream sources of afferent projections to the injection site. Many anterograde and retrograde tracers are available. Few are exclusively transported in one direction, but most show a strong preference. Traditional tracers are not transported trans-neuronally, but a new class of viral tracers is able to travel across synaptic junctions, allowing the mapping of polysynaptic pathways. For an overview of the strengths and weaknesses of different tracers, interested readers are directed to a review by Lanciego and Wouterlood (2011). A consid-eration of the relative utility of different tracers for mesoscale connectomics is provided by Bohland et al. (2009).

Early work used radioactively labeled amino acids as tracers, and measured tracer uptake by autoradiography of tissue sections sampled from selected regions-of-interest that were hypothetically expected to connect to the injected site. The primary aim in this work was to determine whether or not specific brain areas were connected with each other, and estimates of connection strength were largely qualitative (Figure 2.7a). This was painstaking work—each injection required the preparation, dissection, and analysis of a different animal, often involving time-consuming mapping of cytoarchitecture to pre-cisely delineate anatomical areas (nodes) of interest. Any single study thus involved the injection of just a few sites in a few animals.

As discussed in Chapter 1, the first attempt to systematically collate findings across a large number of published tract tracing studies was the work of Felleman and Van Essen (1991). They constructed the first connectivity matrix of an extended brain network, comprising 32 areas of the macaque brain that are involved in vision (Figure 1.1a). This basic idea was later extended by Kötter, Stephan, and colleagues to develop the online, freely accessible Colla-tion of Connectivity data for the Macaque (CoCoMac) database (Stephan et al., 2001; Stephan, 2013), which curates the findings of hundreds of tract-tracing studies (Bakker et al., 2012; Blumenfeld et al., 2014). Parallel efforts led to the construction of a connectome for the cat that was also based on the collation of tract-tracing findings in this animal (Hilgetag et al., 2000; Young, 1992). These two datasets were instrumental in some of the first graph theoretic studies of brain network organization (Chapter 1). A similar online database of tract-tracing findings for the rat and mouse brains, called the brain architecture man-agement system (BAMS), has been developed (Bota et al., 2003, 2005). Tract-tracing findings in the pigeon brain have also been collated in the form of a connectivity matrix (Shanahan et al., 2013).

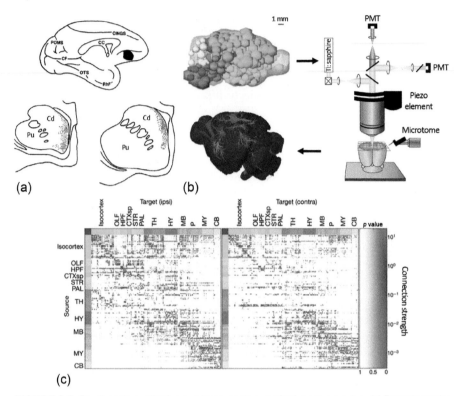

FIGURE 2.7 Small-scale qualitative and large-scale quantitative tract-tracing. **(a)** Example results from a traditional tract tracing experiment. Top shows the location of a tracer injection to the Macaque prefrontal cortex. Bottom panels show sites of tracer uptake in the striatum. The images are hand-drawn, and estimates of connection strength are based on qualitative evaluation of the data. **(b)** Schematic of a modern experimental set-up for whole-brain connectivity mapping using light microscopy. Injections are made at multiple sites distributed throughout the brain. Top left image shows the anatomical distribution of right hemisphere injection sites used to map the mouse connectome. Each brain is imaged one section at a time at the blockface using a light microscope (right panel). A microtome sections off the imaged tissue slice. The sections are digitally stacked and used to reconstruct the spatial distribution of tracer uptake associated with each injection. Bottom left panel shows the tracts mapped following an injection into primary motor cortex. **(c)** The digital images can be spatially registered to an anatomical template, which is used to define network nodes. In the example connectivity matrix shown here case, a statistical model was used to parse the contributions of connectivity between each pair of regions to the measured fluorescence signal. Elements of this matrix correspond to model-based estimates of connectivity between 295 anatomical regions. Variation of connection strength is represented by colors, and confidence in the estimates is represented by opacity. The matrix, generated by multiple injections of an anterograde tracer into the right hemisphere, maps the afferent and efferent projections of all right hemisphere regions, and efferent projections running from areas in the right hemisphere to the left. PMT, photomultiplier tube. *Reproduced **(a)** from Yeterian and Pandya (1991) and **(b,c)** from Oh et al. (2014) and Osten and Margrie (2013), with permission.*

A difficult problem faced by any attempt to synthesize findings across meso-scale tract-tracing studies is node definition. Although functional specialization of neuronal populations is evident at mesoscopic scales, there is ongoing controversy and debate about how exactly these specialized populations should be delineated. For example, Brodmann's initial cytoarchitectonic map of the human brain has since been replaced with many other alternative parcellations (Von Economo, 1929). Such inconsistencies have led to the development of objective, observer-independent methods for cytoarchitectonic mapping (Schleicher et al., 1999), but these have not yet been applied to parcellate entire brains. A further complication is that cytoarchitecture is only one approach to regional parcellation; other approaches that define regions based on spatial variations in myeloarchitecture (Nieuwenhuys, 2013), chemoarchitecture (Zilles et al., 2002), and gene expression (Bohland et al., 2010) may be at least as valid. Most tract-tracing studies have relied on cytoarchitecture, but inconsistencies in the way that these areas are delineated across different investigations complicate the collation and synthesis of results. The development of algorithms for representing tract-tracing findings in a parcellation-independent framework has greatly advanced these efforts (Bota and Arbib, 2004; Stephan and Kotter, 1999; Stephan et al., 2000), and has allowed investigators to use their own parcellations when analyzing the data (Modha and Singh, 2010).

Another limitation of connectomes collated from the published literature concerns the estimation of connection strengths. Most early tract tracing experiments used either binary or qualitative descriptions of axonal connectivity strength (Figure 2.7a). This ambiguity is compounded by variability in the sensitivity of different tracers (Lanciego and Wouterlood, 2011). Analyses of collated connectomes have circumvented this problem by using semiquantitative criteria based on expert judgments about the quality of the evidence and strength of a projection identified in the literature (Bota et al., 2012, 2015). Nonetheless, the weights estimated by these methods will only provide a rough approximation of actual connectivity strength.

Advances in light microscopy and the automation of tissue processing have enabled the construction of whole-brain mesoscale connectomes in mammalian brains using standardized methodologies implemented by the same team of investigators (Osten and Margrie, 2013). In one approach, Oh et al. (2014) acquired data from 469 distinct tracer experiments using injections of a viral anterograde tracer into multiple areas distributed throughout the right hemisphere of the mouse brain (Figure 2.7b). An automated tissue sectioning and blockface-imaging pipeline generated coronally sectioned images with an in-plane resolution of 0.35- and 100-μm slice thickness (Figure 2.7b). Simply acquiring these data posed a significant challenge, with the task being distributed across six scanners, each taking ~18.5 h to complete one brain. The images from each experiment were spatially normalized to a template brain that had been segmented into a hierarchically organized taxonomy of brain structures, as defined in the Allen Institute Reference Atlas (Dong, 2008), and fluorescence signals emanating from the tracer were measured on a voxelwise basis, with each voxel having a volume of 100 μm^3.

The tracer uptake measurements of Oh et al. (2014) were coregistered to a reference atlas, which allowed nodes to be defined based on an established parcellation. However, the tracer injections were not made with respect to any prior method for defining nodes, meaning that the zones of tracer uptake in each experiment did not always conform to the anatomical boundaries of the reference atlas. To get around this problem, a statistical model was used to parse the fluorescence signals emanating from injections that crossed regional borders. An advantage of this modeling approach is that it provides a statistical estimate of the certainty of a projection existing between any two regions (Figure 2.7c). However, the model does make assumptions about how connectivity and the corresponding fluorescence signal of the tracer is distributed across different areas. A different approach to estimating connection weights from these data is described by Rubinov et al. (2015).

In an alternative effort to map the mouse connectome, Zingg et al. (2014) used a double coinjection tracer methodology in over 300 experiments. This approach involved injecting both an anterograde and retrograde tracer into different sites of the same animal, which allowed the authors to resolve the inputs, outputs, and reciprocal connections of each area (Figure 2.8). Injection sites were distributed throughout the neocortex, entorhinal cortex, hippocampus, amygdala, and olfactory areas. Forty-nine regional nodes were defined using the Allen Institute Reference Atlas (Dong, 2008). Connection weights between these regions were defined qualitatively, being classified as absent, weak, moderate, or strong based on an expert judgment of the intensity of anterograde and retrograde labeling for any given pair of regions. For a discussion of other connectome mapping efforts in the mouse, see Mitra (2014) and Osten and Margrie (2013).

The cortical connectome of the macaque monkey has also been mapped using a standardized methodology. Specifically, Markov et al. (2013a) parcelled the cortex into 91 distinct areas using a combination of cytoarchitecture and sulcogyral landmarks (Figure 2.9a). They then injected a retrograde tracer into 29 of these areas. In retrograde tract tracing, the injected site is the projection target and the labeled upstream areas correspond to the sources of the projection (Figure 2.9b). Complete sectioning and inspection of the injected hemisphere allowed projections from all 91 potential sources to be mapped to each injected target area, resulting in a 91 × 29 connectivity matrix of ipsilateral cortical connectivity (Figure 2.9c–e). This matrix thus contains all unidirectional

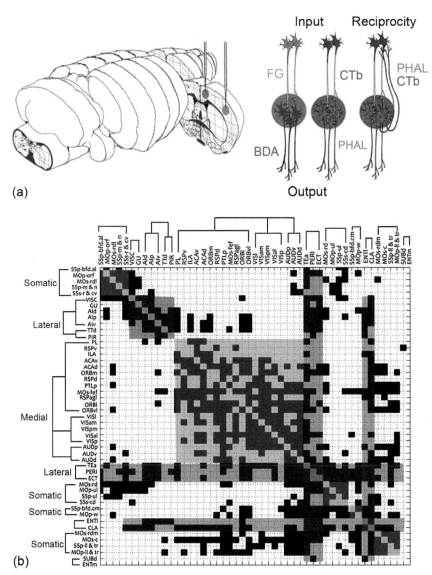

FIGURE 2.8 Mapping the mouse connectome using a co-injection methodology. **(a)** Schematic of the co-injection methodology. An anterograde and retrograde tracer are introduced via injections into two separate areas of the same brain (left). In this case, the anterograde tracer was either *Phaseolus vulgaris* leucoagglutinin (PHAL) or biotinylated dextran amine (BDA), and the retrograde tracer was either cholera toxin subunit b (CTb) or fluorogold (FG). Mapping the uptake of both the anterograde and retrograde tracers allows the identification of the inputs, outputs, and reciprocal connections of an area (right). **(b)** When applied across multiple regions, the method results in separate connectivity matrices for the anterograde and retrograde tracers, which can then be combined into a composite matrix, as shown here for 49 areas of the mouse cortex. The rows and columns have been reordered by a **hierarchical clustering** algorithm and colors represent putative network **modules** (Chapter 9). *Images reproduced from Zingg et al. (2014), with permission.*

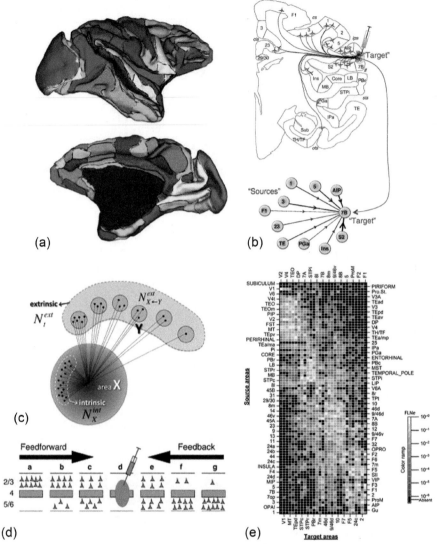

FIGURE 2.9 Mapping the cortical connectome of the macaque. **(a)** Regional boundaries used to parcellate the macaque cortex into 91 areas, defined according to cytoarchitecture and sulco-gyral landmarks. **(b)** Example of mapping interregional connectivity using retrograde tract tracing. The tracer is injected into a target region and is transported retrogradely to the cell bodies making afferent axonal projections to the target region (top). The projection profile can be represented as a simplified graph for each experiment (bottom). **(c)** The strength of connectivity, or edge weight, between an injected target site X and upstream source area Y can be estimated as the fraction of neurons labeled by the tracer (FLN). Specifically, FLN is computed as the number of neurons labeled in any given source area $\left(N^{ext}_{X \leftarrow Y}\right)$ divided by the total number of neurons labeled. The extrinsic FLN (FLNe) is estimated by computing the same fraction after excluding the number of intrinsic neurons labeled within the injected region $\left(N^{int}_{X}\right)$. The FLNe offers a more specific estimate of interregional connectivity strength. **(d)** Hierarchical connectivity can be quantified as the fraction of labeled supragranular neurons (SLN). Most feedforward projections originate in SLN whereas most feedback projections originate in infragranular layers. The schematic shown here is a simplified example of the expected laminar distribution of tracer uptake in different source areas that send either feedforward (left, a–c) or feedback (right, e–g) projections to the injected target site (d). The preponderance of feedforward or feedback projections is strongest for areas on the outer ends of the schematic (areas a and g) and is weaker for areas closer to the middle. **(e)** Connectivity matrix generated from retrograde tracer injections to 29 distinct cortical areas. The result is a comprehensive map of projections running from 91 sources to 29 targets. Colors represent variations in FLNe. *Reproduced (a,e) from Markov et al. (2013a) and (b–d) from Kennedy et al. (2013), with permission.*

projections running from the 91 sources to the 29 targets, and all bidirectional connections between the 29 injected sites. By using a standardized and dense sampling approach to quantify tracer uptake, the authors identified a large number of projections that had not previously been identified in the literature (around 1/3 of all mapped connections), suggesting that reliance on the collation of findings from individual experiments alone may provide an incomplete picture of large-scale brain connectivity (Kennedy et al., 2013).

Injection sites for this macaque connectome were determined based on a prior parcellation of the cortex into cytoarchitectonic and anatomical divisions. The injection zone was small relative to the targeted area, minimizing the uptake of the tracer in adjacent regions (Markov et al., 2013a). As a result, the estimation of interregional connection weights did not require a computational model as used for the mouse connectome (Oh et al., 2014). Instead, Markov and colleagues estimated connection strengths by directly quantifying the fraction of labeled neurons (FLNs) in a given source area; that is, the number of labeled neurons in the source region divided by the total number of labeled neurons in the brain. An estimate of the extrinsic FLN (FLNe) was also computed by first subtracting the number of labeled neurons intrinsic to the injection site from the total number of labeled neurons in the brain, and then dividing the number of labeled neurons in the source region by this value. The FLNe yields a connectivity weight that quantifies only interregional projection strength (Figure 2.9c). Across the connectome, FLNe values varied over five orders of magnitude, a result that concurs with the variability of model-based estimates of connection weights for the mouse brain (Oh et al., 2014; Rubinov et al., 2015). By considering the laminar distribution of the labeled neurons, Markov et al. (2013b,c) were also able to derive a measure of hierarchical connection strength as the fraction of neurons labeled in supragranular layers (SLN). Most feedforward projections in the brain originate in supragranular layers and terminate in layer 4 (or higher) whereas most feedback projections originate from infragranular layers and avoid layer 4. Therefore, a high SLN value suggests a preponderance of feedforward projections whereas a low SLN value suggests a majority of feedback projections (Figure 2.9d).

These large-scale efforts in mammalian brains highlight the potential power and scope of mesoscale tract tracing. However, accurate node definition remains a problem since there is no gold standard parcellation at this scale, and multiple alternative schemes are available (Kennedy et al., 2013; Van Essen, 2013). The choice of a particular parcellation scheme impacts the measurement of connectivity strength. On the one hand, if the locations of the tracer injections are tied to a specific parcellation scheme, connection weights can be estimated using a relatively direct measure, such as the FLN (Markov et al., 2013a). On the other hand, if injections are made across multiple sites without regard for any prior parcellation scheme, it may be necessary to specify a model that parses the contribution of different regions to the measured tracer uptake signal (Oh et al., 2014).

There are some additional considerations when interpreting connection weights inferred from tract tracing. First, axons passing through an injected region may take up the tracer without making a synapse with any neurons in the area, leading to a false positive connection. Second, the volume of tracer uptake within a particular injection site may not necessarily label the full extent of that region, meaning that some projections may be missed. Variability in the extent of the labeled area across experiments may cause fluctuations in any resulting connection weight estimates. Using multiple injections of the same site to assess measurement stability is not practical for large-scale connectome mapping. Markov et al. (2013a) examined this issue via analysis of repeat injections of specific areas in visual and prefrontal cortex. Weight estimates generally varied within an order of magnitude across repeated experiments for one injection site. This variation was lower than the variation in weights observed across regions, which spanned five orders of magnitude, and the estimated connection weight obtained from any single experiment was generally within the 95% confidence intervals of the mean across experiments. However, in some cases, the variation in the estimated weights was large and exceeded two orders of magnitude. These results suggest that some connection weights are estimated with greater reliability than others.

Chiang et al. (2011) employed an alternative to tract-tracing, using genetic labeling and confocal imaging to reconstruct the morphology of around 16,000 individual neurons of the *Drosophila* fruit fly, which corresponds to ~10% of all neurons in the fly's brain. The neurons were then aggregated into putative local processing units (LPUs), which were defined as spatially continuous neuronal populations whose constituent interneurons had fibers that were completely contained within the spatial border of the population. Using this approach, 41 such LPUs were defined, in addition to 6 putative hub areas that did not show the same degree of anatomical segregation. The LPUs were deemed connected if a nerve terminal of a neuron in one unit was located within the borders of another unit. The number of such terminals was used as a measure of connection weight. Estimates of connection strength were thus based on the spatial and morphological features of each neuron within the larger population (Figure 2.10). This database has since been extended to include data from over 20,000 neurons (Shih et al., 2015).

Other promising techniques for mesoscale connectomics include the combination of tract-tracing and transgenic methods to map connectivity between specific, genetically defined, neuronal subtypes (see Jennings and Stuber, 2014, for an overview), and methods for chemically clearing non-neural tissue from postmortem specimens so that they become optically transparent and can be

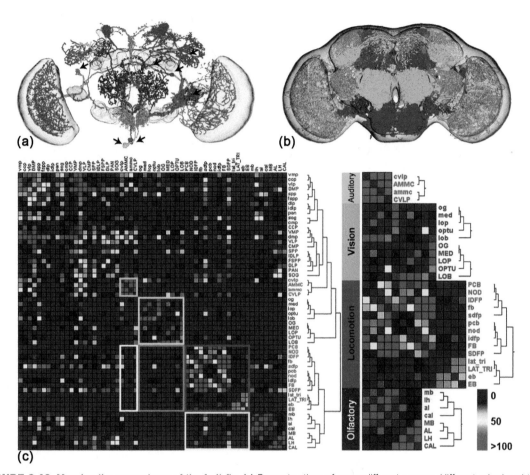

FIGURE 2.10 Mapping the connectome of the fruit fly. **(a)** Reconstructions of seven different neurons (different colors) and their processes. Black arrows indicate the locations of the cell bodies. **(b)** Boundaries of putative local processing units, representing different network nodes. **(c)** Connectivity matrix of axonal connections between local processing units and hub nodes, based on connectivity data in over 16,000 neurons. The rows and columns of the matrix have been ordered using a hierarchical clustering algorithm. The **dendrogram** is shown on the right. The nodes cluster into auditory, visual, locomotion, and olfactory systems. *Figure reproduced from Chiang et al. (2011), with permission.*

imaged as intact, three-dimensional volumes (Chung and Deisseroth, 2013). Tract-tracing techniques for fixed specimens that can be applied to human brains are available, but these methods are difficult to scale for the purpose of whole-brain connectivity mapping (see Lanciego and Wouterlood, 2011, for an overview). Microscopy methods that measure the way in which polarized light interacts with the lipid molecules in myelin show promise for mapping axonal connectivity in human postmortem tissue with submillimeter precision (Axer et al., 2011).

2.2.2   Functional Connectivity at the Mesoscale

Mesoscale analyses of functional connectivity within and between neuronal populations often use invasive electrophysiology. These analyses focus on recordings acquired not from individual cells, but from the extracellular space. The recorded signal, called the local field potential (LFP), reflects the superposition of electric currents arising from multiple cells and neurites in the local vicinity. The most important contribution to the LFP comes from synaptic activity, which creates an electric current via the flux of cations such as Na^+ and Ca^{2+} between the intracellular and extracellular spaces. The LFP also contains contributions from currents created by numerous other cellular processes, such as action potentials, Ca^{2+} spikes, intrinsic cellular currents and resonance, and slow fluctuations in glial function (Buzsáki et al., 2013). These currents superimpose at any given location in neural tissue to generate a potential, measured in volts. The difference in potentials between any two locations generates an electric field that is measured by an appropriately placed electrode (Figure 2.11a).

Synchronized synaptic activity makes a major contribution to fluctuations of the LFP, and the measured signal is often interpreted as an aggregate index of dendritic and somatic inputs to cells situated within a few millimeters of the electrode tip (Logothetis et al., 2001). Synchronous spiking output also makes an important contribution, particularly in higher frequency bands (Buzsáki et al., 2013). The precise location of the electrode, the local distribution of synapses, and the geometry of neurons and their dendritic tree will affect the measured signal. Interested readers are referred to Buzsáki et al. (2013) and Einevoll et al. (2013) for reviews of the biophysical basis of the LFP.

By recording LFPs from multiple electrodes placed in different parts of the brain, we can measure the synchronization of synaptic activity across distributed neural systems (Brovelli et al., 2004; Buschman and Miller, 2007; Gregoriou et al., 2009). However, only a small number of such electrodes can be inserted into the neuropil at one time. A more scalable technique for recording LFPs is **electrocorticography (ECoG)**, also called intracranial electro-encephalography (EEG). ECoG involves the subdural implantation of either a strip or grid of electrodes on the cortical surface (Figure 2.11b). These electrodes are then used to record LFPs from the underlying neural tissue. Dense electrode grids have been developed that can cover nearly the entire lateral and medial surfaces of a cortical hemisphere of the monkey brain (Nagasaka et al., 2011). The technique can also be used to sample activity, in vivo, from relatively large areas of cortex in human neurosurgical patients (Figure 2.11c).

With ECoG, it is possible to construct a functional connectivity network by treating each electrode as an individual node. The recordings acquired from each node thus correspond to the averaged signal arising from the population of neurons beneath and in the local vicinity of the electrode. The position of nodes is dictated by the lattice-like geometry of the electrode array, meaning that the nodes do not necessarily correspond to functionally defined cell populations.

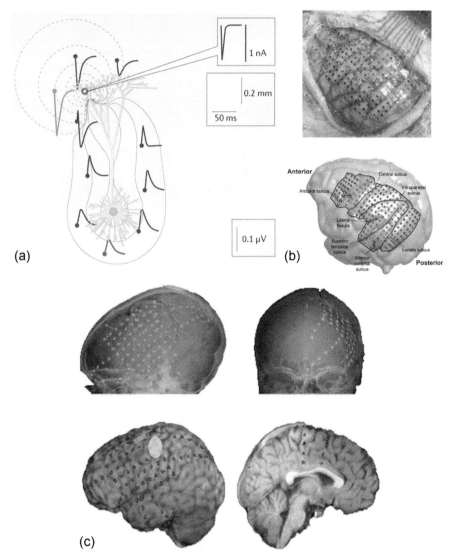

FIGURE 2.11 Functional connectivity networks estimated from local field potentials (LFPs). **(a)** LFPs generated from single neurons depend on the relative positions of the electrode and the stimulated synapse. Shown here are the results of a modeling study in which a single synapse at the apical dendrite (white circle) of a layer V pyramidal neuron was activated. The inset on the right shows the injected current. Thick traces are the extracellular potentials recorded at distinct locations (dots at the start of the traces). Contours indicate the maximal LFP amplitudes on a logarithmic scale (solid and dashed lines indicate positive and negative amplitude, respectively). The current injected to the apical synapse (yellow circle) is negative close to the synapse but positive close to the soma, where the return currents of the synapse are strong. The reverse is found when the synapse occurs at the soma. When considering signals from multiple synapses and other sources, local network organization, neuronal geometry, and the spatial distribution of synapses play an important role in influencing the signal. **(b)** Example ECoG grid implanted on the macaque brain. The grid comprises 252 electrodes. Top is an intraoperative photo. Bottom shows the electrode distribution on a cortical surface rendering generated from an MRI of the monkey. Lines indicate the boundaries and major sulci of the sampled area. Colored dots correspond to electrodes. **(c)** Example ECoG grid used to measure activity in the human brain. Top is a radiograph and bottom is a projection of the electrode locations on the patient's MRI. Colors denote areas of sensorimotor cortex that were being studied by the investigators. *Reproduced **(a)** from Einevoll et al. (2013), **(b)** from Bastos et al. (2015b), and **(c)** from He et al. (2008), Copyright (2008) National Academy of Sciences, U.S.A., with permission.*

Functional connectivity between the nodes of an ECoG array can be estimated using any one of a diverse array of different measures for quantifying the temporal dependence between the recorded signals (Box 2.2). A prominent feature of these recordings (and of neuronal activity in general) is that they oscillate over a range of frequencies that spans several orders of magnitude (Figure 2.2a). Oscillations at each frequency are thought to arise from distinct mechanisms and to make different contributions to behavior. For example, slow 1 oscillations (0.1-2 Hz) of the LFP are prominent in sleep and are thought to reflect the alternation between periods of synchronous depolarization (so-called up-states) and energy conserving periods of hyperpolarization (so-called down-states) of large groups of neurons; in contrast, more rapid oscillations in the gamma band (30-80 Hz), which are present across the brain and in all mental states, are thought to play an important role in local information-processing and are coordinated by fast-spiking interneurons (Buzsáki et al., 2013).

Typically, the spectral power of these oscillations is inversely proportional to frequency, f, giving rise to a characteristic $1/f$-like spectrum (Figure 2.2b and c; Buzsáki and Draguhn, 2004). Similar scaling properties have been observed for measures of brain activity acquired with much lower temporal resolution, such as functional MRI signals (Bullmore et al., 2001; Maxim et al., 2005). The neuronal origins of $1/f$ scaling in these signals is debated, but one contribution arises from network organization. Short time windows allow the recruitment of only a limited number of neurons to a temporally synchronous assembly; longer windows provide sufficient time for more neurons to contribute to the LFP, thereby increasing low-frequency signal power (Buzsáki et al., 2012). As a result, coupled activity in high frequency bands tends to be strongest for neuronal populations separated by short anatomical distances, while long-range synchronization tends to be stronger in low-frequency bands (Kopell et al., 2000; von Stein and Sarnthein, 2000).

The rich temporal organization of LFP signals (or for that matter, any neurophysiological signal recorded with sufficiently high temporal resolution) means that functional connectivity can be measured by Fourier analysis or wavelet analysis within specific frequency bands of interest. The evolution of the networks over time can also be studied by using methods such as sliding window analysis, in which the signal time courses are divided into either overlapping or separate windows of arbitrary length, and network measures are estimated for each window (Box 2.2).

Neuronal synchronization can occur with respect to either the phase or amplitude of the recorded signals (Engel et al., 2013; Siegel et al., 2012). Phase coupling quantifies the consistency of the relative phase between two signals that oscillate with the same frequency, whereas amplitude coupling is typically quantified as the correlation of the amplitude envelopes of the signals (Figure 2.12a and b). These two types of coupling appear to be intrinsic

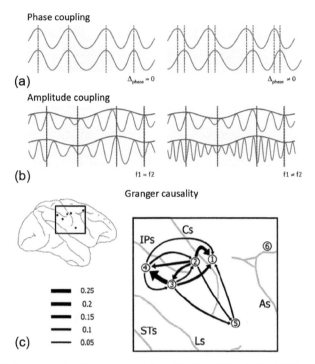

FIGURE 2.12 Measuring functional connectivity with extracellular fields. **(a,b)** Examples of the two principal coupling modes of neurophysiological signals: phase and amplitude coupling. **(a)** Example of phase coupling between two electrophysiological signals (top and bottom). Left shows an example of two signals in perfect phase synchrony with zero lag. The peaks and troughs of the oscillations occur at the same time. Right shows an example of phase synchrony with a nonzero lag. In this case, the phase of one signal is shifted with respect to the other, but the relationship is consistent across the recording period. **(b)** Amplitude coupling measures the correlation in the amplitude envelope (red line) of two oscillating signals. Left panel shows an example of amplitude coupling between two signals with the same underlying carrier frequency. Right panel shows an example of amplitude coupling between two signals with different carrier frequencies. In this case, the bottom signal oscillates at a higher frequency than the top signal, but the amplitude envelopes of these signals are correlated. The left panel shows how phase and amplitude coupling can be dissociated: there is no phase relationship between the two signals, but the amplitude envelopes are strongly correlated. Note that amplitude coupling can be alternatively computed by identifying correspondence between the peaks and troughs of the actual signal, rather than just the amplitude envelope. This often amounts to computing a correlation coefficient between the raw signals, as is typically done in analyses of functional MRI (Section 2.3.3). **(c)** Directed functional connectivity graph computed from LFP recordings in a macaque performing a motor task. Left shows the electrode locations. Right shows the directed influence between electrodes in the beta frequency band. These influences were estimated using Granger causality analysis, which quantifies the degree to which the signal in one electrode is able to predict future variations of signals recorded at other electrodes (see also Box 2.2). *Reproduced (a,b) from Siegel et al. (2012) and (c) from Brovelli et al. (2004), Copyright (2004) National Academy of Sciences, U.S.A., with permission.*

properties of neuronal communication, and are used preferentially by different large-scale neural networks and during different cognitive states (Engel et al., 2013). Cross-frequency coupling can also arise within and between cell populations. In particular, cross-frequency phase-amplitude coupling, where the phase of a low-frequency oscillation is coupled to the amplitude of a high-frequency oscillation, is commonly found in electrophysiological signals recorded from distinct neuronal populations, although phase-phase and amplitude-amplitude cross-frequency coupling have also been identified (Aru et al., 2015; Canolty and Knight, 2010).

The different coupling modes and plethora of methods available for characterizing dependencies between LFP signals afford a rich characterization of dynamic neural interactions, in both the time and frequency domains (Box 2.2). LFPs also allow the estimation of directed measures of functional connectivity that quantify the extent to which signal variations in one site can predict or influence subsequent variations of signals recorded in other sites (Boxes 2.1 and 2.2; Figure 2.12c; Brovelli et al., 2004; Wilke et al., 2010).

The ongoing development of increasingly dense ECoG grids will allow LFPs to be sampled with greater precision and spatial coverage (Fukushima et al., 2015; Nagasaka et al., 2011). ECoG also offers the most direct measure of brain activity that can be acquired in living human participants, but its invasive nature means that it can only be used in neurosurgical patients. It also means that signals can only be sampled from limited, superficial areas of cortex. Deep brain structures remain inaccessible.

2.3   MACROSCALE CONNECTOMICS

The invasive methods for measuring brain connectivity that we have considered thus far offer in-depth access to neural structure and function, but they are difficult to apply across the entire brain, particularly in larger animals and humans. Noninvasive imaging techniques, such as MRI, EEG, and MEG, offer a means for mapping connectivity across the entire brain, in vivo, in model species and humans alike. The advantages of these methods include their clinical safety and tolerability, their coverage of the entire brain (particularly MRI), and their flexibility in allowing studies of whole-brain connectivity across the lifespan and in relation to a wide range of human brain disorders. However, these methods can typically only resolve macroscopic aspects of brain connectivity on the scale of millimeters and centimeters.

The coarse spatial resolution of macroscale techniques demands that we aggregate measures over ever-larger populations of neurons, axons, and synapses. This reduces the precision with which we can define the nodes and edges of a brain network. A typical T1-weighted anatomical MRI has a voxel resolution of about 1 mm^3, which contains an estimated 20,000-30,000 neurons and billions of synapses (Logothetis, 2008). Structural and functional connectivity are typically assessed using diffusion MRI and functional MRI, which have a typical voxel size in the order of 3 mm $\times$ 3 mm $\times$ 3 mm. The spatial resolution of EEG and MEG can be lower still. This low resolution forces us to make several assumptions when defining nodes and edges. In this section, we begin by considering some of the issues associated with defining nodes at the macroscale that are common to analyses of both structural and functional connectivity. We then consider specific issues pertaining to the definition of structural and functional edges separately.

2.3.1 Defining Nodes at the Macroscale

At the microscale, a node can be clearly defined as a single neuron. This definition perfectly aligns with Ramón y Cajal's neuron doctrine, which identifies the neuron as the fundamental element of a nervous system. At the mesoscale, nodes are commonly defined as spatially continuous and often functionally specialized cell assemblies or populations of neurons. There is no gold standard for delineating such populations, so approximations based on cytoarchitecture and anatomical landmarks are used (Bota et al., 2015; Markov et al., 2013a; Oh et al., 2014). This problem is compounded at the macroscale because we generally do not have direct access to cytoarchitecture. Instead, different heuristics are often used to define nodes at this scale.

A simple solution is to treat each measurement point as an individual node. For example, distinct nodes could correspond to individual voxels in an MRI dataset (Eguíluz et al., 2005; van den Heuvel et al., 2008b; for an example, see Figure 2.13a), different electrodes used in EEG, or sensors in MEG (Bassett et al., 2006; Micheloyannis et al., 2006a,b; Stam, 2004). The advantage of this approach is that it does not entail any additional data processing or assumptions—the data are analyzed at the native resolution of the imaging technique and no further aggregation or averaging of measurements is required. However, there is no guarantee that the measurement points coincide with the borders of functionally specialized populations of cells. For example, the tens of thousands of neurons contained within a single MRI voxel may comprise cells from different populations. Alternatively, the borders of a specific, functionally specialized cell population may extend beyond the boundaries of the voxel, a problem that is compounded by the lack of independence between measurements made at spatially neighboring points. In MRI, partial volume effects can smear a focal signal over several nearby voxels, meaning that the effective spatial resolution may be coarser than the dimensions of an individual voxel. In this case, the connectivity between neighboring voxels may be spuriously inflated simply due to the measurement process. EEG and MEG sensors record from even larger populations of cells, and the spatial correspondence between the signals emanating from the underlying neuronal sources and those recorded at the scalp can be distorted as electrical and magnetic fields spread and pass through the cerebrospinal fluid, meninges, skull, and scalp. In other words, the signal recorded at any given scalp location is an aggregate of contributions from a large population of cells that may not even lie directly underneath the sensor. Methods for **source localization** use biophysical models of field spread and tissue conductivity to estimate the locations of neural generators from sensor data (da Silva, 2004; Michel et al., 2004). The accurate modeling of unique sources remains a major challenge.

Rather than relying on measurement points, we could define nodes by co-registering our data to a validated parcellation scheme, such as one based on cytoarchitecture. Stereotactic anatomical and cytoarchitectonic atlases covering the entire brain are available for the mouse and macaque (Figure 2.13b; Dong, 2008; Markov et al., 2013a; Van Essen et al., 2013). In humans, Brodmann's initial cytoarchitectonic parcellation has been mapped to a standard reference space (Van Essen et al., 2013), but does not account for the considerable inter-individual variability in the boundaries of cytoarchitectonic areas (Amunts et al., 2000; Rademacher et al., 2002). Probabilistic maps summarizing the spatial positions of these boundaries across a population of individual brains have been digitized and mapped to reference spaces, but they currently cover only a small portion of the brain (Figure 2.13c; Eickhoff et al., 2005; Van Essen et al., 2013). Variations in myeloarchitecture, which can be quantified with MRI, have shown some promise in delineating regional boundaries (Glasser and Van Essen, 2011), but have not yet been applied to parcellate the entire brain.

An alternative to cytoarchitecture is to use the macroscopic landmarks visible with MRI, such as sulcal and gyral boundaries, to define network nodes (Figure 2.13d). Several such parcellations exist, based on either the delineation of all sulci, gyri, and major subcortical structures in a single individual (Tzourio-Mazoyer et al., 2002), or the aggregation of many such parcellations created across a population of brains to generate a probabilistic atlas—an atlas in which each point in the brain is assigned a probability of belonging to a particular anatomical region (Desikan et al., 2006; Fischl et al., 2004). A limitation of this approach is that cytoarchitectonic areas often cut across sulcal and gyral landmarks (Rademacher et al., 1993; Welker, 1990). Furthermore, the regions that result from this kind of parcellation can show considerable variation in size, which can bias any subsequent estimates of connectivity. For example, larger regions will have a greater surface area to accommodate more incoming and outgoing fiber tracts, and regionally averaged functional MRI time series estimated by averaging all voxel time series in large regions will have higher signal-to-noise than averaged time series representing smaller regions. The size of these larger regions also means that they will tend to correlate with a greater number of

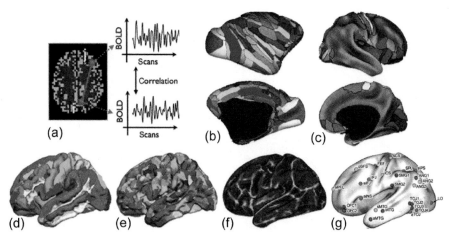

FIGURE 2.13 Defining nodes in macroscale connectomics. **(a)** Example parcellation in which each measurement point (in this case, an MRI voxel) is treated as an individual node. In this functional MRI analysis, an activity time course is extracted from each voxel and functional connectivity between voxel-wise time courses is measured using the correlation coefficient. **(b,c)** Examples of cytoarchitectonic atlases mapped to standard stereotactic space. **(b)** Coverage of the macaque brain is relatively complete. **(c)** Current atlases for the human brain only cover certain parts of cortex. **(d)** An anatomical parcellation of the human cortex based on sulcal and gyral landmarks. **(e)** A random parcellation of the cortex, constructed so that each region is approximately equal in size. **(f)** A parcellation based on variations in voxel-wise profiles of functional connectivity with the rest of the brain. Colored borders correspond to areas where there are abrupt transitions in functional connectivity profiles and which represent putative regional boundaries; dark areas are regions where connectivity profiles are stable. **(g)** Spherical regions centered on stereotaxic coordinates of task-related activation peaks. The regions sample key nodes of interest in the activated networks; in this case, the so-called default mode network (blue) and a network involved in cognitive control (red). Seed-based resting-state functional connectivity analysis was also used to identify some of these foci. *Reproduced **(a)** from van den Heuvel et al. (2008b), **(b,c)** from Van Essen et al. (2013), **(d,e)** from Fornito et al. (2012b), **(f)** from Wig et al. (2011), and **(g)** from Dwyer et al. (2014), with permission.*

other areas. Some of these apparent biases may reflect true biological phenomena, such as some areas being larger in size precisely because they are highly connected. However, if we cannot be sure that our nodes correspond to meaningful functional parcels of the brain, it is difficult to determine whether the measures that we compute in networks constructed with this approach are driven by bona fide variations in connectivity strength or mere differences in node size.

One method for dealing with the problem of unequal node sizes is to parcellate the brain into random parcels, each of which are approximately equal in spatial extent (Figure 2.13e; Fornito et al., 2010; Hagmann et al., 2007; Zalesky et al., 2010b). An advantage of this approach is that we can control node size, but there is no guarantee that the boundaries of the random parcels correspond to the actual borders of functional areas of the brain (Figure 2.14a–d). Generating a random parcellation that is constrained by the boundaries of a coarser, functional, or anatomical parcellation can mitigate this effect, but this method depends on the validity of the coarser parcellation. To ensure that the findings of any network analysis are not driven by a single instantiation of a random parcellation, it is good practice to repeat analyses across few different random parcellations of the same resolution to check for consistency of results. Statistical approaches have been developed for MRI that allow voxelwise inference based on a consensus of results obtained across multiple random parcellations (Da Mota et al., 2014).

We can also use connectivity to define nodes. The logic behind this approach is that each functionally specialized area of the brain possesses a unique connectional fingerprint—a distinct pattern of afferent and efferent connections to other areas that, in large part, defines the function of that area (Passingham et al., 2002). Resolving afferent and efferent connectivity with noninvasive imaging is difficult, but approaches have been developed for MRI that can distinguish between different areas based on profiles of undirected connectivity with other regions. In essence, these methods measure the connectivity of each voxel with all other voxels, and then cluster voxels together if they have a similar profile of connectivity to other areas (Figure 2.13f). Early work showed that diffusion MRI can be used to differentiate between specific regions of the frontal cortex, such as the supplementary motor area and the presupplementary motor area (Johansen-Berg et al., 2002), and to delineate Broca's area from other parts of the frontal operculum (Anwander et al., 2007). The limited capacity of diffusion MRI to track long-range projections (Section 2.3.2) has limited the application of these methods to whole-brain parcellation (for an exception, see Moreno-Dominguez et al., 2014), but similar principles have been used in the analysis of functional connectivity data with considerable success (Cohen et al., 2008; Craddock et al., 2012; Power et al., 2011; Yeo et al., 2011).

Multivariate decomposition of functional MRI signals using techniques such as **independent component analysis** has been used to similar ends (Kiviniemi et al., 2009). In many cases, the boundaries identified with these techniques coincide with functionally specialized areas (Nelson et al., 2010; Power et al., 2011; Wig et al., 2014). A limitation is that some of these methods do not guarantee that putative nodes will comprise a spatially contiguous collection of voxels, since spatially separate regions can show similar connectivity profiles (Power et al., 2011; Yeo et al., 2011). Algorithms with spatial constraints have been proposed to address this problem (Craddock et al., 2012; Moreno-Dominguez et al., 2014).

A final approach is to define nodes according to some a priori criterion. For example, some investigators have mapped patterns of activation during a particular task and generated regions-of-interest based on these maps (Figure 2.13g; Dwyer et al., 2014). Others have used meta-analyses of many different task activation studies to identify the key regions comprising the networks of interest (Dosenbach et al., 2010). The advantage of these methods is that they are informed by measures of brain function and are tailored to address specific hypotheses about neural systems of interest. A limitation is that they are not always translatable across different modalities. For example, in functional MRI, a node can be defined using a spherical region-of-interest identified in a previously mapped pattern of task activation (Figure 2.13g). The region will usually be contained within the gray matter. In diffusion MRI, it can be difficult to track connections to and from such a region because most fiber tracking methods have trouble reconstructing axonal pathways within gray matter (Section 2.3.2; Reveley et al., 2015).

To summarize, methods for defining nodes in macroscopic brain networks use heuristics to account for the limited capacity of noninvasive imaging methods to precisely delineate homogenous cell populations. The heuristic nature of these approaches ultimately reflects the lack of a gold standard for brain parcellation at the macroscale. This is in contrast to the microscale, where each neuron is a natural choice for each node. Consider social networks as an analogy. In these networks, individuals are the most natural choice of nodes, but if the goal is to map a social network for the world's entire population, it may be necessary to consider a coarser scale in which nodes represent groups of individuals. Should such groups be defined according to country, social community membership, hobbies, or some other criterion? The answer largely depends on the kind of inference that will be performed on the social network. For example, if we would like to test whether social ties within Commonwealth nations are stronger relative to ties with the rest of the world, it would be pointless to define nodes based on groups of people with similar hobbies (rather than shared citizenship). Node definition should thus be guided by the focus of the scientific question.

Unlike the example of social networks, the nodal parcellation that is best suited to a given kind of inference in brain networks is usually not obvious. We must therefore be wary of the limitations of the heuristics that are used (Figure 2.14a–d; Butts, 2009; Smith et al., 2011). A common approach in MRI studies is to examine the consistency of results obtained when a low-resolution anatomical template and a high-resolution random parcellation are used in the same dataset (Baker et al., 2015; Hagmann et al., 2008; van den Heuvel and Sporns, 2011). Investigations of the influence of different parcellations on network measures indicate that gross topological properties, such as whether a network is small-world or not, remain consistent across parcellations; however, the quantitative values of these and other topological parameters can show considerable variation (Figure 2.14e and f; Fornito et al., 2010; Hayasaka and Laurienti, 2010; Wang et al., 2009; Zalesky et al., 2010b). Cross-validation is thus important for ensuring that any findings in macroscale analyses are not driven by the choice of a particular parcellation scheme.

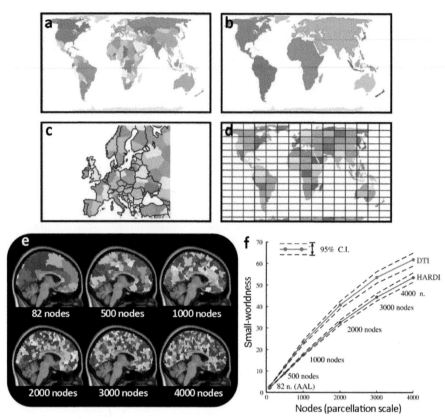

FIGURE 2.14 Different parcellations can influence the estimation of network properties. **(a–c)** Schematic demonstrating how an erroneous parcellation can distort network properties using a geographic map of the world. **(a)** A map of the world's countries, as defined in 2002. **(b)** A parcellation of the world based on large-scale geographic features; specifically, the boundaries between land and water. This coarse-grain parcellation groups many countries into single nodes. For example, all countries in Europe and Asia are treated as one. This approach is similar to parcellations of the brain based on gross anatomical features such as sulci and gyri. **(c)** Random parcellation of Europe. In many cases, the boundaries of the parcels (colors) do not coincide with the actual geographic borders of the countries (black lines). Some parcels contain multiple countries, and other countries are split across more than one parcel. **(d)** Parcellation of the world based on a grid, paralleling a voxel-based approach to node definition. The resulting parcels bear little correspondence to actual national borders. **(e,f)** Example of how parcellations of varying resolutions **(e)** can lead to wide variation in network properties when applied to the same diffusion MRI data; **(f)** in this case, the scalar index sigma, which is a ratio of normalized clustering to normalized path length and is used as a measure of network small-worldness (Chapter 8). All networks are small world, but the precise values of the small-worldness measure vary from less than 5 to over 60 depending on the parcellation scale. DTI and HARDI correspond to different diffusion MRI acquisitions. *Reproduced **(a–d)** from Wig et al. (2011) and **(e,f)** from Zalesky et al. (2010b), with permission.*

A final consideration is individual variability. Much like fingerprints, different brains show marked variation in sulcal and gyral anatomy, cytoarchitecture and axonal wiring, and these differences are related to variations in cognitive ability and other functions (Fornito et al., 2004; Kanai and Rees, 2011; Rademacher et al., 1993, 2002; Van Essen, 2005; Welker, 1990). Such variability will be greatest at microscales, but it also poses a challenge for macroscale analyses. Practically, it means that the correct parcellation for one individual will not necessarily be accurate for another individual.

2.3.2 Structural Connectivity at the Macroscale

The primary method for studying brain connectivity at the macroscale is **diffusion MRI**. As the name implies, diffusion MRI measures the diffusion of water molecules in the brain. In an unconstrained medium, water molecules show *isotropic* diffusion, meaning that they have an equal probability of diffusing in any direction. When there are barriers to the movement of these molecules, their diffusion becomes *anisotropic*, which in this context means that they are more likely to diffuse in the direction running parallel to the barriers.

In neural tissue, axons act as barriers to the diffusion of water molecules and cause the molecules to diffuse preferentially along an axis parallel to the trajectory of the fiber. In diffusion MRI, separate images are acquired to measure the magnitude of water diffusion in different directions oriented around a sphere. The magnitude and direction of water diffusion in any given spatial location (voxel) in the brain is then inferred through a model of the measured signal. In the simplest case, called *diffusion tensor imaging*, the measurements acquired along the different directions are fitted to a three-dimensional ellipsoid, represented mathematically as a 3×3 symmetric matrix, called a **tensor**. The diffusion represented by the tensor can be fully characterized by its three orthogonal **eigenvalues** and their **eigenvectors**. The eigenvalues represent the magnitude of diffusion along the longest, middle, and shortest axes of the ellipsoid, and the eigenvectors represent the orientations of these axes (Figure 2.15a–g; for a primer on eigenvalues and eigenvectors, see Box 5.1).

The putative trajectories of axonal fibers can be reconstructed from diffusion MRI data using **tractography**. The simplest variant, called *deterministic* tractography, involves seeding a specific voxel and then propagating a streamline that follows the principal direction of water diffusion in each voxel until some termination criterion is reached (Figure 2.15h). In *probabilistic* tractography, a probability distribution of the preferred direction of water diffusion is fitted at each voxel and determines the likelihood that a streamline will propagate in any particular direction. Thus, where deterministic tractography uses a point estimate of the preferred direction of diffusion in each voxel, probabilistic tractography models the uncertainty of this estimate, and thus provides confidence

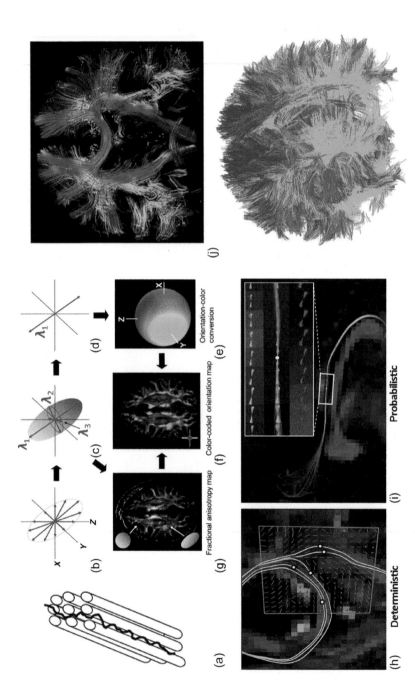

FIGURE 2.15 Macroscale connectomics with diffusion MRI. **(a)** Axons act as barriers to water molecules, causing them to diffuse preferentially in directions that run parallel to the fibers (thick black line). **(b)** In diffusion MRI, water diffusion is measured along multiple axes (blue arrows). **(c)** In diffusion tensor imaging, the measures are fitted to an ellipsoid, which is represented as a tensor; the three eigenvectors of the tensor represent an orthogonal basis for the three directions of water diffusion and the three eigenvalues (λ_1, λ_2, λ_3) represent the amount of diffusion in each direction. **(d)** The principal direction of water diffusion is the direction associated with the largest eigenvalue (λ_1). **(e)** The orientation information can be color-coded and used to map the principal direction of water diffusion in each voxel, as shown in **(f)**. **(g)** We can also quantify the degree to which water diffusion is constrained within a voxel using a measure called fractional anisotropy. If water is free to diffuse in any direction, the modeled ellipsoid (for example, **(c)**) will be spherical; if diffusion is constrained, the ellipsoid will be elongated. Gray shapes are schematic ellipsoids taken from two regions with low (top) and high (bottom) fractional anisotropy. In this map, areas with high fractional anisotropy have a brighter intensity. **(h)** In deterministic tractography, streamlines (white lines) follow the principal direction of water diffusion on a voxel-by-voxel basis. White circles represent seed points. Colored cylinders represent the principal direction of water diffusion in each direction, which is used as a basis for streamline propagation. **(i)** In probabilistic tractography, many streamlines are generated and propagated according to a model of the probability density of fiber orientations at each voxel, allowing for an estimation of the uncertainty in the trajectory of a given tract. In the inset, the probability densities are represented by the cones displayed within each voxel. Accurate estimates of the probability distribution require the sampling of more diffusion directions, and therefore specialized MRI sequences for the acquisition of diffusion-weighted data. **(j)** Tracts can be seeded from all voxels within the white matter volume, or between all pairs of gray matter regions, to generate whole-brain tractograms. Top shows the streamlines passing through a single coronal slice of the human brain. Bottom shows a three-dimensional rendering of all reconstructed streamlines. *Reproduced* **(a)** *from Johansen-Berg and Rushworth (2009),* **(b–g)** *from Mori and Zhang (2006),* **(h–i)** *from Tournier et al. (2011), and* **(j)** *(bottom) from Isenberg (2015), with permission.*

bounds on the trajectory of the reconstructed pathway (Figure 2.15i). Doing so can lead to more robust fiber tracking in noisy data (Behrens et al., 2003).

Alternatives to the tensor model attempt to infer the underlying distribution of fiber orientations within each voxel. Many such models have been proposed and they can be very broadly divided into the following categories: compartmental and fiber models (Alexander, 2005; Behrens et al., 2007; Ferizi et al., 2014), deconvolution methods (Tournier et al., 2004), q-ball imaging (Tuch et al., 2003), and diffusion spectrum imaging (Wedeen et al., 2005). Which of these local diffusion models is best suited for fiber tracking applications and connectome reconstruction remains an active area of research and has motivated fiber tracking competitions among various model proponents (Fillard et al., 2011). The basic motivation for the use of these many and varied alternatives to the tensor model is that they allow the resolution of complex fiber geometries such as crossing, kissing, and forking fiber patterns. In contrast, the tensor model is unimodal and can, strictly speaking, only resolve a single coherent fiber population within any given voxel. Either tensor or crossing-fiber models can be combined with tractography to generate maps of white matter pathways throughout the entire brain (Figure 2.15j). Overviews of the principles and applications of diffusion MRI and fiber tracking have been provided by Mori and Zhang (2006), Tournier et al. (2011), and Johansen-Berg and Rushworth (2009).

Mapping structural connectivity across the entire brain with diffusion MRI requires the selection of appropriate seed points for the initiation of the tracking algorithm. Ultimately, we are interested in mapping connectivity between regions of gray matter, so a logical solution is to seed these areas. However, water diffusion is poorly constrained in gray matter, making fiber tracking difficult. One solution is to seed voxels along the interface of gray and white matter, which allows streamlines to propagate more directly into the white matter volume where fiber tracking is more accurate. However, cross-validation with respect to histological tract tracing has found that this approach is affected by the complex geometry of fibers near the gray/white interface, which causes problems for diffusion tractography algorithms (Reveley et al., 2015). Seeding tracks from voxels in white matter does not overcome this problem, as the fibers near the gray/white matter interface can prevent streamlines from penetrating into cortical gray matter. In this case, streamlines will not be counted as contributing to a connection because they do not terminate in the cortex. A consideration of the strengths and weaknesses of each of these different seeding strategies is provided by Li et al. (2012) and Reveley et al. (2015).

Once streamlines have been seeded and tracked between all pairs of nodes, it is possible to derive several different measures of connectivity strength. The simplest measure is a count of the number of streamlines that connect a pair of regions. This measure assumes that the number of axonal fibers linking two areas is related to the number of connecting streamlines that can be reconstructed from the data. However, streamlines are not tantamount to axons. They are an abstraction of the diffusion model and tractography algorithm that has been applied, and the capacity to track a streamline between two areas can be affected by a number of other factors (discussed below) that are not necessarily related to the strength of axonal connectivity (Jones et al., 2013).

Other measures of connectivity strength are based on measures of water diffusion in each voxel. The most commonly used measure, called *fractional anisotropy*, quantifies the degree to which water diffusion is constrained (in any direction) within a voxel. Fractional anisotropy will be high in areas where there is a dense packing of well myelinated and coherently oriented axonal fibers, because these axons act as strong constraints on diffusion (Figure 2.15g). Conversely, fractional anisotropy can be low when axonal structure has been compromised, when fibers are sparse, or when the fibers are organized in a complex geometry that is not adequately captured by the diffusion model. Fractional anisotropy values can be averaged across all voxels traversed by a set of reconstructed streamlines to derive a putative index of the integrity of that tract. Alternative measures that index water diffusion either perpendicular or parallel to the tract (or in any direction) can also be derived. The disadvantage is that they may lack sensitivity, particularly in clinical applications where a localized disruption in one specific part of an axon may be masked when values are averaged across the extent of an entire tract. Streamline counts may offer greater sensitivity in these cases.

A criticism of both streamline count and diffusion-based measures of axonal structure is that they are sensitive to several influences that are not necessarily related to biologically meaningful characteristics of anatomical connectivity strength, such as axonal number, density, caliber, or myelination. These confounding factors conspire to complicate the interpretation of connection weights derived from diffusion MRI, and include changes in the signal-to-noise of different voxels caused by head motion, physiological noise, or imaging artifacts, as well as complex fiber geometries within a voxel (see Jones et al., 2013 for a thorough discussion). The development of methods to improve the interpretability of the streamline count as a measure of connectivity is an active area of research (Calamante et al., 2015; Smith et al., 2015). One potential solution is to combine diffusion tractography with magnetization transfer imaging, an MRI technique that quantifies the macromolecular content of an imaged voxel (van den Heuvel et al., 2010). The lipid molecules in myelin make a major contribution to this signal in white matter, so the method offers an indirect estimate of myelin content. Some diffusion MRI acquisitions also allow the estimation of axon diameter (Alexander et al., 2010; Assaf et al., 2008), but these measures have not yet been applied to large-scale connectomics.

A further problem of diffusion tractography is that it has difficulty reconstructing long-range pathways, particularly when these pathways intersect other fiber tracts that run in different directions (Zalesky and Fornito, 2009). Streamlines are propagated in discrete steps, typically less than a millimeter in length, and each step presents an opportunity for the streamline to veer off course due to any one of the factors described above that can affect the measured signal. Given that the number of propagation steps required to track a pathway is proportional to its length, long-range pathways present streamlines with more opportunities to veer off course and are thus the most challenging to accurately reconstruct. Principally for this reason, most published human brain connectomes based on diffusion MRI data conspicuously lack long distance, interhemispheric projections between bilaterally symmetrical cortical areas. Such connections are known to exist; indeed they make up the bulk of the corpus callosum, the biggest white matter bundle in the human brain. The absence of these connections in most tractography-based brain graphs is a salient indicator that reconstruction of long distance axonal tracts remains a challenge for diffusion MRI.

Validation studies comparing diffusion tractography results with tract tracing in animal models have found that the various algorithms used for fiber tracking show a trade-off between sensitivity and specificity. Algorithms that attempt to resolve multiple fiber orientations within a voxel provide high sensitivity but are prone to false positives (low specificity) because some of the estimated orientations may be inaccurate or spurious. Other methods that are less adept at resolving crossing fibers show high specificity at the expense of false negatives (low sensitivity; Calabrese et al., 2015; Knösche et al., 2015; Seehaus et al., 2013; Thomas et al., 2014). As a consequence, the chosen method for reconstructing tracts from diffusion MRI will have a major impact on the architecture of the resulting connectome. Sensitive methods will result in dense networks that may contain false positives. These false positives can lead to a randomization of network topology (Chapter 11). Tractography methods with high specificity will result in sparser networks that may include false negatives.

Finally, an important limitation of diffusion tractography is that it cannot resolve the source and target of a projection: it can tell us whether two regions share a connection, but it cannot tell us whether region A projects to B or viceversa. The resulting connectivity matrix is thus symmetric, and yields an undirected network graph.

An alternative approach for studying structural connectivity with noninvasive imaging is to examine interregional covariation of specific morphometric parameters, such as gray matter volume or cortical thickness (Alexander-Bloch et al., 2013a). In this approach, a measure of volume, thickness, or some other relevant metric is extracted for each brain region, and correlations between regions are computed across individuals, after statistically controlling for differences in whole brain volume and other nuisance variables. The resulting connection weight is a correlation estimate, reflecting the degree to which the morphological descriptor covaries between a pair of regions across the sample of brains. Such covariations are thought to be caused by axonal connectivity between areas (Lerch et al., 2006) or coupled developmental or trophic processes (Alexander-Bloch et al., 2013b; Mechelli et al., 2005). The structural MRI covariance method thus provides an indirect estimate of anatomical connectivity.

2.3.3   Functional Connectivity at the Macroscale

Functional connectivity at the macroscale is typically measured using EEG, MEG, or functional MRI. EEG measures the electrical activity of the brain via electrodes placed on the scalp surface. The dominant signal contribution is thought to arise from the synchronized activation of pyramidal neurons; particularly the activation of apical dendrites oriented at a perpendicular angle to the surface. This signal can be thought of as an LFP that has been smoothed through time and space, due to the conduction and attenuation of the electric field as it passes through the cerebrospinal fluid, skull, and scalp (Buzsáki et al., 2012).

MEG uses an array of superconducting quantum interference devices (SQUIDs) placed around the head to measure the weak magnetic fields that are generated by neuronal currents. These magnetic signals do not suffer the same degree of contamination by extracellular tissue as the electrical fields recorded with EEG. As a result, MEG can localize activity with greater spatial precision than EEG. Importantly however, MEG is insensitive to signals emanating from radially oriented neuronal sources whereas EEG is sensitive to both radial and tangential sources (Ahlfors et al., 2010). Several thorough overviews of the principles and applications of EEG and MEG have been published (Hari et al., 2010; Hämäläinen et al., 1993; Lopes da Silva, 2013; Nunez and Srinivasan, 2006; Schomer and da Silva, 2005).

Both EEG and MEG offer high temporal resolution, allowing activity to be sampled on the millisecond scales that are commensurate with the speed of neuronal signaling. This makes them comparable to other electrophysiological signals, such as invasively recorded LFPs, and means that the same analysis techniques can be used to quantify either phase or amplitude coupling between EEG/MEG signals. However, the poor spatial resolution of EEG and MEG poses special problems. The diffuse signal recorded by each sensor, which aggregates activity across large populations of neurons, makes it difficult to localize effects with precision (Figure 2.16a). This problem is compounded by the spread and conduction of the measured fields throughout the various tissue compartments

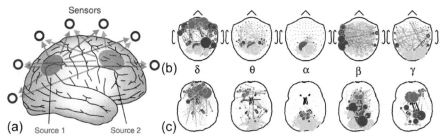

FIGURE 2.16 Macroscale functional connectivity networks measured with EEG and MEG. **(a)** EEG and MEG use sensors outside the head to detect the electrical (EEG) and magnetic (MEG) fields generated by neuronal sources. The signal recorded at any single sensor is an aggregation of multiple neuronal populations distributed throughout the brain. Special methods must be used to model and localize the neural sources generating the measured signals. **(b)** Example resting-state functional connectivity networks measured in distinct frequency bands using MEG. Each of 306 sensors is treated as an individual node. Functional connectivity was quantified using a measure of amplitude coupling, averaged across a sample of 15 healthy human participants. Larger nodes have a higher average connectivity to the rest of the network. Colors indicate different anatomical divisions: red is frontal, blue temporal, green parietal, cyan occipital, and magenta cingulate. **(c)** Functional connectivity networks generated from the same MEG data following source localization. An anatomical parcellation was used to estimate the signals emanating from 148 sources. Note the differences in network topology, particularly in the higher-frequency β and γ bands. *Reproduced **(a)** from Engel et al. (2013) and **(b,c)** from Kitzbichler et al. (2015), with permission.*

of the head (so-called *volume conduction*). Source localization methods infer the neural generators of these signals based on models of how the tissue and geometry of one's head impacts the electrical and magnetic fields (da Silva, 2004; He et al., 2011; Michel et al., 2004; Nunez and Srinivasan, 2006). These techniques allow signals to be estimated for multiple sources distributed throughout the brain, thus enabling an investigation of large-scale functional connectivity networks (Figure 2.16b and c). However, the solutions are only approximate, and often depend on the particular assumptions of the model used for source localization (Darvas et al., 2004). Localizing subcortical sources remains a challenge. Additional preprocessing of the signals, or the use of specific estimators of signal coupling, are also required to deal with volume conduction effects. For example, volume conduction can cause spurious, zero-lag correlations between sensors, so some measures of coupling ignore instantaneous synchronization (Cohen, 2015). The drawback of this approach is that any real zero-lag synchronization, which can play in important role in network dynamics (Gollo et al., 2014; Vicente et al., 2008), will be missed.

Functional MRI generally offers a higher spatial resolution than EEG and MEG, but its temporal resolution is much lower, typically ranging between 0.5 and 1.5 Hz (depending on the acquisition protocol). Most functional MRI studies examine fluctuations of the **blood-oxygenation-level-dependent** (BOLD) signal, a measure of changes in regional levels of cerebral blood oxygenation. Active neurons require oxygen to metabolize glucose—the primary energetic substrate for neuronal signaling. To meet this need, oxygenated blood flows to the activated area. More oxygen is supplied than is needed, causing a net decrease of deoxygenated blood in the region. Oxygenated and deoxygenated blood behave differently when placed in a large magnetic field, such as the 1.5-7 T fields generated by human scanners, or the >9 T fields available for small animal MRI. Deoxygenated blood is weakly attracted to the field (it is paramagnetic) whereas oxygenated blood is weakly repelled by the field (it is diamagnetic). The influx of oxygenated blood to an area that results from neuronal activity thus changes the magnetic properties of that region. This change is measured by the scanner and used as an indirect, hemodynamic marker of local neural activity (Figure 2.17a).

Studies combining functional MRI and invasive electrophysiology have found that the BOLD signal correlates more strongly with LFPs than with the spiking output of an area (Logothetis et al., 2001), but this correlation can vary depending on the psychological context and local neural circuit architecture (Bartels et al., 2008). It is also unclear whether an increase in the BOLD signal reflects a net increase of neuronal excitation in an area, a net increase in inhibition, or some combination of both (Figure 2.17b; Logothetis, 2008). The BOLD response is slow relative to neuronal activity, peaking at around 5 s after stimulus onset. The hemodynamic origin of the BOLD signal also means that it can be affected by local variations in vasculature. For more details on the principles and applications of functional MRI, see Huettel et al. (2014) and Logothetis (2002, 2008). Alternatives to BOLD functional MRI, such as arterial spin labeling (Chen et al., 2015) and functional near-infrared spectroscopy (Scholkmann et al., 2014; Torricelli et al., 2014), can also be used to index regional cerebral blood flow, but have not been used extensively in macroscale connectomics.

Functional connectivity is most commonly measured with functional MRI during task-free resting-states (Fornito and Bullmore, 2010; Fox and Raichle, 2007). In these experiments, individuals lie quietly in the scanner with their eyes either open or closed, without performing any specific task. The signals recorded under such conditions are assumed to reflect spontaneous neural dynamics. These signals have the strongest power at low frequencies (<0.1 Hz), but combined functional MRI and electrophysiological studies suggest that these slow fluctuations correlate with neuronal dynamics unfolding across multiple time scales (He and Raichle, 2009; Mantini et al., 2007). Spontaneous fluctuations of the BOLD signal influence task-evoked activity and behavior (Fox et al., 2006, 2007; Hesselmann et al., 2008); are organized into spatially distributed and temporally correlated networks (Beckmann et al., 2005; Smith et al., 2009); are robust across individuals and time

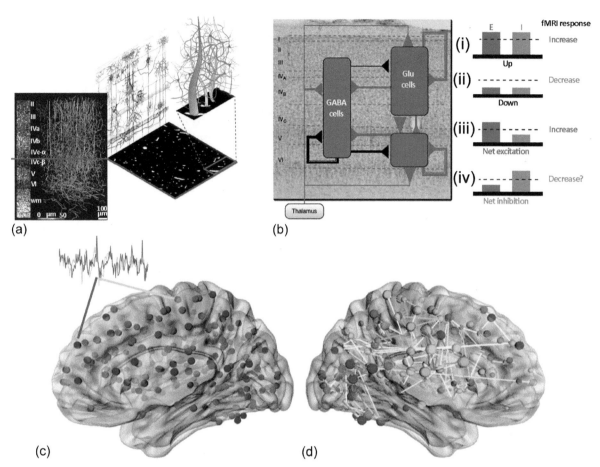

FIGURE 2.17 Mapping functional connectivity networks with functional MRI. **(a)** The neuronal and vascular contents of a voxel. Left panel shows the relative density and laminar distribution of neurons (left column) and blood vessels. Colors represent vessel diameter. Red line corresponds to a section displayed in the right panel. White spots in this section represent cross-sections of the vessels, which account for ~3% of the volume. Neurons, neurites, and glia occupy the space between the vessels, as shown in the color illustration. The top right drawing is of a hypothetical vascular architecture in a small section of tissue (red box). Variations in the levels of oxygenated blood supplied by these vessels cause changes in the magnetic field that is generated by the scanner. These changes form the basis of the BOLD signal. **(b)** A simplified schematic of a canonical cortical microcircuit (left panel). Excitatory cells are shown in red and inhibitory synapses in black. Increases of the BOLD signal relative to a baseline (right panel, dotted black line) can be caused by a combined increase of excitation and inhibition **(i)** whereas a decrease can be caused by a combined reduction of excitation and inhibition **(ii)**. An increase of the BOLD signal may also be caused by an upregulation of excitation alone **(iii)**. Whether or not an isolated increase in inhibition causes a corresponding decrease of the BOLD signal appears to be dependent on the specific neural circuit being investigated. For some neural circuits, an increase of inhibition alone **(iv)**. **(c)** Example parcellation of the brain into functionally defined, spherical regions-of-interest. Functional connectivity is typically measured by the temporal correlation between time series of signal fluctuations extracted from pairs of regions. **(d)** Example brain network graph of functional connectivity generated from functional MRI data acquired in a single human participant. The network has been thresholded to show the top 5% strongest correlations. The colors of the edges encode the strength of the correlation (yellow is lower, red is higher). The colors of the nodes denote the module that each node belongs to (modules are discussed in Chapter 9). In this way, functional connectivity can be mapped across the entire brain. *Reproduced (a,b) from Logothetis (2008), with permission.*

(Damoiseaux et al., 2006; De Luca et al., 2006); and are under strong genetic control (Fornito et al., 2011b; Glahn et al., 2010), indicating that these coordinated fluctuations represent a robust and fundamental component of brain functional organization (Fox and Raichle, 2007). The spatial topography of these so-called resting-state networks is very similar to task-evoked patterns of regional co-activation (Smith et al., 2009), further suggesting that they may reflect the experience-dependent co-activation history of distributed neural systems. This finding is consistent with invasive electrophysiological studies showing that the spatial organization of spontaneous activity resembles stimulus-evoked activity patterns (Kenet et al., 2003; Tsodyks et al., 1999). Resting-state functional connectivity between regions also correlates with underlying anatomical connectivity, as measured with diffusion MRI and tract tracing (Honey et al., 2009; Skudlarski et al., 2008; Vincent et al., 2007; Zalesky and Fornito, 2009).

Functional connectivity in resting-state functional MRI is most commonly measured using the Pearson correlation coefficient between BOLD signal time courses recorded from different regions (Figure 2.17c and d; Box 2.2). The correlation is usually computed using the amplitude fluctuations of the signals themselves, rather than just the amplitude envelope (as is commonly done in analyses of neurophysiological signals recorded with much higher temporal resolution; for an example, see Figure 2.12b). It is common for these functional MRI signals to first undergo preprocessing to remove contributions from nonneuronal physiological sources of noise, such as variations in heart rate and respiration, as well as head motion (Behzadi et al., 2007; Birn et al., 2006; Fox et al., 2009; Kundu et al., 2013; Patel et al., 2014; Power et al., 2012; Satterthwaite et al., 2013). Many other coupling metrics have been used to quantify functional connectivity in these experiments, but studies of BOLD signals simulated on "ground truth" network architectures suggest that the correlation coefficient and partial correlation coefficient perform reasonably well in reconstructing some of the true underlying edge structure of the network (Smith et al., 2011). Nonetheless, these metrics have limitations (Box 2.2). Improvements in the speed of image acquisition (Feinberg et al., 2010) have facilitated the analysis of functional MRI networks as they evolve over time (Chang and Glover, 2010; Zalesky et al., 2014; Box 10.3).

A more general limitation of resting-state functional MRI is that it is difficult to determine whether the spontaneous signal fluctuations are driven by intrinsic neuronal dynamics or transient fluctuations in mental state and cognition. On the one hand, evidence of the robustness of these networks across people and time (Damoiseaux et al., 2006; De Luca et al., 2006), of their persistence across different states of consciousness and in different species (Vincent et al., 2007), and of high heritability of functional connectivity and network topology (Fornito et al., 2011b; Glahn et al., 2010), supports a strong contribution from intrinsic processes and underlying anatomical connectivity (Honey et al., 2007, 2009). On the other hand, several studies have shown that resting-state BOLD signal fluctuations can be influenced by the performance of a prior task (Barnes et al., 2009; Harrison et al., 2008; Lewis et al., 2009) and that they may be associated with transient psychological phenomena such as mind-wandering (Mason et al., 2007). We must therefore be careful in interpreting any functional connectivity estimates that result from such data as directly reflecting an intrinsic property of brain activity or an anatomical connection. Measuring functional connectivity during active task performance can offer more precise control over participants' mental states, and methods are available for isolating task-related functional connectivity from other potential causes of BOLD signal covariation, such as noise, intrinsic connectivity, and coactivation (Cole et al., 2013; Fornito et al., 2012a; Friston et al., 1997).

The low temporal resolution of functional MRI limits the range of functional connectivity measures that can be accurately estimated. For example, we cannot isolate networks within the canonical frequency bands characterized by electrophysiological studies (Figure 2.2a). As a result, most investigators rely on relatively simple coupling metrics, such as the correlation coefficient. The advantage of EEG and MEG is that they allow frequency-specific and time-resolved dynamics to be measured with greater accuracy. These measures can be directly related to invasive recordings. However, this temporal precision comes at the expense of spatial localization. Due to this trade-off between spatial and temporal resolution (see Box 2.1), the edges of functional connectivity networks defined at the macroscale are highly dependent on the measurement technique. The choice of which specific method to use will be dictated by the hypotheses of any individual study: if the temporal structure of neural activity is important, then EEG or MEG is indicated; if the anatomical localization of neural activity is of interest, then functional MRI may be a more appropriate tool.

2.4 SUMMARY

In this chapter, we have considered the problem of node and edge definition with respect to three spatial resolution scales. We have seen that higher resolution measurement techniques allow us to define nodes and edges with cellular and subcellular precision. Coarser mapping technologies inevitably result in a loss of detail and can introduce ambiguities to our definition of nodes and edges. However, the loss of specificity and precision at these coarser scales is counterbalanced by the gain in coverage and greater safety of noninvasive methods, which create opportunities to study human brain network organization in health and disease. The wider network context can be critical for determining microscale structure and dynamics, thus highlighting the benefits of a coarse-grained approach—in the words of the neuroanatomist Valentino

Braitenberg "it makes no sense to read a newspaper with a microscope" (quoted in Logothetis, 2008, p. 871). We thus see that each scale has distinct advantages and disadvantages. The multiscale architecture of the brain (Figures 2.1 and 2.2) means that no single scale is better than the others; rather, connectomes generated at each scale must be analyzed and interpreted with respect to the limitations of the methods used for their construction.

A multiscale architecture has practical consequences for brain network analysis. For microscale connectomics, we must be aware that analyses restricted to small patches of neural tissue will only provide a limited window into network organization. While some microscale methods have been scaled up to study the entire nervous system of relatively small model species such as *C. elegans* and the zebrafish, they cannot explore aspects of brain function that are evident only in larger organisms. For mesoscale and macroscale connectomics, we must be wary of the limits and assumptions of our measurements and models. A thorough validation of the methods used to quantify connection weights, and the cross-validation of findings with respect to different brain parcellations, will enhance confidence in the results obtained at these scales.

This chapter has also demonstrated that the method used to construct a brain network has a major influence on the details included in the connectome. For example, tract-tracing measures yield a directed measure of connectivity between nodes whereas diffusion MRI and correlation-based analyses of functional connectivity yield undirected measures of connectivity. The precise measure used to quantify connectivity strength will also affect the range of values that are in the connectivity matrix and, in turn, influence how we interpret the edge weights of a network. In the next chapter, we consider how these different types of connectivity measures affect network construction, and examine some of the fundamental properties of matrices and graphs that are most relevant for connectomics.

CHAPTER 3

Connectivity Matrices and Brain Graphs

A central goal of **connectomics** is to comprehensively map connections between anatomically distributed neural elements. As we saw in Chapter 2, these elements could be individual neurons, specific neuronal populations, or large-scale brain regions. The number of possible connections between these elements is large; for any network of N **nodes**, the number of possible connections is in the order of N^2. We thus need a method to represent these large data in a succinct and meaningful way.

It is common to represent the connectivity between every pair of nodes in a network as a two-dimensional **matrix**. In this matrix, each row and column corresponds to a different node, and the matrix element positioned at the intersection of the ith row and jth column encodes information about the connection between regions i and j (Figure 1.1). This representation, which is commonly referred to as a connectivity matrix, is fundamental to network analysis (Sporns et al., 2005) and forms the basis of nearly all the analytic techniques discussed in this book. Importantly, the connectivity matrix can be used to generate a graph-based representation of the network, such that each row/column of the matrix is a node in the graph, and each matrix element represents an **edge**. This equivalence between matrix and graph representations means that we can use either in the analysis of brain network connectivity.

In this chapter, we introduce the major types of connectivity matrices that are encountered in connectomics. We discuss the equivalence between matrix and graph-based representations of networks and overview some of the methods used for their visualization.

3.1 THE CONNECTIVITY MATRIX

The connectivity matrix offers a compact description of the pairwise connectivity between all nodes of a network. To build a connectivity matrix, C, for a brain network comprising N nodes, we start by constructing a two-dimensional array, called a square matrix, which comprises N rows and N columns. Each row and corresponding column represents a unique network node. We then populate

89

Fundamentals of Brain Network Analysis. http://dx.doi.org/10.1016/B978-0-12-407908-3.00003-0

the elements of this matrix with the connectivity values that have been estimated for every pair of nodes to construct the $N \times N$ matrix

$$C_{ij} = \begin{bmatrix} C_{11} & C_{12} & \cdots & C_{1N} \\ C_{21} & C_{22} & & C_{2N} \\ \vdots & & \ddots & \vdots \\ C_{N1} & C_{N2} & \cdots & C_{NN} \end{bmatrix}. \tag{3.1}$$

According to this convention the subscripts of C are used to index each element: the first subscript, i, indexes rows and the second subscript, j, indexes columns. This kind of matrix representation is remarkably flexible and can be used to encode a number of different structural properties of a network. We will now consider some of the basic properties of this matrix that are most relevant to brain network analysis.

3.1.1   Diagonal and Off-Diagonal Elements

A square matrix can be separated into diagonal and off-diagonal elements. The diagonal elements, C_{ii}, of the connectivity matrix in Equation (3.1) are shown with black font. Diagonal elements are sometimes interpreted as representing the connectivity of each node with itself, but more generally they can be used to encode some intrinsic property of each node. For example, in a neuronal network, we could enter different values along the matrix diagonal to represent distinct kinds of neurons, such as pyramidal neurons, inhibitory interneurons, and so on. In mesoscopic or macroscopic brain networks, the diagonal could be used to represent some property of the internal circuitry or functional role of each brain region. In **functional connectivity** networks, the diagonal could reflect variations in the local dynamics of each brain region. In practice, the matrix diagonal is seldom used for such purposes in neuroscience. Rather, intrinsic differences between network nodes are commonly ignored and the diagonal is conventionally set to a common value for all nodes. Indeed, many of the graph theoretic measures that have been applied to brain networks ignore the matrix diagonal, often assuming that it contains only zero values. We will also make this general assumption unless specified otherwise.

The off-diagonal elements of the connectivity matrix, C_{ij} (where $i \neq j$), represent the connectivity between pairs of distinct neural elements. The values contained in these off-diagonal elements thus correspond to our measured estimates of pairwise connectivity, and their range will depend on the particular method that we have used to estimate this connectivity (Chapter 2). In general, the values of these elements can be used to represent the type (e.g., excitatory or inhibitory) and strength (e.g., weak or strong) of connectivity between each pair of regions. We can divide the off-diagonal elements of a matrix into a lower triangle, which comprises all values listed beneath the matrix diagonal, and an upper triangle, which comprises all values listed above the matrix diagonal. The lower and upper triangles are shown in Equation (3.1) using blue and red font, respectively.

3.1.2 Directionality

If we have different values in the upper and lower triangle of our connectivity matrix, then the element C_{ij} will not necessarily have the same value as the element C_{ji} and the matrix is asymmetric. In this case, our matrix represents a **directed graph** or network and the asymmetries encode the directions of the connections. Directed networks are also sometimes referred to as *digraphs*. They allow us to map the influence that one network node exerts on another.

Different authors use matrices to represent directionality in different ways. In some cases, efferent projections are listed down the columns and afferent projections along the rows of the matrix; in other cases, this convention is reversed. It is therefore important to understand precisely how a directed connectivity matrix has been constructed before commencing analysis. Unless otherwise stated, in this book we index edges from the second matrix index (column) to the first (row). For example, element $C_{1,4}$ of the connectivity matrix encodes the edge projecting *from* node 4 *to* 1. Conversely, element $C_{4,1}$ encodes the edge projecting in the reverse direction, *from* node 1 *to* 4.

Using this convention, Figure 3.1a shows how we can build a connectivity matrix from a small, directed network comprising six nodes and seven edges (see also Figure 3.1b). In the graph representation, arrowheads represent the direction of connectivity, such that the arrowhead points to the target of each **link** (upper panel). In the matrix representation, directionality is encoded by asymmetries in the matrix elements (lower panel). For instance, node A projects to E, but E does not send a projection back to A. We therefore have a nonzero value in element $C_{E,A}$ and a value of zero in element $C_{A,E}$.

If the upper and lower triangles of the connectivity matrix are identical, we have $C_{ij} = C_{ji}$ and the matrix is symmetric. In this case, the matrix represents an **undirected graph** or network. Such networks allow us to identify which connections exist between specific pairs of network nodes, but they do not allow us to draw any conclusions about the directions of those connections. In brain networks, this means that we cannot make any inferences about possible directions of information flow, or about the causal influences that one neural element may exert on the activity of another. An example of an undirected network is depicted in Figure 3.1c. There are no arrowheads in the graph representation (upper panel), and the upper and lower triangles of the connectivity matrix are mirror images of each other (lower panel).

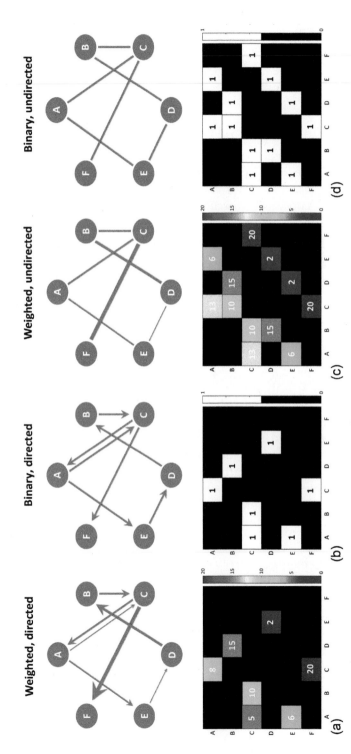

FIGURE 3.1 The equivalence between graphs and matrices. **(a)** A weighted, directed network graph (upper panel), with directionality represented using arrowheads and connection strengths represented as variations in edge thickness. In the corresponding connectivity matrix (lower panel), colors are used to represent variations in weights. **(b)** A binarized or unweighted version of the directed network depicted in **(a)**. All edges have the same thickness (upper panel), and all connectivity weights have a value of zero or one, indicating the absence or presence of a connection. **(c)** An undirected version of the weighted network depicted in **(a)**. Arrowheads are no longer required as there is no directionality in this graph (upper panel). The corresponding **adjacency matrix** is symmetric, such that $C_{ij} = C_{ji}$ (lower panel). The weights of this matrix were computed by summing the weight of connectivity running from node i to j and back again for each pair of regions. **(d)** A binarized and undirected version of the network in **(c)**.

Brain anatomical networks are intrinsically directed, since each axonal projection originates at a cell body and terminates at one or more synapses. Mapping the directionality of neuronal connectivity requires the use of invasive methods, such as **tract tracing** or **electron microscopy**. Noninvasive methods suitable for use in living organisms, such as **diffusion MRI**, do not presently allow us to resolve the direction of an axonal projection. The networks that result from such techniques are therefore undirected. In analyses of brain functional connectivity and **effective connectivity**, directionality can be measured or modeled using specific methods (Chapter 2). In general however, most analyses use the correlation coefficient or related measures in the time domain (e.g., partial correlations) or frequency domain (e.g., coherence) to quantify functional connectivity, resulting in an undirected network. As will be evident throughout this book, resolving the directionality of brain connectivity affords a richer characterization of network organization than analysis of undirected networks. Nonetheless, undirected graphs can still offer important insights into the organization of nervous systems.

3.1.3 Connectivity Weights

Not all connections between neural elements are the same. Some pairs of neural elements may make more synapses with each other, communicate via denser axonal bundles, or have more myelinated fibers that facilitate faster signal transmission. These variations can be described by differences in connectivity weight. Most methods for measuring brain connectivity provide some index of connectivity weight, which for any pair of nodes i and j, is denoted as w_{ij}. We can retain this information in our connectivity matrix by populating the matrix elements with these connection weights, such that $C_{ij} = w_{ij}$. The result is a **weighted graph** or network. In graph representations, it is common to represent differences in connectivity weight between node pairs as variations in edge thickness. In matrix representations, weight variations are often represented using a color scale. Examples of matrix and graph representations of weighted networks are shown in Figure 3.1a and c.

Connectivity weights in brain networks can show wide variation, depending on the method used to estimate interregional connectivity (Chapter 2). For example, if we use electron microscopy to reconstruct each synaptic connection between neurons, the values will range between zero (reflecting no connection) and an upper limit that is determined by spatial and physiological constraints on neuronal wiring (Attwell and Laughlin, 2001; Cherniak, 1990). Alternatively, we may index connectivity as the fraction of neurons in a source area that have been labeled by a retrograde tracer injected into a target region (Markov et al., 2013a), in which case w_{ij} will vary between zero and one, such that $w_{ij} = 0$

indicates an absence of connectivity and $w_{ij} = 1$ indicates that all neurons in the target region were labeled by the tracer.

Weights in functional connectivity networks may span different ranges. For example, correlation-based estimates of functional connectivity will span the interval $-1 < w_{ij} < 1$, where $w_{ij} = -1$ indicates a perfect negative correlation, or anticorrelation, between two regional time courses and $w_{ij} = 1$ reflects a perfect positive correlation. In contrast, if we use an alternative measure to quantify functional connectivity, such as mutual information, our connectivity weights will generally be non-negative. The method we use to estimate connectivity will thus determine the range of weights observed, and how these weights should be interpreted. The weights will also be affected by the sensitivity and resolution of the measurement technique (Chapter 2).

Brain networks can also be represented as unweighted or **binary graphs**. In these networks, $C_{ij} = 1$ if regions i and j are connected and $C_{ij} = 0$ otherwise (Figure 3.1b and d). Binary networks tell us where connections are in the network, but provide no information about variations in connectivity weight between different network nodes. Given that connection weights in brain networks can vary over several orders of magnitude (Markov et al., 2013a; Oh et al., 2014), a reasonable conclusion might be that all analyses of brain networks should be weighted. However, valuable insights into network organization can often be gained by considering the binary **topology** of a connectivity matrix. Most graph theoretic measures used in the analysis of connectomic data have been developed for binary networks. We discuss the advantages and disadvantages of binary representations of brain connectivity in Section 3.2.2.

3.1.4 Sparse Matrices

An alternative network representation to the square matrix is an edge list, which is sometimes also referred to as a **sparse matrix**. This representation is useful when most matrix elements are zero. Each row in this list represents a different edge and typically contains two values, listed in different columns. The first column lists a pair of indices that indicate which pair of nodes is linked by the edge. The second column indicates the connectivity weight of that edge. For example, take the network depicted in Figure 3.1c. The first column of values in a sparse representation of this network indexes the unique edges of this graph, which are {AC, AE, BC, BD, CF, DE}. The second column of values indexes the weights of each edge, which are {13, 6, 10, 15, 20, 2}. For a directed network, separate entries are required for any edges going from node i to j and from j to i.

The name sparse matrix comes from the fact that most of the matrix elements are zero and only the nonzero edges are represented. Any unlisted edges are assumed to have a weight of zero. This compact representation is useful for analyzing large, sparsely connected networks because it eliminates the need to store zero-valued matrix elements in computer memory.

3.2 THE ADJACENCY MATRIX

The connectivity matrix is sometimes also referred to as an **adjacency matrix**, *A*. In **graph theory**, two nodes that are directly connected by an edge are said to be **adjacent** or **neighbors**. The adjacency matrix thus defines the pattern of pairwise adjacencies between nodes.

In connectomics, we often apply some kind of processing or filtering of the connectivity matrix prior to analysis. For example, we might choose to threshold the matrix in order to reduce the influence of low-weight and potentially spurious connections on network topology (Chapter 11). Throughout this book, we will refer to the raw or unfiltered matrix as the connectivity matrix, *C*, and the matrix used for subsequent analysis as the adjacency matrix, *A*. In some cases, the matrices *C* and *A* are equivalent; in other cases, *A* is a processed or filtered version of *C*. In this section, we consider the two most common processing steps applied to the connectivity matrix *C* to yield a final adjacency matrix *A*: thresholding and binarization.

3.2.1 Thresholding

Some brain networks, such as functional connectivity networks based on time series correlations, contain a nonzero value in every off-diagonal element of the matrix (Figure 3.2a). Such networks are referred to as fully connected since there is a nonzero link between every pair of regions. Are all these connections real? Should all of these edges be considered in our analysis? As a general rule, we know that the brain is not a fully connected network. For example, the 302 neurons comprising the central nervous system of *C. elegans* are linked by around 5600 chemical synapses and gap junctions (White et al., 1986), representing approximately 6% of the total possible $N(N-1) = 90,902$ connections that could exist. Connectivity in lower-resolution **connectomes** tends to be higher; for example, the density-estimates of an interareal cortical network of the macaque constructed with viral tract-tracing has been found to be as high as 66% (Markov et al., 2013a). Although dense, this network is still far from being fully connected. These estimates suggest that any method for brain network reconstruction that yields a fully connected connectivity matrix, in which every neural element is connected to every other, is likely to include a substantial percentage of spurious connections. Indeed, most methods for quantifying brain connectivity, whether structural or functional, are associated with measurement noise, complicating the distinction between real and false connections

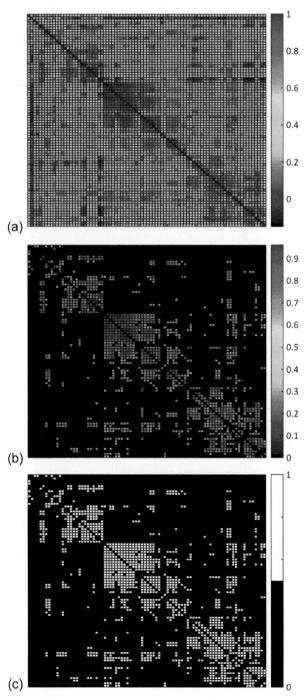

FIGURE 3.2 Thresholding and binarizing an adjacency matrix. **(a)** an unthresholded correlation-based functional connectivity network measured with human **functional MRI**. Each element in this matrix has a nonzero value. **(b)** the same matrix after the network has been thresholded to retain only the 20% strongest weights. **(c)** the thresholded network after binarization.

(Chapter 2). Thus, a small, nonzero value in the connectivity matrix may reflect measurement noise, rather than the presence of an actual connection.

To address this problem, we can apply a cut-off value or threshold, τ, to the connectivity matrix in order to determine which connections should be retained in the adjacency matrix, such that

$$A_{ij} = \begin{cases} C_{ij} & \text{if } C_{ij} > \tau, \\ 0 & \text{otherwise} \end{cases}. \tag{3.2}$$

With this method, the edges C_{ij} are a super-set of the edges A_{ij} (Figure 3.2b). Applying such a threshold can be useful to maximize the separation between signal and noise—between putatively real and spurious links—and to emphasize the topological properties of edges that are less likely to reflect measurement error. It can also improve computational efficiency, since most network measures are faster to compute on networks with fewer edges.

There are two principal disadvantages of thresholding. First, it is often unclear which specific threshold value should be used, so the choice of a final value for τ is left to the investigator. While several different methods for thresholding brain networks have been developed, each of these methods has its own specific limitations (van Wijk et al., 2010). The second problem with thresholding is that it can cause differences in the connection densities of individual networks within a broader population. For example, if we apply a threshold of $\tau = 0.2$ to a sample of human functional connectivity networks, the adjacency matrices of individuals with high average levels of connectivity will have a larger number of edges surviving this threshold than the matrices of people with low average connectivity. This variation introduces a major confound, since most network measures are sensitive to the number of edges in the network. An alternative approach is to adaptively vary the value of τ to achieve the same number of edges in all networks, but this method also has limitations that can complicate analyses. The relative strengths and weakness of different approaches to thresholding brain connectivity data are discussed in detail in Chapter 11.

3.2.2 Binarization

After applying a threshold to the connectivity matrix, the remaining elements can be binarized such that

$$A_{ij} = \begin{cases} 1 & \text{if } C_{ij} > \tau, \\ 0 & \text{otherwise} \end{cases}. \tag{3.3}$$

Analyses of binary connectivity matrices are principally concerned with understanding the topological patterns of connections between nodes, irrespective of variations in their weight (Figure 3.2c). Focusing only on which connections are present, and where they are positioned, can provide insight into the basic architectural plan of a connectome.

Many measures computed on binary networks are simpler and easier to interpret than those computed on weighted networks. This is because variations in some topological properties of weighted networks can be difficult to disentangle from variations in the weights themselves (Alstott et al., 2014b; Barrat et al., 2004; Ginestet et al., 2011; van Wijk et al., 2010). Indeed, variations in weighted network topology may arise from fluctuations in the mean or some other property of edge weights unless some appropriate normalization of the weights is used. Moreover, variations in edge weight can be caused by biases in the methods used to estimate brain connectivity. For example, problems associated with reconstructing long-range fiber bundles in diffusion **tractography** (Zalesky and Fornito, 2009) may spuriously affect the range of edge weights in **structural connectivity** networks constructed using this technique. Binarization can minimize these biases. On the other hand, the large variation in neural connection weights identified in brain networks constructed with high quality, high-resolution tract-tracing (Markov et al., 2013a; Oh et al., 2014) suggests that binarization may eliminate important information about the relative flow of signals between different neural elements. It is therefore often useful to analyze both binary and weighted variants of an adjacency matrix to understand how the connectivity weights of a network relate to its topology.

3.2.3   Network Density and Weight

The adjacency matrix forms the basis for all subsequent network analysis. As we have seen, a basic, yet important measure of this matrix is the total number of edges, E, which is sometimes referred to as the network **degree**. This value is computed as the total number of *unique* nonzero elements of the adjacency matrix. We emphasize the word unique because undirected networks make no distinction between a connection that runs from node i to j or viceversa. Consequently, if we sum all nonzero edges in an undirected, binary network, we count the same edge twice: once from i to j and once from j to i. In other words,

$$2E = \sum_{ij} A_{ij}. \qquad (3.4)$$

The total number of edges in a directed network is thus $E_{\mathrm{dir}} = \sum_{ij} A_{ij}$.

Another way of thinking about the estimation of connection density in directed and undirected networks is to consider each edge as having two ends or stubs. Each pair of nodes connected by an edge is attached to one stub. Any network, whether directed or undirected, will contain $2E$ stubs. If we sum all nonzero elements in the matrix A, we count both stubs of each edge. This makes sense for a directed network, since we want to separately quantify connectivity going from i to j and from j to i. In an undirected network, it means that we count each edge twice.

The **connection density** (sometimes called connectance) of a network refers to the proportion of nonzero elements in A relative to the total possible number of connections that could be formed in the network. The total number of possible connections in an undirected network is given by $N(N-1)/2$, where N is the number of nodes in the network and $N(N-1)$ is simply the number of off-diagonal elements in the adjacency matrix. To understand why, remember that a square, $N \times N$ matrix will have N^2 elements in total, which includes N elements along the diagonal. Since we are interested in the total number of possible connections *between* nodes, we subtract the N diagonal elements of the matrix. The connection density, κ, of an undirected network can thus be quantified as

$$\kappa = \frac{2E}{N(N-1)}. \tag{3.5}$$

Since κ is a proportion, it varies between zero and one, where $\kappa = 0$ indicates that no connections are present, $\kappa = 1$ indicates that the network is fully connected (i.e., all possible connections are present), and $0 < \kappa < 1$ represents the fraction of all possible connections that are present in the network.

When analyzing directed networks, it is critical to distinguish between the ends (stubs) of each edge. Specifically, we must remember that one end is incoming (afferent) and the other is outgoing (efferent). To compute the total degree of the network, we want to count both the edges running from node i to j and vice versa since each stub gives us unique information. Therefore, the total number of edges in a directed network is equal to the total number of nonzero off-diagonal elements in the adjacency matrix (Figure 3.1a and b) and we can write the connection density of a directed network as

$$\kappa_{\text{dir}} = \frac{E_{\text{dir}}}{N(N-1)}. \tag{3.6}$$

Note that Equations (3.5) and (3.6) differ by a factor of 2. Henceforth, we use κ to denote the connection density of both directed and undirected networks.

Let us consider a concrete example to further clarify the distinction between directed and undirected networks. Take a simple, three-node directed matrix:

$$A = \begin{bmatrix} 0 & 1 & 0 \\ 0 & 0 & 1 \\ 1 & 0 & 0 \end{bmatrix}. \tag{3.7}$$

There are three nonzero elements in this matrix: $A_{1,2}$, $A_{2,3}$, and $A_{3,1}$. We have $E_{\text{dir}} = 3$ out of $3(3-1) = 6$ possible connections, so $\kappa = 3/6 = 0.50$.

Now consider the undirected matrix:

$$A = \begin{bmatrix} 0 & 1 & 0 \\ 1 & 0 & 1 \\ 0 & 1 & 0 \end{bmatrix}. \tag{3.8}$$

There are four nonzero elements in this matrix, which represent the presence of two unique connections out of a maximum possible number of three. We make no distinction between incoming or outgoing connections, so $A_{ij} = A_{ji}$ and the number of nonzero elements is equal to twice the number of *unique* connections (i.e., $2E = \Sigma_{ij}A_{ij}$). We can therefore calculate the total number of edges in an undirected network as the number of nonzero elements in *either* the upper or lower triangle of the matrix, or more generally as $\frac{1}{2}\Sigma_{ij}A_{ij}$. The total number of possible connections in this undirected network is thus $3(3-1)/2 = 3$ and the connection density is $\kappa = 2/3 = 0.67$.

For weighted networks, we can also compute the total weight of the edges in the adjacency matrix. Depending on the method used to measure brain connectivity, it is possible to have positively and negatively signed weights (e.g., a correlation coefficient may be either negative or positive). If we denote positive weights as w_{ij}^+ and negative weights as w_{ij}^-, the total weight across both signs, $W^\pm$, of an undirected graph is given by half the sum of all its positive and negative edge weights,

$$W^\pm = \frac{1}{2}\sum_{ij} w_{ij}^\pm. \tag{3.9}$$

We can also compute total weight separately for positive and negative weights as $W^+ = \frac{1}{2}\Sigma_{ij}w_{ij}^+$ and $W^- = \frac{1}{2}\Sigma_{ij}w_{ij}^-$, respectively. For directed networks, we do not halve the sum of weights because once again, each matrix element provides unique information about network connectivity. In some cases, such as correlation-based networks, the mean weight of the connectivity matrix may offer a more intuitive summary than the total weight (see Chapter 4).

The connection density of a brain network can vary as a function of species, resolution scale and measurement technique. Across species, brains with more cortical neurons send proportionally fewer axons through white matter (Herculano-Houzel, 2012; Herculano-Houzel et al., 2010; Ventura-Antunes et al., 2013). In other words, larger brains make fewer long-distance connections, which may be a consequence of larger brains requiring greater metabolic and material resources to maintain connectivity (Ringo, 1991).

The impact of resolution scale on connection density is such that networks mapped at lower spatial resolutions tend to have a higher density of connectivity. This is because the probability of finding a connection between a pair of small neural elements (e.g., two neurons amongst many billions) will be lower than the probability of finding at least one axon linking a pair of larger elements (e.g., two cytoarchitectonically defined cortical areas amongst several hundred). This effect will be amplified when there is a mismatch in the resolutions used for node definition and connectivity mapping—if we try to measure connectivity between two large populations of neurons, a high-resolution connectivity mapping technique such as invasive tract-tracing will have a much higher likelihood of identifying at least one connection compared to a lower-resolution mapping technique, such as diffusion MRI. Thus, we should expect the connection density of a network to be particularly high when a low-resolution parcellation is combined with a high-resolution method for mapping connectivity.

The measurement technique for quantifying connectivity can also affect connection density. Structural connectivity networks derived using tract tracing and deterministic diffusion tractography (Chapter 2) often yield connectivity matrices that contain some zero-valued elements. These zero values could be a true negative (i.e., an absent connection), or a false negative caused by limitations and/or biases of the measurement technique. Other techniques, such as diffusion tractography based on high-order crossing-fiber models and probabilistic tracking methods typically yield much denser connectivity matrices (in the order of >80% for ~100 nodes). It is likely that these dense matrices include false positive connections—connections mapped between pairs of regions that are not truly interconnected (Thomas et al., 2014). Filtering and thresholding methods can potentially eliminate some of these false positives (Smith et al., 2013; Sherbondy et al., 2010). Brain networks constructed with these methods thus have an intrinsic density imposed by both the actual connectivity of the brain and the biases of the measurement technique.

In other networks, particularly correlation-based networks such as those studied in analyses of functional connectivity and morphometric covariance, there is no intrinsic density: all edges have a continuously varying nonzero weight and the network is thus fully connected by construction. In these cases, thresholding determines network density and becomes a critical processing step prior to analysis (Chapter 11).

3.3 NETWORK VISUALIZATION

Visualization is critical in network analysis, and there are many different ways of visualizing both adjacency matrices and brain graphs. In this section, we consider how different types of matrix and graph visualizations can be used to emphasize distinct aspects of network organization.

3.3.1   Visualizing the Adjacency Matrix

Visualizing an adjacency matrix can provide useful insight into network organization. As we have seen, each row and column represents a different network node, and each element represents the pairwise connectivity between nodes. When visualizing this matrix, the elements are often colored to represent variations in edge weight (e.g., Figure 3.2). Other properties can be encoded by treating the different elements as independent tiles that can vary in size and shape (e.g., Figure 1.1).

We can gain insight into different organizational properties of the network by reordering the rows and columns of the matrix. Some common examples are shown in Figure 3.3. All four matrices represent the same dataset—a human structural connectivity matrix estimated between 82 anatomically defined regional nodes, averaged over diffusion MRI data acquired in a sample of 40 healthy individuals. In Figure 3.3a, rows and columns are ordered according to anatomical criteria, such that all left hemisphere regions are listed first, followed by all right hemisphere regions. Within each hemisphere, the ordering of the regions is the same, so that the first-listed left hemisphere region and first-listed right hemisphere region correspond to homologous areas. With this ordering, we see two clear blocks of increased connectivity: one in the upper-left quadrant and one in the lower-right quadrant. These two blocks represent the intrahemispheric connectivity of the left (upper left quadrant) and right (lower right quadrant) hemispheres. Interhemispheric connectivity is depicted in the upper right and lower left quadrants. The white boxes in Figure 3.3a highlight subdiagonal elements of the matrix that encode connectivity between homologous regions in the left and right hemispheres. From this anatomically informed matrix representation, we see that intrahemispheric connectivity is stronger than interhemispheric connectivity, and that there is a tendency for homologous regions in opposite hemispheres to be connected with each other.

In Figure 3.3b, rows and columns are ordered by ranking nodes in descending order from the most highly connected to the most weakly connected. Highly connected areas appear in the top left corner. These areas have broad connectivity with most other brain regions, whereas weakly connected regions generally have low connectivity across the entire brain (bottom right corner). This structure is consistent with the brain being organized around a core of highly interconnected areas with more weakly connected nodes in the periphery (see also Chapter 6).

Figure 3.3c orders rows and columns of the adjacency matrix to emphasize the modular organization of the network (see Chapter 9). All nodes belonging to the same topological **module** (delineated by white lines) are listed together, giving the matrix a block-diagonal appearance. We can see that connectivity between members of the same module is higher than connectivity between modules.

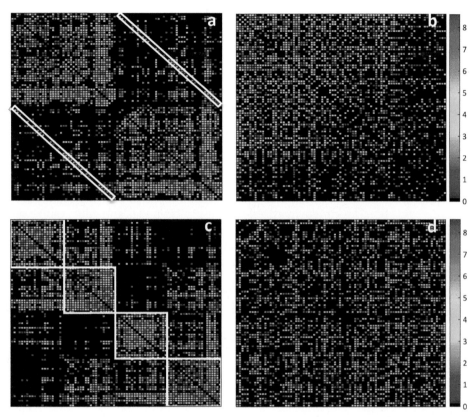

FIGURE 3.3 Visualizing the adjacency matrix. Different ways of visualizing an 82-region human structural connectivity matrix, constructed by averaging diffusion MRI data across 40 healthy participants. Edge weights are plotted on a logarithmic scale. **(a)** Rows and columns are ordered such that all left hemisphere regions appear first, followed by all right hemisphere regions. The densely connected blocks in the upper-left and lower-right quadrants of the matrix thus represent strong intrahemispheric connectivity, whereas the darker colored lower-left and upper-right quadrants represent weaker interhemispheric connectivity. White boxes highlight subdiagonal elements that correspond to the connectivity of each region with its homologue in the other hemisphere. **(b)** Rows and columns have been ordered from most to least connected, as determined by summing each node's connectivity weights with the rest of the network. The most highly connected nodes appear in the upper left corner of the matrix. **(c)** Rows and columns are ordered to emphasize modular structure. Regions belonging to the same topological module (Chapter 9) are clustered together along the diagonal, giving a block-diagonal structure to the matrix. Modules are delineated by white lines. Connectivity within each module is strong and relatively weak between modules. **(d)** Rows and columns are in random order. The structure in the network is no longer clearly visible by eye.

In Figure 3.3d, the rows and columns are ordered at random and no clear structure is discernible. Comparing these visualizations shows how reordering the adjacency matrix can emphasize different network properties. Many other types of ordering are possible (Henderson and Robinson, 2013). In some analyses, node ordering is more than cosmetic—it can also affect the computation of certain network properties (Blondel et al., 2008).

3.3.2 Visualizing Brain Graphs

As shown in Figure 3.1, we can trivially alternate between a matrix-based and graph-based representation of a network. Indeed, generating a graph from an adjacency matrix is simple. Each brain region, corresponding to a given row (and column) of the adjacency matrix, is represented in a graph as a distinct node (commonly depicted as a circle or sphere), and each connection is represented as an edge linking two nodes. In graph theory, a node is sometimes also called a **vertex** and an edge is sometimes called an **arc** or **link**.

Variations in connection weights can be represented in a graph by the thickness of each edge. Directionality can be represented by arrowheads that are attached to the ends of the edges, such that the arrow points to the target of the connection. Much like matrices, we can project nodes and edges onto a graph in different ways in order to emphasize distinct network attributes. Here, we consider three classes of graph layouts: anatomical, circular, and force-directed. Examples of each are presented in Figure 3.4.

Anatomical network projections position nodes according to their physical location in the brain. For mammalian brains, this is most commonly done within the three-dimensional coordinate space of the intracranial volume, such that each node is plotted according to its stereotactic $\{x, y, z\}$ coordinates and edges are drawn as straight lines between nodes. Colors can be used to highlight different groupings of nodes and/or edges (e.g., Figure 3.4a). Anatomical projections are excellent for understanding spatial effects in the data and for localizing specific findings. However, the visualization can become cluttered, particularly for large and dense networks. Moreover, we can only project two-dimensions of the three-dimensional space onto a printed page at any one time, so multiple views are required to "escape from flatland" and obtain a comprehensive visual depiction of the data (Tufte, 1990). Edge bundling and other algorithms can reduce clutter in these projections (Böttger et al., 2014).

Circular projections offer a succinct characterization of network connectivity. In this projection, nodes are positioned around the perimeter of a circle (ring) and edges are drawn either as straight lines or curves cutting through the circle interior. This type of visualization was used by Watts and Strogatz (1998) in their classic illustration of **small-world** network properties (Figure 1.3). We can also illustrate specific properties by ordering the nodes along the perimeter of the circle according to different criteria, such as cortical lobe or topological module. Edges can vary in width or color to depict variations in specific edge-wise properties, such as connectivity weight. Additional properties can be encoded as colored glyphs in concentric rings around the perimeter of the circle (e.g., Figure 3.4b). These visualizations are sometimes called *connectograms* (Irimia et al., 2012). This format can also accommodate hierarchical network structures (Modha and Singh, 2010; Samu et al., 2014).

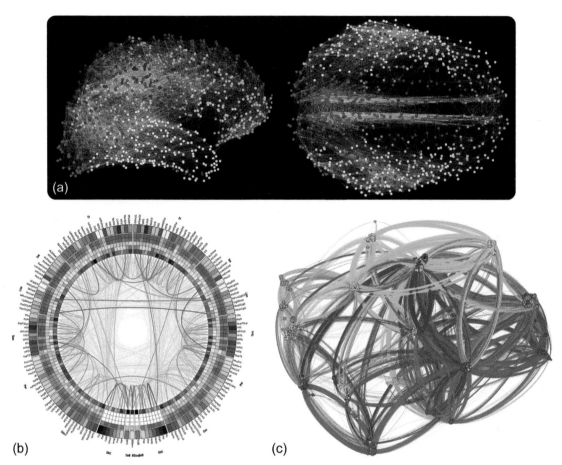

FIGURE 3.4 Visualizing brain graphs. **(a)** An anatomical projection of a 988-region human structural connectivity network constructed using diffusion tractography. In this projection, each node is located according to its anatomical position, defined by a three-dimensional coordinate system. White edges represent links between nodes. Node colors denote the anatomical division to which each region belongs. **(b)** A circular projection, also called a connectogram, of human cortical network structural connectivity. Each node is represented in the outer-most circle as a tile and is ordered according to broad anatomical division, such as frontal, parietal, limbic, and so on. The other concentric circles encode variations in different node-level properties—from outside to inside, they represent regional (1) volume; (2) average cortical thickness; (3) total surface area; (4) cortical curvature; and (5) connectivity degree. Edges crossing the diameter of the circle represent structural connections between nodes. Blue represents connections with low, green moderate, and red high connectivity weight. **(c)** A force-directed projection of a voxelwise human functional MRI functional connectivity network comprising over 22,000 nodes and nearly 60 million edges. Nodes are clustered into topological modules, and edges linking different modules have been clustered together using a bundling algorithm. Even with such a massive network, topological relations between nodes can be discerned with this projection. See also Box 3.1. *(a) Reproduced from Samu et al. (2014), (b) from Van Horn et al. (2012), and (c) from Zuo et al. (2012) with permission.*

Force-directed projections model the graph as a system of nodes that are attracted or repelled according to a physical force that is proportional to some measure of pair-wise node distance, such as the connectivity weight or topological **path length**. With this approach, an optimal solution typically strives for some combination of the following attributes (Gibson et al., 2013): (1) minimal edge crossings, to minimize clutter and allow edge **paths** to be followed easily; (2) symmetry, which aids an understanding of network structure; (3) uniform edge lengths, to avoid distortions; (4) uniform spatial distribution of nodes, to minimize clutter and yield a regular structure; and (5) a positioning of nodes such that their spatial proximity in the layout is correlated with their topological adjacency, in order to illustrate pair-wise node relations.

The optimum solution for a force-projected layout is commonly found by initiating the algorithm with a preliminary placement of nodes, followed by iterative node displacement in order to minimize an objective function that defines the total energy of the layout, as determined by the physical forces assigned to the edges. The objective function is defined so that low-energy layouts are those in which the graphical location of topologically adjacent nodes is near some predefined ideal distance, whereas nonadjacent nodes are distant from each other. Some of the most popular approaches define node attraction and repulsion using a spring-like force (see Box 3.1). Alternative techniques, such as those based on multidimensional scaling, find a projection of the data that maximally corresponds to the observed topological distances between nodes (Dwyer et al., 2006; Kruskal, 1964).

BOX 3.1 SPRING-EMBEDDED NETWORK VISUALIZATIONS

When displaying a network on a two-dimensional page, we should ideally position the nodes and edges in a way that is both clear and informative about some specific aspect of network organization. Spring-embedding is a popular method designed to address this problem. Spring-embedded projections treat each node as a steel ring linked by springs (edges), thereby forming a mechanical system. The optimal layout is defined as one that minimizes the total energy of the system.

The idea of using spring-like forces to determine the distance between nodes in a graph visualization was first proposed by Eades (1984), who built on an earlier approach based on gravitational forces described by Tutte (1963). A widely used refinement of Eades' method was proposed by Fruchterman and Reingold (1991). Each node in the Fruchterman and Reingold projection is subjected to attractive and repulsive forces. The aim is to position nodes as uniformly in space on the page as possible. Attractive forces bring nodes that are linked by an edge closer together on the page, but do not act between pairs of nodes that are not connected. In contrast, repulsive forces act between all pairs of nodes, forcing apart pairs that are drawn close together on the page. This repulsion prevents nodes from occupying the same space. For large networks, the algorithm is iterated several times, starting from an initial random placement of nodes and then re-positioning each node based on its computed attractive and repulsive forces until a stable equilibrium is reached. To ensure convergence, the distance by which nodes can be repositioned is moderated as the number of iterations grows; that is, the degree of displacement becomes smaller as the layout improves.

Another popular spring-embedded algorithm, proposed by Kamada and Kawai (1989), models the spring force between nodes according to their topological distance, which is determined by their shortest path length (Chapter 7). The method thus assumes that the ideal spatial location of

BOX 3.1 SPRING-EMBEDDED NETWORK VISUALIZATIONS—CONT'D

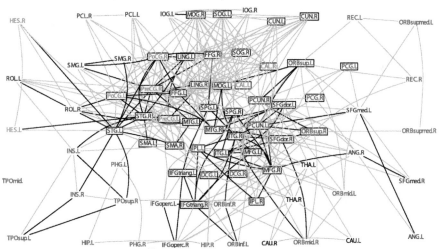

FIGURE 3.1.1 A spring-embedded projection of a human functional connectivity network. The Kamada-Kawai projection was used for this image. The 90-region network was reconstructed from resting-state functional MRI data. Nodes are positioned such that their location in the graph approximates their topological distance. Nodes near the center of the graph thus have low average path length to other regions. As a result, topologically central nodes are located near the center of the graph. Different colored text represents different anatomical divisions. Short-distance connections are in gray, long distance connections in black. *Reproduced from Achard et al. (2006) with permission.*

the nodes should be determined by their pair-wise path length. The optimal layout is identified by minimizing an objective function that quantifies the difference between the sum of Euclidean distances (on the page) and the sum of topological distances between pairs of nodes. In other words, this method attempts to project nodes on the page such that their spatial distance corresponds closely with their topological distance (Figure 3.1.1).

Spring-embedded projections are simple, flexible, and have an intuitive theoretical basis. They can also be useful for revealing the modular organization of networks (Fair et al., 2009; Fornito et al., 2012a; Zuo et al., 2012; see also Figure 3.4c). However, finding a globally optimal solution can be challenging, especially for large networks, and the final result can depend on the initial placement of the nodes. A multistep approach that starts with a Kamada-Kawai projection to generate an initial layout and then identifies a final optimum with the Fruchterman-Reingold algorithm can alleviate some of these problems (Gibson et al., 2013; Kobourov, 2013). Alternative multilevel algorithms are also available, as are techniques for projecting the data into non-Euclidean spaces (see Kobourov, 2013 for a review).

Force-directed approaches often provide a useful visual summary of topological relationships between nodes, and can be particularly useful for identifying strongly connected subsets, such as modules (Figure 3.4c). Force-directed approaches can also be combined with other constraints to aid the visualization of certain network properties. For example, in Figure 1.9b, a force-directed projection is combined with ring constraints to illustrate the modular organization of a functional connectivity network.

There are many other methods for visualizing networks. Here, we have only considered those techniques that have received widespread application in neuroscience. Interested readers are directed to Margulies et al. (2013) for an overview of different methods for visualizing brain connectivity. For a more general treatment, the classic text is Tufte (1990). Interesting discussions are also provided by Gibson et al. (2013) and Krzywinski et al. (2012). Developing algorithms for succinct and informative visualization of dense and complex datasets such as those used in network analysis is an active area of research and many different software platforms are available for generating different projections of brain network data. Ultimately, there is no single correct way of visualizing the data—different projections can be used to emphasize different aspects of network organization.

3.4   WHAT TYPE OF NETWORK IS A CONNECTOME?

We have considered in this chapter several different ways to represent a brain network. Is one type of representation better than others? Can any of them really describe a connectome accurately? To answer these questions, it is useful to contrast the connectomic maps that are currently being constructed across different species with an idealized version of such a map (Fornito et al., 2013; see also Lichtman et al., 2014).

What are the critical properties of a connectome that should be captured by an ideal map? First, let us consider how we define our nodes. Chapter 2 considered the complexities associated with appropriately defining valid nodes for certain types of connectomic analysis. In general, our nodes should be spatially embedded, since nervous systems exist in physical space and spatial constraints have an important influence on network topology (Bullmore and Sporns, 2012; Chapter 8). Our nodes should also be intrinsically homogeneous, meaning that they represent a coherent functional entity (Butts, 2009). This is important because a network node that encompasses many different functional units of the brain will complicate interpretation of any subsequent analyses. Our nodes should also be extrinsically distinct or heterogeneous—we should be able to differentiate them based on some relevant characteristic, such as cytoarchitecture, myeloarchitecture, gene expression, neurophysiology, or other relevant parameter. This heterogeneity can be readily represented in the diagonal elements of the connectivity matrix, or by using different colors or shapes to represent nodes in a brain graph (e.g., Figure 1.8d).

Next, let us consider the edges of an ideal brain graph. What do we know about connectivity in the brain? First, we know that connections are inherently directed. Structurally, each connection has a source and a target. Functionally, activity in one part of the brain exerts a causal influence on the dynamics of other areas. We also know that connections vary in terms of weight and type. Structurally, the weight or strength of connectivity between two regions can vary in terms of the number of connecting fibers, their cross-sectional area, myelination, conduction speed and numerous other parameters describing axonal integrity and function. Functionally, different pairs of brain regions vary over time in their propensity for synchronization. The connectome is thus an intrinsically weighted network. Moreover, we know that brain regions form different types of connections. Neurons can communicate via either chemical synapses or gap junctions, synapses can be dendritic, axonal or somatic, some connections are excitatory, some inhibitory, and so on. Neuronal connectivity is thus directed, weighted, and heterogeneous. In this chapter, we have seen how directionality and weight can be modeled with graph theory. Graph theory also allows for the representation and analysis of different classes of edges. For example, signed weights can be used to differentiate excitatory and inhibitory interactions or positive and negative correlations in regional activity.

A final important characteristic of brain networks is that they are dynamic. In structural connectivity networks assessed at microscopic scales, synaptic plasticity unfolds over submillisecond timescales (Song and Abbott, 2001). At the macroscopic level, interregional structural connectivity changes over periods spanning days to years in accordance with developmental programs or experience-dependent plasticity (Zatorre et al., 2012). Functionally, neuronal ensembles rapidly synchronize and desynchronize (Varela et al., 2001), but also display slower dynamics that persist across multiple temporal scales (Buzsáki and Draguhn, 2004; Figure 2.2). Faster timescales are readily accessible with invasive recording techniques or noninvasively with **EEG** and **MEG**. They are not directly accessible with functional MRI, although computational modeling suggests that rapid dynamics do contribute to **BOLD** signal fluctuations (Deco and Jirsa, 2012). More generally, this dynamism means that any cross-sectional snapshot of brain connectivity—be it structural or functional—will only offer a limited picture of brain organization.

To summarize, an ideal brain graph should account for the following properties:

1. Spatial embedding.
2. Heterogeneity of node properties.
3. Weight of connectivity.

4. Directionality of connectivity.
5. Heterogeneity of edge types.
6. Dynamic changes in network organization.

How well do our current maps and models capture these properties? Figure 3.5 offers a summary schematic illustrated at the level of a macroscopic human brain network. The top row is a representation of an ideal brain graph. The bottom row shows the kinds of graphs that have been produced thus far, and lists the techniques that have been used for producing such graphs. Many of these techniques were considered in Chapter 2.

The first property of an ideal brain graph—spatial embedding—is trivially accounted for by considering the physical locations of each network node. This information is often readily available, as most connectomes are constructed within a standard stereotactic coordinate reference system.

The second property—node heterogeneity—is commonly ignored in neuroscience. Instead, the focus has been on understanding interactions *between* nodes, and on defining the role of each brain region purely on the basis of its extrinsic connectivity. This simplification may be a reasonable approximation, since the functional specialization of a neural element is, to a large extent, determined by its connectivity profile with other areas (Passingham et al., 2002; Song et al., 2014). However, we know that the nodes of a brain graph can vary along a number of important dimensions that are related to the function of that node. There is an extraordinary heterogeneity of cell types at microscales, with potentially several thousand different types of neuron in mammalian brains (Bota et al., 2003). At mesoscales and macroscales, regions can vary in terms of cell density, chemoarchitecture, and the presence of specific cell types, such as the Purkinje cells of the cerebellum and large Betz cells of motor cortex.

Ideally, we would incorporate these variations into our brain network model. In practice, this heterogeneity can be difficult to capture, as it requires knowledge of the functional specialization of each and every individual network node. For some well-characterized organisms with a relatively small number of nodes, such as *C. elegans*, it is possible to draw such distinctions. For example, Figure 1.8d shows the neuronal circuitry controlling chemosensory organs (amphids) of the nematode, with different classes of neurons (sensory, motor, and interneurons) represented as different shapes. Distinguishing between different classes of nodes in this way is more challenging for larger brain networks, particularly those assessed at mesoscopic and macroscopic resolutions. At these coarser resolutions, it is difficult to delineate boundaries between nodes in a biologically meaningful way. Moreover, the number of nodes that are delineated can be arbitrary given the lack of clearly defined spatial borders between

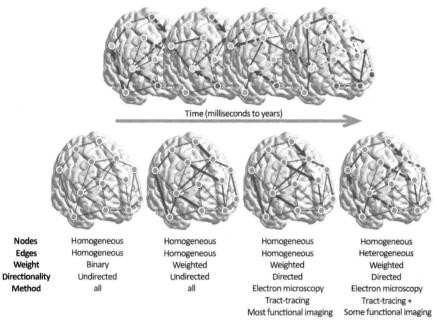

Time (milliseconds to years)

Nodes	Homogeneous	Homogeneous	Homogeneous	Homogeneous
Edges	Homogeneous	Homogeneous	Homogeneous	Heterogeneous
Weight	Binary	Weighted	Weighted	Weighted
Directionality	Undirected	Undirected	Directed	Directed
Method	all	all	Electron microscopy	Electron microscopy
			Tract-tracing	Tract-tracing +
			Most functional imaging	Some functional imaging

FIGURE 3.5 Ideal and actual brain graphs. Upper panel illustrates the key features of an ideal graph model of the brain, comprising spatially embedded, heterogeneous nodes (heterogeneity is represented by different colors) and edges that are directed (arrows), weighted (edge thickness), and variable in type (dashed vs. solid edges). The network is also dynamic, varying over time-scales ranging from milliseconds to years, depending on the resolution and type of connectivity (structural or functional). Lower panel shows the various types of brain graphs analyzed thus far and lists the methods that can be used for their construction, with reference to the connectivity mapping techniques discussed in Chapter 2. The entry 'all' means that all such methods can be used to generate the graph. For example, a binary, undirected graph can be generated using any method for connectivity mapping, given appropriate processing of the native data. The term 'tract-tracing +' refers to the combination of tract-tracing with genetic labelling and/or other techniques to distinguish different types of connections (see Chapter 2). The term 'most functional imaging' means that most methods for measuring functional connectivity can be used to measure directionality, although care should be taken when interpreting such measures (see Boxes 2.1 and 2.2). The term 'some functional imaging' means that some functional connectivity measures can be used to distinguish different types of edges. For example, the correlation coefficient can be used to distinguish between pairs of nodes with positively and negatively correlated activity. Note that the list of methods for measuring connectivity becomes smaller as more details are represented in the graph. For example, diffusion MRI cannot be used to resolve directionality. Few studies have considered node heterogeneity, although this can, in principle, be captured using methods such as electron microscopy (see Chapter 2 for more details). *Adapted from Fornito et al. (2013) with permission.*

large populations of neurons (Chapter 2). The precise functional role that each node plays within the wider network context is also often unclear.

The third property of an ideal brain graph—weighted connectivity—is readily captured using most connectome mapping methodologies. The validity of the weights that result from the mapping process as an index of the weight of neuronal connectivity depends on the method used (Chapter 2). The fourth property—directionality of connectivity—can be mapped via invasive methods such as electron microscopy and tract tracing, but cannot be resolved with non-invasive studies of human structural connectivity with current imaging technology. For functional networks, high-resolution and invasive temporal recordings, or models of effective connectivity for less direct neurophysiological measures, can be used to resolve directed interactions (Chapter 2). However, the technical challenges of acquiring invasive recordings for large-scale brain networks, and the computational burden of models of effective connectivity, have thus far limited their application to relatively small networks of brain regions.

The fifth property—edge heterogeneity—can be difficult to characterize without invasive techniques. Studies have examined the separate properties of chemical synapse and gap junction networks in *C. elegans* (Varshney et al., 2011). Others have used variations in the laminar organization of inter-regional projections in the macaque to distinguish between feed-forward and feed-back projections (Felleman and Van Essen, 1991; Markov et al., 2013b). The combination of tract-tracing and genetic labeling of neurons in the *Drosophila* fruit fly has allowed distinctions between excitatory and inhibitory connections (Shih et al., 2015). In human MRI, dynamic causal models that include modulatory interactions between regions can be specified (Stephan et al., 2008). Others have distinguished between positive and negative correlations in regional activity time courses (Fornito et al., 2012a; Rubinov and Sporns, 2011). Distinctions between other connection types are possible, but are technically challenging for large-scale neural systems.

The final property of an ideal brain graph—dynamism—has traditionally been ignored in the literature, with most studies focusing on understanding network organization at a single snapshot in time. In the human literature, longitudinal and cross-sectional studies of topological variations in structural connectivity are emerging (Baker et al., 2015; Khundrakpam et al., 2013), and modeling dynamic changes in the organization of functional connectivity networks is a rapidly developing area (Bassett et al., 2011; Hutchison et al., 2013; Zalesky et al., 2014; Kitzbichler et al., 2011).

This brief discussion highlights how our current models of brain networks do not account for the full complexity of neuronal connectivity. As discussed in Chapter 1, simple models are still useful, provided that they are not too simple. An advantage of graph theory is that it provides scope for an increase in model complexity as technical advances allow us to resolve increasingly finer details of brain connectivity across different measurement scales. The challenge for people working in the field of connectomics will be to determine the minimal level of detail required to gain a sufficiently comprehensive understanding of brain structure and function.

3.5 SUMMARY

In this chapter, we have seen that matrices and graphs, while being formally equivalent, offer flexible and diverse methods for representing a broad range of features pertaining to neuronal connectivity. Two important distinctions for neuroscience are between directed and undirected graphs and weighted and unweighted graphs. Directed graphs tell us about the directionality of neuronal connectivity and are based on knowledge of the source and target of each connection. Undirected graphs do not make this distinction and only tell us which pairs of neural elements are connected. Weighted graphs encode variations in the strength of connectivity between node pairs whereas unweighted graphs encode connectivity in a binary format as either present or absent. Analysis of each of these representations can be useful. However, as will be evident throughout this book, directed and weighted graphs often allow for a richer characterization of brain connectivity.

Whether directionality and weight can be accurately resolved depends on the method used to measure connectivity. Current representations of brain networks do not account for all relevant aspects of neuronal connectivity, but graph theory is sufficiently flexible to allow additional properties, such as node and edge heterogeneity, to be incorporated once appropriate data are available. Graph theory thus offers a powerful framework for succinctly summarizing the complex interconnectivity of nervous systems. In the chapters that follow, we will see how it can be used to characterize a diverse array of structural and functional properties of brain networks.

CHAPTER 4

Node Degree and Strength

Understanding how connectivity varies across **nodes** is a fundamental step in network analysis. Perhaps the simplest quantity that we can compute in this context is a count of the number of connections that each node has to the rest of the network, a measure called node **degree**. In many real-world networks, the distribution of degree values across nodes is heterogeneous—we often find that many nodes have a few **links** while a smaller number of key nodes take the lion's share of connectivity, marking them as putative **hubs** that facilitate integration across the network.

The importance of highly connected hub nodes to network function can be illustrated by considering the global air transportation network, in which the nodes represent airports and the **edges** are the nonstop flights that directly interconnect the airports (e.g., Figure 1.4a). Major airports such as London Heathrow take on much more traffic and are connected to a wider variety of airports than smaller, local airfields such as the one located in Newmarket, near Cambridge in the UK. The consequences for global air traffic flow arising from a disruption at Heathrow will be more severe than any disruptions that occur at Newmarket. Heathrow thus acts as an important hub of the network.

The search for hubs of the human brain has a long history. In antiquity, this search concentrated on finding the location of the soul. Democritus (c. 460-c. 370 BCE) and Plato (c. 427-c. 347 BCE) were among the first to place the soul in the head, before Herophilus (c. 335-c. 280 BCE) specifically localized it to the ventricles (Rose, 2009). The Roman physician Galen (c. 129-c. 216 CE) echoed these thoughts when he characterized the ventricles as the production and distribution center of psychic pneuma, the hypothetical fuel for the nervous system (Rocca, 1997). Approximately 1000 years later, the French polymath René Descartes (1596-1650) shifted the seat of the soul to the pineal gland, which he viewed as a convergence point for psychic spirits flowing throughout the brain's nervous pathways.

Neuroscientists working after Descartes gradually eschewed spiritual accounts of brain function and instead concentrated on understanding how different parts of the brain were specialized to perform different functions (Fodor, 1983; Gibson, 1962; Hubel and Wiesel, 1959; Penfield and Jasper, 1954). This

115

Fundamentals of Brain Network Analysis. http://dx.doi.org/10.1016/B978-0-12-407908-3.00004-2

work identified a remarkable functional specificity of anatomically localized neurons and neuronal populations, suggesting that there might not be a central point of convergence in the brain and that cognition, emotion, and behavior are the product of specialized, distributed, and parallel processes (Rumelhart et al., 1986). Such models are distinctly decentralized and somewhat egalitarian, in that no single region has a privileged role. Nonetheless, the idea that some brain regions may be more important than others has persisted. For example, psychological models of executive cognitive processes often posit the existence of a supervisory or central system that is thought to reside in the prefrontal cortex and which controls the functions of other areas (Baddeley, 1996; Shallice, 1988). These views have been criticized for assuming the existence of a homunculus-like overseer of neural processing (Hazy et al., 2007).

Network-based accounts have shed new light on the distinct functional roles played by different brain regions. Specifically, analyses of brain network **topology** have revealed that, much like the global air transportation network, not all elements of a **connectome** are equal. In fact, there is wide variation in the degree to which individual neural elements connect with others, with some nodes showing a particularly high degree of connectivity that implies a hub-like function redolent of the role played by London Heathrow Airport. In mammals, these hubs often localize to higher-order association cortices (Achard et al., 2006; Buckner et al., 2009; van den Heuvel and Sporns, 2011), consistent with a model proposed by Mesulam (1998) nearly 20 years ago, in which these association areas act as convergence zones for the integration of unimodal, polymodal, limbic, and paralimbic processes (see Chapter 1). This work suggests that no single brain region controls others; rather a collection of areas distributed throughout the brain play a vital role in integrating the functions of other, segregated network elements.

Given these considerations, the distinction between hub and non-hub nodes is important in many network analyses. In this chapter, we consider some basic methods for measuring how connected an individual node is to the rest of the network, and for investigating how connectivity is distributed across the brain. In particular, we will see how characterizing the distribution of brain connectivity across nodes allows us to understand the potential influence of highly connected hub areas on the function of nervous systems.

4.1   MEASURES OF NODE CONNECTIVITY

We first consider some basic measures for summarizing the connectivity of each node in a network. Many other measures of network topology are based on these relatively simple quantities. We examine cases for binary and weighted, undirected and directed networks.

4.1.1 Node Degree

Node degree is one of the most basic measures that we can calculate from a graph. In a binary undirected network, the degree, k_i, of node i is the number of edges connecting node i with all other $j = 1 \ldots N - 1$ nodes,

$$k_i = \sum_{j \neq i} A_{ij}. \tag{4.1}$$

This definition is equivalent to adding all the nonzero elements of the ith row (as above) of the binary connectivity **matrix**. Equivalently, we could sum across the ith column, since A is symmetric for an undirected network. Figure 4.1a demonstrates how node degree can be computed from an **adjacency matrix** and network graph. The mean degree of an undirected network is the average of all the node degrees,

$$\langle k \rangle = \frac{1}{N} \sum_{i=1}^{N} k_i. \tag{4.2}$$

In directed networks, the adjacency matrix is asymmetric, and we can distinguish between the incoming and outgoing connections of a node. We therefore have two types of degree: (1) the **in-degree**, which counts the number of incoming edges (i.e., edges that point to or terminate at a node); and (2) the **out-degree**, which counts the number of outgoing edges (i.e., edges that point away or originate from a node). For a directed, binary adjacency matrix, in which each element contains a one if and only if there is an edge from node j to i and zero otherwise, the in-degree of node i is given by the number of non-zero edges from all nodes j to i,

$$k_i^{in} = \sum_{j \neq i} A_{ij}, \tag{4.3}$$

and the out-degree of node j is given by the number of nonzero edges from node j to all nodes i,

$$k_j^{out} = \sum_{i \neq j} A_{ij}. \tag{4.4}$$

In brain networks, k^{in} measures the number of afferent connections and k^{out} the number of efferent connections of each node. Panels b and c of Figure 4.1 demonstrate how the in-degree and out-degree can be calculated from a directed adjacency matrix.

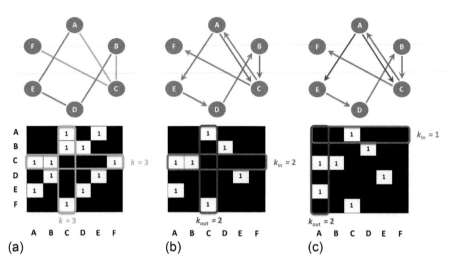

FIGURE 4.1 Computing node degree in undirected and directed binary networks. **(a)** Example of a binary, **undirected graph** (top) and its corresponding adjacency matrix (bottom). The degree of node C is illustrated here. Since the matrix is symmetric, we obtain the same value when summing across the rows or columns of the matrix (green highlights). **(b)** A directed variation of the graph depicted in **(a)**. Here, node C has two incoming projections: one from node A and one from node B. It also has two outgoing projections: one to node A and the other to node F. In this matrix, the element (i,j) denotes the edge running from node j to i. We therefore calculate the in-degree, k_{in}, of node C by summing the corresponding row of the matrix (red highlight). The out-degree, k_{out}, is computed by summing the corresponding column (blue highlight). In this case, the in-degree and out-degree of node C both equal 2, and the total degree is $k_{tot} = k_{in} + k_{out} = 4$. In contrast, we have $k = 3$ for node C in the undirected network in **(a)**. This discrepancy arises because the connection between nodes C and A is reciprocal—an edge runs in both directions between these two nodes. In a directed network, we count both edges; in an undirected network, we only count one connection since we do not distinguish between incoming and outgoing links. **(c)** Calculation of k_{in} and k_{out} for node A in the same directed network as depicted in **(b)**. Node A has one incoming connection from node C and two outgoing connections: one to node C and one to node E. Once again, we compute k_{in} by summing row A of the matrix and k_{out} by summing column A. In this case, $k_{in} \neq k_{out}$, and is an example of how asymmetries can arise in directed networks.

4.1.2 Node Strength

Node degree quantifies the total number of edges incident to a node in a binary network. In binary networks, edges assume a value of either zero or one. In weighted networks, edges can assume a range of different values, allowing us to capture variations in the strength of connectivity between pairs of neural elements. In some cases, the range of values in weighted networks can be large. For example, in **structural connectivity** networks generated with **diffusion MRI**, connectivity weights are sometimes measured as the number of reconstructed streamlines linking pairs of brain regions (Chapter 2). These values are necessarily integer and can range anywhere between zero and an arbitrarily large number.

In other cases, such as correlation-based networks, the values of the edge weights range between -1 and 1.

The analog of node degree in weighted networks is node **strength**, which we denote with s. In an undirected network, this quantity can be computed as the sum of the connectivity weights of the edges attached to each node i,

$$s_i = \sum_{j \neq i} w_{ij},$$

(4.5)

where w_{ij} is the strength or weight of the edge linking nodes i and j. In directed networks, Equation (4.5) gives the in-strength of node i. The out-strength can be computed by instead summing over the rows of the weighted adjacency matrix.

It is sometimes useful to normalize strength values. Say that we have a correlation-based network comprising six nodes, one of which is connected to the other five with weights 0.2, 0.3, 0.4, 0.6, and 0.7. If we sum these values as per Equation (4.5), we obtain 2.2. The average of these values is 0.4, and is a more intuitive summary of connectivity in this context. We can thus define a normalized measure of node strength as

$$s_i' = \frac{1}{N-1} \sum_{j \neq i} w_{ij}.$$

(4.6)

In some weighted networks, such as those based on correlations, the edge weights can be signed. In a brain network, a positively signed connection may imply an excitatory interaction or cooperation between neural elements, whereas a negative sign may imply inhibition or antagonism. In this case, it is useful to retain a distinction between these qualitatively different types of interaction when computing nodal summary measures such as strength. We can thus compute the positive strength, s_i^+, and negative strength, s_i^-, of the ith node separately by summing over its positive edge weights, w_{ij}^+, and negative edge weights, w_{ij}^-, respectively,

$$s_i^+ = \sum_{j \neq i} w_{ij}^+, \quad \text{and} \quad s_i^- = -\sum_{j \neq i} w_{ij}^-.$$

(4.7)

As per Equation (4.6), we can compute normalized variants of these measures by dividing either sum by $N-1$.

Rubinov and Sporns (2011) have proposed a unitary, normalized strength measure for signed networks that scales the relative contribution of positive and negative edge weights,

$$s_i^* = s_i'^+ - \left(\frac{s_i^-}{s_i^+ + s_i^-} \right) s_i'^-,$$

(4.8)

where s'^+_i and s'^-_i correspond to the normalized positive and negative strength, respectively, of node i. The expression $s^-_i / (s^+_i + s^-_i)$ is the relative proportion of negative weight that contributes to node i's connectivity. It thus scales the contribution of s'^-_i to s^*_i by the proportion of negative weight in the network. If the total weight of negative edges is low, s'^-_i will make less of a contribution to the final value of s^*_i. For example, if 50% of the total edge weight is negative, the contribution of s'^-_i to s^*_i will be half the contribution of s'^+_i. This scaling factor ensures that positively weighted edges make a stronger contribution than negative edge weights to s^*_i. This asymmetry between the contributions of positive and negative weights is sensible if we think of hubs in brain networks as integrating diverse network elements. Integration is implied by positive rather than negative edge weights, suggesting that we should preferentially emphasize the contribution of positively weighted edges. Alternatively, we could index a node's global connectivity by taking the sum of the absolute edge weights attached to that node. In this case, we quantify the total connectivity of a node, irrespective of the polarity of its edge weights.

4.1.3   Node Degree and Network Density

Recall from Chapter 3 that the total number of edges in a network, E, is the total number of *unique* nonzero elements in the adjacency matrix, A. In directed networks, this value is calculated as $E_{\mathrm{dir}} = \sum_{ij} A_{ij}$, assuming the convention that A_{ii} equals zero. In undirected networks, we halve this number to avoid double-counting edges, such that $E = \frac{1}{2}\sum_{ij} A_{ij}$. Equivalently, we can compute E for an undirected network as half the sum of node degree values,

$$E = \frac{1}{2}\sum_{i=1}^{N} k_i. \tag{4.9}$$

Using this relation, we can also compute the mean degree of an undirected network as $\langle k \rangle = 2E/N$.

For directed networks, the number of edges, E_{dir}, can be calculated as the total number of incoming edges, or the total number of outgoing edges,

$$E_{\mathrm{dir}} = \sum_{ij} A_{ij} = \sum_{i=1}^{N} k_i^{\mathrm{in}} = \sum_{j=1}^{N} k_j^{\mathrm{out}}. \tag{4.10}$$

Henceforth, we use E to denote the total number of unique edges in either a directed or undirected network. Note that the mean in-degree, $\langle k^{\mathrm{in}} \rangle$, and mean out-degree, $\langle k^{\mathrm{out}} \rangle$, of a directed network are equal, so we have

$$\langle k^{\mathrm{in}} \rangle = \frac{1}{N}\sum_{i=1}^{N} k_i^{\mathrm{in}} = \frac{1}{N}\sum_{j=1}^{N} k_j^{\mathrm{out}} = \langle k^{\mathrm{out}} \rangle. \tag{4.11}$$

To see why the mean in-degree and out-degree are equivalent, consider the network in Figure 4.1b. The in-degrees of this network are $\{1, 1, 2, 1, 1, 1\}$ and the out-degrees of this network are {2, 1, 2, 1, 1, 0}. Despite these sets of numbers being different, they both sum to 7 and the mean of both is 1.67. We thus see that even though k_{in} and k_{out} may vary for any individual node, the mean in-degree and out-degree of a network are equivalent.

4.2 DEGREE DISTRIBUTIONS

Degree and strength are useful summary measures of the connectivity of each node in a network. Once we have estimated degree or strength for all nodes, it is useful to understand how these measures are distributed across the network. In particular, characterizing the **degree distribution** is an important element of network analysis. In **connectomics**, degree distributions allow us to determine whether our network contains hubs and to understand the potential influence that these hubs have on brain function. The degree distribution can also provide insight into the developmental or growth processes that have shaped network topology (Barabási and Albert, 1999; Newman, 2005a; Box 4.1).

A simple way to understand the degree distribution of a network is to map its **degree sequence**: the set of degree values for all network nodes. Consider the undirected network depicted in Figure 4.1a, which has the degree sequence {2, 2, 3, 2, 2, 1}. One node (node C) has degree $k = 3$, four nodes (nodes A, B, D, and E) have degree $k = 2$, and one node (node F) has degree $k = 1$. Using this information, we can compute the probability of finding a node with degree k in the network by calculating the fraction of nodes with that degree. In our example, we have $P(\text{degree} = 1) = 1/6$, $P(\text{degree} = 2) = 4/6$, and $P(\text{degree} = 3) = 1/6$.

For larger networks, it is often instructive to plot the fraction of nodes with degree k, for $k = 0, \ldots, N - 1$. The resulting histogram is called the *probability density* of the node degree, since the height of the histogram at point k is the probability of finding a node with degree k. Confusingly, this histogram is sometimes also called the degree distribution. However, the degree distribution technically corresponds to the *cumulative distribution* of degree, which is the probability of finding a node in a network with a degree that is less than or equal to k. This is equivalent to summing the height of the histogram at points 0 through k. Subtracting this probability from one, we obtain the complementary cumulative distribution function (cCDF), which is commonly used to visualize degree distributions in network analysis. If we sort the node degree values of a network in descending order and assign each value a rank ranging from 1 to N, with 1 corresponding to the node with lowest degree and N to the node with highest degree, we can define the probability of finding a node with degree greater than k as

$$P(\text{degree} > k) = 1 - \left(\frac{r_k}{N}\right), \tag{4.12}$$

where r_k is the maximum rank of a node with degree k.

Some illustrative probability density and cCDF plots for relevant classes of networks are illustrated in Figure 4.2. The properties of these distributions are discussed in the following sections. In particular, we focus on three major classes of networks, as defined by the form of their degree distribution, that are particularly useful for understanding brain organization. These classes are called **single scale**, **scale free**, and **broad scale** (see Amaral et al., 2000). Our discussion will focus on the simple case of undirected, binary networks, although it is possible to analyze the in-degree and out-degree distributions of directed networks in the same way. Technically, the degree distribution of a directed network is characterized by the two-dimensional joint distribution of the in-degree and out-degree, although joint distributions are seldom analyzed in connectomics (see Varshney et al., 2011 for an exception).

4.2.1   Single-Scale Distributions

The **Erdős-Rényi graph** is a classic example of a single-scale network. Recall from Chapter 1 that an Erdős-Rényi graph is constructed by independently drawing a connection between nodes with a probability p (for more detail, see Chapter 10). The degree distribution of this network follows a binomial distribution.

Binomial distributions characterize the outcome probabilities of a series of independent binary outcomes. In the network context, the binary outcome is the presence or absence of a connection at each node. In an Erdős-Rényi graph, these connections are formed independently at each node by sampling from a uniform distribution. Formally, the probability of finding a node with degree k in such a network is

$$P(\text{degree} = k) = \binom{N-1}{k} p^k (1-p)^{N-1-k}, \quad k \leq N - 1, \qquad (4.13)$$

where p is the probability of finding an edge between a pair of nodes; $(1-p)$ is the probability that the pair is not connected; $N-1-k$ is the remaining number of possible connections that an individual node can make after we have chosen k edges; and $\binom{N-1}{k} = \dfrac{N-1!}{k!(N-1-k)!}$ is called the binomial coefficient and quantifies the number of possible ways to choose k positive outcomes (in this case, the presence of a connection) from $N-1$ observations. We use $N-1$ here to avoid counting self-connections. The binomial coefficient thus tells us the total number of different combinations in which any individual node can connect to k other vertices.

To develop an intuition for Equation (4.13), consider a fair coin toss. Say that we toss the coin four times and want to compute the probability of obtaining two

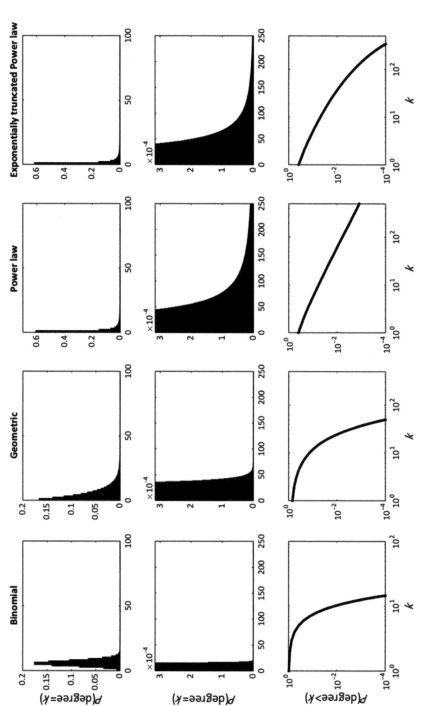

FIGURE 4.2 Single-scale, scale-free and broad-scale degree distributions. Top row shows the probability densities for binomial ($p=.01$), geometric (also sometimes called exponential, with $\lambda = 1/\langle k \rangle$, where $\langle k \rangle = 5$), power-law ($\gamma = 2$), and exponentially truncated power-law ($\gamma = 2$, $k_c = 200$) distributions for a network with 500 nodes. The horizontal axis in these plots has been truncated to better illustrate the shape of the distributions. The middle row focuses on the tails of the distributions shown in the top row. To better demonstrate the shape of the tails, the vertical axes have been truncated and the horizontal axes have been extended. These plots clearly show an abrupt truncation of the tails of the binomial and geometric distributions, in comparison to the power-law and exponentially truncated power-law distributions. Looking closely, it is also possible to see that the tail of the power-law distribution is 'fatter' than the tail of the truncated power-law distribution, corresponding to a higher probability of finding high degree nodes. These differences are clearly seen in the bottom row, which presents the corresponding complementary cumulative distribution functions, plotted on logarithmic axes. The full shape of the distributions and their scaling behavior is more apparent in these plots. Specifically, we see that the probability of finding a node with $k > 10^2$ is several orders of magnitude higher in the power law network (just under 10^{-2}) compared to the binomial and geometric networks (less than 10^{-10}). In fact, the power law distribution appears as a straight line on the logarithmic axes, consistent with scaling behavior over several orders of magnitude (true scale-free networks demonstrate this behavior across all scales). The exponentially truncated power law scales over a limited regime, before showing a rapid decay in the probability of finding nodes with degree above the cut-off value, k_c

"heads" in any order. The binomial coefficient gives the number of possible ways to obtain two "heads" in four coin tosses: $\binom{N}{k} = \frac{4!}{2!2!} = 6$. Note that we use N rather than $N-1$ in this example because we are counting all possible outcomes. If H denotes the outcome "heads" and T denotes the outcome "tails," these possibilities are: HHTT, HTHT, HTTH, TTHH, THHT, and THTH. The quantity p^k in Equation (4.13) is the probability of obtaining the outcome "heads" k times. We thus have $p^k = 0.5^2 = 0.25$, since there is a 50% chance of obtaining "heads" on any single coin toss ($p = 0.5$), and we are interested in computing the probability of obtaining two heads in a set of four coin tosses ($k = 2$). Because we are interested only in obtaining exactly two heads, the other two coin tosses in the sequence of four must be tails. These represent instances of the outcome that we are not interested in, the probability of which is $(1-p)^{N-k} = 0.5^2 = 0.25$. Substituting the values into Equation (4.13), we obtain $P(2) = 0.38$.

In this coin toss example, we determine the probability of obtaining k outcomes from N coin tosses, where $k = 2$, $N = 4$, and the outcome is "heads." If we extend this analogy to networks, our question becomes: what is the probability of finding a node with degree k in a network of N nodes? The analog of the coin toss in the network context is determining whether a connection is present between an arbitrary node and each of the other $N-1$ vertices. So, we have $N-1$ possible "coin tosses," or instances where a connection is either present or absent. The value p in the network context corresponds to the probability of finding a connection at any given node. In an Erdős-Rényi graph, each node has an equal probability of receiving a connection, so p is equivalent to the **connection density** of the network.

An example binomial degree distribution for an Erdős-Rényi network is shown in Figure 4.2. Note the similarity between the shape of the probability density and the well-known bell-shaped curve that is characteristic of Gaussian (normal) distributions. In fact, for sufficiently large N, the binomial distribution is well approximated by a Gaussian. For this reason, networks with binomial degree distributions are sometimes called Gaussian. They are also sometimes referred to as *Poisson random graphs* in the limit of large N (Newman, 2010). In these networks, most nodes have a degree that is similar to the mode of the distribution, and the probability of finding a node with lower or higher degree decays rapidly as we move further from this average. Since the degree of most nodes in these systems converges to a single modal value—a characteristic scale—these networks are classed as single scale.

Other degree distributions are also consistent with a single scale. For instance, exponential degree distributions characterize some real-world networks and have the form

$$P(\text{degree} = k) \sim e^{-\lambda k}, \qquad (4.14)$$

where λ is a positive scaling constant (sometimes called the rate parameter) that defines the rate of decay in the probability of finding a node with increasing degree k. Technically, exponential distributions are continuous probability distributions. For discrete variables, such as node degree, the corresponding distribution is called geometric although the terms exponential and geometric are sometimes used interchangeably in this context. Geometric degree distributions are highly skewed, such that most nodes have very low degree and the probability of observing deviations from these low values decays exponentially. As in the binomial case, this sharp decay means that the distribution has a single, characteristic scale (see Figure 4.2). The defining feature of single-scale systems is thus a rapidly diminishing likelihood of observing large deviations from a characteristic modal value, consistent with an egalitarian or homogeneous distribution of connectivity across nodes.

4.2.2 Scale-Free Distributions

One of the major discoveries to emerge from network science in the past few decades is that connections in many real-world networks—including the brain—do not arise from a process of randomly interconnecting pairs of nodes, as assumed by the Erdős-Rényi model (Barabási and Albert, 1999; Sporns, 2011b; Watts and Strogatz, 1998). This finding implies that the degree distributions of these networks do not follow a binomial. In fact, Barabási and Albert (1999) showed that the degree distribution of many real-world networks is better approximated by a power law with the form

$$P(\text{degree} = k) \sim k^{-\gamma}, \tag{4.15}$$

where $P(k)$ is the probability of finding a node with degree k, and γ denotes the power-law scaling exponent.

Equation (4.15) indicates that the probability of finding a node with increasing degree k diminishes as a power of γ. This rate of decay is slower than that observed in binomial or exponential networks (Figure 4.2). This slower decay means that the probability of finding nodes with large k is higher in power law compared to single-scale distributions. In most power-law systems, $2 < \gamma < 3$ (Newman, 2005a). Higher values of γ indicate a sharper slope in the distribution, a more restricted power-law regime, and a lower probability of finding highly connected nodes.

We can appreciate the difference between single-scale and power-law networks by comparing their probability densities, as shown in Figure 4.2 (top row). In the power-law distribution, most nodes have very low-degree. A similar property is observed in the exponential case. However, the power-law distribution also shows an extended tail, comprising nodes with very high values of k. This long tail, sometimes called a "fat" tail, represents the higher probability of finding high degree nodes in power-law degree distributions compared to single-scale systems. In other words, power-law distributions contain nodes with a very large number of connections. These nodes represent putative network hubs.

A second important difference between single-scale and power-law distributions is that the former shows a clear peak. As we have seen, this peak defines the characteristic *scale* of the network: the modal value that is representative of the degree of most other nodes in the network and from which large deviations are rare. In contrast, there is no clear peak in power-law distributions (Figure 4.2). Rather, we observe a continuous progression from frequent, low-degree nodes to rare, very highly connected elements. This progression means that there is no single modal or representative node that we can select to estimate the average degree of the other nodes. In other words, these networks have no characteristic scale. For this reason, networks with power-law degree distributions are also called scale-free networks (Barabási and Albert, 1999).

An important feature of scale-free systems is that the shape of the power-law distribution will be the same regardless of our measurement scale. A simple demonstration of this property is offered by Newman (2005a). Suppose that we find computer files of size 2 kB are ¼ as common as files of size 1 kB. If we switch the measurement scale to megabytes, we also find that files of size 2 MB are ¼ as common as files of size 1 MB. In this example, the shape of the distribution is not contingent on the scale used to measure file size. For comparison, consider the duration of telephone calls, which has traditionally (and somewhat simplistically) been modeled as an exponential random variable (Dunlop and Smith, 1994). Under this model, if a call of 1 min duration is four times more likely than a call of 2 min, we cannot say that a 1-h call is four times more likely than a 2-h call. In this case, the shape of the distribution depends on whether we measure call duration on the scale of minutes or hours. It can be shown mathematically that scale-invariance is a unique property of power-law distributions (Newman, 2010). For this reason, power-law scaling is characteristic of self-similar or **fractal** systems—systems that show similarity across different scales of measurement. Power laws are thus found across the diverse range of manmade and natural phenomena, including the brain, that demonstrate fractal or self-similar organization (Box 4.1).

A preliminary visual test of whether a degree distribution might follow a power law is to plot its cCDF on logarithmic axes. Recall that the cCDF is the probability that a node has degree greater than k. A plot of $\log P(\text{degree} > k)$ as a function of $\log k$ will approximate a straight line if the distribution follows a power law. In exponential distributions, $\log P(\text{degree} > k)$ shows a sharper cut-off as $\log k$ increases. In a log-linear plot, an exponential cCDF will appear as a straight line. Some examples are presented in Figure 4.2.

Plotting the cCDF on logarithmic axes offers a more robust visualization of power-law scaling than the probability density. When plotting a probability density, we must bin the data into different intervals of k. The binning

BOX 4.1 THE UBIQUITY OF POWER LAWS

The Italian economist Vilfredo Pareto (1848-1923) was among the first to draw attention to the importance of power laws. He found that they described the distribution of income in Italy at the time, in which just 20% of the population owned 80% of the land. A keen gardener, he also discovered that 80% of his peas were produced by only 20% of pods. This so-called 80/20 rule later came to be known as Pareto's Law. Zipf's law is another famous example of a power law. Named after the linguist George Zipf, it describes the power-law scaling of the frequency with which words appear in text (Zipf, 1936, 1949).

There is evidence to suggest that power-law distributions characterize a wide range of physical, biological, and man-made phenomena. Examples include citations of scientific papers, the number of hits received by web sites, book sale volumes, the magnitude of earthquakes, the diameter of moon craters, the intensity of solar flares, the frequency of family names, and the population of cities, to name but a few (Clauset et al., 2009; Newman, 2005a). An intuitive characterization of power-law scaling was provided by management consultant Joseph Juran, who called it the principle of the "vital few and trivial many" (Juran, 2005). He later favored the term "useful many" to avoid down-playing the importance of the "many."

In the brain, power laws describe the distribution of periods of phase synchrony between activity time courses recorded using either **MEG** or **functional MRI** (Kitzbichler et al., 2009), the power spectrum of **BOLD**, **EEG**, and **ECoG** signal fluctuations (Bullmore et al., 2004; He, 2011; He et al., 2010; Miller et al., 2009), the propagation of neuronal avalanches both in vitro and in vivo (Beggs and Plenz, 2003; Petermann et al., 2009), and the allometric scaling of brain size across species (Bassett et al., 2010; Bush and Allman, 2003).

Several known mechanisms can generate power-law distributions (Frank, 2009; Newman, 2005a). The two most relevant for our purposes are **preferential attachment** and self-organized criticality. Preferential attachment, often also referred to as a rich-get-richer or Yule process (Newman, 2005a; Simon, 1955; Yule, 1925), occurs when some property of an existing system element affects its probability of interacting with a new element that is added to the system. For example, a scientific article with a high citation rate gains prestige, increasing the likelihood that it will be cited by new articles. Barabási and Albert (1999) were able to grow networks with power-law degree distributions by adding connections preferentially to nodes with higher degree (see also Chapter 10).

Power-law distributions are also observed in critical systems. A system is in a critical state when it is on the verge of a phase transition; that is, as it shifts from one state to another. For example, each atom in a ferromagnetic piece of metal has a magnetic moment or *spin*. The well-known Ising model characterizes the behavior of these spins as the temperature of the magnet is varied. If we cool the magnet until some critical point, the spins of its atoms undergo a phase transition from an initial random alignment to uniform orientation. Close to the critical point, the system vacillates between the two states, expressing complex patterns of spin correlations (Figure 4.1.1). In this regime, several important characteristics of the system show a power-law distribution, such as the correlation length (the distance over which atoms communicate with each other) and the strength of the magnet (determined by the fraction of atoms with aligned spins).

A self-organized critical system is a system that has self-organized around the critical point. A simple example is the forest fire model of Drossel and Schwabl (1992). Imagine that we add trees to an initially bare forest. Early on, the forest will be sparse and any random lightning strikes will be associated with a low chance of a fire spreading. As we add more trees, clusters of trees form, thereby increasing the probability of larger fires. If we add trees until the entire forest is filled, a single lightning strike could wipe out the entire forest. If we start growing trees again and repeat the process many times, the system fluctuates around a certain critical point that ranges from being very resilient to being very vulnerable to perturbation by lightning. The distribution of fire sizes in this model follows a power law.

Self-organized critical systems are theoretically claimed to be widespread in nature (Bak, 1996), and power-law scaling of several different properties of brain function have been interpreted as evidence of critical neural dynamics (Beggs, 2008; Kitzbichler et al., 2009). Very simple dynamical models operating in a critical regime can closely reproduce the spatial topography and other properties of the canonical resting-state **functional connectivity** networks observed in human functional MRI (Haimovici et al., 2013), suggesting that criticality may play a central role in shaping the functional organization of the brain.

Continued

BOX 4.1   THE UBIQUITY OF POWER LAWS—CONT'D

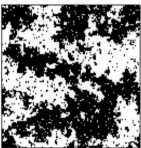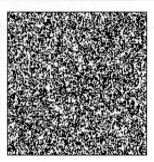

FIGURE 4.1.1 Criticality in the Ising model of atomic spins in a magnet. Here, the spins are modeled as a **lattice** comprising 128 nodes. Each spin can be in one of two states: +1 (shown in white) or −1 (shown in black). The model describes how these spins interact as the temperature of the magnet is varied from low, in which nearly all spins are aligned and thus perfectly ordered (left), to hot, in which the spins are randomly oriented (right). At some critical temperature (middle), the spins display a complex, fractal-like organization in which many properties, such as the correlation length of the spins, show a power-law distribution. *Reproduced from Kitzbichler et al. (2009) with permission.*

aggregates adjacent data points, causing a loss of information. This loss is especially problematic in the tails of the distribution, where samples are often limited. These tails are the most relevant part of the distribution for power laws, as they contain the highly connected hubs. It is therefore important to characterize the tails as accurately as possible. The cCDF does not require binning of the data since $P(\text{degree} > k)$ can be defined for any value of k. As a result, the cCDF will have as many points along the x-axis as there are values of k in the network and provides a more complete representation of the data (see Figure 4.2).

In most cases, simply showing that the cCDF approximates a straight line when plotted on a logarithmic scale is not sufficiently rigorous to demonstrate power-law scaling. Instead, we require quantitative criteria to determine whether a degree distribution follows a power law. A straightforward solution is to estimate the best-fitting straight line via least squares regression of $\log k$ on $\log P(\text{degree} > k)$. However, this approach assumes independence of observations, which does not hold for a cCDF: the cCDF is a cumulative index, meaning that its values are not independent. Least-squares regression also suffers from other biases that can lead to the erroneous conclusion of power-law scaling when in fact there is none. These problems are thoroughly discussed by Clauset et al. (2009), who propose an alternative framework for evaluating power-law scaling based on maximum-likelihood estimation of the scaling parameter γ and likelihood ratio tests of the probability that the observed data

BOX 4.2 TESTING FOR POWER-LAW SCALING IN NETWORK DEGREE DISTRIBUTIONS

Clauset et al. (2009) propose a rigorous statistical framework for evaluating the existence of power-law scaling in empirical data. There are three main steps in this approach, which we briefly summarize here. Technical details can be found in the original article.

Step 1: Estimate γ using maximum-likelihood estimation. As discussed in Section 4.2.2, least-squares regression of $\log P(\text{degree} > k)$ on $\log k$ is not the most rigorous method for determining the parameters of the best-fitting power-law function. Maximum likelihood estimation is a more robust and accurate approach to parameter estimation for this purpose. For discrete distributions, this method can be used to approximately estimate the scaling parameter γ as

$$\hat{\gamma} \simeq 1 + N \left[\sum_{i=1}^{N} \log \frac{k_i}{k_{\min} - \frac{1}{2}} \right]^{-1} \quad (4.2.1)$$

where $\hat{\gamma}$ is an estimate of γ derived from the data, $k_{\min}$ is a cut-off above which power-law scaling occurs, and k_i are the observed values of k that are greater than the cut-off $k_{\min}$. The precise value of $k_{\min}$ is arbitrary but important, as it determines where power-law scaling emerges in the distribution. Clauset et al. suggest choosing a value that maximizes the similarity between the cumulative distribution functions of the observed data and the best-fit power-law model. This similarity can be tested using the Kolmogorov-Smirnov (KS) statistic, which quantifies the maximum distance between the cumulative distributions of the data and fitted model.

Step 2: Estimate the goodness-of-fit between the data and a power law. Step 1 fits a power-law model to the data. Step 2 tests the goodness of this fit. The KS statistic is again used for this purpose. Specifically, Clauset et al. recommend the following procedure: (a) compute the KS statistic for the best fit model; (b) generate a large number of synthetic power-law distributed datasets with the same estimated scaling parameter $\hat{\gamma}$ and $k_{\min}$ cut-off; (c) fit a power-law model to each synthetic dataset as per the methods described in step 1; and (d) use the distribution of the fitted values to estimate a *p*-value for the observed fit. If the *p*-value is below a predetermined threshold ($p < 0.10$ is suggested), there is a difference between the goodness-of-fit in the observed data and the fits obtained in the synthetic power-law networks, indicating that a power law is not an appropriate model.

Step 3: Compare the power law with alternative distributions via a likelihood ratio test. The outcome of step 2 may suggest that a power law describes the data well, but other distributions may fit equally well. Testing the relative fits of competing models allows us to determine which distribution best models the data. Relative fits can be examined by computing the log-likelihood of the data under two competing distributions. In this way, we can compare the fit of a power-law distribution to other key distributions, such as binomial, geometric, or other forms.

do indeed conform to a power law (Box 4.2). A practical guide to fitting power-law distributions to empirical data is provided by Alstott et al. (2014a).

4.2.3 Broad-Scale Distributions

To summarize, we have thus far considered two classes of networks: (1) single-scale networks, which have a modal, characteristic degree and a low probability of observing large deviations from this mode; and (2) scale-free networks, which have no representative degree or characteristic scale and a higher probability of finding highly connected hub nodes. An intermediate class of networks that is relevant to neuroscience is referred to as broad scale (Amaral et al., 2000).

Broad-scale networks show power-law scaling over some limited range of degree, but the probability of finding very highly connected nodes falls off more rapidly than scale-free systems. A broad-scale degree distribution is often found in networks where the addition of new links is subject to external constraints (Amaral et al., 2000; Mossa et al., 2002). Examples of such constraints include the financial cost of adding a link in a technological network, the physical distance of a new connection in a spatially embedded system, or the metabolic resources required to support interactions between the nodes of a biological network. Such systems contrast with virtual networks, such as the WorldWide Web, where the **cost** of adding hyperlinks is relatively low and there are few physical constraints on network size (Barabási and Albert, 1999).

Nervous systems are physically embedded, and both spatial and metabolic constraints limit the number of connections that any single neural element can possess. Accordingly, an early functional MRI study of thresholded, binary whole-brain functional connectivity networks (Achard et al., 2006) found evidence for an exponentially truncated power-law degree distribution of the form

$$P(\text{degree} = k) \sim k^{-\gamma} e^{-k/k_c}, \tag{4.16}$$

where k_c is the cut-off degree at which the power-law regime gradually transitions to an exponential decay. Low values of k_c correspond to a lower probability of finding highly connected network nodes; that is, a quicker transition from a power law to exponential decay. An example probability distribution and complementary distribution function for an exponentially truncated power law is shown in Figure 4.2.

Subsequent analyses of human structural connectivity (Gong et al., 2009; Iturria-Medina et al., 2008), **structural covariance** (He et al., 2007), and functional connectivity (Fornito et al., 2010; Hayasaka and Laurienti, 2010; Lynall et al., 2010) brain networks have supported the initial functional MRI study of Achard and colleagues (2006) by reporting evidence for a broad-scale degree distribution. Evidence of an exponential decay in the tail of the degree distributions of the *C. elegans* nervous system (Amaral et al., 2000) and the macroscale macaque cortical network (Modha and Singh, 2010) has also been reported, although a truncated power-law hypothesis was not formally tested in these analyses. The form of some of these distributions is illustrated in Figure 4.3.

Other analyses have reported conflicting findings. One study found that the tail of the in-degree and out-degree distribution of the *C. elegans* chemical synapse network approximates a power law (Varshney et al., 2011), and some human studies, particularly those examining high-resolution voxelwise functional MRI networks, have also reported evidence of scale-free properties (Eguíluz et al., 2005; van den Heuvel et al., 2008b). One report that systematically varied network resolution in the same functional MRI dataset found that power-law scaling was more prominent as the resolution shifted from the order of 10^2 regions to $>10^4$ voxels

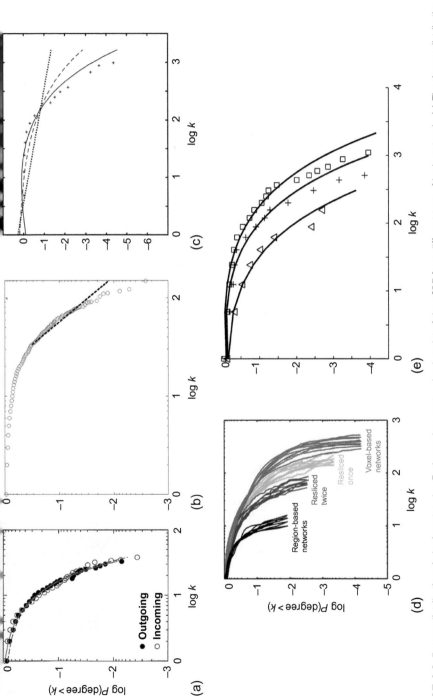

FIGURE 4.3 Degree distributions in brain networks. Each panel presents a log-log plot of the cCDF for a different type of brain network. **(a)** The degree distribution of incoming and outgoing links of the *C. elegans* neuronal network. **(b)** The degree distribution of a 383-region macaque connectome collated from published **tract-tracing** studies. Red circles represent the empirical distribution and the black line is the best-fitting power law, which does not adequately characterize the shape of the empirical distribution. In both **(a)** and **(b)**, we see evidence of an exponential decay in the tail of the distribution. **(c)** The degree distribution of a human 78-region cortical connectome reconstructed with diffusion MRI. The symbol + represents the empirical data, the solid line is the best-fitting exponentially truncated power law, the dashed line is the fit of an exponential (single scale) distribution and the dotted line is a power law. The truncated power law had the best fit in this analysis. **(d)** The degree distribution of human resting-state functional connectivity networks assessed with functional MRI at different resolutions, ranging from a 90-region anatomical parcellation (black lines) to a 16,000 node voxel-level network (red lines). Blue and green lines represent intermediate resolutions. Different lines represent different datasets. Note how the power-law regime broadens as the resolution (and network size) is increased. There was also a corresponding reduction in the parameters γ and k_c (Equation 4.16). **(e)** The degree distribution of an anatomical covariance network constructed from correlations in the average cortical thickness of 54 anatomically defined regions. The different symbols represent the empirical data thresholded at different correlation thresholds. The solid lines are the best-fitting exponentially truncated power-law distributions. The natural logarithm of node degree, k, is presented on the horizontal axes of panels **(c)** and **(e)**. All other panels use the base 10 logarithm. **(a)** *Reproduced from Amaral et al. (2000), Copyright (2000) National Academy of Sciences, U.S.A.;* **(b)** *from Modha and Singh (2010);* **(c)** *from Gong et al. (2009);* **(d)** *from Hayasaka and Laurienti (2010); and* **(e)** *from He et al. (2007) with permission.*

(Hayasaka and Laurienti, 2010; Figure 4.3d). This resolution effect may be caused by statistical problems associated with estimating power laws. A true scale-free network will demonstrate scaling over all degrees and scales beyond some minimum cut-off of degree (Clauset et al., 2009). To be confident of this scaling in empirical data, power-law scaling should be evident over several orders of degree magnitude. In a small, low-resolution network, the available observations typically do not scale more than two or three orders of magnitude. This limited range will result in noisy estimates of the tail of the degree distribution. A wider range of k provides more samples to better characterize the tail.

Another reason for the inconsistent descriptions of degree distributions across different connectomic data may be that relatively few studies have used systematic and rigorous methods for comparing the fits of a wide variety of different distributions to brain network data (Box 4.2). This is an important limitation, since an exponentially truncated power law can be difficult to distinguish from a pure exponential when the truncation point or cut-off degree for the power law is low.

4.3   WEIGHT DISTRIBUTIONS

Degree distributions describe a property of the binary topology of a network and offer no information about the strength of connectivity between nodes. The analog of node degree in weighted networks is node strength (Section 4.1.2), and we can analyze how strength is distributed across nodes in a similar way to the analysis of degree distributions. Indeed, analysis of human structural connectivity networks constructed with diffusion MRI indicates that both the strength distribution and degree distribution of brain networks have a similar form (Hagmann et al., 2007).

In weighted networks, we can also examine the distribution of edge weights in the connectivity matrix. In this case, we can examine how connection weights are distributed across all pair-wise links, rather than how summary measures such as degree and strength are distributed at the nodal level. Independent analyses of high-quality tract-tracing datasets that describe connectivity of the mouse whole-brain (Oh et al., 2014) and visual system connectomes (Wang et al., 2012), as well as the macaque interregional cortical connectome (Markov et al., 2014; Oh et al., 2014), have shown that connection weights vary over several orders of magnitude, and that their distribution approximates a lognormal.

4.3.1   The Lognormal Distribution

In a lognormal distribution, the logarithms of the edge weights are normally distributed, regardless of the base of the logarithm function. Lognormal distributions often arise when there is a low mean with large variance, and when values cannot be less than zero. The distribution of raw values is thus skewed, with an extended tail similar to the tail observed in scale-free and broad-scale systems. Indeed, lognormal and power-law distributions are mathematically related and can be generated by similar processes (Mitzenmacher, 2004).

Figure 4.4 plots the distribution of raw and logarithmically scaled connectivity weights of a cortical connectome constructed for the macaque using invasive tract tracing (Markov et al., 2013). In the distribution of raw values, we see an extended tail consisting of a small number of very strongly weighted links (Figure 4.4a). In contrast, the distribution of the logarithms of the weights has a symmetric, Gaussian-like form (Figure 4.4b).

Under a Gaussian distribution, the edge weight probability density is

$$P(\text{weight} = w) = \frac{1}{\sigma\sqrt{2\pi}} e^{-\frac{(w-\mu)^2}{2\sigma^2}}, \qquad (4.17)$$

where μ and σ are the mean and standard deviation of the distribution, respectively. The distribution of edge weights in a signed network is often well characterized with the Gaussian distribution. For example, the connectivity weights associated with correlation-based functional networks can be roughly approximated with the Gaussian distribution once the correlation coefficients have been transformed to a z-score.

In networks where all the edge weights as necessarily positive, such as structural connectivity networks, the Gaussian distribution is not an accurate model since it provides support for both positive and negative values. For unsigned networks, the lognormal distribution may provide a better characterization of the edge weights. The probability density under a lognormal distribution is

$$P(\text{weight} = w) = \frac{1}{w\sigma\sqrt{2\pi}} e^{-\frac{(\ln w - \mu)^2}{2\sigma^2}}, \quad w > 0. \qquad (4.18)$$

There are two differences between Equations (4.17) and (4.18). First, the denominator in Equation (4.18) is multiplied by w due to the change in variables $w \rightarrow \log(w)$, and ensures that the density sums to one over all values of w. Second, by taking the natural logarithm in the exponent in Equation (4.18), we contract very large deviations from the mean and bring the extended tail of extremely large weights closer to the mode. We also expand small weights below the mode, thus giving rise to a bell-shaped distribution of logarithmically transformed values. The normal distribution is characterized by its mean, μ, and variance, σ, whereas the lognormal distribution is characterized by its geometric mean and geometric standard deviation, which are computed as $\mu' = e^{\mu}$ and $\sigma' = e^{\sigma}$, respectively.

A normal distribution often arises from the additive effect of many independent processes whereas lognormal distributions arise from multiplicative

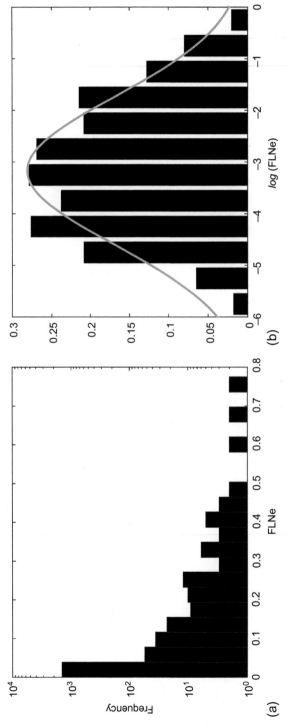

FIGURE 4.4 The weight distribution of the cortical connectome of the macaque. Data were taken from a connectivity matrix constructed via retrograde viral tracer injections into 29 cortical regions of the macaque. The result is a matrix encoding the connectivity between 29 cortical target regions (injection sites) and 91 sources (sites of tracer uptake). The connection weights represent the fraction of neurons labeled by the tracer in each source area, relative to the number of labeled neurons outside the injection site (the extrinsic fraction of labeled neurons, or FLNe; see Chapter 2). **(a)** A histogram of raw edge weights in this connectivity matrix. The vertical axis is plotted on a logarithmic scale to facilitate visualization of the shape of the distribution. All weights are positive, the mean is low and there is an extended tail where some connections have very high weight. **(b)** The probability density of the base 10 logarithm of the same connectivity weights. This distribution approximates a Gaussian (gray line). Also note that the connection weights span five orders of magnitude. Data are displayed in panel **(b)** as in Ercsey-Ravasz et al. (2013). Further details about this connectome are provided by Markov et al. (2013a).

processes (Limpert et al., 2001). Let us consider what this distinction means for network organization. The binomial degree distribution of an Erdös-Rényi graph, which approximates a Gaussian in the limit of large N, arises from the random *addition* of links. Since links are added at random, and independently of each other, they have an *additive* effect on the degree distribution. In contrast, neuronal networks are thought to engage in a large number of multiplicative computations (see Silver, 2010 for a review). In such computations, driving inputs to a neuron are multiplied by a constant, which alters the sensitivity or gain of a neuron to changes in its afferent signals. This multiplication can rise from modulatory or inhibitory influences. It allows greater flexibility to changing environmental conditions, facilitates signal amplification, and enhances the **efficiency** of information transmission (Brenner et al., 2000; Silver, 2010). Changes in the gain of coupled subsets of neurons will alter their synchronization dynamics and synaptic efficacy (Azouz, 2005; Silver, 2010). In turn, these changes will alter the connection weights of the network and may give rise to a lognormal distribution.

Physical and metabolic constraints on network wiring may also give rise to a lognormal distribution of connectivity weights. It has been shown that the lognormal distribution of weights in the interregional cortical connectome of the macaque is a mathematical consequence of two factors: (1) the physical layout of different regions in the brain, in which the distribution of interregional distances is approximately Gaussian; and (2) a distance-dependent cost on forming long-range connections, which causes a rapid (exponential) decay in the probability of finding a strong connection between two regions with increasing spatial separation (Ercsey-Ravasz et al., 2013). In other words, the lognormal distribution of connectivity weights in this network results from both the geometry of the cortex and a pressure to minimize the wiring costs of the brain (see also Chapter 8).

Connectivity weights are not the only aspects of brain organization that show a lognormal distribution. In fact, such distributions characterize many diverse structural and functional processes of the brain, and are thought to support increased dynamic range, dynamical stability, robustness to perturbation, energy efficiency, and computational simplicity (see Buzsáki and Mizuseki, 2014 for a review). The prevalence of lognormal distributions for many neural properties also suggests that a small number of highly active and connected neural elements (the tail of the distribution) may account for a large bulk of neuronal activity, forming a putative network "backbone" that is supported by a large number of less active and more weakly connected elements (the mode of the distribution; Buzsáki and Mizuseki, 2014). Under this model, the weak links are thought to facilitate the flexible engagement of diverse cell assemblies and the fine-tuning of neuronal activity. This organizational motif is consistent with a bipartite division of neural elements into a strongly connected core and weakly connected periphery. We review graph theoretic methods for characterizing core-periphery structure in Chapter 6.

4.4   SUMMARY

In this chapter, we have considered some of the most basic measures that can be computed for a graph: namely, the number of connections attached to each node (degree) and the total weight (strength) of these connections. We have seen how characterizing the distribution of these measures across the network allows us to understand the potential influence of certain nodes on network function. In particular, we considered three important classes of networks that can be defined by the form of their degree distribution: single scale, scale free and broad scale. Studies attempting to determine the class to which brain networks belong have yielded conflicting findings. We expect highly connected hub nodes to exert a stronger influence in long-tailed scale-free or broad-scale topologies, since a large proportion of network connectivity will be concentrated on these nodes.

The high connectivity of putative network hubs suggests that these nodes have a large capacity to influence, or be influenced by, many other network elements. In other words, they are topologically *central*. Importantly, degree and strength are not the only measures available for characterizing the importance of individual network nodes. As we will see in the next chapter, **graph theory** provides a wide range of such measures, and our notion of "importance" in the network context very much depends on the specific functional properties of the network that we wish to understand.

CHAPTER 5

Centrality and Hubs

Connectivity is not distributed uniformly across the elements of a nervous system. Instead, like many other real-world networks, nervous systems show a skewed, long-tailed **degree distribution** that points to the existence of highly connected network **hubs**. This heterogeneous distribution implies that different **nodes** serve distinct topological roles in the network, with highly connected nodes exerting a particularly important influence over network function. However, measures of **degree** or **strength** provide only a partial account of the functional significance of individual nodes. In fact, a much broader set of measures can be used to quantify the importance or influence of a node on network function. These measures capture distinct aspects of topological **centrality**—the capacity of a node to influence (or be influenced by) other network elements by virtue of its connection **topology**.

In this chapter, we overview some key measures of topological centrality and consider their interpretation in neuroscientific applications. We show that these measures can lead to different ways of defining hubs, and that they often assume a particular dynamic of information flow on the network. Understanding these assumptions is an important step in selecting an appropriate centrality measure for connectomic analysis.

5.1 CENTRALITY

The concept of centrality was introduced by Bavelas (1948) to understand how the structural position of an individual within a social network determines the influence of that person in group-wide processes. A large number of centrality measures have been proposed in the intervening years. Each of these measures captures distinct aspects of what it means to be "central" in a network. The diversity of measures available partly reflects a difficulty in formulating a specific definition of network centrality. This difficulty arises because our precise notion of centrality depends on the kind of influence that we aim to measure.

Freeman (1979) proposed that a common theme across different measures of centrality is that they arrive at the same conclusion when applied to a star-like (also called hub-and-spoke) network: namely, that the central node in this

137

Fundamentals of Brain Network Analysis. http://dx.doi.org/10.1016/B978-0-12-407908-3.00005-4

configuration will be identified as having the highest centrality. An example star network is shown in Figure 5.1. The node located in the center of the graph is clearly the most topologically central: it connects to all other nodes, which are otherwise not connected to each other. Any communication in this network must pass through the central node.

Freeman (1979) observed that three fundamental properties can be ascribed to the central node in Figure 5.1: (1) it has the maximum possible *degree*, since it is connected to all other nodes; (2) it falls on the shortest possible topological **path** *between* all pairs of nodes; and (3) it is located at the shortest topological distance from—or is maximally *close* to—all other nodes. Topological distance refers to the shortest **path length**, or the minimum number of **edges** required to link any two nodes in a network, and makes no assumptions about the spatial relationships or physical distances between nodes.

Based on this analysis of a simple star-like graph, Freeman (1979) argued that **degree**, **betweenness** and **closeness** are three cardinal aspects of topological centrality. In the following sections, we consider different measures of each characteristic. We focus on those measures that have been applied to neurobiological data, or which are relevant for understanding critical concepts

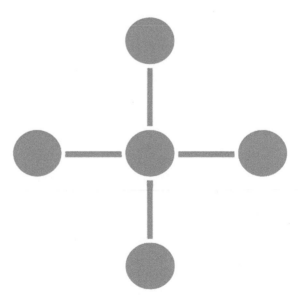

FIGURE 5.1 Three cardinal aspects of centrality. The central node in this network displays the three primary characteristics of centrality proposed by Freeman (1979). It has maximum degree, since it connects to all other nodes; it has maximal betweenness, since it falls on the shortest path between all possible pairs of nodes; and it has maximal closeness, being located at the shortest topological distance to all other nodes.

pertaining to centrality in the brain. More general treatments of network centrality are provided by Freeman (1979), Borgatti (2005), and Borgatti and Everett (2006).

5.1.1 Degree-Based Measures of Centrality

Degree centrality, C_D, is the simplest measure of centrality. It is equivalent to the node degree (Chapter 4). Thus, for an undirected network we have

$$C_D(i) = k_i = \sum_{i \neq j} A_{ij}. \tag{5.1}$$

This definition of centrality assumes that nodes with many connections exert more influence over network function and thus have higher topological centrality compared to nodes with fewer connections.

One limitation of degree centrality is that all connections are treated equally. Consider two nodes, i and j, each of which has a degree of three. Based on this equivalence, degree centrality will assign equal importance to these two nodes. Now suppose that the three **neighbors** of node i have degrees equal to 2, 3, and 4, whereas the neighbors of node j have degrees equal to 8, 12, and 14. The high-degree neighbors of node j are presumably more influential than the neighbors of node i. By association, they also make node j more influential than node i, since j is able to reach a larger number of network elements in a smaller number of steps. In this example, the influence of each node is determined by the degree of its neighbours, rather than its own degree. Put simply, the "quality" of a node's neighbors may be just as important as the quantity. Degree centrality counts quantity, but does not consider quality.

Eigenvector centrality is a measure that accounts for the quantity *and* quality of a node's connections. It considers both the degree of that node and the degree of its neighbors. To estimate this measure, we must consider the **eigenvectors** and **eigenvalues** of the **adjacency matrix**, A. An eigenvector of A is a nonzero vector x that, when multiplied by A, satisfies the condition $Ax = \lambda x$. The scalar λ is an eigenvalue of A and represents a factor which, when multiplied with x, is equal to Ax (Box 5.1). The eigenvector centrality, $C_E(i)$, of node i is defined as the ith entry in the eigenvector belonging to the largest eigenvalue of A, which is denoted λ_1. Alternatively, we can think of the eigenvector centrality of node i as being equivalent to the summed centrality of its neighbors (Newman, 2010). Given that $Ax = \lambda x$, we can write

$$C_E(i) = x_i = \frac{1}{\lambda_1} \sum_{j=1}^{N} A_{ij} x_j. \tag{5.2}$$

BOX 5.1 A PRIMER ON EIGENVALUES AND EIGENVECTORS

Eigenvalues and eigenvectors play an important role in networks and in **matrix** theory more generally. Eigenvectors and eigenvalues are found by solving the equation

$$Ax = \lambda x, \qquad (5.1.1)$$

where A is an $N \times N$ matrix, the column vector $x = [x_1, \ldots, x_N]$ is an eigenvector of A, and the scalar λ is the corresponding eigenvalue. Setting $x = 0$ trivially solves Equation (5.1.1), but we seek a nontrivial solution such that $x \neq 0$. We can think of Equation (5.1.1) as asking the following question: if we multiply a matrix A by a vector x, and then multiply that same vector x by a constant value λ, can we get the same result?

To solve Equation (5.1.1), we can rewrite it as

$$(A - \lambda I)x = 0, \qquad (5.1.2)$$

where I is an $N \times N$ identity matrix with diagonal elements equal to one and off-diagonal elements equal to zero. Its inclusion in Equation (5.1.2) allows us to subtract the scalar λ from the diagonal elements of the matrix A. Apart from the trivial case, $x = 0$, Equation (5.1.2) will only have a solution when $|A - \lambda I| = 0$, where $|X|$ denotes the determinant of the matrix X. The expression $|A - \lambda I|$ is also called the characteristic polynomial of A.

To take a concrete example, consider the 2×2 matrix

$$A = \begin{bmatrix} 2 & 2 \\ 5 & -1 \end{bmatrix}. \qquad (5.1.3)$$

We then have

$$A - \lambda I = \begin{bmatrix} 2 & 2 \\ 5 & -1 \end{bmatrix} - \lambda \begin{bmatrix} 1 & 0 \\ 0 & 1 \end{bmatrix} = \begin{bmatrix} 2-\lambda & 2 \\ 5 & -1-\lambda \end{bmatrix}. \qquad (5.1.4)$$

Next, we find the values of λ satisfying $|A - \lambda I| = 0$. Since A is a 2×2 matrix, its determinant is given by $A_{11}A_{22} - A_{12}A_{21}$, so we have

$$|A - \lambda I| = (2-\lambda)(-1-\lambda) - 10. \qquad (5.1.5)$$

The right hand side of Equation (5.1.5) reduces to $\lambda^2 - \lambda - 12$. The eigenvalues will thus be the values of λ that allow us to solve the quadratic equation $\lambda^2 - \lambda - 12 = 0$, which are -3 and 4. We can then find the eigenvectors by solving the equation

$$\begin{bmatrix} 2-\lambda & 2 \\ 5 & -1-\lambda \end{bmatrix} \begin{bmatrix} x_1 \\ x_2 \end{bmatrix} = 0, \qquad (5.1.6)$$

after substituting each eigenvalue. For example, if we solve Equation (5.1.6) for $\lambda = -3$, we get $x_1 = 2$ and $x_2 = -5$. These values constitute the eigenvector for $\lambda = -3$.

We then repeat this process for every unique eigenvalue. A matrix with N columns has N eigenvalues, so the method illustrated here is not practical when N is large. There are several alternative and more efficient algorithms available. An accessible introduction to eigendecomposition for networks provided by Newman (2010).

The largest eigenvalue is called the principal eigenvalue and is denoted λ_1. The leading eigenvector is the vector associated with this eigenvalue. The full spectrum of eigenvalues is called the *eigenspectrum* of A.

In directed networks, we can define left and right eigenvectors separately by considering outgoing and incoming connections, respectively. The right eigenvectors, x_{right}, are given by $Ax_{\text{right}} = \lambda x_{\text{right}}$ whereas the left eigenvectors, x_{left}, are given by $(x'_{\text{left}}A)' = A'x_{\text{left}} = \lambda x_{\text{left}}$. Here we use x' and A' to denote the transpose of x and A, respectively.

To understand why Equation (5.1.1) is important, imagine that the matrix A describes a system that effects some linear, geometric transformation of a vector, such that its direction, scale, or both can be altered. The eigenvectors of A are those vectors that change only in magnitude but not direction. The change in magnitude is determined by λ. For this reason, eigenvalues are sometimes thought of as corresponding to the natural "modes"—often called eigenmodes—of a system. For instance, the eigenvalues of an oscillating system correspond to the natural or resonant frequencies of the oscillations, and the eigenvectors correspond to their shapes. In statistics, the eigendecomposition of a covariance matrix is equivalent to a principal component analysis; the eigenvectors correspond to orthogonal directions of variance (principal components) and the eigenvalues to the amount of variance explained by each direction.

Eigendecomposition has been used extensively to characterize and model **functional connectivity** in human brain imaging signals (Bullmore et al., 1996b; Friston, 1994; Friston et al., 2014). In **graph theory**, the principal eigenvector of the adjacency matrix gives us a measure of the topological centrality of each node (Section 5.1.1), although eigendecomposition has also proven useful for understanding other aspects of network topology, such as its modular organization (Newman, 2013). Spectral graph theory is specifically concerned with understanding how the eigenspectrum of a graph relates to its topology and dynamics (Brouwer and Haemers, 2012; van Mieghem, 2012; Box 7.2).

This definition, first proposed by Bonacich (1987), endows a node with high centrality if it has many neighbors, if its neighbors are highly connected, or both.

Equation (5.2) can be used to compute centrality in weighted networks, subject to certain conditions (Newman, 2004a, 2010). In particular, the largest eigenvalue of A must be unique and positive, which is mathematically guaranteed for any square matrix with strictly positive entries (the Perron-Frobenius theorem) but is not guaranteed for matrices with negative weights, such as those representing correlation-based networks. One solution, proposed by Lohmann et al. (2010), is to remap edge weights to a positive range by simply adding 1, such that $r_{ij} \leftarrow r_{ij} + 1$. As the authors acknowledge, one problem with this approach is that a strong negative correlation will be given lower weight than a weak positive correlation. An alternative is to take the absolute value of the correlation coefficient, in which case a strong negative correlation will be given an equal weighting to a strong positive correlation. The decision of which transformation to use depends on how we choose to interpret the polarity of the edge weights.

In directed networks, the adjacency matrix, A, is asymmetric and it has two leading eigenvectors, one left and one right (Box 5.1). If edges are listed in the matrix such that A_{ij} is the edge from node j to i, the right eigenvector defines centrality according to the **in-degree** of each node and its neighbors and the left eigenvector defines centrality according to the **out-degree**. The correct choice between these two measures depends on the properties that we are interested in extracting from our network. For example, in the online social network of Twitter users, a person with many followers (high k^{in}) will be more central than someone who follows many people (high k^{out}) but who is not followed by many others (low k^{in}). In this context, we are more interested in properties related to the in-degree of each node, so we should estimate centrality using the right eigenvector. In brain networks, either the right or left eigenvectors may be informative: the left eigenvector will allow us to identify central sources of information whose efferent connections can influence many other nodes; the right eigenvector will allow us to identify putative sinks of information that receive a large amount of afferent information. Note that the definition of eigenvector centrality presented in Equation (5.2) is based on the in-degree of each node (i.e., right eigenvectors) when applied to a directed network. We retain this convention below.

A limitation of eigenvector centrality in directed networks is that it can assign a centrality score of zero to some nodes, even if their in-degree and out-degree are greater than zero. For example, assume that we estimate $C_E(i)$ using the right eigenvector because we are interested in centrality related to incoming connections. Any node i with an out-degree greater than zero, but with no incoming **links** (i.e., $k^{out} > 0$ and $k^{in} = 0$), will have $C_E(i) = 0$. This is because we only sum over nodes that project to i when computing $C_E(i)$, and there are no such nodes if i has zero in-degree. In this scenario, it may be sensible that a **vertex** with zero in-degree should have zero centrality, but it means that any node with all of its incoming edges from nodes with zero centrality will also have zero centrality.

One way around this problem is to add constants to Equation (5.2) that endow all nodes with nonzero centrality. This approach is taken by Bonachich and Lloyd's *alpha centrality* measure (Bonacich and Lloyd, 2001) and the closely related measure of *Katz centrality* initially proposed for the analysis of social networks (Katz, 1953). However, like eigenvector centrality, these measures can assign high centrality to low-degree nodes simply because they are connected to a high-degree hub. To understand why this is problematic, consider the World Wide Web. It does not make sense to assign high centrality to an individual page because it receives a link from the search engine Google, since this link will only be one of the many millions of links that Google makes. One solution is to scale the contribution that Google makes to the centrality of each individual site by Google's overall connectivity. In more general terms, we can scale the contribution that a neighbor j of node i makes to i's centrality by the out-degree of j (Newman, 2010),

$$x_i = \alpha \sum_{j=1}^{N} A_{ij} \frac{x_j}{k_j^{out}} + \beta, \tag{5.3}$$

where β is a constant (typically $\beta = 1$) that ensures a nonzero centrality score even for nodes with zero in-degree. The free parameter α weights the contribution of network topology to the centrality score, such that high values of α will increase the contribution of the sum in Equation (5.3).

One remaining problem with Equation (5.3) is that the first term is undefined when a node has no outgoing links, since we have $x_j/0$ when $k_j^{out} = 0$. One solution is to artificially set the out-degree of such nodes to $k_j^{out} = 1$. Since $A_{ij} = 0$ for all i when $k_j^{out} = 0$, this makes no difference to the calculation. We can then rewrite Equation (5.3) in matrix form as

$$x = \alpha A D^{-1} x + \beta \mathbf{1}, \tag{5.4}$$

where D is a diagonal matrix of node out-degrees such that $D_{ii} = \max\left(k_i^{out}, 1\right)$, and $\mathbf{1}$ is a column vector of ones of length N. If we set $\beta = 1$ and rearrange Equation (5.4), we get the definition for the well-known *PageRank centrality*,

$$C_{PR} = x = \left(I - \alpha A D^{-1}\right)^{-1} \mathbf{1} = D(D - \alpha A)^{-1} \mathbf{1}, \tag{5.5}$$

where I is an $N \times N$ identity matrix that contains ones along the diagonal and zeros elsewhere.

Initially proposed by Google cofounders Sergey Brin and Larry Page (Brin and Page, 1998), PageRank centrality is one of the algorithms used to rank websites in the results of the popular search engine. The measure scales the contributions that the neighbors of node i make to its centrality by the degree of those neighbors, thereby accounting for any potential biases associated with links to highly connected nodes. The parameter α is sometimes called a damping factor. Its precise value is arbitrary, but is constrained to the range $\alpha \in [0, 1/\lambda_1]$, where λ_1 is the largest eigenvalue of the matrix AD^{-1} (Newman, 2010). When α is high, C_{PR} will be more strongly determined by network topology. When α is low, the contribution of topology to C_{PR} will be reduced, resulting in more uniform centrality scores across nodes.

Figure 5.2 demonstrates some important differences between degree, eigenvector, and PageRank centrality in an example network comprising fourteen nodes. In this network, node 6 has the highest degree (six links). The next most connected is node 10 (five links), followed by node 7 (four links). All other nodes have only one edge. If we compute the eigenvector centrality of each node in this network, we find that node 6 is again the most central. We also find that node 7 has higher eigenvector centrality than node 10. This is because most of node 10's neighbors (nodes 11, 12, 13, and 14) have low degree, whereas node 7 links to nodes 6 and 10—the highest degree nodes in the graph. Node 7 is therefore better positioned to influence (or be influenced by) a larger number of network elements.

When we compute PageRank centrality, we find that the relative rankings of nodes 7 and 10 have swapped again, with node 10 being identified as more central. This switch occurs because the connections that endow nodes 7 with high eigenvector centrality—the links to nodes 6 and 10—must compete with the other neighbors of these high-degree vertices. Considering these links in relation to the total connectivity of nodes 6 and 10 reduces their contribution to the PageRank centrality of node 7.

What happens when we apply these centrality measures to brain networks? Figure 5.3 shows differences between degree, eigenvector, and PageRank centrality in an analysis of a large sample (1003 individuals) of high-resolution (>22,000 voxels) human functional connectivity networks measured with resting-state **functional MRI** (Zuo et al., 2012). Figure 5.3a shows the spatial correspondence between regions identified as being the most central nodes across the three measures. The anatomical distribution of the degree and PageRank centrality measures is very consistent: both highlight areas of association

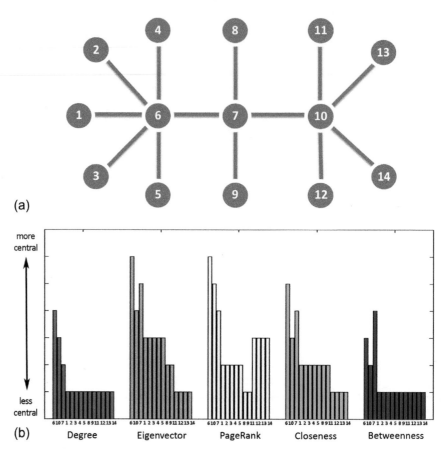

FIGURE 5.2 Measuring centrality by degree, closeness, and betweenness. **(a)** An example undirected network comprising fourteen nodes. **(b)** The degree, eigenvector, PageRank, closeness, and betweenness centrality of each node in this network. To facilitate comparison across measures, each node was assigned a rank based on its score on that measure, such that a higher rank is associated with higher centrality (vertical axis). For each measure, nodes have been ordered from left to right in descending order of degree centrality rank. PageRank centrality was computed with $\alpha = 0.85$, $\beta = 1$ (Brin and Page, 1998).

cortex, particularly in lateral and medial parietal regions, as being the most central. In contrast, eigenvector centrality shows a different pattern, with regions of primary and somatosensory cortex being the most central. Figure 5.3b shows the scatter plots of the associations between these measures. Degree and PageRank centrality are highly correlated. Both measures also correlate with eigenvector centrality, but there are two different types of association. One group of

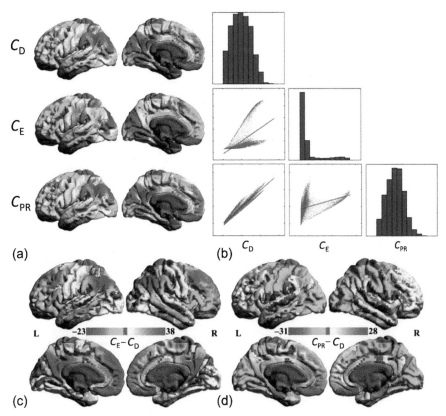

FIGURE 5.3 Degree-based centrality measures in human functional connectivity networks. Networks were constructed using resting-state functional MRI in 1003 healthy individuals at voxel-wise resolution (>22,000 nodes). Functional connectivity was measured as the Pearson correlation between regional time courses. **(a)** Spatial maps showing regional variations in degree centrality, C_D, eigenvector centrality, C_E, and PageRank centrality, C_{PR} for the left hemisphere (the pattern was similar in the right hemisphere). Centrality scores have been rescaled to z-scores, with red colors indicating higher centrality and blue colors indicate lower centrality. **(b)** Scatterplots for the bivariate associations between the centrality measures. The best-fitting straight line is shown in red. The histograms of each measure are shown along the diagonal. **(c)** Cortical map of regions showing a statistically significant difference in eigenvector and degree centrality scores. Blue coloring highlights regions where $C_D > C_E$, red coloring highlights the opposite trend. **(d)** Cortical map showing differences between PageRank and degree centrality scores. Blue coloring highlights regions where $C_D > C_{PR}$, red coloring highlights the opposite trend. *Images reproduced from Zuo et al. (2012) with permission.*

voxels shows a strong correlation, with a rise in either degree or PageRank centrality being associated with a rapid increase in eigenvector centrality. The other group of voxels shows a weaker association, such that an increase in degree or PageRank centrality is associated with only a small change in eigenvector centrality. Voxels in this second group are likely to have many neighbors of low degree.

Panels c and d of Figure 5.3 map regions showing a statistically significant difference between eigenvector and PageRank centrality relative to degree. Degree emphasizes the centrality of cortical association areas relative to the eigenvector measure. PageRank centrality emphasizes the centrality of association areas to an even greater extent. The eigenvector and PageRank measures were also found to emphasize the centrality of subcortical regions, suggesting that these areas are linked to network hubs (Zuo et al., 2012). Correlations between centrality measures are likely to vary across different types of brain networks, but these data usefully illustrate how distinct metrics can emphasize different characteristics of the topological role played by individual brain regions. One caveat is that degree-based measures of centrality can be difficult to interpret in the analysis of correlation-based networks. These difficulties are discussed in Box 5.2.

BOX 5.2 IDENTIFYING HUBS IN CORRELATION-BASED NETWORKS

Many of the measures discussed in this chapter use degree, strength, or path length to characterize the topological centrality of a node. Path-based measures suffer from particular problems when applied to correlation networks. First, it can be difficult to define shortest paths in signed networks, complicating the way in which we interpret putative paths of information flow. Second, the shortest path length between two nodes in a correlation network is already captured by the correlation coefficient. In particular, the correlation coefficient captures the effect of both direct and indirect paths in the underlying structural network, meaning that everything about the interaction between a pair of nodes is conveyed by the correlation coefficient itself. This suggests that an analysis of indirect paths, performed on a thresholded (and sometimes binarized) correlation matrix, offers little additional information. These issues are considered in more detail in Chapter 7.

Degree-based and strength-based measures are also limited when analyzing correlation networks. A detailed examination of this issue by Power and colleagues (2013) revealed that, for brain functional connectivity and other correlation networks, there is a strong association between the degree of an individual node and the size of the **module** to which that node belongs (Figure 5.2.1). This association arises because of the transitivity of the correlation coefficient. This property mandates that a given node i cannot be strongly correlated with node h and node j if these two nodes are weakly correlated themselves. A typical modular decomposition of a correlation network will group nodes such that they show high average correlation within modules and low average correlation between modules, meaning that any individual node will rarely show a strong correlation with other nodes outside its own module. As a result, the degree and/or strength of each node will be determined primarily by the number of other nodes in the same module.

Networks that are not based on correlations, such as structural connectivity networks, do not show this bias, since node i is free to connect to both nodes j and h independently of the connectivity between j and h. Measures that consider each node's connectivity in relation to the modular organization of the brain may provide more appropriate measures of centrality for correlation-based networks (Power et al., 2013). Some of these measures are discussed in Chapter 9.

BOX 5.2   IDENTIFYING HUBS IN CORRELATION-BASED NETWORKS—CONT'D

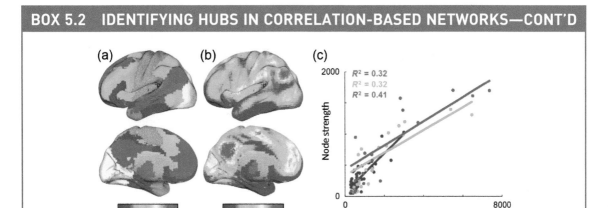

FIGURE 5.2.1 The association between node strength and module size in human correlation-based functional connectivity networks. **(a)** Cortical surface illustrating the module membership of each node in a voxel-wise functional connectivity network constructed from healthy human resting-state functional MRI data. Each color corresponds to a different module, and the color scale depicts variations in module size, calculated as the number of voxels in that module (e.g., the red module is the largest). **(b)** Cortical surface depicting variations in node strength in the same voxel-wise network. The highest node strength is observed for voxels belonging to the red module, which is also the largest module (compare with **(a)**). **(c)** The correlation between node strength and module size. Different colors correspond to results obtained after thresholding the functional connectivity network at different levels. The association between strength and module size is consistent across thresholds. *Figure reproduced from Power et al. (2013) with permission.*

5.1.2   Closeness Centrality

In graph theory, the terms **neighbors** and **adjacent** refer to node pairs that are directly connected by at least one edge. This usage implies a sense of closeness: we can think of two nodes being maximally close—in a topological sense—if they share a direct connection, whereas two nodes that are only linked indirectly through many intermediate nodes are topologically distant. Based on this logic, we can define the topological distance between nodes i and j as the number of edges on the shortest path between them. This measure, denoted l_{ij}, is commonly called the shortest path length or **geodesic** distance. For example, in the network depicted in Figure 5.2, the shortest path length between nodes 1 and 6 is $l_{1,6} = 1$, between nodes 1 and 3 is $l_{1,3} = 2$ and between nodes 1 and 14 is $l_{1,14} = 4$. A node with short average path length is able to interact with many network elements via only a few links, suggesting that it is topologically central. In particular, we might assume that messages emanating from such a central node will spread to other network elements in a relatively short period of time. Conversely, signals originating from noncentral elements of the network will only take a short time to reach the central node. This is the intuition, first articulated by Bavelas (1950), behind defining the centrality of a node based on closeness.

More formally, the *closeness centrality* of a node can be defined as the inverse of its average shortest path length (Beauchamp, 1965),

$$C_C(i) = \frac{N-1}{\sum_{j \neq i} l_{ij}},$$ (5.6)

where l_{ij} is the shortest path length, or topological distance, between nodes i and j. Distance here is defined purely in topological terms, and is not related to physical proximity in any way. In brain networks, this means that two areas at opposite ends of the brain could be topologically "close" simply because they share a direct connection.

In directed networks, if we adopt the convention that l_{ij} is the shortest path from node j to node i, then Equation (5.6) defines the closeness centrality according to the shortest paths that are incoming to node i. We can also compute closeness centrality based on the paths outgoing from node i, in which case we would instead sum over l_{ji} for $j=1,\ldots,N$ in Equation (5.6).

In weighted networks, closeness centrality can be estimated by finding the shortest *weighted* path length between regions, where the weight of the path is determined by the sum of the edge weights on that path. Using this approach, it is possible for a node with many low weight edges to have the same centrality as a node with only a few high-weight edges. To address this ambiguity, Opsahl et al. (2010) have proposed generalized definitions for degree, closeness, and betweenness centrality that incorporate a tunable parameter to differentially emphasize the contribution of the number of edges (also called the **hop count**) and weight of those edges (Box 5.3). Methods for finding weighted and directed shortest paths in networks are covered in Chapter 7. Box 5.2 considers some of the ambiguities associated with measuring path length in correlation-based networks.

Closeness centrality cannot be computed for fragmented networks. Fragmented networks comprise disconnected subsets of nodes, called components (Chapter 6). The shortest path length between nodes in disconnected components is defined as infinite and thus $C_C(i) = 0$ for all nodes in a fragmented network. One way to deal this problem is to treat each component as a separate network and compute closeness centrality only for nodes within the same component. However, this method is biased as geodesic distances will be shorter in smaller components. An alternative is to take the average of the inverse shortest path lengths (Opsahl et al., 2010; Rochat, 2009),

$$C'_C(i) = \frac{1}{N-1} \sum_{j \neq i} \frac{1}{l_{ij}}.$$ (5.7)

BOX 5.3 COMBINING DEGREE AND STRENGTH TO QUANTIFY CENTRALITY

Most of the centrality measures discussed in this chapter use either degree or path length to quantify node centrality. These measures have a clear definition and interpretation in binary networks. In weighted networks, we often compute variants of these measures by using a sum of edge weights. When we sum edge weights, it is possible for a node with many low weight edges to have the same centrality as a node with only a few high-weight edges. Should one be favored over the other?

To get around this ambiguity, Opsahl et al. (2010) propose generalized definitions for degree, betweenness, and closeness centrality that use a tuning parameter, α, to weight the relative contribution of the number of edges (hop count), and weight of those edges, to node centrality. They thus define a scaled degree centrality measure as

$$C_D^{w\alpha}(i) = k_i^{1-\alpha} s_i^{\alpha}, \tag{5.3.1}$$

where α is the positively valued tuning parameter. If $\alpha=0$, $C_D^{w\alpha}$ is determined only by degree, k_i, and if $\alpha=1$, $C_D^{w\alpha}$ is determined only by node strength, s_i. When $0 < \alpha < 1$, we favor the contribution of node degree and when $\alpha>1$, we favor node strength on a small number of edges. Figure 5.3.1 illustrates how the scaling parameter α impacts scaled degree centrality.

Scaled closeness centrality is defined as

$$C_C^{w\alpha}(i) = \frac{N-1}{\sum_{j\neq i} l_{ij}^{w\alpha}}, \tag{5.3.2}$$

where $l_{ij}^{w\alpha}$ is the scaled weighted shortest path computed after incorporating the scaling parameter.

In a **weighted graph**, the shortest path between nodes i and j is the path that has the lowest sum of edge weights (Chapter 7). We take the lowest sum because weights have traditionally been interpreted as a measure of the **cost** or "distance" of a connection. In brain networks, edge weights typically index the strength of connectivity between nodes, and we expect that paths with high weights will have greater topological proximity. It is therefore common to remap the edge weights, using an inverse (or other) transformation, such as $w_{ij} \leftarrow 1/w_{ij}$ (see Chapter 7).

The scaled, weighted shortest path is defined as the path with the lowest sum of remapped edge weights, where each edge weight has been raised to the power of α. Once again, when $\alpha=0$, the shortest path will be determined purely by binary topology. When $\alpha=1$, the shortest path will be found only by considering the total weight of each path. When $\alpha<1$, less emphasis is placed on edge weights, with the shortest paths mainly influenced by the need to minimize hop counts. When $\alpha>1$, any heterogeneity in the edge weights is amplified and thus shortest paths are more influenced by the need to minimize the total edge weight of the path. Using this scaled approach to finding weighted shortest paths, we can also define a scaled measure of betweenness centrality by substituting appropriately scaled quantities into Equation (5.8). Methods for finding weighted shortest paths are considered in Chapter 7.

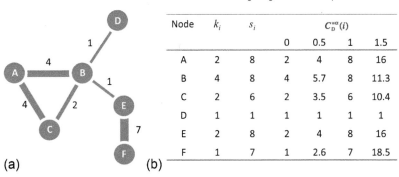

Node	k_i	s_i	$C_D^{w\alpha}(i)$			
			0	0.5	1	1.5
A	2	8	2	4	8	16
B	4	8	4	5.7	8	11.3
C	2	6	2	3.5	6	10.4
D	1	1	1	1	1	1
E	2	8	2	4	8	16
F	1	7	1	2.6	7	18.5

(a)   (b)

FIGURE 5.3.1 Scaled degree centrality for weighted networks. (a) An example weighted network, where the values next to each edge represent connectivity weight. **(b)** Table showing, for each node in the network, the degree, k_i, strength, s_i, and scaled, weighted degree centrality, $C_D^{w\alpha}(i)$ for different values of the scaling parameter α. As the value of this parameter increases, we assign more weight to nodes with fewer, higher weight edges. For example, node A has $k_i=2$ and $s_i=8$ whereas node F has $k_i=1$ and $s_i=7$. For $C_D^{w\alpha}$, the centrality of node A is higher for all values of α except $\alpha=1.5$. At this value, node F has higher centrality despite having one less link, because the contribution of strength is weighted heavily. *Example adapted from Opsahl et al. (2010) with permission.*

When this measure is estimated in a disconnected network and $l_{ij} = \infty$, the inverse $1/l_{ij} = 0$, ensuring that the "infinite" path makes no contribution to $C_C(i)$. Calculated in this way, closeness centrality is identical to the definition for nodal **efficiency** (Achard and Bullmore, 2007), which we discuss in Chapter 7. The intuition here is that a node with a short average path length to all other vertices has a higher efficiency of connectivity (Latora and Marchiori, 2001, 2003). To summarize, Equation (5.6) defines closeness centrality as the inverse of the average shortest path length of an index node to all others, whereas Equation (5.7) defines closeness centrality as the average of the inverse distances. The advantage of Equation (5.7) is that it can be computed in fragmented graphs because the inverse of an infinite path length between nodes is zero.

5.1.3   Betweenness Centrality

Betweenness centrality is a popular measure that was independently proposed by Freeman (1977) and Anthonisse (1971). It measures the proportion of shortest paths between all node pairs in the network that pass through a given index node. As the name implies, we can think of this measure as indexing the extent to which a node lies "between" other pairs of nodes. If information travels through a network along the shortest path, then nodes that lie on many shortest paths will mediate a high proportion of traffic, and thus represent central elements of the network. In this sense, such a node might play a controlling role in the passage of information through the network (Freeman, 1979) or act as a traffic bottleneck.

To compute the betweenness centrality of a node i, we calculate the proportion of shortest paths between nodes j and h that pass through i,

$$C_B(i) = \frac{1}{(N-1)(N-2)} \sum_{h \neq i,\, h \neq j,\, j \neq i} \frac{\rho_{hj}(i)}{\rho_{hj}}, \qquad (5.8)$$

where $\rho_{hj}(i)$ is the number of shortest paths between h and j that pass through i, ρ_{hj} is the number of shortest paths between h and j, and $(N-1)(N-2)$ is the number of node pairs that does not include node i. The normalization by ρ_{hj} accounts for the possibility that multiple shortest paths may exist between any pair of nodes.

Betweenness centrality can be computed in the same way for weighted and directed networks. In these cases, the shortest paths are determined using methods appropriate for weighted and/or directed networks (Chapter 7; Box 5.3). Betweenness centrality can also be computed for each individual edge (Girvan and Newman, 2002). To do this, we calculate the proportion of shortest paths between node pairs that traverse a given edge.

One interesting property of nodal betweenness centrality is that a vertex does not need to have a high degree to be considered topologically central. This property is demonstrated in Figure 5.2. In this network, node 7 has the highest betweenness centrality despite having only the third highest degree. This is because any path between nodes in the group on the left of the graph (i.e., nodes 1, 2, 3, 4, and 5) and nodes in the group on the right of the graph (i.e., nodes 11, 12, 13, and 14) must pass between node 7. We also find that the closeness centrality of node 7 is higher than the closeness of node 10, even though node 10 has higher degree. This is because node 7 is topologically closer to node 6 and its many neighbors.

Both closeness and betweenness centrality are based on shortest paths. Therefore, the identification of a node as topologically central using one of these measures is based on the assumption that information is routed along the shortest paths of the network. As we discuss in Section 5.1.5, this assumption may not be appropriate for the brain. Finding shortest paths requires knowledge of global network topology. This knowledge is not likely to be accessible to any individual neuron or neuronal population if we assume that there is no central control system or homunculus in the brain. Alternative measures that do not make such assumptions have been proposed. Typically, these measures incorporate information about all possible paths between brain regions. For example, the *random walk betweenness* of a node is quantified as the number of times a **random walker** traveling between any pair of network vertices passes through the index node (Newman, 2005b). An alternative measure, called **communicability**, accounts for all possible paths between nodes, with shorter paths given higher weight (Estrada and Hatano, 2008). Other measures can be derived by assuming specific models of network dynamics (Goñi et al., 2013; Mišić et al., 2014). Some of these measures are considered in Chapter 7.

5.1.4 Delta Centrality

A more general way to think about centrality that does not specifically depend on either degree, closeness or betweenness is based on the effect that removal of a given node has on the structure and function of the rest of the network (Latora and Marchiori, 2007). The intuition behind this measure is that inactivation or removal of highly central nodes will exert a disproportionate impact on remaining network elements. Formally, the *delta centrality* of a node is defined as

$$C_\Delta(i) = \frac{\Delta M_i}{M} = \frac{M(G) - M(G')}{M(G)}, \tag{5.9}$$

where M is any measure used to index the functioning or topological integrity of the graph, G, ΔM_i is the change in M observed after the removal of node i and its incident edges, and G′ denotes the graph obtained after this removal. Any measure of network topology can be substituted for M, depending on the specific research questions being asked of the data.

The logic of delta centrality has been used to identify a putative scaffold of **structural connectivity** in the human brain, the removal of which has a major impact on a large number of network properties (Irimia and Van Horn, 2014; Figure 5.4). Delta centrality can also be computed at the level of individual edges. In this case, we compare the difference in a network property before and after the removal of each edge (van den Heuvel and Sporns, 2011). Delta centrality is closely related to the analysis of network robustness, which we discuss in Chapter 6.

5.1.5   Characterizing Centrality in Brain Networks

The interpretation of some centrality measures requires that we assume a particular dynamical principle that governs the way signals propagate in the network. For example, betweenness centrality is an appropriate measure for networks in which information travels along shortest paths, whereas random walk betweenness is most appropriate for networks in which signals propagate via an unguided diffusion process (Chapter 7). Given the link between centrality measure and system dynamics, are some measures of topological centrality more closely related to actual neural communication processes than others?

Borgatti (2005) suggests that different centrality measures can be categorized along two dimensions. These dimensions describe the different forms of communication dynamics that are assumed by each measure (see also Borgatti and Everett, 2006). The first dimension describes how information flows between nodes. Borgatti distinguishes between serial transfer, serial duplication, and parallel duplication processes. Serial transfer refers to a process by which packets of information move from one node to the next (Figure 5.5a). The packet is indivisible and can only be in one place at any time, so it is transferred from node to node much like when a letter passes through multiple post offices before reaching its final destination. In a duplication process, the information packet is copied at each node. In serial duplication, only one replication is performed per unit of time (Figure 5.5b). For example, a viral infection spreads by infecting a host, replicating, and then infecting another host through direct contact. In parallel duplication, the replicated information can affect multiple targets simultaneously (Figure 5.5c). For example, a computer virus that spreads via email takes control of a host computer and sends a copy of itself to affect other computers (Telesford et al., 2011a).

Borgatti's second dimension concerns the trajectory that information can follow through the network. The trajectory could follow a shortest path, which

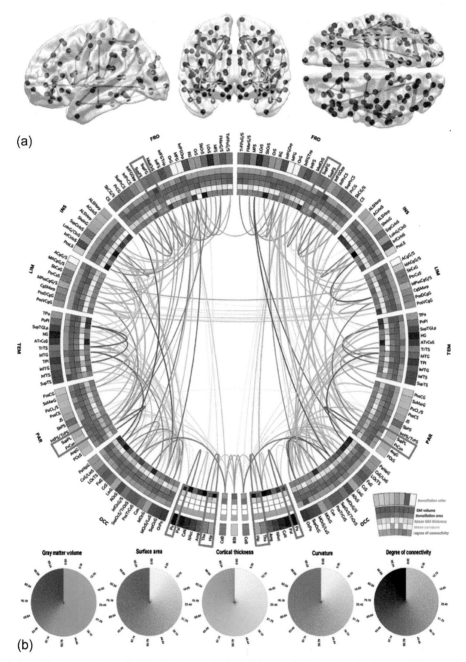

FIGURE 5.4 A putative structural scaffold for the human brain. In this analysis, human structural connectivity networks constructed with **diffusion MRI** were artificially lesioned by removing individual nodes and their incident edges. A multivariate extension of delta centrality analysis was used to characterize the effect of each simulated lesion on a large number of topological measures. Here we see an anatomical projection **(a)** and connectogram **(b)** of connections whose removal leads to a statistically significant change in global network topology. The anatomical projection has been thresholded at a more stringent level than the connectogram to facilitate visualization. Edge colors in both **(a)** and **(b)** represent links with high (red), average (green), or low (blue) fractional anisotropy, a marker of white matter microstructure. In the connectogram, concentric circles highlight variations in different nodal properties. Color wheels demonstrate the color scales for each ring. Nodes highlighted by red boxes correspond to brain areas previously identified to represent a densely connected core of hub areas (a **rich club**; see Chapter 6) in the human brain. *Figure reproduced from Irimia and Van Horn (2014) with permission.*

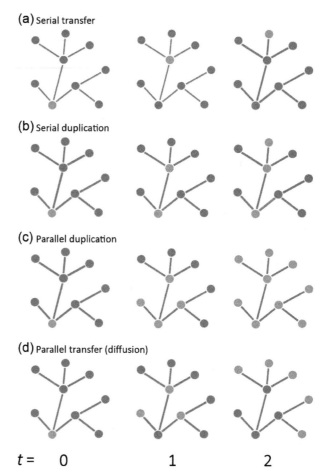

(a) Serial transfer

(b) Serial duplication

(c) Parallel duplication

(d) Parallel transfer (diffusion)

$t =$ 0   1   2

FIGURE 5.5 Information flows in networks. **(a)** Example of serial transfer, where an indivisible information packet (red) moves from one node in the network to another at each time step, t. **(b)** Example of serial duplication, where information is duplicated as it passes from node to node, but only ever follows a single path. **(c)** Example of parallel duplication, where the information is duplicated and simultaneously propagates to all neighbors of each node at each time step. **(d)** Parallel transfer may emulate neuronal communication. In this model, a signal propagates from one node to its neighbors in parallel, but does not remain at the originating node. This mode of communication is similar to a diffusion model of signal propagation. Diffusion models are discussed in more detail in Chapter 7. Note that only a fraction of a given node's neighbors are likely to be activated at each time step. Time is shown in discrete units for illustrative purposes.

implies that the nodes of the system have a global knowledge of network topology. Postal systems offer an example of such a trajectory: each post office (node) knows the geographic location of every other post office, and is thus able to efficiently route the letter through the shortest possible path to the target destination. Alternatively, the trajectory might not be influenced by knowledge of network topology. Such a trajectory could follow either a **path**, **trail** or **walk**. A walk is an unrestricted trajectory throughout a network—it is free to propagate along any direction and may return to previously traversed nodes and edges. A trail is more restricted—it can return to an already visited node but cannot reuse an edge. A path is the most restricted: it is a trajectory in which neither nodes nor edges can be visited more than once. In summary, paths are a subset of trails and trails are a subset of walks (more details on paths, walks, and trails are provided in Chapter 7).

Where do nervous systems fit within this classification scheme? Telesford et al. (2011a) propose parallel duplication as a model for neural communication. They argue that an action potential travels along the axon of a spiking neuron to activate a diverse number of synapses in an apparently stochastic manner. Neurons thus send an output to multiple targets, consistent with parallel transmission. However, a neuron, upon spiking, will return to its resting membrane potential after a short refractory period and await further incoming signals. This contrasts with a more typical example of a parallel duplication process, such as a computer virus, where the source stays infected while also infecting other targets (see Figure 5.5c). A parallel transfer model, in which information is transferred (rather than duplicated) to multiple targets simultaneously (Figure 5.5d), may provide a more accurate description of information flow in the brain. This model is similar to a diffusion process, in which a signal propagates to affect multiple neighbors simultaneously. There are many different network models of communication by diffusion, where the trajectory of information flow is often modelled as a trail or walk. An overview of these models is presented in Chapter 7.

What does a parallel transfer model tell us about how to measure centrality in brain networks? Consider how some of the centrality measures examined in this chapter can be interpreted under this model. Degree centrality may be characterized as a measure of the immediate influence that a node can have on the network (Borgatti, 2005). To understand why, consider a hypothetical network of neurons linked by synapses. If an individual neuron spikes at a given time, the effects will be felt in its neighbors after a short axonal conduction delay (Figure 5.5d). Thus, spiking neurons with more neighbors (i.e., neurons with high degree centrality) will have a larger immediate effect on the network.

Closeness centrality extends this concept beyond immediate effects of spiking neurons to account for effects on topologically distant network elements. Specifically, a neuron with high closeness centrality will affect a larger number of other neurons within a shorter period of time, because it can impact those neurons via a relatively short number of synaptic links. Similarly, eigenvector and PageRank centrality also characterize the global spread of an event occurring at a specific node, since nodes more closely connected with other central nodes will also impact a larger number of other elements in a shorter time. This effect is illustrated by comparing Figure 5.5a and d. In this simplified example, information that propagates either via serial or parallel transfer will reach topo-logically distant network elements at a similar time, suggesting that degree-based and closeness-based measures can both offer useful characterizations of network centrality.

It is important to emphasize that the examples depicted in Figure 5.5 are sim-plified, heuristic models of neuronal communication. In reality, signal propa-gation under a diffusion or parallel transfer model can be affected by many physiologically relevant parameters, such as variations in axonal conduction velocity, differences in the fraction of neighbours that are activated by a source node, and so on. Each of these factors, which are often difficult to quantify in large-scale networks, may affect the validity of any given centrality measure as an index of neural communication (see also Box 7.3). Mišić et al. (2015) and Goñi et al. (2013) have proposed a number of alternative measures of centrality that are appropriate for networks in which diffusion is a likely mechanism of communication. In brain networks, many of these measures correlate with more traditional centrality measures based on degree, closeness and betweenness. However, these correlations are not perfect, suggesting that diffusion-based mea-sures can reveal aspects of network centrality that are not captured by the more traditional metrics. Some of these diffusion-based measures are discussed in Chapter 7.

5.2   IDENTIFYING HUB NODES

In Chapter 4, we discussed the importance of a network's degree distribution. For many real-world networks, this distribution is characterized by a fat tail that points to the existence of highly connected nodes. These highly connected nodes are often called hubs because they are hypothesized to act as focal points for the convergence and divergence of information in the network. This section considers some of the different approaches that have been used to define hubs in brain networks.

5.2.1   Classifying Hubs Based on Degree and Centrality

An intuitive definition of hub nodes is that they have high degree. According to this definition, we could rank nodes based on their degree and set a specific threshold that defines a cut-off above which all nodes are considered hubs (van den Heuvel et al., 2008b). Unfortunately, this choice of threshold is arbi-trary and means that hubs can be defined in networks with all types of degree distribution (e.g., a **single-scale** distribution), provided that a sufficiently lib-eral threshold is chosen. Ideally, cutoff values should be used in conjunction with an analysis of the degree distribution when trying to identify putative hub nodes. If the distribution is fat-tailed and supports the existence of hubs, then a cut-off may be appropriate.

Focusing just on degree to define hubs may be too simplistic or restrictive. As we saw in Section 5.1, not all connections are the same—some nodes may connect to other important nodes, implying that they play a more influential role in the network. From this perspective, nodes with high eigenvector centrality may be more appropriate candidates for hub nodes. Alternatively, we may view high betweenness or closeness centrality as a desirable property for candidate hubs, because such nodes can bridge together different subgroups within a network. We could also view nodes with high delta centrality as putative hubs, since their removal will have a major impact on the topological integrity of the network.

These considerations highlight how the definition of a hub node is closely related to the concept of centrality, and that different types of hubs can be defined depending on the specific aspects of centrality that we wish to prioritize. In principle, we could choose any of the numerous centrality measures proposed in the literature as a basis for identifying hubs and assign the label of "hub" to nodes with the highest scores on that particular metric. In practice, node rankings on many of these properties are often correlated, so it can be useful to aggregate rankings across measures to devise a more robust, consensus classification of individual nodes as either hubs or nonhubs.

5.2.2   Consensus Classification of Hubs

In an early attempt to identify hubs in brain networks, Sporns et al. (2007) analyzed the anatomical connectivity matrices of the macaque and cat brains, as collated from the synthesis of a large number of published **tract-tracing** studies. They examined the relationship in each network between measures of node degree, closeness centrality, betweenness centrality, clustering, and three-node **motif** fingerprints. A three-node motif is a pattern of connectivity between a triplet of nodes. For example, the three nodes could all be connected to each other (a closed triangle), or two nodes may be connected to the third without being connected to each other (an open triangle). In a directed network, there are 12 possible ways to connect three different nodes (Figure 8.1). Specific types of motifs are particularly prevalent in brain networks (Figure 8.2), suggesting that they represent basic building blocks of network topology. Motifs are discussed in detail in Chapter 8.

Sporns et al. (2007) found that both the macaque and cat **connectomes** showed an overrepresentation of the so-called m_9^3 motif, which is characterized by a single apex node that reciprocally links two other nodes that are not directly connected to each other (Figure 5.6). Hub regions should be

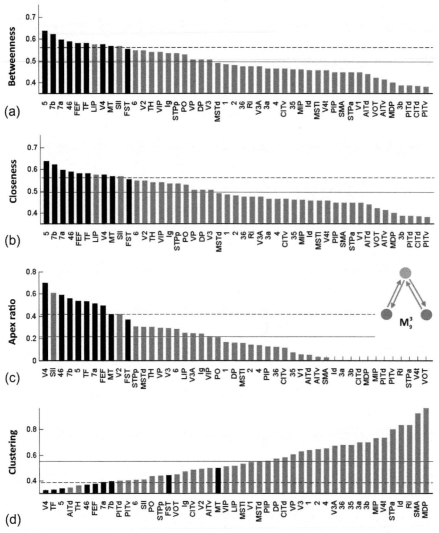

FIGURE 5.6 Centrality and clustering in the macaque cortical connectome. A 47-region structural connectivity matrix for the macaque was constructed by collating the results of a large number of published tract-tracing studies. In all plots, nodes have been ranked according to their betweenness centrality **(a)**, closeness centrality **(b)**, apex ratio **(c)**, and **clustering coefficient (d)**. Black bars highlight nodes with a degree greater than one standard deviation above the network mean. High-degree nodes have high betweenness, closeness, and apex ratio, and low clustering. The inset in **(c)** illustrates the m_9^3 connectivity motif. Nodes with a high apex ratio are more likely to occupy the apex position (red vertex) in this motif. In all panels, solid horizontal lines correspond to the mean across nodes and broken lines to one standard deviation above the mean. Abbreviations correspond to different brain regions. *Images were adapted from Sporns et al. (2007) with permission.*

overrepresented as apex nodes in this motif, since hubs are thought to bridge or integrate otherwise segregated network elements. To test this hypothesis, Sporns and colleagues defined a measure called the apex ratio—computed as the fraction of times a node was located at the apex position in all instances of the m_9^3 motif—and found that all nodes with high degree also ranked highly on the apex ratio measure. Moreover, the authors found that measures of degree, closeness centrality, and betweenness centrality were all moderately correlated ($r > 0.50$), and that high degree nodes also ranked highly on these other centrality metrics (Figure 5.6).

The high apex ratio of hub nodes suggests that they will have a low clustering coefficient. Recall from Chapter 1 that the clustering coefficient of a node is the fraction of its neighbors that are also connected to each other (see also Chapter 8). In the m_9^3 motif, the neighbors of the apex node are not directly connected with each other. A node with a high apex ratio is thus likely to have a low clustering coefficient. Indeed, it was found that brain regions with high degree, apex ratio, closeness centrality, and betweenness centrality were also among the nodes with the lowest clustering coefficient (Sporns et al., 2007; Figure 5.6). This relationship between centrality and clustering is consistent with a hierarchical organization of brain connectivity in which hub nodes link otherwise unconnected elements (Ravasz and Barabasi, 2003). Human MRI has shown that this hierarchical organization is disrupted in brain disorders such as schizophrenia (Bassett et al., 2008).

The analysis of Sporns and colleagues (2007) suggests that hubs in brain networks have several topological characteristics. First, they have high degree and connect to many other nodes. Second, they have high closeness centrality, and are thus able to communicate with many other nodes via only a small number of links. Third, they have high betweenness and low clustering, by virtue of their role in integrating otherwise unconnected nodes. These characteristics were used by van den Heuvel et al. (2010) to develop a consensus-based definition of hubs in a diffusion MRI analysis of human structural connectivity networks. Specifically, these authors reasoned that hubs should display high connectivity, high betweenness centrality, low average path length (which is inversely related to closeness centrality), and low clustering. They ranked each of 108 anatomically defined brain regions on each measure. Each region was then assigned a score of one each time any of the following conditions were met: (1) the node was ranked in the top 20% of nodes with highest degree; (2) it was in the top 20% of nodes with the highest betweenness; (3) it was in the bottom 20% of nodes with the lowest clustering; or (4) it was in the bottom 20% of nodes with lowest average path length. The result was a node-specific hubness score, ranging from zero to four, where a score of four indicates that the node demonstrates the expected characteristics of a hub across each of the four measures. This method identified the following brain regions as having a maximum

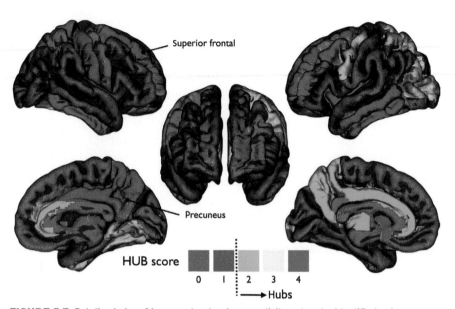

FIGURE 5.7 Putative hubs of human structural connectivity networks identified using a consensus approach. Brain regions were assigned a hub score ranging from 1 to 4, based on whether they were ranked amongst nodes with the highest degree and betweenness centrality and/or lowest average path length or clustering coefficient. *Figure reproduced with permission from van den Heuvel et al. (2010).*

hub score of four, thus marking them as putative hub regions of the human connectome: left and right caudate, left and right superior frontal gyrus, right middle cingulate gyrus, right precuneus, left and right putamen, and left thalamus (Figure 5.7).

5.2.3   Module-Based Role Classification

Different types of hubs can be identified in relation to the modular organization of a network. Recall from Chapter 1 that modules are strongly connected subsets of nodes. Their strong interconnectivity implies some degree of functional specialization, or topological segregation, from other network elements. Guimerá and Amaral (2005) proposed that different kinds of hubs can be identified based on how the connections of each node are distributed across different topological modules. For example, the links of so-called *connector* hubs have a relatively even distribution across different modules, suggesting that these nodes play an important role in integrating different subsystems within the network. In contrast, *provincial* hubs are strongly connected to other nodes in the same module, suggesting that they contribute to specialized information processing. We discuss methods for characterizing the topological roles of individual nodes in relation to network modules in Chapter 9.

5.3 SUMMARY

Topological centrality is a multifaceted concept, and there are many different graph theoretic measures for quantifying the centrality of a network node. Here, we have focused on those measures most commonly applied in the analysis of brain networks. Many of these measures may be categorized according to the extent to which they index aspects of centrality that are related to degree, closeness, and betweenness (Freeman, 1979). Other approaches, such as delta centrality, identify key nodes based on the structural or functional consequences of their removal from the network.

Different measures of centrality make different assumptions about how information flows on the network. These assumptions should be borne in mind when interpreting the results of a centrality analysis. For example, measures based on shortest paths have proven popular in the characterization of brain networks, but they assume that nodes have a global knowledge of network topology that allows information to be efficiently routed along the shortest path. It is unlikely that any individual neuron, or population of neurons, possesses such knowledge. Alternative measures based on a diffusion model of communication may be more appropriate in the analysis of brain networks.

The choice of centrality measure has important implications for how we think about the role of hubs in a brain network. A more general strategy is to compute a range of relevant topological measures that we think are characteristic of hubs and use a consensus approach to distinguish them from non-hub nodes. This approach ensures that our findings are not biased by any single concept of centrality.

In considering any of these measures, it is important to examine their distribution across network nodes. In so doing, we can determine whether a specific subset of nodes has an unusually high level of centrality, which would support a distinctive topological role for these network elements.

CHAPTER 6

Components, Cores, and Clubs

A **centrality** analysis aims to identify individual **nodes** that are topologically positioned to exert a strong influence on brain function. In this chapter, we extend this idea to consider *collections* of nodes and **edges** that play a central role in network organization and dynamics. These subsets of nodes and edges, which are also called **subgraphs**, represent putative backbones of network **topology**.

To see why network backbones are important, consider the global air transportation network, in which airports are nodes and direct flights between airports are edges (Figure 1.4a). The **hubs** of this system, such as London Heathrow, Los Angeles LAX, and Dubai International, are connected to many other airports. These hubs also have a large number of direct flights between them (Alstott et al., 2014b; Colizza et al., 2006). Many travelers will be familiar with this phenomenon: long-haul flights between regional airports that are located in different continents often involve a leg that links a hub in the continent of departure to a hub located in the destination continent. In other words, many routes between different nodes in the network involve a flight between two hub airports. More generally, we can say that the subgraph of **links** between hub airports defines a densely connected topological core that facilitates travel between regional airports.

Turning to the brain, we can similarly ask: are certain subsets of neural elements densely interconnected, forming an information-processing core akin to the hubs of the global air transportation network? This question is central to understanding how the brain integrates information across anatomically distributed subsystems. As we will see, there are a numerous graph methods for identifying topologically central subgraphs of nodes that represent candidate network cores, at different scales of topological resolution. We begin at the broadest scale, by trying to understand whether or not a network is fragmented, and consider how such an analysis can be used to understand the robustness of a network to attack or damage. We then focus on techniques that allow more refined distinctions between the core and periphery of a network, and examine how these methods have contributed to our understanding of integrative processes in brain networks.

Fundamentals of Brain Network Analysis. http://dx.doi.org/10.1016/B978-0-12-407908-3.00006-6

6.1   CONNECTED COMPONENTS

One of the most basic questions that we can ask about a network is whether all of its constituent nodes are interconnected such that they form a single component. In the parlance of **graph theory**, we ask whether the network is **node-connected**, or simply connected (Figure 6.1a). A node-connected network is one in which a **path** can be traced between any pair of nodes by following the edges of the graph. The minimal set of edges required to form a node-connected network is called the **minimum spanning tree** (MST; Box 6.1). Networks containing subsets of nodes that cannot be linked by a traversable path are called fragmented or disconnected, and each subset is called a **connected component** or, more simply, a component (Figure 6.1b). Fragmented networks can contain isolated or singleton nodes—nodes with zero **degree** that are not connected to any other system element (Figure 6.1b).

Most real-world networks contain one large component that encompasses a majority of nodes. For example, in scientific collaboration networks, where scientists are nodes and coauthorship of papers is represented by edges, the largest component for the discipline of biology contains 92% of nodes; similarly, it contains 85% in physics and 82% in mathematics (Newman, 2004b). The largest connected component of the film actor network, in which actors are nodes and links represent a costarring role in a movie, includes 90% of nodes (Watts and Strogatz, 1998). Due to its extent, the largest component is sometimes also referred to as a *giant component*. However, this term has a more specific meaning in the theory of random graphs such as those described by the Erdős-Rényi model, where a giant component refers to a component whose size grows in proportion to the number of nodes in the graph (Box 6.2).

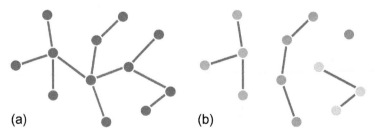

(a)   (b)

FIGURE 6.1 Node-connected and fragmented networks. **(a)** An example of a node-connected network. A path can be traced between any pair of nodes by following the edges of the graph. **(b)** This particular network can be fragmented after removing any one of the edges. Here, three edges have been removed to yield a network comprising four separate components: two consist of four nodes (blue and purple), one consists of three nodes (green), and the fourth comprises a single, isolated node with zero degree (red). A path exists between any pair of nodes within a component, but no paths can be found between pairs of nodes located in two distinct components.

BOX 6.1 THE MINIMUM SPANNING TREE

In a binary network, the MST is a subgraph comprising the minimum number of edges that ensures a path can be found between all pairs of nodes. Since we require at least $N-1$ edges to connect N nodes, the MST will always comprise $N-1$ links. In weighted networks, each edge is assigned a distance penalty that is some function of its weight, and the MST is the subset of edges that minimizes the total sum of these penalty values subject to a path existing between all node pairs. In brain networks, these values will typically represent some function of the reciprocal of the connectivity weight, such as $1/w_{ij}$, since we want strongly weighted edges to index greater topological proximity (see also Chapter 7).

The MST is not necessarily unique for a given network, unless that network is binary. Several different algorithms are available for finding the

MST of a graph. One popular method, called Kruskal's algorithm, is demonstrated in Figure 6.1.1.

By construction, the MST lacks many properties known to be important for the brain, such as clustering (Chapter 8) and **modularity** (Chapter 9). Examining its properties can nonetheless be useful, as it may represent a foundational backbone of the network (Alexander-Bloch et al., 2012; Stam et al., 2014). The MST is also useful when thresholding networks. By first finding the MST and adding edges to this backbone, we ensure that our resulting networks will be node-connected (Alexander-Bloch et al., 2010). Thresholding procedures are covered in Chapter 11.

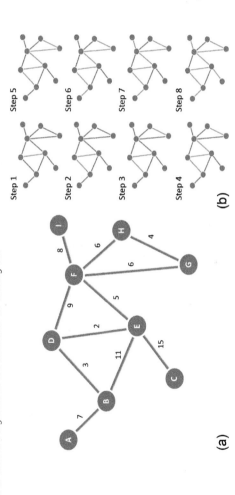

(a)

(b)

FIGURE 6.1.1 Kruskal's algorithm for finding the minimum spanning tree in a weighted network. **(a)** An example weighted network. The numbers next to each edge correspond to the edge weights. These weights are the result of a similarity-to-distance remapping (Chapter 7), which ensures the MST comprises strong connections. **(b)** The stepwise progression of Kruskal's algorithm. In step 1, the lowest weighted edge is selected first, which in this case is edge *DE* (red). In the remaining steps, the algorithm adds edges in ascending order of their remapped weight, provided that the edges do not form any **cycles** in the graph. For example, at step 5 there are two edges with a weight of 6: edge *FG* and edge *FH*. We only add edge *FG* because *FH* would close a triangle between nodes *F*, *G* and *H*, and thus form a cycle. The algorithm proceeds until no more edges can be added without closing a triangle. The MST corresponds to all the edges marked in red in the final step.

We should not be surprised that the largest connected component of many real-world networks includes a large fraction of nodes since connectedness between system elements is a prerequisite for integrated function. The Internet would fail in its goal of connecting people around the world if it comprised multiple fragmented subnetworks. The same can be said for the air transportation network, electric power grids, and most other networks.

BOX 6.2 COMPONENTS AND PERCOLATION IN RANDOM GRAPHS

In graph theory, a giant component is a connected component that contains a large proportion of the total number of nodes, and whose size grows in proportion to N. A giant component usually emerges rapidly as connections are added to a network. Imagine a network that has no connections between nodes. In this system, each node forms its own component, so the number of components in the network is N. As we add edges to the graph, nodes coalesce to form small, fragmented components that gradually link up with each other until a giant component emerges that encompasses a large fraction of nodes. The emergence of this large component occurs when the **connection density** passes the **percolation** threshold (Section 6.1.3).

In a seminal paper, Erdős and Rényi (1960) established conditions in which random graphs are likely to contain a giant component (further details on the Erdös-Rényi model are provided in Chapter 10). If p is the probability of finding a connection between a pair of nodes, a giant component is likely to exist when $p > 1/N$. In contrast, when $p < 1/N$, the graph is likely to be fragmented and contain many small components, the largest of which is in the order of $\log(N)$ nodes (Figure 6.2.1). At the transition point of $p = 1/N$, the largest component is of order $N^{2/3}$ nodes. These results are asymptotic, meaning that they only hold true for very large graphs.

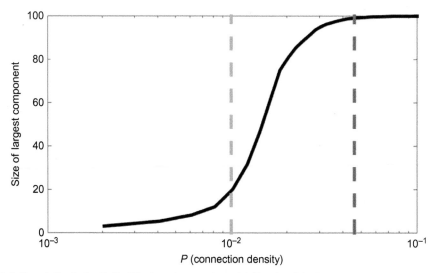

FIGURE 6.2.1 Percolation in the Erdős-Rényi random graph model. The size of the largest connected component (black) in an **Erdős-Rényi graph** ($N = 100$) shows a sharp increase with the connection density, p, corresponding to a phase transition at $p \sim 1/N$ (light gray dashed line). The graph is likely to be node connected when $p > \log(N)/N$ (dark gray dashed line).

BOX 6.2 COMPONENTS AND PERCOLATION IN RANDOM GRAPHS—CONT'D

To build an intuition for these results, we can equivalently state that a giant component is likely to exist when $\langle k \rangle > 1$, where $\langle k \rangle$ is the mean degree. In other words, each node need only possess one link on average to form a well-connected network. In a random graph, $\langle k \rangle$ and p are related as $\langle k \rangle = p(N-1)$. For sufficiently large N, the condition $\langle k \rangle > 1$ is satisfied when $p > 1/N$ (i.e., when it is likely that every node has at least one connection). Note that p corresponds to the connection density of an Erdős-Rényi graph.

Erdős and Rényi also established that random graphs are likely to be node connected, in the large N limit, when $p > \log(N)/N$ (Figure 6.2.1). Conversely, they are likely to be fragmented when $p < \log(N)/N$, with the largest of these fragmented components having a size in the order of $\log(N)$ nodes. When $1/N < p < \log(N)/N$, a giant component emerges, but the network will still contain smaller components, with the size of the largest of these other components being in the order of $\log(N)$ nodes.

Recall from Chapter 4 that the binomial **degree distribution** of Erdős-Rényi graphs does not accurately represent the distributions seen in many real-world networks. Percolation theory has been extended to random graphs with arbitrary degree distributions. The theory behind these graphs is discussed by Newman et al. (2001) and an analysis of percolation processes on these graphs is given by Callaway et al. (2000). Percolation theory for k-cores (Section 6.2.2) has been developed by Dorogovtsev and colleagues (Dorogovtsev et al., 2006a,b; Goltsev et al., 2006). A detailed overview of percolation and other critical processes on complex networks has been provided by the same group (Dorogovtsev et al., 2008).

Nervous systems are also characterized by a single large component. The largest component of the chemical synapse network of the *C. elegans* nervous system contains all neurons (Varshney et al., 2011). Similarly, the nodes of the 49-region Drosophila fruit fly **connectome** (Shih et al., 2015) and the 29-region cortical connectome of the macaque monkey (Markov et al., 2013a) constitute a single connected component. This is compatible with our expectation that any individual neural element (or subset of elements) in a brain network will not be isolated from other elements. In this section, we consider how components are defined in undirected and directed networks, how they can be used to understand the brain's robustness to injury, and how they can facilitate the analysis of population differences in brain networks.

6.1.1 Components in Undirected Networks

A component of an undirected network is a subgraph of vertices in which each member can be linked to every other member via one or more paths (Figure 6.1). The size of a component can be measured in terms of the number of nodes or edges it comprises. If size is measured in terms of nodes, it can attain any integer value between 1 and N. When the size of the largest component is equal to N, the network is node connected. When the size of the largest component is less than N, the network is fragmented and is comprised of multiple components.

Figure 6.1b depicts an example network that comprises four components. Within each component, it is possible to find a path between each pair of nodes, but no paths can be found between pairs of nodes located in two distinct components. Components of size one define isolated nodes that are not connected to any other network element and thus have degree of zero. Various algorithms are available for finding the components of a graph. One of the more popular methods, the breadth-first search, is described in Box 6.3.

6.1.2 Components in Directed Networks

In directed networks, we can distinguish between different types of components. One distinction is between weakly and strongly connected components. A *weakly connected component* is defined in the same way as a component is defined in an undirected network: as a subset of nodes in which every pair of members can be linked by one or more interconnecting paths, regardless of the direction of those edges. In this case, the directed network is analyzed without any regard for edge direction. A *strongly connected component* is defined with respect to the directions of the edges. Specifically, a strongly connected component is a subset of nodes in which there exists a directed path that runs in *both* directions between every pair of its constituent elements.

The differences between weakly and strongly connected components are illustrated in Figure 6.2. We see that every node in a strongly connected component must belong to at least one cycle—a closed loop of edges that points in the same direction around the loop. For example, take the strongly connected component shown in dark blue in Figure 6.2, which forms a cycle comprising four nodes and four edges. For each node i in this component, we can find a directed path that runs out of i and reaches every other node. Similarly, we can find a path starting in any other node that runs into node i—we can thus find paths between node pairs that run in both directions. If a node cannot be linked to any others in such a way, it forms its own strongly connected component of size one.

Directed networks also allow us to distinguish between out-components and in-components. An out-component is a set of nodes that can be linked to an index node i (and which includes i) via one or more directed paths emanating from that node. The definition of an out-component is thus node centric and depends on the choice of a specific index node. This means that any single node may belong to multiple out-components. Moreover, all members of a strongly connected component will necessarily have identical out-components. This is because all nodes that can be reached by any index node in a strongly connected component are, by definition, also reachable by other nodes in that component. The members of the strongly connected component of an index node are thus a subset of that node's out-component. These properties are illustrated in Figure 6.3a and c.

BOX 6.3 FINDING CONNECTED COMPONENTS IN NETWORKS

The connected components of a network can be found using the breadth-first search (BFS) algorithm. The BFS is an iterative technique for visiting all the connected elements of a graph. The algorithm starts at an arbitrary node, visits its **neighbors**, and then visits the unvisited neighbors of those neighbors and so on. The set of nodes traversed by the BFS will thus comprise a connected component. Figure 6.3.1 offers a step-by-step example of how the algorithm proceeds in a small, undirected network of eight nodes. If the network contains multiple components, we initialize the algorithm again in any node that was not visited by an earlier run of the algorithm. Alternative algorithms for finding connected components in undirected graphs (Dulmage and Mendelsohn, 1958), and to find strongly connected components in **directed graphs** (Sharir, 1981; Aho et al., 1983), are also available.

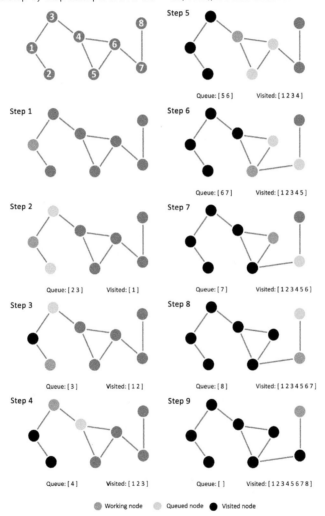

Queue: [5 6]   Visited: [1 2 3 4]

Step 1

Step 6
Queue: [6 7]   Visited: [1 2 3 4 5]

Step 2
Queue: [2 3]   Visited: [1]

Step 7
Queue: [7]   Visited: [1 2 3 4 5 6]

Step 3
Queue: [3]   Visited: [1 2]

Step 8
Queue: [8]   Visited: [1 2 3 4 5 6 7]

Step 4
Queue: [4]   Visited: [1 2 3]

Step 9
Queue: [ ]   Visited: [1 2 3 4 5 6 7 8]

● Working node   ● Queued node   ● Visited node

FIGURE 6.3.1 Example of the breadth-first search algorithm. Top left panel shows an example network. The remaining panels show the step-by-step progression of the breadth-first search algorithm. In step 1, an index node (here, node 1) is arbitrarily selected as a 'working' node. In step 2, this node is marked as visited and its neighbours—nodes 2 and 3—are placed in a queue. The enqueued nodes are then sequentially selected as the next working nodes. In step 3, we select node 2 first. The only neighbor of this node is node 1, which has already been visited. In step 4, we mark node 2 as visited and proceed to node 3. Node 4 is the sole neighbor of node 3, so node 4 is enqueued. In step 5, node 3 is marked as visited and node 4 becomes the working node. In step 6, node 4 is removed from the queue and marked as visited. We arbitrarily select one of node 4's neighbours as the working node (here, node 5) and enqueue the other neighbor (node 6). In steps 7 to 9, the algorithm proceeds to visit each remaining node of the network in the same way, enqueuing only unvisited nodes at each step, until all nodes in the component have been visited.

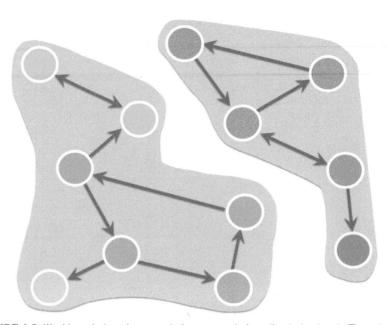

FIGURE 6.2 Weakly and strongly connected components in a directed network. The network depicted here comprises two weakly connected components, represented by gray shading, and five strongly connected components, represented by different node colors. A path can be found between each pair of nodes comprising a weakly connected component. A strongly connected component requires that bidirectional paths exist between each pair of nodes. The green and red nodes form their own strongly connected components because they are only linked to other nodes via a path that goes in one direction.

An in-component is the set of nodes from which there exists one or more directed paths that link to an index node *i*. In-components have similar properties to out-components; namely, they depend on the specific element selected as the index node; an individual node can belong to more than one in-component; and all members of the same strongly connected component will have the same in-component. Nodes comprising the strongly connected component of the index node are a subset of its in-component. These properties are illustrated in Figure 6.3b and d.

Given these characteristics, we can relate out-components, in-components, and strongly connected components in the following ways (Newman, 2010):

1. All members of a strongly connected component will be part of each other's out-component and each other's in-component.

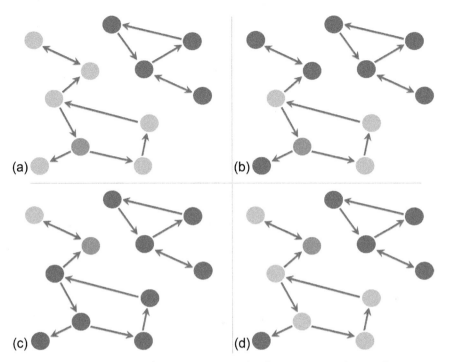

FIGURE 6.3 In-components and out-components of a directed network. Top row illustrates the out-component **(a)** and in-component **(b)** of an index node (red), within the same network depicted in Figure 6.2. The out-component (orange, **a**) is the set of all vertices that can be linked via an outward path emanating from the index node. The in-component (orange, **b**) is the set of all vertices that can be linked via a path that points towards the index node. Bottom row shows the out-component **(c)** and in-component **(d)** for a different index node. The out-component for the node in the top row includes all nodes in the same weakly connected component, while the in-component only includes nodes in the same strongly connected component (see also Figure 6.2). In contrast, the in-component for the index node in the bottom row contains nearly all vertices in the same weakly connected component, whereas the out-component of this node only contains one other node. Comparing with Figure 6.2, we also see that the strongly connected component of an index node is given by the intersection of its in-component and out-component.

2. Any node that is in both the out-component and the in-component of an index node will be in the same strongly connected component, since paths between the two nodes exist in both directions.
3. The intersection of a node's out- and in-components defines the strongly connected component of that node.

6.1.3   Percolation and Robustness

The sizes of connected components within a network can be used to understand the robustness of the system to node damage or failure. This is done by computationally and selectively "lesioning" individual nodes by removing all of their connections from the **adjacency matrix** and recomputing the size of the largest component. If we repeat this process one node at a time, we can examine the capacity of a network to maintain connectedness under different kinds of structural perturbation or "attack". We can "lesion" individual edges in a similar way. This virtual lesion methodology does not seek to model any pathophysiological details of the lesion. Rather, it aims to understand how the focal disruption of nodes and edges affects the broader brain network.

In graph theory, **percolation** refers to processes that occur on a network as nodes and/or edges are added or deleted. Percolation theory emerged from the physical study of how a fluid passes, or percolates, through a porous material. The material is often modeled as a **lattice**. The nodes in this lattice correspond to different sites or locations on the material and the edges represent bonds. These bonds can either be open, allowing passage of the fluid, or closed, obstructing fluid passage. Percolation in this system occurs if there is a path of open bonds that exists from one side of the lattice to the other. This process is the same as the lattice forming one large, connected component.

As we remove more nodes or edges from a network we often find that there is some critical percolation threshold, p_c, at which the network fragments. Above this threshold a large component emerges that includes a major fraction of nodes. Below this threshold, the network is fragmented into multiple smaller components. The same is true if we move in the other direction, by starting with an empty network and gradually adding connections between nodes. In fact, Erdös and Rényi were the first to use this approach to study percolation in random graphs, finding that a giant component suddenly appears at a certain connection density (Box 6.2). Figure 6.4 illustrates similar behavior in the connectomes of the fruit fly, mouse, and macaque. In each case, we start with an empty adjacency matrix and gradually add connections by populating the elements of our adjacency matrix in descending order of connection strength. At each iteration we compute the size of the largest component. Across all networks, only a small proportion of connections ($\leq$10%) need to be added before we observe a large component that encompasses most network nodes.

This rapid shift from **fragmentation** to the emergence of a large component at the percolation threshold implies a phase transition: a rapid change in the state of a system at some critical value of a tunable parameter. A classic example of a phase transition is the change in state of water from liquid to gas as temperature is increased: once the temperature (tunable parameter) reaches the boiling

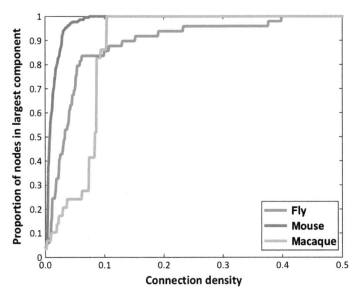

FIGURE 6.4 Connectedness in brain networks. This figure plots the size of the largest connected component in the 49-region connectome of the Drosophila fruit fly (Shih et al., 2015), the 213-region connectome of the mouse (Oh et al., 2014), and the 29-region cortical connectome of the macaque (Markov et al., 2013a), as a function of connection density. For each network, we gradually add edges in decreasing order of weight and plot the size of the largest weakly connected component. In all networks, the size of the largest component increases rapidly as a function of increasing connection density and we only need to add a small proportion of connections before the largest component spans all network nodes and the network is node connected.

point of water, the volume of liquid dramatically reduces as liquid turns to gas (change in state); another classic example is the Ising model of ferromagnetism, which is discussed in Box 4.1. In the Erdös-Rényi graph, the tunable parameter is the probability of connection between nodes (i.e., the connection density), and the change in state corresponds to an abrupt change in the size of the largest component (Box 6.2).

We can analyze percolation processes on networks to simulate the effect of progressive damage to individual brain regions or connections that may result from disease or injury. Network topologies that withstand fragmentation as a greater proportion of nodes and connections are removed have a greater topological robustness. The ability to withstand fragmentation may thus explain how individual brains vary in their resilience to pathological attack. This kind of analysis may also allow us to model which specific disruptions best capture the brain changes seen in a clinical population, and make predictions about the level of impairment that might be expected following insult to specific network elements (Fornito et al., 2015; Lo et al., 2015).

Two types of attack on the network are often simulated in these analyses: random and targeted. Random attack involves the removal of nodes or edges with uniform probability. Targeted attack involves the removal of network elements according to some prespecified criterion, such as node degree or some other index of topological centrality (Chapter 5). The attacks can be limited to a specific subset of nodes or edges, or they can proceed incrementally to examine how network organization changes as an increasing fraction of nodes and/or edges is deleted.

In communication and transportation networks, it is easy to see how random or targeted attack of a network *in silico* can be used to model real-world failures or attacks. Random attack models stochastic processes that may cause failure in individual network elements, such as component malfunction, operator error, or natural disaster. Targeted attack can be used to predict the consequences of criminal or terrorist sabotage of network hubs, such as attacks on the New York Stock Exchange or London Heathrow Airport. It is interesting to note that the first ever Wi-Fi network, called ALOHAnet, built in 1971, was designed as a star-like topology with a central hub node that allowed for efficient transmission of data between nodes at the periphery (see Figure 5.1). These days, computer networks rarely use star-like architectures because the central node represents a single point of failure. If the switch, hub, or computer at this node fails or is attacked, the entire network is disabled.

In neuroscience, simulating random attack allows us to understand how the brain might respond to unexpected pathological perturbations that affect specific network elements, such as those caused by stroke, tumor, or traumatic injury. Targeted attack can be used to understand how removal of specific elements, which may be preferentially affected by certain diseases (e.g., the caudate nucleus in Huntington's disease or the substantia nigra in Parkinson's disease), impacts the remaining network.

A common method for characterizing network robustness is to compute the size of the largest connected component after removal of each node and its incident connections, or after removal of each individual edge. The extent of network fragmentation after each "lesion" can be quantified as the fraction of nodes in the largest component of the graph (Albert et al., 2000). A measure of fragmentation that takes into account all components (not just the largest one) is given by the ratio of the sum of the number of node pairs in each connected component to the total number of node pairs in the entire network (Chen et al., 2007),

$$F = 1 - \frac{\sum_{j=1}^{m} N_j(N_j - 1)}{N(N-1)}, \tag{6.1}$$

where m is the number of components in the fragmented network, and N_j is the number of nodes in component j. The ratio F equals zero for a node-connected network and equals one for a completely fragmented network that contains no edges, such that each node forms its own cluster with $m = N$ and $N_j = 1$.

The robustness of a network to damage depends strongly on its degree distribution. Recall from Chapter 4 that a network's degree distribution describes how connectivity is distributed across its constituent nodes. Also recall that we can use this distribution to categorize networks into one of three broad classes: **single scale**, **scale free**, and **broad scale**. Albert et al. (2000) examined the resilience of single-scale and scale-free systems to random and targeted attack. Compared to single-scale networks, scale-free topologies showed a greater resilience to random failure, maintaining connectedness until about 75% of nodes were removed. However, scale-free networks demonstrated a greater vulnerability to targeted attack of the highest degree nodes, fragmenting at a lower percolation threshold (i.e., after removal of just 18% of nodes, compared to 28% for single-scale networks). The authors found that the Internet and World Wide Web displayed a robustness to random failure and vulnerability to targeted attack that was comparable to the scale-free model (see also Figure 6.5).

To understand why the robustness of single-scale and scale-free networks differed in the analysis of Albert et al. (2000), consider what the degree distributions of these classes of networks tell us about their topology. In scale-free systems, most nodes have low degree. Only a relatively small number of nodes, represented in the tail of the degree distribution, have a very high degree. This means that the probability of choosing such a highly connected node at random is very low, and we have a much higher chance of selecting a low-degree node whose deletion will have a limited effect on network integrity. In contrast, the scale-free topology is vulnerable to targeted attack precisely because many of the network's links are attached to the hubs. Selective attack of these high-degree nodes will rapidly fragment the network. In a single-scale network, most nodes have degrees that are close to a single, modal value, and large deviations from this mode are very rare. This means that the disparity in the connectivity of a high-degree node and low-degree node in a single-scale network is much smaller than in a scale-free system. Therefore, a targeted attack of high-degree nodes in a single-scale system will make a smaller impact on network integrity. On the other hand, the homogeneity of degree across the nodes of a single-scale system makes it more vulnerable to random attack.

It is possible to use the relationship between robustness and degree distribution to derive analytic percolation thresholds for random graphs with various degree distributions (Callaway et al., 2000). For example, it has been shown that for scale-free networks with a power-law exponent $\gamma < 3$, the fraction of nodes that must be removed before the network fragments, p_c, approaches one as $N \to \infty$ (Cohen et al., 2000). In other words, in the limit of large N, scale-free networks with $\gamma < 3$ will rarely (if ever) fragment under random attack. A large number of scale-free networks found in the real world are characterized by a power-law exponent that ranges between 2 and 3 (Barabási and Albert, 1999; Newman, 2005a), underscoring the practical relevance of this result. Indeed, even though real-world networks have a finite size, we have seen that they display remarkable robustness to random failure. For example, it has been estimated that $p_c > 0.99$ for random attack of a network model of the Internet with $N > 10^6$ and power-law exponent $\gamma \sim 2.5$ (Cohen et al., 2000). Theoretical studies also confirm that in a targeted attack of power-law networks, only a small fraction of nodes must be removed before the graph fragments (Callaway et al., 2000; Newman, 2003a).

In Chapter 4, we considered evidence that brain networks are neither single scale nor scale free, but may instead show a broad-scale degree distribution. What does this mean for the robustness of nervous systems? Achard and colleagues (2006) were the first to address this question in their study of human whole-brain resting-state **functional connectivity** networks. They found that these networks showed comparable resilience to random attack, when compared with random and scale-free networks. However, the brain was more resilient to targeted attack than the scale-free network: removal of only 20% of the most highly connected nodes in the scale-free network reduced the size of the largest connected component by 50%, whereas the brain showed the same level of fragmentation only after the top 40% most connected nodes had been removed (Figure 6.5a). This differential vulnerability occurs because the concentration of links on hub nodes is weaker in a broad-scale network compared to a scale-free system.

Fragmentation is not the only measure that can be used to examine network resilience to attack. Other studies have used average **path length** (Chapter 7) and related measures (Achard et al., 2006; Albert et al., 2000; Kaiser et al., 2007b; van den Heuvel and Sporns, 2011), as well as computational models of neuronal dynamics simulated on structural network architectures to examine the dynamical consequences of lesions to specific nodes (Alstott et al., 2009; Honey and Sporns, 2008; Young et al., 2000). In this work, model-based estimates of functional connectivity are compared before and after specific elements of the structural network are lesioned, allowing the identification of nodes whose removal has a major impact on network function (Figure 6.5b). Models of cascading failures in networks, in which dysfunction of one node can propagate to affect other nodes, have also been developed (Watts, 2002). These models have not been extensively applied to the brain but they may be particularly useful for understanding how disease processes spread throughout a connectome (de Haan et al., 2012; Raj et al., 2012).

We can also use robustness analysis to understand which specific network elements are important in promoting certain network properties, such as topological integration or segregation. In this case, we remove individual nodes or edges and determine the degree to which their removal impacts some global topological or functional property of the network. This analysis is closely related to the concept of delta centrality discussed in Chapter 5, and allows us to identify central neural elements whose removal may exert a disproportionate impact on network integrity – putative Achilles' heels of a nervous system (Fornito et al., 2015; see also Figure 5.4).

6.1.4 Components and Group Differences in Networks

The analysis of network components has several other uses in neuroscience. One application is to guide the selection of an appropriate thresholding strategy when constructing an adjacency matrix. If we reasonably assume that the brain is a node-connected system and that no neural element or subset of elements should be isolated from any other, we can constrain our approach so that we do not use a threshold that fragments the network. For example, we could find the MST of the graph (Box 6.1) to ensure connectedness, and then add connections in descending order of weight to achieve a desired connection density (Chapter 11).

Another fruitful application of component analysis has been in statistical methods for dealing with the **multiple comparisons problem** in connectome-wide analyses. In this work, statistical tests are applied at each network edge, and an appropriate threshold for determining statistical significance must be chosen to maintain control over **familywise error** while retaining adequate statistical power. Traditional methods, such as the Bonferroni or **false discovery rate** corrections are overly conservative because they treat each test independently. An alternative method, termed the **network-based statistic**, evaluates the null hypothesis at the level of connected components of edges showing a common effect. This method can provide a considerable gain in statistical power (Zalesky et al., 2010a). The network-based statistic and related approaches are discussed in Chapter 11.

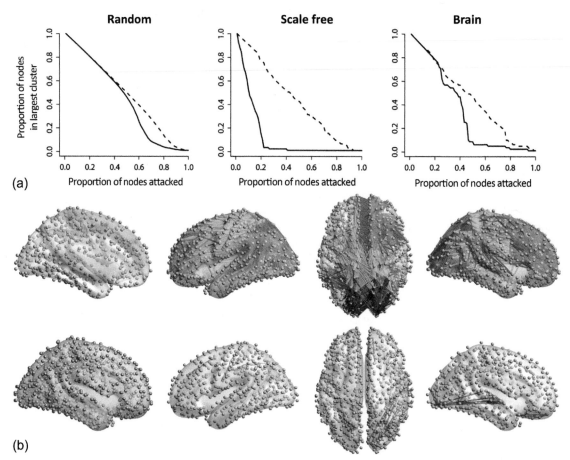

(a)

(b)

FIGURE 6.5 The effects of random and targeted attack on brain network structure and dynamics. **(a)** Largest component size as a function of the proportion of nodes (and their incident edges) removed via random or targeted attack (dashed and solid lines, respectively) in Erdös-Rényi (left) and scale-free (middle) graphs compared to a 90-region human functional connectivity network measured with resting-state **functional MRI** (right). In this study, the degree distribution of the brain was best characterized by an exponentially truncated power law, consistent with a broad-scale distribution (Chapter 4). Here, we see that, relative to the scale-free network, the brain shows greater robustness to targeted attack and comparable resilience to random node deletion. **(b)** The effects of focal lesions on model-based estimates of network functional connectivity. A computational model of large-scale brain dynamics was simulated on a 998-region human **structural connectivity** network empirically derived from **diffusion MRI**. Lesions were simulated by removing a set of spatially contiguous nodes, and model-based estimates of functional connectivity were compared before and after the lesion. Top row shows the location (left, green) of the lesioned nodes and the resulting effects on functional connectivity (right). The lesioned area had high **betweenness** centrality (Chapter 5). Edges in red correspond to links where functional connectivity was decreased following the lesion; edges in blue correspond to links where functional connectivity was increased after the lesion. Bottom row shows the results of the same analysis after lesioning an area with low betweenness centrality. Lesioning the central region (top) results in widespread and complex changes involving both increased and decreased functional connectivity. In comparison, lesioning the noncentral area (bottom) results in spatially circumscribed functional changes. *(a) Reproduced from Achard et al. (2006) with permission.* *(b) Reproduced from Fornito et al. (2015), using data and methods reported in Alstott et al. (2009), with permission.*

6.2 CORE-PERIPHERY ORGANIZATION

Connected components offer a fairly coarse description of the core of a network. In particular, since most real-world networks, including the brain, comprise one large component that spans most nodes, component analysis does not allow us to identify subsets of network elements that act as a critical backbone or information-processing core—a densely connected and topologically central subgraph that exists within the node-connected component. Returning to the example of the global air transportation network, we cannot deduce anything about the central role of hub airports such as Dubai International in routing airline traffic if the only thing we know about the network is that it forms a single, connected component.

In general, a network with a clear distinction between core and topologically peripheral nodes will show the following properties (Borgatti and Everett, 2000):

1. Core nodes should occupy a topologically central position in the network.
2. Core nodes should be highly interconnected with each other.
3. Peripheral nodes should be (at least) moderately connected to core nodes, but sparsely interconnected with each other.

Many real-world networks show evidence of such a core-periphery organization (Csermely et al., 2013; Kitano, 2004). In particular, it has been suggested that a core-periphery architecture emerges in a network when resources are scarce and there is some **cost** involved in forming connections between nodes (Csermely et al., 2013). The core may thus provide a cost-effective solution for the integration of distributed network elements in the periphery. A strong core-periphery organization optimizes robustness to random node failures (Peixoto and Bornholdt, 2012), because the core becomes a focal point for network integration. The high interconnectivity of core nodes also promotes **degeneracy** and thus robustness to node failures (Kitano, 2004). In this context, degeneracy refers to the capacity of different elements of a system to perform the same function, and can be distinguished from **redundancy**, which involves the repetition of identical system elements (Tononi et al., 1999). Degeneracy supports resilience to perturbation, but in a network where degeneracy is driven by a topological core, this robustness may come at the cost of vulnerability to targeted attack (Priester et al., 2014). These two characteristics of core-periphery organization—susceptibility to resource limitations and robustness to perturbation—are critical constraints on brain network topology (Bullmore and Sporns, 2012). It is thus logical to consider whether brain networks are organized into one or more cores surrounded by a sparsely connected periphery.

In this section, we consider some of the main approaches used for defining the core and periphery of brain networks. If network components describe the global or macroscale topology of a network, and measures of node centrality (Chapter 5) characterize the topological microscale, we can think of the dichotomy between core and periphery as a mesoscale topological property (Rombach et al., 2014).

6.2.1   Maximal Cliques

Core-periphery analysis has a long history in social network science, where many algorithms have been proposed to identify cohesive subgraphs within a broader network of social ties. Early definitions focused on maximally connected subsets of nodes, called **maximal cliques** (Luce and Perry, 1949). A maximal clique is a set of nodes in a network that is fully connected—each node in this set has a direct link to every other node in the clique, and the set is not a subgraph of any other clique. In other words, the connection density of the nodes in a clique is maximal.

There is some evidence for strong clique organization in brain networks. For example, an analysis of a dense 29-region macaque interregional cortical connectome constructed from invasive **tract-tracing** experiments identified 13 different cliques, each comprising 10 nodes (Ercsey-Ravasz et al., 2013; Figure 6.6a). Each clique comprised a set of nodes that was linked by every possible directed connection—it was a fully connected subgraph. These cliques spanned 17 unique areas, which were collectively interconnected with a density of 92% (Figure 6.6b). This density was much higher than the 54% connection density found between the 17 core nodes and the remaining 12 regions in the periphery, and the 49% connection density of the edges linking the 12 peripheral areas (Figure 6.6c). We thus see evidence of each of the three characteristics of core-periphery structure outlined at the start of this section: a densely connected core that is moderately connected to peripheral nodes, and a periphery that is sparsely connected with itself. However, defining the core of a network based on an analysis of maximal cliques will result in a very narrow definition, since maximal cliques tend to be small in most real-world networks, particularly if these networks are sparse. In the following sections, we consider less restrictive definitions of core and periphery.

6.2.2   *k*-Cores and *s*-Cores

To enable a more general characterization of network core structure than a consideration of its cliques, Seidman (1983) proposed a method called k-core decomposition. In this approach, nodes with degree less than k are removed until all remaining nodes in the network have at least k connections. That is,

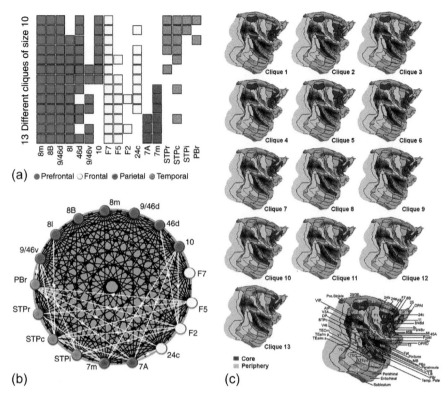

FIGURE 6.6 Maximal cliques of the macaque cortical connectome. **(a)** Cortical regions implicated in each of 13 cliques of size 10. Each node within a clique is directly connected to every other node. Each row in the plot corresponds to a different clique and columns represent the 17 different cortical areas implicated in the network core. The network comprised a total of 29 regions. Squares indicate the membership of each region to each clique. **(b)** Connectivity between the 17 cortical regions belonging to the core of the macaque cortical connectome. The connection density of this network is 92%. White edges represent links where there was no connection between a region pair. **(c)** Cortical flat maps showing the anatomical distribution of areas belonging to each clique, and the network core and periphery. *Images reproduced from Ercsey-Ravasz et al. (2013) with permission.*

the k-core is the subgraph comprising nodes with a degree of at least k (Eidsaa and Almaas, 2013). This definition conforms to the intuition that a network core should comprise a set of highly connected, and mutually interconnected, vertices.

The k-core decomposition algorithm proceeds as follows. In the first iteration, all nodes with $k = 1$ and their incident edges are removed from the graph. The degree of the remaining nodes is recalculated and only the remaining nodes with degree greater than 1 are retained to define the 2-core of the graph. On the second iteration, nodes with $k \leq 2$ (and their edges) are removed to define the 3-core. In the third iteration, nodes with $k \leq 3$ (and their edges) are removed to define the 4-core and so on. At each step, if removal of a node and its edges causes the degree of one of its neighbors to drop below k, then that neighbor is also removed from the graph. The algorithm proceeds until all nodes have a degree that is less than k (Figure 6.7). In general, multiple disconnected cores may be present at any given level of k, the $(k+1)$ core is a subgraph of the k-core, and any maximally connected cliques of the graph are contained within a given k-core.

Based on this procedure, we can assign a core index to each node, defined as the value of k corresponding to the highest level k-core to which the node belongs. Higher values point to a more central role within the network core. This index is also sometimes referred to as the shell of a node, since we can imagine a k-core decomposition as unpacking a network into a series of layers, much like an onion, where we have a densely interconnected core (the subgraph present at the highest level of k) surrounded by concentric shells of less central nodes (Garas et al., 2012; Figures 6.7 and 6.8a and b). The core index of a node also provides a measure of its topological centrality (Chapter 5). In general, any

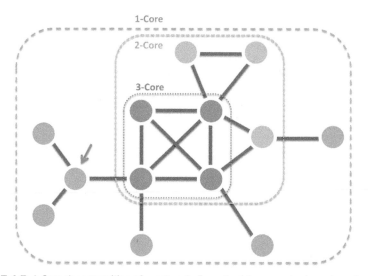

FIGURE 6.7 *k*-Core decomposition of a network. Example of the core structure of a network, with different colors indicating the *k*-core to which each node belongs. Arrow highlights a node that belongs to the 1-core despite having a degree of 3 because removal of other nodes in the 1-core (the blue neighbors) reduces the degree of this node to 1. *Image adapted from Alvarez-Hamelin et al. (2008) with permission.*

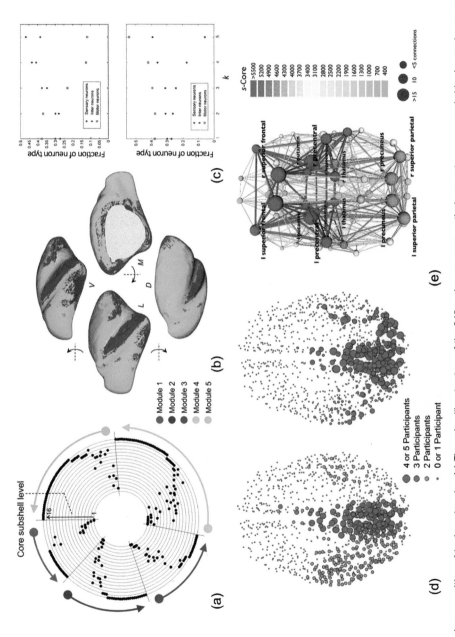

FIGURE 6.8 Core decomposition of brain networks. (a) The onion-like structure of the 242-region macaque cortical connectome, as revealed by *k*-core decomposition of a connectivity **matrix** reconstructed from the collation of published tract-tracing findings. Each concentric circle corresponds to a different *k*-core or shell, and dots represent the shell to which each region belongs. Regions are ordered according to their topological **module**, as indicated by the different colors shown around the perimeter of the ring. **(b)** The anatomical distribution of each module for the network analyzed in **(a)**. **(c)** *k*-Core decomposition of the *C. elegans* nervous system performed separately for synaptic **out-degree** (top) and **in-degree** (bottom) revealed that sensory neurons constitute the innermost cores based on out-degree and motor neurons occupy the inner-most cores based on in-degree. **(d)** *K*-Core (left) and *s*-core (right) decomposition of human cortical structural connectivity networks reconstructed with diffusion MRI reveals remarkable consistency in the brain regions showing the highest core index scores. These regions are principally located in medial posterior areas. Red circles represent the consistency with which each of 998 regions was implicated in the innermost core across five different people. **(e)** An *s*-core decomposition of human structural connectivity networks measured in a larger sample of healthy individuals replicated the prominence of medial posterior areas, while also implicating a more distributed network of brain regions. *Images in (a) and (b) reproduced from Harriger et al. (2012), (c) from Chatterjee and Sinha (2008), (d) from Hagmann et al. (2008), and (e) from van den Heuvel and Sporns (2011), with permission.*

measure that quantifies the participation of a node in the network core—that is, a measure of node *coreness*—is also an index of centrality since the core is, by definition, a subset of topologically central vertices. However not all centrality measures index coreness (Borgatti and Everett, 2000); for example, a node with high betweenness centrality but low degree (see Figure 5.2 for an example) is unlikely to belong to an inner k-core of a network (a different perspective on the role of nodes with high betweenness centrality in the core of a network is considered in Section 6.2.4).

The k-core decomposition algorithm can be generalized to weighted networks by replacing k with node **strength**, denoted with s (Eidsaa and Almaas, 2013; Garas et al., 2012; node strength is defined in Chapter 4). An s-core decomposition defines the s-core of a network as comprising all nodes with strength greater than s, where s is a threshold value defined by the minimum strength in the subgraph. Because this definition does not account for node degree, s-core and k-core decompositions may diverge, and the innermost layers of an s-core decompositions may not necessarily be subgraphs of lower order k-cores (Eidsaa and Almaas, 2013).

Both k-core and s-core decomposition can be generalized to directed networks by separately considering cores for in-degree/strength and out-degree/strength. For example, Chatterjee and Sinha (2008) performed separate k-core decompositions of the incoming and outgoing synaptic connections of the *C. elegans* nervous system. They found that the innermost core for synaptic in-links comprised mostly motor neurons whereas the innermost core for synaptic out-links consisted mainly of sensory neurons (Figure 6.8c). This result was consistent with their finding that sensory neurons have low in-degree whereas motor neurons have low out-degree. Neurons belonging to the lateral ganglion were implicated in the innermost in-cores and out-cores, suggesting that these neurons represent critical hubs that link sensory and motor function in the worm.

Hagmann and colleagues (2008) were the first to perform a core decomposition of the human connectome in their analysis of undirected structural connectivity networks measured with diffusion MRI. They found that the regions most consistently implicated in the core across the five participants studied were located in the posterior medial cortex, particularly the posterior cingulate and precuneus. Similar results were obtained using s-core decomposition (Figure 6.8d). This finding was later replicated and extended to include frontal and lateral cortical regions, as well as subcortical areas (van den Heuvel and Sporns, 2011; Figure 6.8e). The same analysis also revealed that regions with a high core index have high topological centrality, and that they play a critical role in linking nodes that belong to different modules. This result is consistent with the hypothesis that these areas act as a communication backbone for the network.

6.2.3 Model-Based Decomposition

Clique and core decompositions offer a data-driven method for identifying putative network cores. An alternative approach is to pose a model of an idealized core-periphery structure, and quantify the extent to which this model structure is expressed in the observed network. Borgatti and Everett (2000) were the first to develop such an approach, using a simple **block model** to define an idealized core-periphery architecture. Block modeling is conducted at the level of the adjacency matrix of a network. It involves permuting the rows and columns to fit, as much as possible, some ideal form (model) of the matrix.

Figure 6.9 shows the adjacency matrices for some possible block models of core-periphery structure using a hypothetical network of 50 nodes. The left panel depicts the most extreme form of core-periphery organization: nodes 1-10 form the core, being maximally connected to each other as well as all other nodes. Nodes 11-50 occupy the periphery, being connected to the core yet unconnected with each other. This structure corresponds to a generalization of a simple star or hub-and-spoke graph, where each node of the core is connected to all others, and nodes in the periphery do not connect with each other.

The right panel of Figure 6.9 depicts a less extreme model of core-periphery organization. In this case, the core is still maximally connected and the nodes in the periphery do not connect with each other, but connectivity between the core and periphery is not complete (i.e., not all connections are present). In the

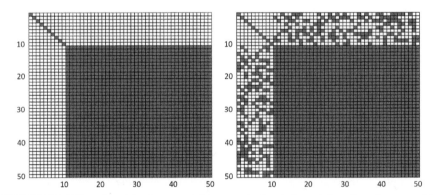

FIGURE 6.9 Different models of core-periphery organization. Left panel shows the adjacency matrix of a 50-node network with strong core-periphery organization, in which 10 nodes (the first 10 rows and columns) are maximally connected with each other and with all other nodes, whereas the other nodes are not connected with each other. White elements denote a connection; gray elements denote an absent connection. Right panel shows an adjacency matrix depicting moderate core-periphery organization, in which the 10 nodes comprising the core are maximally connected, and are only moderately connected with peripheral nodes. Peripheral nodes are not connected to each other. *The models have been adapted from Borgatti and Everett (2000).*

language of block modeling, the core-core region is called a 1-block (all $A_{ij}=1$); the core-periphery region is an imperfect 1-block; and the periphery-periphery region is a 0-block (all $A_{ij}=0$). This is perhaps a more realistic network model than the one shown in the left panel of Figure 6.9, but the model still makes strong assumptions about network structure; namely, that all core nodes are maximally connected and all periphery nodes are unconnected. It is also possible to specify variants of these models that do not make such strong assumptions.

Once a model is selected, our goal is to see how well the model fits the experimental data. To this end, we must find a class vector, c, of length N, that defines the class to which each node belongs. For example, using the model shown in the left panel of Figure 6.9, we can define the class vector such that $c_i=1$ if node i belongs to the core and 0 otherwise. From this vector, we can construct a so-called pattern matrix in which the elements $\delta_{ij}=1$ if $c_i=1$ or $c_j=1$, and $\delta_{ij}=0$ otherwise (note that this rule can be varied to accommodate other models of core-periphery organization). We can then compute the similarity between our pattern matrix and empirical adjacency matrix as

$$\rho = \sum_{ij} A_{ij}\delta_{ij}. \tag{6.2}$$

The element-wise product $A_{ij}\delta_{ij}$ will only equal one if two conditions are met: (1) there is a direct connection between nodes i and j in the original network (i.e., $A_{ij}=1$); and (2) either i and/or j belong to the core, as determined by the class vector c (i.e., either $c_i=1$ and/or $c_j=1$). The similarity index ρ counts the number of edges for which both conditions are met. High values indicate a strong correspondence between the pattern matrix and the real data. In fact, ρ corresponds to an unnormalized Pearson correlation coefficient applied to matrices, so we could also say that a network displays core-periphery structure to the extent that its topology correlates with the architecture of the idealized model (Borgatti and Everett, 2000).

To fit the idealized model of core-periphery structure to an empirical network, we need to find a class vector c that maximizes the quantity ρ. This vector can be found using combinatorial optimization techniques, such as **simulated annealing** or genetic algorithms. The basic idea is to shuffle the vector c, such that the total number of one and zero entries is preserved, but their order is randomized. A good optimization technique will shuffle c again and again in such a way that each successive shuffle incrementally increases the similarity between the pattern matrix and adjacency matrix.

In principle, we could determine the statistical significance of ρ by repeating the process in an ensemble of matched random graphs to generate an empirical null distribution of fit statistics (see Chapters 10 and 11). This is important because we can always find an optimal permutation of c that maximizes the similarity between our pattern matrix and adjacency matrix, regardless of whether or not the empirical data contain a clear core-periphery structure. Generating a null distribution for the similarity measure ρ allows us to statistically quantify the probability of obtaining a particular level of similarity by chance. Approaches for generating such null distributions for networks are covered in Chapter 10.

This model-based approach can be extended to handle directed and weighted networks, to accommodate continuous rather than discrete distinctions between core and periphery, and to describe different models of interaction between core and periphery (Borgatti and Everett, 2000). However, one limitation is that it assumes a fixed core size. We must therefore make a strong prior assumption about the number of vertices that we expect will constitute the network core. We must also make some assumptions about the sharpness of the distinction between core and periphery, such as whether the distinction is discrete or continuous.

To overcome these limitations, Rombach et al. (2014) developed a general method to account for cores of different sizes and quality, where quality refers to the sharpness of the transition between core and periphery. Specifically, they define the class vector c^* according to a transition function,

$$c_i^*(\alpha, \beta) = \begin{cases} \dfrac{i(1-\alpha)}{2\lfloor \beta N \rfloor}, & i \in \{1, \ldots \lfloor \beta N \rfloor\} \\ \dfrac{(i-\lfloor \beta N \rfloor)(1-\alpha)}{2(N-\lfloor \beta N \rfloor)} + \dfrac{1+\alpha}{2}, & i \in \{\lfloor \beta N \rfloor + 1, \ldots N\} \end{cases} \tag{6.3}$$

The first case in this equation runs over the peripheral nodes, which are the nodes with index $i = 1, \ldots, \beta N$. The second case runs over the core nodes. The tunable parameter β takes values in the range [0, 1] and sets the size of the core, such that lower values result in a more inclusive definition of the core (i.e., when $\beta = 0$, all N nodes are in the core). It can be seen that the first case sums over an empty set when $\beta = 0$, and thus there are no peripheral nodes for this parameter choice. The other tunable parameter, α, sets the size of the difference in scores between the lowest-ranked core node and highest-ranked peripheral node. When $\alpha = 1$, nodes will be discretely assigned to either the core or periphery. In particular, when $\alpha = 1$, the upper case (periphery) in Equation (6.3) is zero, while the lower case (core) simplifies to one. If $\alpha < 1$, we obtain a more graded or fuzzy categorization. We can think of this value as a measure of core quality, since a sharp distinction between core and periphery nodes will be associated with high values of α. We can therefore tune α to account for both discrete and continuous core-periphery architectures.

Figure 6.10 shows the pattern matrix resulting from various combinations of α and β values. In each matrix, the vector c has been ordered from core to peripheral nodes, so that core nodes appear in the top left corner. For constant α, we see that higher values of β result in a larger core. Similarly, for any given value of β, higher α will result in a sharper boundary between nodes defined as core and periphery. For example, at $\alpha = 1$, nodes are either assigned to the core or periphery, with no intermediate values. Rombach et al. (2014) also discuss other possible transition functions that can be substituted for Equation (6.3).

For a specific combination of α and β, we can define a quality function that quantifies the fit between the model and data in a similar fashion to Equation (6.2),

$$\rho_{\alpha,\beta} = \sum_{i,j} A_{ij} c_i^* c_j^*. \qquad (6.4)$$

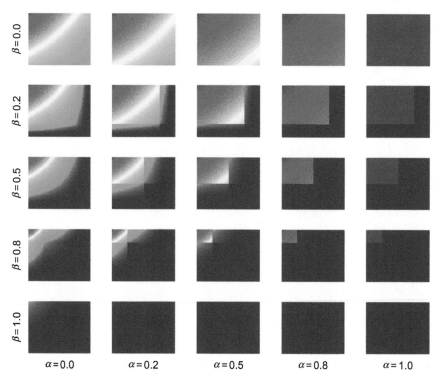

FIGURE 6.10 Idealized models of core-periphery network structure. Influence of the parameters α and β on the idealized model of core-periphery structure proposed by Rombach et al. (2014). Shown here are pattern matrices for various combinations of these parameter values. Red denotes a core node, blue a peripheral node, and intermediate colors are nodes interposed between the two when a continuous classification is used. For visualization, the rows and columns of each matrix, corresponding to the vector c, have been ordered from core to periphery, such that core nodes appear in the top left of each matrix.

The main difference between Equations (6.2) and (6.4) is as follows. Equation (6.2) weights the contribution of all core-core and core-periphery connections equally. Equation (6.4) is a generalization that weights the contribution of all connections according to how well they fit the gradual transition between core and periphery, as determined by specific the choice of α and β for the transition function defined in Equation (6.3). When using Equation (6.4), our goal is to permute the order of the elements in the class vector c_i^* to maximize $\rho_{\alpha,\beta}$ for specific values of α and β, using an optimization scheme, such as simulated annealing. Once this optimal permutation is found, we can define an aggregate core score for each node across a range of values of α and β as

$$CS(i) = Z\sum_{\alpha,\beta} c_i^*(\alpha, \beta)\rho_{\alpha,\beta},\qquad(6.5)$$

where c^* and $\rho_{\alpha,\beta}$ are values computed for the optimal node ordering and Z is a normalization constant chosen to ensure $CS(i)$ attains a maximum of one for $i = 1,\ldots N$. Using this definition, we can assign a core score to each node that considers a wide range of core definitions that vary according to size (β) and quality (α).

This approach was used to understand the temporal dynamics of core and periphery regions in human functional connectivity networks, measured with functional MRI during a motor learning task (Bassett et al., 2013b). Specifically, the authors split the signal time courses of each of 112 different brain regions into distinct temporal windows. They first distinguished between a *temporal* core and periphery based on the flexibility with which each region dynamically changed its module affiliation across temporal windows (Figure 6.11a). The temporal core, which maintained a consistent module affiliation over time, was largely comprised of unimodal somatosensory-motor areas; the temporal periphery, which showed more variable module affiliation across time, comprised association cortices (Figure 6.11b). The authors then computed *topological* coreness scores using a variant of the transition function in Equation (6.3), and values of α and β averaged across the optimal values for different participants, scan sessions, and task conditions. They found that brain regions with high topological coreness scores showed low flexibility in module affiliation, meaning that such regions were also likely to belong to the somatosensory-motor temporal core (Figure 6.11c). In contrast, regions with low topological coreness scores showed higher flexibility, and were more likely to be in the temporal periphery, which was comprised mainly of association cortex.

6.2.4 Knotty Centrality

The core decomposition methods discussed thus far emphasize degree as a defining feature of the network core. This definition may overlook the contribution of other nodes that are topologically central despite having low degree.

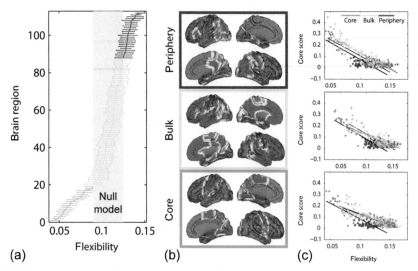

FIGURE 6.11 Dynamics and topology of core-periphery organization in human brain functional connectivity networks. Shown here is the association between the temporal and topological cores of human brain networks assessed with functional MRI. Dynamic network analysis of functional connectivity during a motor learning task was used to identify a temporal core as a set of nodes showing consistent module affiliation over time (dynamic modularity analysis is discussed in Chapter 9). The topological core was defined using the core-decomposition method described by Rombach et al. (2014). **(a)** The flexibility in module affiliation for each of 112 different brain regions. Areas with significantly lower module flexibility than expected by chance comprised a temporal core of the network (cyan); areas with high flexibility corresponded to a temporal periphery (maroon nodes); and areas with chance-expected levels of flexibility formed a temporal bulk (gold). **(b)** The anatomical distribution (red) of the temporal core, periphery, and bulk. The temporal core comprised mainly somatosensory-motor areas and the temporal periphery included regions of association cortex. **(c)** Flexibility in dynamic module affiliation was negatively correlated with topological coreness scores across minimal (top), moderate (middle), and extensive (bottom) training conditions of the motor learning task. Thus, regions with high *topological* coreness scores showed less flexibility in dynamic module affiliation and were more likely to belong to the *temporal* core. *Images adapted from Bassett et al. (2013a, 2013b) under a creative commons license.*

To illustrate this point, consider the example networks depicted in Figure 6.12. The network in panel (a) contains a subset of highly interconnected, high-degree nodes, consistent with the various definitions of network cores that we have already considered. The network in panel (b) contains no nodes with unusually high degree. It is modular, which we know is an important characteristic of brain networks (Chapter 9), and each module links only to a central module. Vertices belonging to the central module represent a putative core of the network because they act as a conduit for the flow of information between

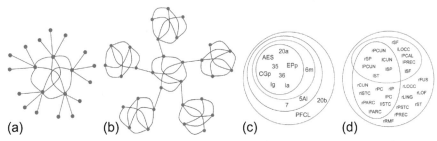

FIGURE 6.12 The knotty center of brain networks. **(a)** Example of a typical core-periphery organization, in which high-degree nodes are strongly connected with each other. **(b)** Example of a knotty center. In this network, all nodes have the same degree, but the five nodes in the middle of the graph are clearly more central (specifically, they show high betweenness centrality), as they provide the only route for other subsystems in the network to communicate. **(c)** Knotty center and other types of network cores of the 52-region cat cortical connectome, collated from published tract-tracing studies. The knotty center is in blue; high-degree nodes (**rich-club** (RC); see Section 6.3) are in purple; the compact knotty center is in green; and nodes with the highest betweenness centrality are in red. **(d)** Knotty center of the 66-region human cortical connectome constructed with diffusion MRI. The knotty center is in blue; structural core as revealed by *k*-core decomposition in red; and the compact knotty center is in green. In panels **(c)** and **(d)**, the acronyms correspond to abbreviated region names. See Shanahan and Wildie (2012) for more details. *Figures reproduced from Shanahan and Wildie (2012) with permission.*

all other pairs of modules. However, the nodes in this core have the same degree as the nodes in the other modules, and would not be identified as a core using *k*-core or model-based decomposition.

The example in Figure 6.12 suggests that betweenness centrality (see Chapter 5) might be a useful alternative measure to define the network core. Accordingly, Shanahan and Wildie (2012) defined the *knotty centrality* of a subgraph, *S*, in a network as

$$KC(S) = \frac{E_S}{N_S(N_S - 1)} \sum_{i \in S} C_B'(i), \tag{6.6}$$

where N_S and E_S denote the number of nodes and edges in the subgraph *S*, respectively, and $C_B'(i)$ is the betweenness centrality of node *i* normalized with respect to the whole graph, *G*,

$$C_B'(i) = \frac{C_B(i)}{\sum_{j \in G} C_B(j)}. \tag{6.7}$$

The term $E_S/N_S(N_S - 1)$ in Equation (6.6) gives the proportion of connections in *S* relative to the total number of possible edges in the subgraph. It should also be apparent that this term counts edges in both upper and lower triangles of the adjacency matrix, and thus Equation (6.6) is defined for a directed network and $C_B'(i)$ is the normalized *directed* betweenness centrality. The measure can be easily adapted for undirected networks by computing an undirected measure of betweenness and counting only the edges in one triangle of the matrix (or multiplying the numerator in Equation (6.6) by a factor of two).

By emphasizing betweenness, knotty centrality defines the core of a network as a highly interconnected subset of nodes that mediates a high proportion of shortest paths in the network. Values of $KC(S)$ vary between zero and one: $KC(S)=0$ if none of the nodes in S are connected (i.e., $E_S=0$) and $KC(S)$ is large when S if a fully connected graph (i.e. a maximal clique) and most shortest paths pass through S. The knotty center of a graph G is defined as the subgraph S associated with the highest value of $KS(S)$. It is possible for multiple subgraphs to have the same value of $KS(S)$, in which case the network has multiple knotty centers. We can define a measure of how "knotty" a graph is as the knotty centrality of its knotty center(s). To facilitate comparison across networks, $KC(S)$ can be normalized with respect to equivalent values computed in an ensemble of randomized graphs matched to the observed network for number of nodes, edges, and degree distribution (see Chapter 10).

Finding the knotty center of a graph is computationally expensive in large networks, since we must examine every possible subgraph of nodes. Shanahan and Wildie (2012) recommend restricting the search to the subset of nodes showing the highest betweenness centrality and/or the nodes that are connected to such vertices since, by definition, the knotty center will incorporate nodes scoring highly on this measure. Interested readers can find more details about search algorithms in Shanahan and Wildie (2012).

A variant of knotty centrality that favors smaller subgraphs, termed compact knotty centrality, can also be computed by considering the proportion of nodes excluded from S,

$$KC_C(S) = (1 - N_S/N)\frac{E_S}{N_S(N_S - 1)}\sum_{i \in S} C_B'(i). \tag{6.8}$$

Shanahan and Wildie (2012) found that knotty centrality offers a more inclusive definition of the core of a brain network compared to other methods. For example, in the 54-region cortical connectome of the cat, nodes with the highest betweenness centrality comprised a subset of the compact knotty center, which was a subset of high-degree hub nodes, which in turn were a subset of the noncompact knotty center (Figure 6.12c). In a 66-region human cortical network constructed using diffusion MRI, the structural core of the network, defined using k-core decomposition, partially overlapped with the compact knotty center, and both sets of nodes were subsets of the noncompact knotty center (Figure 6.12d). A similar pattern was evident in a separate analysis of the pigeon connectome (Shanahan et al., 2013). This consistent overlap was not found in a range of other real-world networks and model graphs, where the different core definitions showed greater divergence (Shanahan and Wildie, 2012). This convergence across different methods applied to brain networks suggests that the structural core of a connectome comprises a set of high-degree nodes that also lie on a large proportion of shortest paths (van den Heuvel et al., 2012).

6.2.5 Bow-Tie Structure

Bow-tie structure describes the core-periphery organization of directed networks. In such a structure, peripheral nodes are categorized as either feed-in or feed-out. Feed-in nodes act as sources or initiation points for edges and paths that project into the core, whereas feed-out nodes act as targets or sinks for edges and paths that project out of the core. Nodes in the core tend to be reciprocally connected.

Bow-tie architecture is a common characteristic of directed biological networks and is thought to support robustness and adaptability (Kitano, 2004). Robustness is supported by the core, which mediates a set of highly conserved and degenerate processes. Adaptability, which encompasses both functional diversity and evolvability, is supported by weak linkages between the core and the wings of the bow-tie (i.e., the feed-in and feed-out nodes). Nodes in the feed-in wing enable flexible adaptation to diverse inputs; nodes in the feed-out wing facilitate the execution of appropriate outputs (Csermely et al., 2013; Kitano, 2004; Figure 6.13a).

The hierarchical organization of brain networks, which includes both feed-forward and feed-back projections (Felleman and Van Essen, 1991), suggests that a bow-tie topology is an attractive model of the connectome. Markov et al. (2013b) presented the first characterization of such an organization for the 29-region cortical connectome of the macaque. As discussed previously, this network contains a dense, reciprocally connected core of 17 nodes (Ercsey-Ravasz et al., 2013; Figure 6.6). By considering the laminar organization of neuronal connectivity, the team was able to distinguish between feed-back and feed-forward projections into and out of the core (see Figure 2.9). Feed-forward projections were defined as pathways originating in supragranular layers and terminating in layer 4. Feed-back projections were defined as those originating in the infragranular layers of higher cortical areas and avoiding layer 4 in lower areas. This categorization allowed the authors to construct a preliminary bow-tie map of this network (Figure 6.13b). The core includes areas of frontal and parietal association cortex, the feed-forward wing

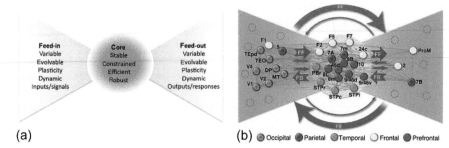

(a)   (b) ⊙ Occipital ● Parietal ◍ Temporal ○ Frontal ● Prefrontal

FIGURE 6.13 The bow-tie architecture of directed networks. **(a)** Generic bow-tie structure for complex networks. The periphery tends to be highly variable, evolvable, and dynamic whereas the core is more stable, less evolvable, highly integrated, and supports robustness. The feed-in wing provides inputs to the core; the feed-out wing affects outputs. **(b)** Bow-tie model of the 29-region cortical connectome of the macaque. The dense core of this network, comprising 17 regions connected at 92% density, is shown in the center (see also Figure 6.6). The feed-forward (FF) and feed-back (FB) wings were defined based on asymmetries in edge weight for connections projecting to and from the core, as determined by the fraction of neurons in supragranular cortical layers that were labeled by viral tracer injections. *(a) Adapted from Csermely et al. (2013) and Kitano (2004). (b) Reproduced from Markov et al. (2013b) with permission.*

comprises mainly primary sensory and motor areas, and the feed-back wing involves a smaller number of frontal and parietal areas that putatively coordinate the functions of other network nodes. In general, the methods for characterizing bow-tie structure in directed networks are not developed to the same extent as the other core-periphery decomposition methods discussed thus far, although some algorithms for characterizing bow-tie properties have been proposed (Supper et al., 2009).

6.3   RICH CLUBS

Many of the core decomposition techniques discussed in this chapter focus on degree as the defining feature of a network core. They thus identify a highly interconnected set of hub nodes that feed, and are fed by, other network elements. Extending this logic, it is possible to define a core based on the density of connectivity between the hubs of a given network. For instance, we could ask: do high-degree nodes, which represent putative network hubs, show higher interconnectivity with each other than expected by chance? To take some specific examples, we could ask whether CEOs of Fortune 500 companies show a statistically significant tendency to socialize or do business with each other, whether the most highly cited scientists in a particular discipline are more likely to cite each other, or whether major air transportation hubs are highly interconnected. In brain networks, we can ask whether candidate hub regions are also strongly connected with each other. These are precisely the questions addressed by an analysis of the **rich-club** properties of a network.

6.3.1 The Unweighted Rich-Club Coefficient

Rich-club organization in complex networks was first reported by Zhou and Mondragón (2004) in their study of Internet topology, as defined at the level of autonomous servers. Their approach involves ranking nodes in descending order of degree, such that the node with highest degree is assigned a rank of one. For each rank, r, they define a rich-club coefficient, $\phi(r)$, as a ratio of the number of edges in the subgraph defined by nodes with rank less than r, relative to the total possible number of edges in this subgraph. In other words, the rich-club coefficient is the connection density of the subgraph of nodes with rank less than r or, conversely, the subgraph of nodes with degree higher than the level specified by r. The authors found that the Internet showed a higher rich-club coefficient over a broad range of r when compared to different random graph models of scale-free topologies. For example, they found that the top 1% most highly connected nodes in the Internet were linked by 32% of total possible connections. By comparison, the density of connections between these nodes in random graphs was $\leq 18\%$.

Colizza et al. (2006) modified the definition of ϕ to avoid ranking nodes. They define the rich-club coefficient at each level of degree, k, as

$$\phi(k) = \frac{2E_{>k}}{N_{>k}(N_{>k}-1)},\tag{6.9}$$

where $E_{>k}$ is the number of edges in the subgraph comprising only nodes with degree greater than k, and $N_{>k}(N_{>k}-1)/2$ is the total possible number of edges in this subgraph. Equation (6.9) defines $\phi(k)$ for an undirected network, so we multiply the numerator by two because the denominator double-counts edges. To compute $\phi(k)$ in a directed network, we do not double the numerator and take k as the total degree. Alternatively, we can compute ϕ separately for in-degree and out-degree (see also Opsahl et al., 2008; Smilkov and Kocarev, 2010).

If the rich-club coefficient $\phi(k) = 0$ for any given k, we have no edges present in the subgraph of nodes with degree greater than k. If $\phi(r) = 1$, all such edges are present. In general, we can think of the value k as defining a cut-off value of "richness," where richness is defined in terms of degree. Equation (6.9) thus quantifies the connectivity between the richest nodes in a graph, for a given level of k. In a network with rich-club organization, we expect that the richest nodes will be densely connected, which points to an interlinked core of high-degree nodes. We thus expect that $\phi(k)$ will increase as a function of k.

In their analysis, Colizza et al. (2006) found that even networks with random topologies such as Erdős-Rényi graphs show this characteristic increase of $\phi(k)$ with increasing k. This increase occurs simply because nodes with high degree are more likely to share a connection, even in a random graph. To get around this problem, Collizza and colleagues proposed a normalized rich-club coefficient,

$$\phi_{\mathrm{norm}}(k) = \frac{\phi(k)}{\langle \phi_{\mathrm{rand}}(k)\rangle}, \tag{6.10}$$

where $\langle\phi_{\mathrm{rand}}(k)\rangle$ is the average rich-club coefficient computed in an ensemble of randomized networks matched for size, connection density and degree distribution (see Chapter 10). Values of $\phi_{\mathrm{norm}}(k)$ that are greater than one suggest rich-club organization; values around one suggest no evidence of rich-club organization and values below one indicate a distinct tendency for high-degree nodes to be disconnected from each other. Using this metric, Colizza and coworkers (2006) found no evidence for rich-club behavior in the autonomous server Internet map studied by Zhou and Mondragón (2004), but did find evidence for rich-club organization in air transportation and scientific collaboration networks. In contrast, the protein interaction network of the yeast *Saccharomyces cerevisiae* showed decreasing values of ϕ_{norm} for increasing k, suggesting that the hubs of this network are less connected than expected by chance.

We can adopt a more rigorous approach and explicitly perform statistical tests to assess the significance of ϕ_{norm} at each level of k (Jiang and Zhou, 2008). If we generate a sufficient number of randomized networks (typically using algorithms that preserve the degree distribution of the empirical network; see Chapter 10), we can compute an empirical null distribution of $\phi_{\mathrm{rand}}(k)$ values that can be used to estimate the statistical significance of our observed metric. A p-value to test the null hypothesis of chance-level rich-club organization can be estimated as the proportion of $\phi_{\mathrm{rand}}(k)$ samples in this null distribution that are greater than the observed $\phi(k)$. If we repeat this analysis over all levels of k, the range of k in which $\phi(k)$ differs significantly from $\phi_{\mathrm{rand}}(k)$ is where we see significant rich-club organization: the so-called rich-club regime. The results of this analysis are typically plotted in the form of a rich-club curve, which shows how $\phi(k)$ and $\langle\phi_{\mathrm{rand}}(k)\rangle$ vary as a function of k. These curves are plotted for the *C. elegans* and mouse connectomes in Figure 6.14.

Since ϕ is computed as a function of k, there is no single degree at which the rich club is defined. This means that the nodes comprising the rich-club will depend on which specific level of k is chosen for analysis. A simple approach might be to take the level of k corresponding to some criterion for defining network hubs, such as nodes with k greater than one standard deviation above the mean

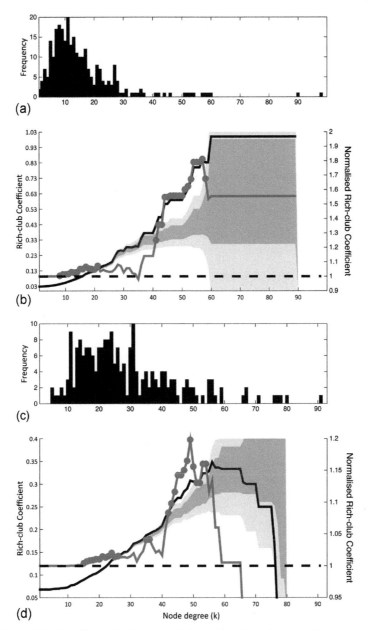

FIGURE 6.14 Rich-club organization of the microscale nematode connectome and mesoscale mouse connectome. **(a)** Degree distribution of the *C. elegans* network of chemical synapses between 297 neurons (Varshney et al., 2011), as mapped with **electron microscopy**. The distribution has an extended tail, consistent with the existence of a small number of hub neurons (see also Chapter 4). **(b)** Rich-club curve for the *C. elegans* chemical synapse network. The observed (black) and normalized (red) rich-club coefficient is plotted at each level of degree, *k*. The gray area corresponds to the range of rich-club coefficients obtained in an ensemble of 5000 networks randomized to preserve size, connection density, and degree distribution (Chapter 10). Dark gray shading corresponds to the mean value ± one standard deviation; light gray corresponds to the mean ± two standard deviations. Filled circles on the red line indicate levels of *k* where the observed rich-club coefficient was significantly greater than the corresponding values obtained in the ensemble of randomized networks ($p < 0.05$). **(c)** Degree distribution for the mesoscale mouse connectome, mapped using viral tract tracing between 213 anatomically defined regions (Oh et al., 2014). **(d)** Rich-club curve for the mouse connectome. In both the worm and mouse datasets, there is a sharp rise in the normalized rich-club coefficient at around $k = 40$. The range of *k* in which this increase is significant is called the *rich-club regime*. At very high levels of degree (around $k > 60$ for both networks), the small number of nodes that remain in the networks makes it difficult to accurately quantify rich-club properties, and leads to large variance in the null distributions.

(Sporns et al., 2007; van den Heuvel and Sporns, 2011), or some proportion of the most connected vertices (Ball et al., 2013). Alternatively, we could examine putative rich-club nodes using more or less inclusive definitions by considering subgraphs at different levels of k that show evidence of rich-club organization, much as we would when analyzing a node's coreness (e.g., Figure 6.8a).

6.3.2   The Weighted Rich-Club Coefficient

The rich-club effect can be generalized to weighted networks. The basic procedure is as follows (Opsahl et al., 2008):

1. Rank nodes according to some measure of prominence or "richness," such as degree or strength.
2. Apply a threshold to define a subgraph that retains only nodes (and their connections) that are higher than a certain rank.
3. Compute the total sum of weights on the links between the nodes comprising this subgraph.
4. Compute the sum of weights for the same number of edges, but this time for the most highly weighted edges in the entire network.
5. Take the ratio of the quantities computed in steps 3 and 4.

Formally, we can define the weighted rich-club coefficient as (Opsahl et al., 2008)

$$\phi^{w}(r) = \frac{W_{>r}}{\sum_{l=1}^{E_{>r}} w_l^{\text{rank}}}, \tag{6.11}$$

where $W_{>r}$ is the sum of weights on the edges in the subgraph of nodes with rank greater than r, $E_{>r}$ is the number of edges in this subgraph, and w^{rank} is a vector of all edge weights in the network, ranked from largest to smallest weight. The denominator in Equation (6.11) sums over the same number of edges in w^{rank} that are contained in the subgraph used to compute $W_{>r}$. If $\phi^{w}(r) = 1$, the sum of the weights of the edges between the richest nodes is maximal; that is, the most highly weighted edges are the edges of the rich-club. Values of $\phi^{w}(r) < 1$ quantify the fraction of the strongest weights in the network that exist on the links between the richest nodes. The metric used to define node "richness" is arbitrary and either node degree or node strength are suitable. Most analyses of brain networks have used node degree (Harriger et al., 2012; van den Heuvel et al., 2012; van den Heuvel and Sporns, 2011).

The unweighted rich-club coefficient defined in Equation (6.9) measures whether the richest, high-degree nodes of the network are also densely connected with each other. The weighted coefficient defined in Equation (6.11) assesses whether the most highly weighted edges in the network are found on the links between the richest nodes. Alternative methods for computing weighted rich-club coefficients have been proposed, each of which varies in the assumptions made about the relationship between the binary topology of the network and the weights on the edges (for an overview, see Alstott et al., 2014b). We focus here on the definition given in Equation (6.11) because it is the most widely applied in neuroscience.

In some networks, weighted and unweighted measures of ϕ may diverge, and these discrepancies offer insight into the relationship between topology and weights in the system. For example, Opsahl et al. (2008) examined the scientific collaboration network of researchers working in condensed matter physics, as extracted from the online database *arXiv* from 1995 to 1999. In this network, an edge was drawn if two authors coauthored a paper, and edge weights indexed the intensity of the collaboration, such that a larger number of coauthors was associated with a lower weight. The analysis revealed a rich-club effect for the unweighted measure but not for the weighted metric, suggesting that while authors with many collaborators tend to work together, the intensity of their collaboration does not differ from chance expectations.

As with the unweighted rich-club coefficient, the weighted index must also be compared to an ensemble of randomized surrogate graphs to generate a weighted normalized coefficient, $\phi^{w}_{norm}(r) = \phi^{w}(r)/\langle \phi^{w}_{rand}(r) \rangle$. Alstott and colleagues (2014b) discuss how the choice of an appropriate randomization strategy is critical in this context. Specifically, randomization can occur at two levels: (1) at the level of binary topology, by rewiring the edges that connect different node pairs; and (2) at the level of the weights attached to those edges (see Chapter 10). Randomization strategies that shuffle both these properties at the same time will conflate the contributions of binary topology and connectivity weight to the weighted rich-club coefficient. To disentangle these contributions, Alstott and colleagues suggest analyzing both the unweighted and weighted rich-club coefficient in the same network. The unweighted measure is normalized with respect to networks randomized at the level of binary topology (i.e., by rewiring edges). The weighted measure is normalized with respect to randomized networks in which the binary topology is preserved but the weights are shuffled.

Using this approach, Alstott and colleagues (2014b) found that human structural connectivity networks constructed using diffusion MRI only showed evidence of rich-club organization using the unweighted coefficient. In fact, the authors found that $\phi^{w}_{norm} < 1$ for the brain, suggesting that the weight on the links between rich nodes was lower than expected by chance. Alstott et al. suggest that this divergence of unweighted and weighted rich-club coefficients may be due to the brain's need to balance integration with segregation: dense connectivity between hub nodes at the level of binary topology enables integration across diverse neural subsystems, whereas a low total weight on these edges facilitates segregation of function. Indeed, we could imagine a scenario in which both high connection density and high strength of connectivity between hub nodes could lead to excess integration, reduced segregation of normally specialized processes, and increased vulnerability to hypersynchronized, seizure-like dynamics (Kaiser et al., 2007a). This balance between integration and segregation contrasts with other networks, such as the global air transportation network, where the primary objective is integration. Accordingly, Alstott and colleagues found that the air transportation network shows rich-club organization with both unweighted and weighted measures, offering a quantitative confirmation of the intuition expressed at the start of this chapter that hub nodes such as London Heathrow and Dubai International airports are more likely to be connected by a direct route than explained by chance (unweighted rich-club effect), and to have a higher number of flights between them (weighted rich-club effect). It is also possible that the low values of ϕ^{w}_{norm} that Alstott and colleagues observed in the human brain were driven by an inability of diffusion **tractography** to accurately reconstruct long-range fiber tracts (Chapter 2). A predictable result of this known bias of the technique is a reduction in the weight of long-range projections (Zalesky and Fornito, 2009). It is precisely the links between brain network hub nodes that tend to be long-range (van den Heuvel et al., 2012; Fulcher and Fornito, 2016).

6.3.3   Rich-Clubs in Brain Networks

The first evidence of rich-club organization in a brain network was reported by Zamora-López et al. (2010). These authors identified significant rich-club organization over a broad range of degrees in the 53-region cortical connectome of the cat. Focusing on the level of k where the strongest rich-club effect was observed ($k=23$), they found that the rich-club comprised 11 hubs from all four major topological modules of this network: visual, auditory, somatomotor, and frontolimbic. The density of connectivity within the rich-club was higher than for any other subsystem. Further, the average topological overlap (see Chapter 8) of connections attached to rich-club nodes was comparable to the values obtained for pairs of nodes in the visual, auditory, somatomotor, or frontolimbic modules. Thus, while hubs are distributed across different modules, the high interconnectivity and topological similarity of these hub areas suggests that they also work together as a closely coordinated, module-like group that integrates the functions of other areas. In support of this view, an independent analysis of the cat cortical connectome found that rich-club regions belong to more modules, on average, than other areas (de Reus and van den Heuvel, 2014; see also Figure 6.15 and Chapter 9).

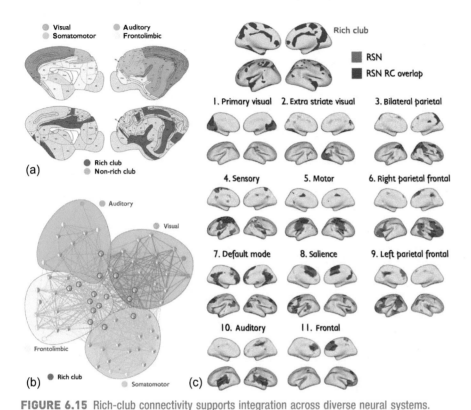

FIGURE 6.15 Rich-club connectivity supports integration across diverse neural systems.
(a) Anatomical distribution of modules (top) and rich-club (bottom) of the cat cortical connectome.
(b) Topological projection of connectivity in the cat connectome, showing the position of rich-club areas with respect to the modular organization of the network. Colors within each node indicate the number of modules to which each region is assigned, using a method for overlapping modular decomposition. Rich-club regions were involved in a larger number of modules than other areas. They were also more likely to act as connector hubs, having links distributed across different modules (see Chapter 9).
(c) Anatomical overlap of rich-club members of the human brain with 11 canonical, large-scale resting-state functional connectivity networks (RSN). The rich-club was mapped via an analysis of structural connectivity networks reconstructed with diffusion MRI. The functional connectivity networks were mapped by analysis of functional MRI data. Top image shows the anatomical distribution of rich-club hubs. Bottom images show the overlap of these areas with regions of each functional connectivity network. The rich-club was implicated in each functional system. *(a) and (b) Reproduced from de Reus and van den Heuvel (2013b) and (c) from van den Heuvel and Sporns (2013a, 2013b) with permission.*

Rich-club organization in human structural connectivity networks reconstructed with diffusion MRI was discovered by van den Heuvel and Sporns (2011; Figure 6.16a). In this work, putative hubs forming the brain's rich-club included regions of frontal and parietal association cortex, as well as subcortical regions such as the thalamus and striatum (Figure 6.16d and e). These regions also

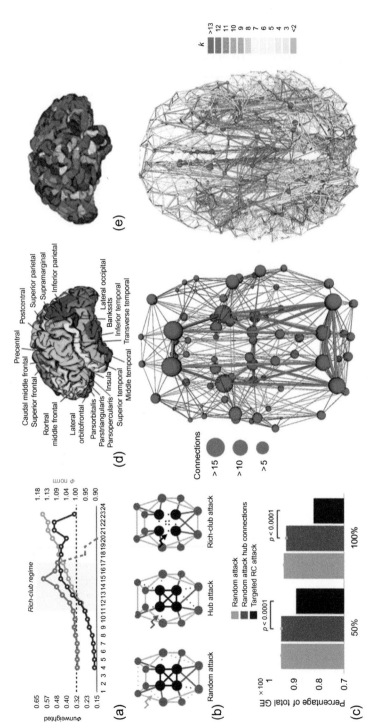

FIGURE 6.16 Rich-club organization of the human connectome. **(a)** Example rich-club curve for an 82-region structural connectivity network reconstructed using diffusion MRI. The observed values for the unweighted rich club coefficient, ϕ^w, are in dark gray, the corresponding values in randomized networks, ϕ^w_{rand}, in light gray, and the normalized coefficient, ϕ^w_{norm}, in red. **(b)** Schematic illustrating the effects of different kinds of simulated attack on the connectome. Three types of attack were simulated by randomly reducing the weight of individual edges that either (1) were outside the rich club (left); (2) linked hub regions in the rich club to nonhub areas outside the rich club (middle); or (3) linked two hub areas in the rich club (right). **(c)** The effects of the three different types of attack shown in **(b)** on the global **efficiency** (GE) of the network, which is a measure of topological integration (Chapter 7). The effects of down-weighting 50% and 100% of each connection type are shown on left and right, respectively. **(d)** Anatomical location of the rich club using a low-resolution 82-region anatomical parcellation (top). Rich-club connections are shown in thick blue (bottom) and node sizes are scaled according to degree. **(e)** Rich-club connections identified using a high-resolution 1170-region random parcellation of the same data (top). Nodes and edges are colored according to their participation in the rich club, as determined by the highest level of k at which they were included in the club (bottom). *Images reproduced from van den Heuvel and Sporns (2011) with permission.*

had high coreness values (compare Figures 6.16d and e and 6.8e), as determined by an *s*-core decomposition, pointing to a convergence between these two methods in determining the putative structural core of the human connectome. Furthermore, these hubs had high betweenness centrality, being located on 89% of all shortest paths linking non-rich-club regions. As with the cat cortex, midline cortical rich-club areas acted as connector hubs, linking topologically distinct modules, whereas subcortical rich-club members were provincial hubs, playing a central role in their own modules (Figure 6.15c; see also van den Heuvel and Sporns, 2013b).

Computational models of communication dynamics simulated on empirically derived connectomes have confirmed that the rich-club nodes mediate the majority of information flow in the brain (Mišić et al., 2014) and that they integrate the functions of topologically segregated modules (Gomez-Gardenes et al., 2010). Robustness analyses also support the centrality of these areas, showing that a targeted attack of hub-hub connections leads to a dramatic reduction in the topological integration of the network compared to random removal of other types of links (van den Heuvel and Sporns, 2013b; Figure 6.16b and c). Finally, connections between members of the brain's rich club account for a disproportionate fraction of the brain's wiring cost, extending over longer distances, on average, than other types of connections (van den Heuvel et al., 2012; Fulcher and Fornito, 2016). This result is consistent with metabolic imaging evidence suggesting that the energetic demand of putative hub regions in the human brain is higher than for other areas (Bullmore and Sporns, 2012; Collin et al., 2014; Tomasi et al., 2013; see also Chapter 8). It is also consistent with a study of the mouse connectome showing that the transcriptional signature of hub connectivity is driven by coordinated expression of genes regulating the synthesis and breakdown of adenosine triphosphate (ATP), the energetic substrate of neuronal signalling (Fulcher and Fornito, 2016). Analyses of *C. elegans* (Towlson et al., 2013) and macaque connectomes (Harriger et al., 2012) have confirmed that rich-club connectivity is a topologically central and costly aspect of brain network organization.

Collectively, these findings indicate that RCs in nervous systems comprise topologically central yet costly cores of highly interconnected high-degree nodes, which integrate the functions of anatomically disparate subsystems or modules. In support of this view, hub neurons of the *C. elegans* nervous system are all born early in development, prior to the first evidence of movement in the animal (Towlson et al., 2013). This developmental timing suggests that the genesis of these neurons is a necessary substrate for the integrated nervous function that is required to support coordinated and adaptive behavior. Functional networks derived from **multielectrode array** recordings of neuronal cultures in vitro also show evidence of an early emergence of hub nodes, which become more connected by a "rich-get-richer" rule (Box 4.1) so that a rich-club is consistently represented within 28 days of in vitro culture (Schroeter et al., 2015). Moreover, early development of rich-club topology in these cellular systems plays a key role in facilitating synchronized bursting dynamics of the network. Some preliminary evidence suggests that brain network hubs also emerge early in human development, with rich-club organization being apparent from as early as 30 weeks of gestation (Ball et al., 2013). Research in adolescents indicates that rich-club connectivity continues to mature later in life, demonstrating a shift from high connectivity between subcortical hubs at around 16 years of age to greater integration of frontal hubs by 18 years (Baker et al., 2015). These changes may support a gradual transition in brain network organization from one dominated by spatially proximal functional subsystems to a more spatially distributed and globally integrated network architecture (Fair et al., 2009). The ongoing development of connections between hub areas may have clinical implications, given that many psychiatric disorders have their peak age of onset in late adolescence (Paus et al., 2008). Indeed, accumulating evidence indicates that pathology tends to accumulate in hub regions across a broad spectrum of brain disorders (Buckner et al., 2009; Crossley et al., 2014; Fornito and Bullmore, 2014; van den Heuvel et al., 2013a).

6.3.4   Assortativity

Assortativity (also called assortative mixing or homophily) is a topological property that is sometimes confused with rich-club organization. Assortativity refers to the tendency of nodes in a network to link with other similar nodes. This concept is easily understood when considering social networks, where assortativity can be intuitively understood as the tendency of like to associate with like. Imagine that we categorize people in a social network according to some characteristic such as ethnicity, gender, or socioeconomic class. Assortativity occurs when people in the same category tend to associate with each other; for instance, when wealthy people are friends with other wealthy people, middle-class individuals associate with other members of the middle-class, and people in poorer classes nominate each other as friends. Conversely, disassortativity (or heterophily) occurs when there are links between people belonging to different categories; for example, when wealthy people are friends with poorer people.

Network assortativity is typically examined in relation to degree, although in principle it can be measured in relation to any other nodal property. The degree assortativity of an undirected network is defined as the Pearson correlation coefficient of the node degrees of connected pairs of vertices j and k (Newman, 2002), and can be written as

$$a = \frac{E^{-1}\sum_i j_i k_i - \left[E^{-1}\sum_i \frac{1}{2}(j_i + k_i)\right]^2}{E^{-1}\sum_i \frac{1}{2}(j_i^2 + k_i^2) - \left[E^{-1}\sum_i \frac{1}{2}(j_i + k_i)\right]^2}, \qquad (6.12)$$

where E is the number of edges in the network and j_i and k_i are the degrees of nodes that are linked by the i th edge, with $i = 1,\ldots,E$. As with the Pearson correlation coefficient, $-1 \le a \le 1$, where positive values reflect a tendency for high-degree nodes to connect to each other while low-degree nodes link to each other (assortativity); negative values suggest that high-degree nodes tend to connect to low-degree nodes (disassortativity). In directed networks, it is also possible to compute separate assortativity coefficients by considering only the in-links of nodes j and k, only the out-links, the in-links of j and out-links of k, and vice versa (Foster et al., 2010; Newman, 2003b).

It is easy to see how rich-club organization can be confused with assortativity: the rich-club coefficient measures the density of connectivity between high-degree nodes, whereas assortativity measured the tendency of nodes with similar degree to connect to each other. The difference between these measures is that assortativity, when specified in terms of degree, describes a relationship between nodes across all levels of k. The rich-club coefficient is only concerned with connectivity amongst nodes with degree greater than k and makes no statement about the connectivity of nodes with lower degree. Another way of understanding this distinction between assortativity and the rich-club coefficient is to think about how these measures relate to Borgatti and Everett's (2000) model of a network with strong core-periphery structure (Figure 6.9). In this model, core nodes are strongly connected to each other and to peripheral nodes. The peripheral nodes, in turn, are only weakly connected amongst themselves. This organization is consistent with a rich-club, to the extent that the core will comprise the highest degree nodes in the network, and these nodes will also be densely connected with each other. However, it is not characteristic of a network with high assortativity, since an assortative network will have strong connectivity between high-degree nodes, strong connectivity between low-degree nodes, and only weak connectivity between high-degree and low-degree nodes. In fact, the model depicted in the left panel of Figure 6.9 is consistent with a disassortative organization, since hubs are strongly connected with low-degree nodes, which in turn are not connected with each other.

This chapter has presented considerable evidence supporting a strong core in brain networks (Bassett et al., 2013b; Hagmann et al., 2008; van den Heuvel and Sporns, 2011), but there is little evidence for strong assortativity of neural connectivity: estimates in human structural connectivity networks are low (Klimm et al., 2014; Samu et al., 2014) and the neuronal network of C. *elegans* actually shows evidence of disassortativity ($a = -0.163$) (Newman, 2002), despite having a rich-club structure (Figure 6.14; see also Towlson et al., 2013). Moreover, seizure states may be associated with an increase in the assortativity of functional connectivity networks, suggesting that assortative mixing between different brain regions may lead to pathological dynamics (Bialonski and Lehnertz, 2013). We can thus conclude that a network with strong core-periphery organization will also have rich-club structure, and may or may not be positively assortative. Moreover, assortativity is not a necessary condition for formation of a rich-club.

6.4   SUMMARY

In this chapter, we have covered different methods for understanding the topological macrostructure and mesostructure of brain networks. We have seen how analysis of connected components helps us to understand network robustness to disease or injury, and how a brain network can be divided into a densely connected core that is surrounded by a sparsely connected periphery. This core overlaps extensively with a set of topologically central and highly connected hub regions—collectively called a rich-club—that plays an important role in integrating information across diverse nervous systems.

Our discussion has principally focused on degree-based measures as these have been the most widely used in the literature. Alternative measures of centrality can be used to define network cores and peripheries (Batagelj and Zaveršnik, 2002; da Silva et al., 2008; Holme, 2005). There are also many other methods for characterizing the core-periphery structure of a network. For example, some methods define the core based on the behavior of a **random walker** on the network, which implicitly incorporates a model of system dynamics (Rossa et al., 2013). Other approaches have proposed more general block models of core-periphery organization (Pavlovic et al., 2014; Zhang et al., 2015). In the analysis of functional connectivity networks, we have also seen how cores can be defined based on their pattern of functional interaction with different networks over time (Bassett et al., 2013b; de Pasquale et al., 2012).

The important role of core nodes in supporting integrated network function draws attention to precisely how the topological integration of a network can be quantified. Many of these measures are based on tracing paths through the network, which is the topic of the next chapter.

CHAPTER 7

Paths, Diffusion, and Navigation

A key question for **connectomics** is how topological properties of brain graphs relate to the signaling and information-processing of nervous systems. Neurons communicate via electrical impulses known as action potentials. These discrete, short-lasting depolarizations of the axonal membrane are propagated by axons at a speed known as the conduction velocity. Axons that have larger cross-sectional areas, and axons that are coated with an electrically insulating myelinated sheath, have faster conduction velocities than thinner or unmyelinated axons. The digital code of action potentials (spiking or firing) must be converted to analog, chemical, or electrical signals to traverse the synaptic cleft and allow information to propagate from one neuron to another (Debanne et al., 2013). Network **topology** constrains the way in which these signals propagate throughout a system. **Graph theory** can be used to gain insight into the flow of neural information between the elements of a nervous system.

Which specific graph properties can be used to understand information flow in the brain? It is well known that synaptic transmission is more time-consuming, metabolically expensive, and prone to noise than axonal action potential propagation, so it is reasonable to expect that the minimum number of synaptic **edges** that must be traversed to send a signal between any pair of **nodes** should be low. To put this more simply, the number of synapses on a **path** between two neurons should be minimized for the sake of speed and fidelity of signal transmission. In the language of graph theory, we might hypothesize that the network should have a **characteristic path length** that is close to the theoretically minimum characteristic path length of a comparable random graph. Indeed, this was precisely what Watts and Strogatz (1998) found in their seminal analysis of the neuronal network of *C. elegans*. A near-minimal characteristic path length has since been reported in analyses of other brain networks studied in different species across micro, meso, and macro scales (for a review, see Bullmore and Sporns, 2012).

Given these findings, **path length** and related metrics based on shortest paths seem like reasonable measures of the capacity of a network to support rapid, integrated, and efficient communication. All of these metrics are based on quantifying the shortest path length or **hop count**, which respectively define the minimum sum of edge weights, or the minimum number of edges,

207

Fundamentals of Brain Network Analysis. http://dx.doi.org/10.1016/B978-0-12-407908-3.00007-8

comprising a path between a pair of nodes. There is an intuitively straightforward connection between the shortest paths in a network and information processing. For example, in cell networks, low hop count directly predicts shorter minimum conduction time of action potentials between neurons. Similar arguments have justified the use of measures based on shortest paths in mesoscale **connectomes** derived from axonal **tract tracing** experiments (Rubinov et al., 2015) and in macroscale connectomes derived from **diffusion MRI** (van den Heuvel et al., 2012).

However, this apparently simple topological analysis of network integration and communication **efficiency** makes an important assumption; namely, that signals propagate in brain networks via the shortest paths between network elements. From the Olympian perspective of a neuroscientist gazing down on the wiring diagram of *C. elegans*, we can easily identify the shortest path between any pair of neurons. Knowing that fewer synaptic edges mean faster, higher-fidelity signaling, we may readily assume that physiological signaling will follow the shortest path. However, a worm does not have access to such a map of its own nervous system. How does one neuron "know" that the best and cheapest way for it to send a signal to another neuron is via a two-hop path mediated by a third neuron? Put another way, the routing of information along shortest paths mandates that each node has global knowledge of the network to find the appropriate route. This assumption may be sensible for engineered or manmade systems such as the worldwide air transportation network, where an individual traveler has access to flights between all countries and can identify a route that reaches the target destination via the smallest number of flights. As we saw in Chapter 5, individual nodes in biological networks, such as the brain, never have such global knowledge, motivating work on alternative models for neural signal transmission that do not depend on homunculus-like global knowledge of optimal routes.

We can articulate the distinction between different models of neural communication by introducing the concept of a "walker," an agent that traverses a network by "walking" along its edges in a sequential manner. In this way, the walker traces out a path in the network. Let's begin by supposing our walker is a tourist called Alice. She is visiting Paris and wants to walk to the Eiffel Tower from her hotel. How might Alice decide which direction to take next when she arrives at an intersection in her walking path? Alice wants to reach the Eiffel Tower by walking the shortest possible distance, so her choice of direction is determined by the use of a map of Paris, which she can use to estimate the shortest path between any pair of locations in the city.

Now consider a second tourist named Bob, who does not necessarily want to visit the Eiffel Tower. He instead explores the streets of Paris without a destination in mind. Bob randomly chooses a direction whenever he arrives at an intersection, but soon realizes that doing so has him walking around in circles. Bob therefore biases his random choices so that he is more likely to exit an intersection along a wide or busy road, as opposed to a narrow alleyway. In a weighted network, this is equivalent to Bob biasing his random choice to highly weighted edges. Bob is called a **random walker**. In many ways, Bob's walk is the polar opposite of Alice's. Alice has a clear destination and knows the optimal route to get there; Bob has no clear target and there is no apparent order to his journey, apart from the fact that it is constrained by the underlying network.

Our third and final tourist, Jane, wants to meet Alice at the Eiffel Tower, but does not have a map of the city—she knows her destination but does not know how to get there. To compensate for this lack of global knowledge, Jane makes decisions based only on the information at hand—the local properties of her current location. For example, she might choose the busiest or widest road at each intersection, or she may see the Eiffel Tower from afar and choose the road that is most closely oriented in that direction. These locally informed decisions cannot guarantee that Jane will ever reach the Eiffel Tower, but she makes an informed guess based on the available information.

Our three tourists—Alice, Bob, and Jane—are metaphors for three different models of neural information flow. Alice is *routed* via a (shortest) path based on global knowledge of the network. Her path is completely determined by a map. Jane attempts to *navigate* to her desired destination according to local information. Our walker Bob does not have a particular destination in mind, preferring to *diffuse* through the network in a random manner. In this chapter, we will revisit Alice in the context of shortest path routing, Jane in the context of network navigation and locally greedy schemes for communication, and Bob in the context of diffusion models of information flow. Which of these three models—routing, diffusion, or navigation—best describes the flow of neural information remains to be ascertained empirically. Different topological metrics are theoretically predicted to be more relevant to one model than another.

Most analyses of network integration in nervous systems to date are based on metrics related to the shortest paths in the connectome. This is likely a reflection of the apparent technical and interpretational simplicity of shortest path routing, as well as the possibly misleading analogy that this routing strategy is fundamental to the design and reliability of many engineered networks. Because of the importance of efficient flow in such networks, the metrics for shortest path routing were established before the field of connectomics emerged. They were then rapidly translated to brain network analysis without much regard for their validity as models of neural information flow.

Recent work challenges shortest path routing as a model of neural information flow, mainly due to the fact that all shortest paths must be computed centrally (Goñi et al., 2013). Therefore, network measures that examine the efficiency of navigation (Jane) and diffusion (Bob) models of information flow have increasingly been used to investigate nervous systems. We will see that these measures include **communicability** (Estrada and Hatano, 2008), search information, and path transitivity (Goñi et al., 2014), and locally greedy navigation schemes that have their origins in social (Kleinberg, 2000) and wireless telecommunications networks (Stojmenovic, 2002). The term "locally greedy" comes from the fact that each node is responsible for greedily choosing the next outgoing **link** in a path, based on what it believes will lead to the desired destination most efficiently. With navigation, paths are therefore defined locally in a decentralized manner and do not require any single node to know the topology of the network as a whole.

In this chapter, we consider network measures quantifying the efficiency of neural information flow under shortest path routing, navigation, or diffusion. We will consider the physiological plausibility of these different measures as models for neural communication, and highlight examples of their application to characterize signaling and network integration in human brain networks. Throughout this chapter, the concept of efficiency will be considered under several different guises and quantified using a variety of network measures. These measures quantify the efficiency with which information can be communicated between the distributed elements of a network under different models of information flow. We will see that global, nodal, and local efficiency quantify the efficiency characteristics of shortest path routing, whereas diffusion and resource efficiency measure the efficiency of communication under a diffusion process. We begin our discussion with consideration of some basic graph properties and processes that are essential to the estimation of these measures; namely, **walks**, **trails**, **paths**, and **cycles**.

7.1   WALKS, TRAILS, PATHS, AND CYCLES

Walks, trails, and paths are fundamental concepts in graph theory (see also Chapter 5). Consider a set of edges in an **undirected graph** that is ordered to form a sequence in which any pair of successive edges shares a common node. Such a sequence defines a walk. It represents a feasible sequence of edges that an entity can "walk" along, in a contiguous manner, without having to ever "jump" from one node to another and thereby break the contiguity of the sequence. A walk can traverse multiple edges and nodes; that is, an entity that is "walking" along a walk can visit the same node or edge on multiple occasions. The definitions of trail and path follow trivially: a trail is a walk in which all edges are unique, while a path is a trail in which all edges *and* nodes are

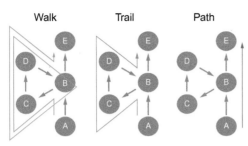

FIGURE 7.1 Walks, trails, and paths in a directed network. The walk (left; *A-B-C-D-B-C-D-B-E*) traverses three of the same edges and nodes at least twice. The trail (center; *A-B-C-D-B-E*) traverses node *B* on two occasions, but does not traverse any edges more than once. Embedded within both the walk and trail is a cycle (*B-C-D-B*). The path (*A-B-E*) never traverses an edge or node more than once. Polysynaptic pathways are modeled as paths in a graph and are the simplest kind of pathways that can be studied in brain networks. Walks and trails can potentially characterize the flow of information in neural feedback circuits.

unique. For a given graph, the set of all paths is therefore a subset of the set of all trails, which is in turn a subset of the set of all walks. A closed walk, or cycle, is a walk in which the first and last nodes are the same. Cycles can be embedded within a larger walk. In an undirected network, a single edge can even be considered a cycle, since the edge can be traversed back and forth. In some of the literature, the term "path" is used as a generic description of trails, walks, paths, and cycles. In this case, a *simple* path is used to mean a path in which all nodes and edges are unique. Examples of a walk, trail, and path are shown in Figure 7.1.

7.1.1 Shortest Paths

Polysynaptic neural pathways are the simplest to understand and quantify in brain networks using graph analysis. Unlike walks and trails, paths do not allow neural information to loop over the same neuron or axon on multiple occasions. Given that traversing the same network elements is clearly inefficient, it might seem that walks and trails are not useful in the analysis of brain networks. However, neural feedback circuits are an example where they can potentially find utility. Later in this chapter, we will also see that walks and trails are intrinsic to diffusion models of neural communication and measures used to quantify the efficiency of a diffusion process. For now though, we focus only on paths and thereby assume neural information flows in a feed-forward manner along sequences of successive axons and their synaptic contacts.

The definition of a path can be easily extended to encompass weighted and **directed graphs**. In directed networks, a sequence of edges defining a path must be ordered such that all edges are aligned in the same direction, and thus the head of one edge is always followed by the tail of the next. The length of a path in a **binary graph**, or path length, is defined as the total number of edges comprising the path. This is also called the hop count in many applications, with the traversal of each edge representing the completion of a hop along the route.

In **weighted graphs**, the path length can either be defined in terms of the hop count, or more typically the sum of the edge weights (or some function thereof) traversed by the path. When the path length is defined in terms of edge weights, it is possible for paths with long hop counts to have short path lengths. In this case, the path length can quantify the optimality of the path or the likelihood that it will be traversed. For example, if the edge weights are proportional to the extent of congestion on each arterial of a road network, then the optimal route may involve using many lightly congested backroads, as opposed to a single, crowded highway. In this case, a short path length with a long hop count may be desirable. Approaches are available that allow one to tune the relative contributions of hop count and edge weights when defining path length (Box 5.3).

Shortest paths have been fundamental to brain network analysis. Many studies have focused on shortest paths as the principal routes along which information is communicated in the brain. Most notably, this assumption has been crucial in establishing the widely accepted view of the brain as an efficiently wired, **small-world** network (Bassett and Bullmore, 2006; Latora and Marchiori, 2001). It seems intuitively plausible that information in nervous systems is routed along shortest paths, since this minimizes conduction delays, susceptibility to interference by noise, and the amount of energy required to transmit information between network elements. Whether or not neural information is routed via the shortest paths of a nervous system nonetheless remains an unresolved question in neuroscience.

What exactly do we mean by a shortest path? In a graph, a shortest path, or **geodesic**, is an ordered set of edges linking two nodes in a network for which the sum of the weights of its constituent edges is minimal. There may be multiple shortest paths for any given pair of vertices, especially in binary networks. In binary networks, all edges effectively have a weight of one, and thus a shortest path is any path between the pair of vertices that traverses the fewest number of edges, or hops. This is not necessarily the case in weighted networks, where shortest paths preferentially traverse edges with *low* weights, possibly at the expense of an increased hop count.

In weighted brain networks, it is usually the edges with the *highest* weights that are assumed to delineate the strongest and most reliable connections, and by extension the most likely communication pathways. For example, when **tractography** is performed using diffusion MRI data, the weight of an edge is often estimated by the number of streamlines reconstructed in the associated axonal fiber bundle, otherwise known as the streamline count. The edges with the highest weights are therefore assumed to be the strongest and most reliable. Similarly, in axonal tract tracing studies, the fraction of neurons labeled by the tracer can serve as the edge weight (Markov et al., 2014). A high fraction of labeled neurons indicates a strong and reliable axonal connection from the injection site to the target.

Studying shortest paths in weighted brain networks might therefore seem anomalous, since shortest paths will preferentially avoid the very edges that are thought to comprise the most likely routes along which it is assumed information flows. In fact, it might seem that our focus should be on the *longest* paths in weighted brain networks. However, longest paths would seem to be implausible, since they imply that neural information could be routed along an unnecessarily large number of hops just for the sake of passing through a few highly weighted edges. Figure 7.2a provides an example.

Longest paths are not studied in neuroscience for this reason. Instead, we apply a similarity-to-distance remapping to the edge weights *before* computing shortest paths, so that the largest weights become the smallest and the smallest weights become the largest. Shortest paths can then be sensibly defined under such a monotonic remapping. The simplest and most common remapping is,

$$w_{ij} \leftarrow 1/w_{ij}, \tag{7.1}$$

where w_{ij} denotes the edge weight (Rubinov and Sporns, 2010; Goñi et al., 2014). Figure 7.2 shows an example of this remapping. It can be seen that the

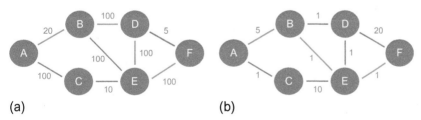

(a)   (b)

FIGURE 7.2 Edge weights can be remapped before computing shortest paths. The reciprocal of each edge weight in network **(a)** is computed and multiplied by 100 to yield the edge weights in **(b)**. The multiplication by 100 does not affect the calculation of shortest paths and is performed here to avert fractional values. It may be tempting to consider the longest paths in network **(a)** based on the rationale that highly weighted edges (i.e., strong and reliable connections) are preferentially traversed. However, longest paths yield physiologically implausible routes. For example, the longest simple path between nodes A and F in **(a)** is shown in green and diverts to node D, rather than traversing the direct connection between nodes B and E. It has a path length of $20 + 100 + 100 + 100 = 320$. The shortest path in the remapped network **(b)** is shown in green and has a path length of $5 + 1 + 1 = 7$. It represents a more efficient communication route between nodes A and F.

longest path in the original network is not the same as the shortest path in the network with remapped edge weights.

There is no theoretical justification for a remapping based on the reciprocal of the edge weights, and we might alternatively consider $w_{ij} \leftarrow 1/w_{ij}^{\beta}$, where $\beta > 0$ is a constant. Larger values of β accentuate the confidence we have in edges with large weights representing strong and reliable connections. If w_{ij} is between 0 and 1, other alternatives that may be considered are $w_{ij} \leftarrow -\log w_{ij}$ (Goñi et al., 2014) and $w_{ij} \leftarrow 1 - w_{ij}$.

7.1.2   Finding Shortest Paths

Shortest paths can be efficiently computed using many algorithms. Dijkstra's algorithm is one of the oldest and perhaps the most well-known and widely used. The algorithm is efficient and versatile. It is applicable to both directed and undirected networks, and deals with weighted edges without any modifications. The only caveat is that all edge weights must be non-negative. This is rarely a limitation for measures of structural brain connectivity. For correlation-based functional networks, negative edge weights are common, complicating the computation of shortest paths. For now, we only consider graphs with non-negative weights.

Dijkstra conceived the algorithm in 1956, apparently while he was shopping with his fiancée and thinking about the shortest way to travel from Rotterdam to Groningen. He published the algorithm in a two-page note three years later (Dijkstra, 1959). The paper remains within the top 1% of most cited papers ever published across all fields of science. While Dijkstra's algorithm is the most widely used for computing shortest paths, it was not the first. **Matrix** methods for binary networks were developed in the early 1950s by Shimbel (1953) and Luce (1950). These methods rely on the fact that if A is a binary **adjacency matrix**, then A^n characterizes all the walks in the network that traverse exactly n hops. Hence, if the (i,j)th element of A^n is nonzero, a walk exists between nodes i and j. Orden (1956) showed that the shortest path problem in a weighted graph can be solved using linear programming, which is a general method for solving optimization problems. Leyzorek et al. (1957) and Schrijver (2012) developed an algorithm that is more or less identical to Dijkstra's algorithm. The interested reader is referred to Schrijver (2012) for a historical perspective on these developments.

Dijkstra's algorithm is iterative. At each iteration the algorithm updates the path lengths from a user-defined initial node to all other nodes in the graph. In particular, each node is assigned a single value that represents its path length from the initial node. The algorithm seeks to decrease the path length values assigned to each node. To begin, the initial node is assigned a path length of zero (the path length from a node to itself is zero), while all other nodes are assigned infinity. The algorithm maintains a list of the nodes that have been visited. A table storing the *predecessor* of each node is also maintained. A node's predecessor is the previous node along a given path.

To begin with, all nodes are marked as unvisited, except for the initial node, and the predecessor of each node is undefined. At each step, the algorithm visits an unvisited node and in doing so tests whether the path length can be decreased from the initial node to that unvisited node. The choice of node to visit at each step is crucial (see below). The user can terminate the algorithm once a desired target node is visited, at which time the shortest path length from the initial node to the desired target node is known. Alternatively, the algorithm can be allowed to run until all nodes have been visited, in which case the shortest path lengths from the initial node to all other nodes in the graph are available. The algorithm proceeds by repeating the following five steps:

1. Consider the unvisited **neighbors** of the current node. Neighbors are nodes that are directly connected to the current node. (At the first step, the current node is the initial node and its unvisited neighbors are any nodes that can be reached from it using a single edge.)
2. Compute *tentative* path lengths from the initial node to each of the unvisited neighbors of the current node. This is done by adding the path length assigned to the current node to the weight of the edge between the current node and its unvisited neighbor. For example, if the current node has a path length of x relative to the initial node, and the current node is linked to an unvisited neighboring node via an edge with weight y, the tentative path length for that unvisited neighbor becomes $x + y$.
3. For each unvisited neighbor, compare its newly calculated tentative path length to its currently assigned path length. Replace the currently assigned value with the newly calculated one, if and only if the new value is smaller than the current value. If this replacement is performed, the predecessor of the unvisited neighbor is set to point to the current node.
4. Add the currently visited nodes to the set of visited nodes. A visited node will never be visited again.
5. Select the unvisited node with the *smallest* assigned path length as the next node and repeat Steps 1-5 until all nodes have been visited.

If the algorithm is terminated once all nodes have been visited, then any node with a path length of infinity cannot be reached from the initial node. In other words, no paths exist from the initial node to any node with a path length of infinity. The algorithm yields the shortest path length from the initial node to all other nodes, but not the sequence of edges comprising the shortest paths. To determine the sequence of edges, we can backtrack through the graph, starting from the target node and ending at the initial node. At each backtracking step, we proceed to the node that is listed as the predecessor of the current node. In this way, we can trace out (in reverse) the sequence of edges comprising the shortest path. The same backtracking process can be repeated for other target nodes, thereby yielding the shortest path between the initial and desired target node.

The mechanics of Dijkstra's algorithm can be understood with a worked example. We step through the computation of the shortest path shown in Figure 7.2b. This requires us to maintain a list of unvisited nodes, and for each node, its predecessor node, current path length and whether or not it has been visited. Each step of the algorithm is shown in Figure 7.3. The path length of every node, other than node A, is initialized with infinity in Step 1. In Step 2,

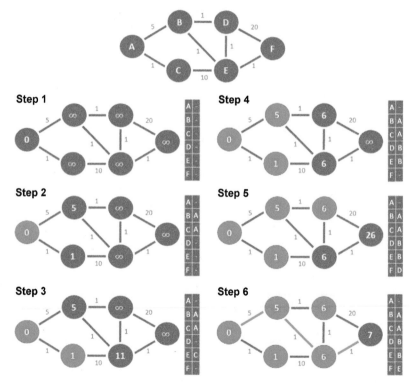

FIGURE 7.3 Using Dijkstra's algorithm to compute the shortest path between a pair of nodes in a weighted, undirected network. The aim is to identify the shortest path between nodes A and F. The value assigned to each edge is its remapped weight (see Figure 7.2). For each of the six steps of the algorithm, the value residing within each node is the current path length from node A to that node. Visited nodes are colored in blue. Unvisited nodes are gray. The table shown to the right of each network tabulates the predecessor of each node. Specifically, the second column lists the predecessor of the node in the first column. See main text for an explanation of each step.

the path lengths of the two neighbors of node A are updated and node A is marked as visited (colored blue). In Step 3, node C is identified as the unvisited node with the smallest path length. The path length to its only unvisited neighbor, node E, is updated to 11 and node C is marked as visited. In Step 4, node B is selected and the path length to its two unvisited neighbors (nodes D and E) is updated. Node D is marked as visited. Note that the path length for node E is reduced from 11 to 6 and its predecessor node is accordingly changed to B. In Step 5, we can select either node D or E, since they both have a path length of 6. We select node D arbitrarily, update the path length to node F from infinity to 26 and mark node D as visited. Finally, in Step 6, we select node E, and update the path length of node F to 7. We could include another step in which we visit the only unvisited node (node F), but this would be superfluous here. The shortest path is shown in green in Step 6 and can be determined by backtracking through the predecessor table. We begin at node F and see from the sixth row of the table that its predecessor is node E. We then see that the predecessor of node E is node B, and the predecessor of node B is our initial node (i.e., node A). The shortest path is thus A-B-E-F. The algorithm would be applied in the same way for directed networks, but in this case we must respect the direction of edges when determining the neighbors of each node. Specifically, if an edge is present from node A to node B, but not in the reverse direction, then B is considered a neighbor of A, yet A is not a neighbor of B. Dijkstra's algorithm can also be used to determine the shortest paths in binary graphs (weight of each edge is one), although faster algorithms are available in this case, such as the breadth-first search (Box 6.3).

An important observation in this example is that the path length assigned to an unvisited node can change as the algorithm proceeds to that node. For example, in Step 4, the path length assigned to node E is reduced from 11 to 6 because the algorithm realizes at this step that its initial path to node E via node B was suboptimal. The example also demonstrates that the naïve approach of traversing the lowest weight edge outgoing from each successive node (i.e., the lowest weight edge after applying the remapping in Equation (7.1)) does not necessarily identify the shortest path. This locally greedy approach identifies the path A-C-E-F, which has a path length of 11, whereas the shortest path, A-B-E-F has a path length of 6.

7.1.3   Shortest Paths and Negative Edges

The correlation coefficient and other measures of **functional connectivity** can validly span a range of negative and positive values. Since Dijkstra's algorithm cannot handle negatively weighted edges, it cannot be directly used in correlation-based functional networks. How can shortest paths be delineated in these networks?

The concept of shortest paths in networks with negatively weighted edges is tricky. In the graph theory literature, the presence of a negative edge in an *undirected* graph is usually taken to mean that all shortest path lengths are negative infinity and thus the concept of a shortest path becomes ill-defined. As long as a negative edge can be reached, all shortest paths will make their way to that negative edge and then cycle back and forth along it for eternity. In other words, the length of any path can be reduced simply by cycling back and forth one more time along a negative edge, and thus the shortest path length is always negative infinity. Recall that a single undirected edge can define a cycle. The term shortest *path* is rather misleading here since a path by definition cannot traverse the same edge more than once, and thus cycling back and forth along the same edge in this way results in a walk, not a path. However, in the network literature, a "path" is often used in a generic sense to mean any kind of route, whether it is a walk, trail, or path.

To avoid shortest paths having a length of negative infinity, it may be feasible to allow only noncyclic (acyclic) shortest paths; that is, shortest paths that do not traverse the same edge or node more than once. These are sometimes known as *simple* shortest paths and can be determined by solving what is known as the longest path problem (Zamfirescu, 1976). Computing simple shortest paths in networks with negative edges is unfortunately a very difficult problem to efficiently solve. Moreover, even if the simple shortest paths in networks with negative edges could be computed, their meaningfulness in the context of routing information in brain networks is unclear. For example, a shortest path may traverse many edges with small positive weights just for the sake of reaching a single edge with a large negative weight. In this case, the reduction in path length afforded by traversing the edge with large negative weight must outweigh the increase incurred by traversing the many positive edges required to reach the negative weight. Whether this is plausible from a physiological perspective depends on the interpretation given to negatively weighted edges. If they are interpreted as inhibitory connections, for example, it would seem implausible for information to be routed along a path that traverses substantially more hops than necessary simply for the sake of reaching an inhibitory edge.

Two scenarios are possible in *directed* networks with negatively weighted edges. If the network comprises a negative cycle that can be reached from a node—an example of which is shown in Figure 7.4a—the concept of a shortest path is once again ill-defined. In particular, a negative cycle can be traversed for an eternity, yielding shortest path lengths of negative infinity. If no negative cycles are present, shortest paths are well-defined, but an algorithm that is more computationally expensive than Dijkstra's is required to compute them, and for the same reasons given above, it is unclear whether these shortest paths characterize physiologically meaningful pathways via which neural information is routed.

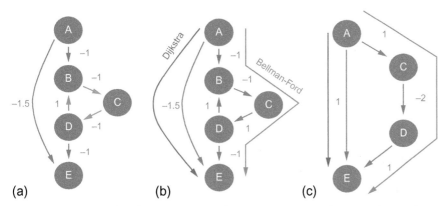

FIGURE 7.4 Computation of shortest paths in directed networks comprising negatively weighted edges. The three edges colored in green form a negative cycle **(a)**. The presence of this negative cycle indicates that the shortest path length from node A to node E is negative infinity. In particular, the length of any path from node A to node E can be reduced by one infinitely many times by traversing the negative cycle again and again. When negative cycles are detected, an alternative is to consider noncyclic (simple) shortest paths, such as the path A-B-C-D-E. However, these are generally difficult to compute in large networks. The network in **(b)** does not contain any negative cycles (sign of the edge from node C to node D has been reversed), meaning the Bellman-Ford algorithm can be used to compute the shortest path from node A to node E. The shortest path identified by the Bellman-Ford algorithm is colored in blue and is of length -2. Dijkstra's algorithm incorrectly declares the path colored in red as the shortest path from node A to node E. In particular, Dijsktra's algorithm visits node E first and assigns it a path length of -1.5. Node E is then marked as visited, and thus this suboptimal path length of -1.5 becomes fixed. While the path identified by Dijkstra's algorithm is suboptimal in the sense that it is not the shortest in length, it arguably represents a more direct and biologically plausible route from node A to node E. The shortest path identified with the Bellman-Ford algorithm deviates to node C just for the sake of traversing an extra negatively weighted edge. This highlights a potential lack of biological plausibility with respect to shortest paths as routes of information flow in the brain. Rather than using the Bellman-Ford algorithm, a more typical approach for dealing with negatively weighted edges is to add a constant to each of the edge weights so that the smallest edge weight becomes zero. Dijkstra's algorithm can then be used on the remapped edge weights. For the network shown in **(c)**, this amounts to adding 2 to each edge weight, since the smallest edge is of weight -2. Dijkstra's algorithm then identifies the path colored in red as the shortest. But it is trivial to see that the true shortest path is of length 0 and colored in blue. This example demonstrates that remapping the edge weights to eliminate negatively weighted edges does not necessarily ensure that the true shortest paths can be identified with Dijkstra's algorithm.

Figure 7.4b shows a simple example in which Dijkstra's algorithm is unable to determine the shortest path in a directed network comprising negatively weighted edges, but no negative cycles. The Bellman-Ford algorithm (Bellman, 1958) can be used to determine shortest paths in directed networks with negatively weighted edges as long as no negative cycles are present. The Bellman-Ford algorithm can also be used to detect the presence of negative cycles. The algorithm is quite similar to Dijkstra's algorithm. The main difference is that nodes are no longer marked as visited and unvisited. This means each node's path length can be updated at any iteration of the algorithm, whereas Dijkstra's algorithm only allows the path length of unvisited nodes to be updated. To begin, the initial node is assigned a path length of zero, while all other nodes are assigned infinity, just like Dijkstra's algorithm. The following two steps are then repeated $N-1$ times, where N is the number of nodes in the network:

1. Consider an edge in the network. Suppose that edge is from node j to node i. If the path length assigned to node i relative to the initial node is greater than the path length assigned to node j plus the remapped edge weight w_{ij}, then update the path length of node i to equal the path length of node j plus w_{ij}. If an update is performed, set the predecessor of node i to node j. (Remember that in directed networks, we use the convention w_{ij} to mean the weight of the edge from node j to node i.)
2. Repeat Step 1 for all edges in the network.

As with Dijkstra's algorithm, the shortest path from the initial node to a desired target node can then be delineated by backtracking through the predecessor of each node, beginning with the target node. To detect the presence of any negative cycles, an additional iteration of the above two steps can be performed, bringing the total number of iterations to N. If the path length of *any* node is updated during this additional iteration, it can be concluded that a negative cycle is present, and thus the concept of a shortest path is rendered ill-defined for any node that can reach the negative cycle.

The Bellman-Ford algorithm is rarely used in the connectomics literature because it is most relevant to *directed* graphs, whereas the vast majority of brain networks that have been analyzed to date are undirected, such as those considered in studies of human brain networks. The most common approach to deal with negatively weighted edges simply involves removing them. This can be achieved by thresholding away the negatively weighted edges (Chapter 11) or by remapping the edge weights such that all negatively weighted edges become positive. Taking the absolute value of each edge weight is the simplest remapping that achieves this goal. According to this remapping, an edge with a large negative weight becomes an edge with a large positive weight. This remapping essentially discards any distinction in edge polarity, which is undesirable if the polarity of an edge is assumed to distinguish between inhibitory and excitatory connections, or some other distinct from of interaction between nodes. An alternative mapping involves identifying the edge with the most negative weight and then *adding* the absolute value of its weight to the weights of all edges in the network. This remapping can be written as,

$$w_{ij} \leftarrow w_{ij} + |\min_{uv} w_{uv}|, \tag{7.2}$$

where $|\min_{uv} w_{uv}|$ denotes the absolute value of the most negative edge weight. This remapping ensures that the smallest edge weight becomes zero and simply involves adding a constant to each edge weight. Dijsktra's algorithm can then be used to compute shortest paths in the set of remapped edge weights.

It is important to note that the shortest paths yielded by applying Dijkstra's algorithm to the edge weights remapped according to Equation (7.2) or some other remapping that ignores edge polarity is not necessarily the same as the shortest paths yielded by applying the Bellman-Ford algorithm to the original edge weights. It is the Bellman-Ford algorithm that yields the correct set of shortest paths; namely, the paths that minimize the sum of edge weights. Let's consider a simple example to demonstrate this point. Suppose we have two distinct paths between a pair of nodes. The first path comprises a single edge of weight 1. The second path comprises three edges, two of weight 1 and the other of weight -2. Which of these two paths has the shortest path length? The path length of the first path is obviously 1, while the second is $1 + 1 - 2 = 0$. The second path therefore has the shortest path length and is the path that will be identified by the Bellman-Ford algorithm. Now consider what happens if we apply the remapping given by Equation (7.2) to the edge weights and then use Dijkstra's algorithm to compute shortest paths. This amounts to adding 2 to each of the edge weights. After this remapping is applied, the path length of the first path becomes 3, while the second becomes $3 + 3 + 0 = 6$. Hence, the first path is incorrectly identified as the shortest path. This example is illustrated in Figure 7.4c. While the remapping given by Equation (7.2) can yield incorrect shortest paths from the perspective of minimizing the sum of edge weights, these shortest paths might still be meaningful from a neurobiological perspective. For example, while the first path in the above example is not the shortest path, it traverses fewer hops (one vs. three hops) and might therefore be advantageous in terms of maximizing communication efficiency and fidelity.

In summary, analyzing shortest paths in brain networks with negatively weighted edges is challenging. Most of the connectomics literature deals with negative weights by eliminating them through thresholding, reversing their polarity or adding a constant to all the edge weights so that the smallest edge weight becomes zero. The Bellman-Ford algorithm is a more principled approach to the computation of shortest paths in networks with negatively weighted edges, although it is limited to directed networks without negative cycles. It remains to be determined which of these approaches yields the most plausible routes in brain networks, or whether neural information is indeed routed via the shortest paths of a nervous system. These aspects are considered in greater detail in Box 7.1, particularly with respect to the validity and interpretation of shortest paths in correlation-based functional networks.

BOX 7.1 INTERPRETING SHORTEST PATHS IN BRAIN NETWORKS

The analysis of shortest paths is pervasive in connectomics, but empirical evidence supporting the physiological significance of shortest paths in nervous systems has not been rigorously established. The study of shortest paths in correlation-based functional networks is especially susceptible to interpretational difficulties, beyond the challenges posed by the potential for negative edge weights. Numerous authors have noted that a strong functional connection between a pair of regions does not exclusively indicate the presence of a direct anatomical connection between those regions, since strongly correlated activity between distinct neural elements can result from one or more indirect, multi-synaptic (multi-hop) paths (Rubinov and Sporns, 2010; Smith et al., 2011; Zalesky et al., 2012b). Consequently, the weight of a functional connection inherently captures both *direct* connections and *indirect* paths that traverse one or more intermediate nodes. Given that the edge weight itself already encapsulates the contribution of paths, the logic of studying paths in correlation-based functional networks is rather ambiguous. Paths in functional networks represent sequences of statistical associations (Rubinov and Sporns, 2010), but they are unlikely to characterize meaningful routes of information flow. Despite these challenges, shortest path lengths in functional networks have proven to be a useful biomarker. For example, they are an accurate predictor of an individual's intelligence (van den Heuvel et al., 2009) and other cognitive abilities (Bassett et al., 2009), they are significantly heritable (Fornito et al., 2011b; van den Heuvel et al., 2013a), and they differ significantly in patients with psychiatric or neurological disorders, compared to the general population (Xia and He, 2011).

Measures of functional connectivity have been proposed to exclusively capture direct connections between pairs of regions by suppressing the contribution of indirect paths. The most well-known of these is the partial correlation coefficient (Marrelec et al., 2009; Salvador et al., 2005; Brier et al., 2015; Varoquaux and Craddock, 2013). Shortest paths in partial correlation-based brain networks may better characterize

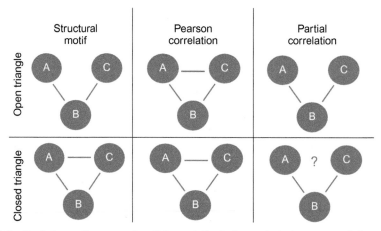

FIGURE 7.1.1 Distinction between Pearson and partial correlation in the context of an open and closed triangle motif. The first column shows two structural **motifs**. The second and third column show the functional networks estimated by computing the pairwise Pearson and partial correlation, respectively, between the neural dynamics arising from these structural motifs. Pearson correlation estimates a direct connection between nodes *A* and *C* of the open triangle motif. This "spurious" connection is shown in red and represents a false positive with respect to the underlying structural motif. The shortest path in the structural motif between nodes *A* and *C* must traverse two hops, via node *B*, whereas a direct connection is available in the Pearson correlation network. Shortest paths in correlation-based functional networks are therefore unlikely to characterize meaningful routes of information flow, since the correlation coefficient inherently captures the presence of any paths. Shortest paths computed in partial correlation functional networks may better characterize information flow, although in some circumstances this measure of functional connectivity can mistake closed triangle motifs for open ones, as suggested by the question mark.

BOX 7.1 INTERPRETING SHORTEST PATHS IN BRAIN NETWORKS—CONT'D

signaling routes and information flow because the contribution of indirect paths (shared variance) is essentially regressed from the edge weights, leaving only the contribution of the direct connections. The downside of partial correlation as a measure of functional connectivity is that it can incorrectly suppress direct connections that are supported by an underlying axonal connection. In other words, true functional connections can be overlooked with partial correlation. This typically results in closed triangle motifs being spuriously "broken open" and mistaken for open triangles, leading to a drastic reduction in measures such as the **clustering coefficient** in partial correlation functional networks (Zalesky et al., 2012b). Figure 7.1.1 shows the distinction between Pearson and partial

correlation in the context of a closed and open triangle motif. While some studies advocate the use of partial correlation, they have evaluated this measure using test networks that are devoid of any closed triangle motifs (i.e., the clustering coefficient is zero), and thus the important effect of closed triangles being mistaken for open ones is not evident (Smith et al., 2011). The graphical LASSO (Friedman et al., 2008) and other regularization approaches yield promising measures of functional connectivity that provide a balance between the benefits of Pearson and partial correlation (Varoquaux and Craddock, 2013). Their future use may facilitate better interpretability of shortest paths and other networks measures computed in functional brain networks.

7.2 SHORTEST PATH ROUTING

It is common to average the path length over all shortest paths in a network to yield a summary measure of the efficiency with which information can be routed via shortest paths. A network with a short average path length between nodes is considered topologically integrated and an efficient network (Latora and Marchiori, 2001). In this section, we consider related measures that quantify network integration and the efficiency of shortest path routing. Much like Alice, the tourist that we encountered at the start of this chapter who was able to find her destination with the use of a map, these measures assume that neural information is routed along the globally optimal (i.e., shortest) path between nodes.

7.2.1 Characteristic Path Length

The characteristic path length, L, is the average shortest path length between all possible pairs of nodes in a network,

$$L = \frac{1}{N} \sum_i l_i = \frac{1}{N(N-1)} \sum_{i \neq j} l_{ij}, \tag{7.3}$$

where l_i is the average shortest path length from node i to all other nodes and l_{ij} is the shortest path length from node j to node i, which is computed with Dijkstra's algorithm or one of its variants. This formulation is applicable to both directed and undirected networks.

The characteristic path length was popularized by Watts and Strogatz (1998) in their seminal work on small-world networks. Together with network efficiency (Section 7.2.2), it is the simplest and most widely used measure of integration in brain networks (Rubinov and Sporns, 2010). Brain networks with a short characteristic path length are thought to integrate information more efficiently between nodes. The intuition here is simple: a short characteristic path length means that information can, on average, be routed between pairs of nodes using only a few edges. In this way, short path lengths minimize the metabolic **cost** associated with routing action potentials across axons and synaptic contacts, and hence could provide faster, more direct, and less noisy information transfer (Bullmore and Sporns, 2009).

The characteristic path length is ill-defined in fragmented networks, since by definition, a path will not exist between at least one pair of nodes. The path length for such a node pair is assumed to be infinite, meaning the characteristic path length will also be infinite. Various alternatives can be used to meaningfully define the characteristic path length of a fragmented network. These include constraining the average to only those node pairs between which a path exists (i.e., estimating the average separately for each **connected component** of the network); setting the path length of all node pairs between which a path does *not* exist to the longest (finite) path length in the network; or applying a specialized thresholding method that necessarily yields connected graphs (Chapter 11). A common approach is to compute the average shortest path length between all possible node pairs using the harmonic mean (Newman, 2003a),

$$L' = N(N-1) \left[\sum_{i \neq j} \frac{1}{l_{ij}} \right]^{-1}. \qquad (7.4)$$

The harmonic mean formulation of the characteristic path length can be meaningfully computed in fragmented networks because the reciprocal of the path length, $1/l_{ij}$, is summed. The reciprocal of the path length between disconnected nodes is zero, under the assumption that the reciprocal of infinity is zero, and thus the harmonic mean is necessarily finite (unless the graph contains no edges at all), irrespective of whether the path length for a particular node pair is infinite. In this way, considering the reciprocal of the path length simplifies the numerical issues in estimating the path length between disconnected nodes in fragmented graphs.

Apart from the fact that the harmonic mean formulation of the characteristic path length remains meaningful in fragmented networks, another advantage is that it is less susceptible to the influence of node pairs with exceptionally long shortest paths. Such node pairs might represent outliers. It can be argued that their contribution to the average should be down-weighted. Based on the same rationale, it has also been argued that the harmonic mean is a **hub**-centric measure of integration because it emphasizes the contribution of path lengths originating from hub regions and down-weights peripheral nodes (Achard and Bullmore, 2007).

Let's consider a concrete example to demonstrate some of these aspects of the harmonic mean. Suppose the shortest path lengths between all possible node pairs in a small network are 2, 2, 3, 1, and 12. The path length of 12 appears to be an outlier and might be due to a network reconstruction anomaly or associated with a weakly connected peripheral node. The characteristic path length is $L = 4$, whereas the harmonic mean is $L' = 2.1$. The latter is clearly more characteristic of the majority of path lengths in this network. The exceptionally long shortest path length of 12 should not diminish integration from a global perspective, since many other node pairs remain well integrated with short path lengths and therefore shortest path routing will, on average, remain efficient.

Path lengths can also be computed for individual nodes. To determine the extent to which a specific node is integrated within its network, the path length between that node and all other $N - 1$ nodes is averaged to yield a node-specific measure of integration. The harmonic mean can be used to compute this average.

In summary, the characteristic path length provides a global measure of a network's capacity to integrate information using shortest path routing. Using the harmonic mean to compute the characteristic path length is desirable in fragmented networks and yields a measure of global integration that is not unduly influenced by a small proportion of node pairs with exceptionally long shortest paths.

7.2.2 Global and Nodal Efficiency

Network efficiency is an alternative measure of integration that is closely related to the characteristic path length. In fact, a network's global efficiency, E_{glob}, is the reciprocal of the harmonic mean of its path lengths,

$$E_{\text{glob}} = \frac{1}{L'} = \frac{1}{N(N-1)} \sum_{i \neq j} \frac{1}{l_{ij}}. \tag{7.5}$$

Global efficiency was proposed by Latora and Marchiori (2001) in their attempt to characterize small-world behavior in terms of a single variable. They contend that E_{glob} is the efficiency of information exchange in a parallel system in which all nodes are capable of concurrently exchanging information via shortest paths, whereas $1/L$ is a better measure of the efficiency of a sequential system in which information is processed serially. Since the brain is thought to be a massively parallel system, it has been argued that E_{glob} provides a truer characterization of connectome integration (Achard and Bullmore, 2007). This definition is valid both for unweighted and weighted graphs, and can also be applied to directed networks. Network **fragmentation** does not present a problem.

In practice, $1/L$ and E_{glob} are usually highly correlated, unless the distribution of shortest paths is skewed, in which case E_{glob} is less influenced by the long tail of a skewed distribution. Analyzing a network's characteristic path length in addition to its global efficiency is therefore often redundant.

Efficiency can also be defined for an individual node. Specifically, the nodal, or regional, efficiency of the jth node is given by,

$$E_{nodal}(j) = \frac{1}{N-1} \sum_i \frac{1}{l_{ij}}. \tag{7.6}$$

The nodal efficiency is the normalized sum of the reciprocal of the shortest path lengths from a given node to all other nodes in the network. We can use this measure to localize global efficiency effects to specific nodes. Nodal efficiency quantifies how well a given region is integrated within the network via its shortest paths. Hub nodes tend to have the highest nodal efficiency (van den Heuvel and Sporns, 2013a), facilitating their central role in efficiently integrating and distributing information. Note that this definition is identical to the measure of **closeness** centrality defined in Equation (5.7).

Latora and Marchiori (2001) also define a node-specific measure known as the local efficiency. They define local efficiency of the ith node, $E_{loc}(i)$, such that,

$$E_{loc}(i) = \frac{1}{N_{G_i}(N_{G_i} - 1)} \sum_{j,h \in G_i} \frac{1}{l_{jh}}, \tag{7.7}$$

where G_i denotes the **subgraph** comprising all nodes that are immediate neighbors of the ith node. In other words, the local efficiency of node i is defined with respect to the subgraph comprising all of i's neighbors, after removal of i and its incident edges. The local efficiencies for each node can be averaged over all nodes to estimate the mean local efficiency of a network.

Nodal and local efficiency should not be confused. Whereas nodal efficiency characterizes the extent to which a node is integrated within its entire network, local efficiency reflects the extent of integration between the immediate neighbors of the given node. In this way, local efficiency can be considered a generalization of the clustering coefficient that explicitly takes into account paths. The clustering coefficient only counts the direct connections between the neighbors of a node (see Chapter 8). Both indirect paths and direct connections are considered by local efficiency. Since the node for which local efficiency is computed is not an element of the subgraph G_i, it has been argued that local efficiency also provides a measure of fault tolerance that indicates how efficiently the neighbors of a node are able to communicate when that node is disrupted (Achard and Bullmore, 2007; Latora and Marchiori, 2003).

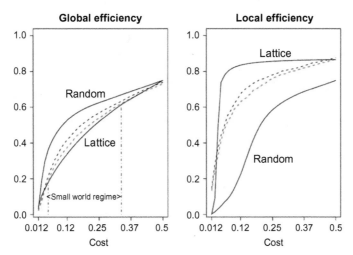

FIGURE 7.5 The global and local efficiency of correlation-based functional brain networks. Global and local network efficiency plotted as a function of network **cost** in correlation-based functional networks mapped in groups of young (dashed black line) and old (dashed red line) healthy individuals. The solid lines represent the efficiencies in appropriately matched **lattice** and random networks. Random networks are the most globally efficient networks, whereas lattices are the most locally efficient. Cost here is approximated by the **connection density**. Small-world organization is evident over the range of costs for which the empirically measured brain networks exceed the global efficiency of appropriately matched lattice networks as well as the local efficiency of random networks. *Image reproduced from Achard and Bullmore (2007), with permission.*

Latora and Marchiori (2001) argue that small-world networks can be parsimoniously defined as networks that are both globally and locally efficient. In other words, small-world networks are efficient integrators of information over a range of scales. Figure 7.5 shows the global and local efficiency plotted against the connection density (considered as a proxy for wiring cost) for correlation-based functional brain networks. Small-world organization is evident over the range of connection densities for which the brain network exceeds the global efficiency of appropriately matched **lattice** networks, as well as the local efficiency of random networks (Chapters 8 and 10). Using this approach, network efficiency can be used to identify small-world organization. As we will see in Chapter 8, small-world networks such as the brain are able to achieve this simultaneously high level of global and local topological efficiency for relatively low wiring cost, giving rise to a cost-efficient or economical organization.

Most of the literature to date has focused on characterizing the efficiency of static networks mapped over a single, extended period of time. Network efficiency has also been analyzed dynamically in a time-resolved manner both with respect to spontaneous fluctuations measured with resting-state **functional MRI** (Zalesky et al., 2014) and task-evoked activity measured with **MEG** (Kitzbichler et al., 2011). In this work, **sliding window analysis** was used to map changes in functional network topology as a function of time. The findings of both studies suggest that functional connectivity networks transition between states characterized by high and low global efficiency, which may reflect the emergence of an integrated but costly workspace configuration when the demand for cognitive processing is high, and a more segregated, locally efficient and less metabolically demanding configuration when task demands are low (Bullmore and Sporns, 2012). Examples are shown in Figure 7.6 (see also Figure 1.10).

7.3 DIFFUSION PROCESSES

We have discussed how it is unlikely that any individual node within a brain network possesses the global knowledge required to find the shortest path to another network element. For this reason, a process of diffusion has been suggested as a more appropriate model of neural information flow, since it can operate without global knowledge of the network topology (see also Chapter 5).

Whereas navigation (described later in this chapter) and shortest path routing imply that information flows along a single path to a single destination, a diffusion process implies that the flow of information is dispersive, propagating simultaneously along multiple "fronts," without being directed towards a single destination (similar to the parallel transfer model discussed in Chapter 5). As shown in Figure 7.7, shortest path routing is associated with a single path to a single destination, whereas with diffusion, information sent from a single node can reach any of a number of destinations via any of a number of paths. In this way, diffusion can be considered a process of broadcasting information, whereas routing is akin to sending a letter in the mail to a single destination (see also Chapter 5).

In this section, we will see that shortest paths are still useful in measuring the efficiency of information flow under a diffusion model. The analysis of diffusion on a graph is best understood with analogy to a random walker, such as our tourist Bob, who was introduced at the start of this chapter. Bob was content to walk the streets of Paris without a map by randomly picking one of the various roads available to him whenever he reached an intersection. If we follow Bob for a sufficient amount of time, he will eventually reach the Eiffel Tower. What is the probability of Bob reaching this destination by the shortest path—the same path followed by Alice, who used a map of Paris to get there? If the probability is high, the shortest path is accessible, and there will be little difference in the efficiency of the paths taken by Bob and Alice. If this probability is low, the shortest path is "hidden," and Alice's route will be much more efficient than Bob's because the distance Alice walks to reach the Eiffel Tower is shorter. In this way, understanding how easy it is to find the shortest path via a random walk allows us to understand the efficiency of communication on a network in which signals propagate according to a diffusion process.

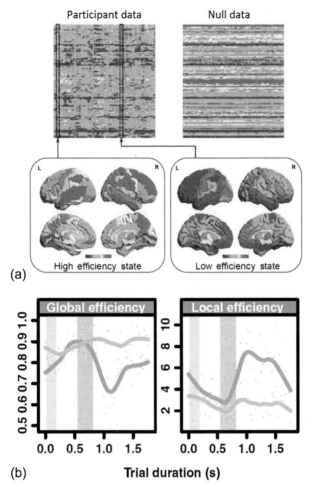

(a)

(b) Trial duration (s)

FIGURE 7.6 The dynamics of network efficiency in time-resolved functional brain networks mapped in healthy human volunteers. **(a)** Nodal efficiency is plotted as function of time using sliding window analysis, measured with resting-state functional MRI in a single participant. Each row represents one of 116 anatomically defined regions and time progresses from left to right on the horizontal axis. Warmer colors indicate higher values of efficiency. Fluctuations in nodal efficiency are coordinated across multiple regions and these fluctuations are absent in null data generated with a stationary **vector autoregressive** process (see Chapter 10). Multiple spatially distributed regions simultaneously increase, for brief intervals, their nodal efficiency and, by inference, their capacity to transfer information. However, these intervals of high efficiency are supported by long anatomical connections and are thus likely to carry an extra metabolic cost. Nodal efficiencies were rendered onto the cortical surface for representative high- and low-efficiency states. **(b)** Global and local efficiency were mapped in a group of individuals during performance of a working memory task. Functional brain networks were mapped dynamically with MEG and a sliding window approach to investigate network changes with an approximate time resolution of 10 ms over the course of memory trials lasting 1.6 s each. In an easy version of the task (red line), working memory is only required for the first half (800 ms) of the trial; there are no active cognitive demands in the second half of the trial. In the difficult version of the task (blue line), working memory performance is required continuously throughout the trial. Global efficiency (integration) is high and local efficiency (clustering) is low in networks recorded under both task conditions for the first half of the trials. In the second half of the trials, global efficiency remains high in networks recorded during performance of the difficult task, but there is a rapid decrease in the global efficiency (and increase in local efficiency) of networks recorded under easy task conditions. These data are compatible with the theory that effortful cognition depends on a highly integrated, global workspace architecture, and that release of cognitive effort is associated with relaxation of the functional network architecture to a less globally efficient but locally clustered and metabolically less expensive configuration (Figure 1.10). *Reproduced (a) from Zalesky et al. (2014) and (b) from Kitzbichler et al. (2011), with permission.*

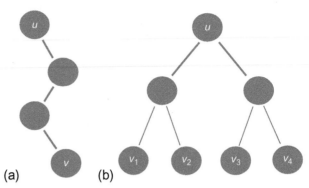

FIGURE 7.7 Shortest path routing and diffusion are two models describing the flow of neural information. **(a)** Under a routing model, information is routed along a single shortest length path from node *u* to a single destination, denoted node *v*. While this is the most widely studied model of neural information flow, it assumes that information is intended for a single destination and that a shortest path to that destination can be readily identified. **(b)** A diffusion process is such that the flow of information is dispersive, propagating simultaneously along multiple "fronts," without being directed towards a single destination. Four possible destination nodes are shown, $v_1, \ldots, v_4$. If the flow comprises discrete entities, an outgoing edge is chosen randomly at each fork in the network, possibly biased by the edge weight. The flow of information is therefore distributed among multiple destinations using multiple paths (see also Chapter 5).

7.3.1   Search Information and Path Transitivity

Goñi and colleagues (2014) were among the first to empirically challenge the untested assumption that information transfer in nervous systems necessarily follows shortest paths. They did this by examining how two specific characteristics of the shortest path between two brain regions, as identified using **structural connectivity** networks generated from diffusion MRI, related to the functional connectivity of those areas. These two characteristics are called *search information* and *path transitivity*.

To understand these measures, we can imagine that neural information diffuses through a network much like the flow of fluids in a network of pipes. Fluid flow is in fact a prime example of a diffusion process operating on a network. The fluid can be imagined to comprise discrete particles (random walkers), where each particle randomly selects one of the outgoing pipes at each branch point. The pipes outgoing from each node traversed by a path play an important role in governing the flow of the fluid, because they represent branch points at which a portion of the flow can be diverted. Each of these diversions presumably weakens the strength of the fluid flow on the underlying shortest path between the source and destination. The extent of this dispersion is quantified using the search information. Paths associated with high search information are characterized by many branch points. Although fluid can be diverted at each branch point, thereby weakening the underlying flow, this diverted fluid may eventually return to the underlying shortest path at an upstream node. In this case, the flow of information on the underlying path is therefore restored and the branch point acts like a local detour. The number of local detours along a path is measured using path transitivity.

Search information (Sneppen et al., 2005; Trusina et al., 2005) measures the difficulty encountered by a random walker like Bob in identifying the shortest path between a pair of nodes as a matter of chance. Consider a path from node u to node v that traverses the ordered sequence of nodes $\Omega_{vu} = (u, a_1, a_2, ..., a_J, v)$. The search information associated with Ω_{vu} is determined by the probability, $P(\Omega_{vu})$, that a random walker beginning at node u reaches node v via the ordered sequence of nodes Ω_{vu}. More specifically, the search information of Ω_{vu} is given by $S(\Omega_{vu}) = -\log_2 P(\Omega_{vu})$. The logarithmic remapping of $P(\Omega_{vu})$ ensures that the search information is high when the probability of a random walker traversing the path Ω_{vu} is low, and vice versa.

The probability that a random walker beginning at node u reaches node v via Ω_{vu} is not necessarily equal to the probability that a random walker beginning at node v reaches node u via $\Omega_{uv} = (v, a_J, ..., a_2, a_1, u)$. In other words, $P(\Omega_{uv}) \neq P(\Omega_{vu})$. This does not present a problem for directed networks. However, for undirected networks, this implies that the search information associated with a bidirectional path is dependent on which end of the path the random walker is initiated. To eliminate this dependence, the search information associated with a bidirectional path is defined as the average of the search information associated with Ω_{uv} and Ω_{vu} (Goñi et al., 2014); that is,

$$S(\Omega_{vu}) = \frac{-\log_2 P(\Omega_{uv}) - \log_2 P(\Omega_{vu})}{2}. \tag{7.8}$$

The probability, $P(\Omega_{vu})$, that a random walker beginning at node u traverses the path defined by the sequence of ordered nodes $\Omega_{vu} = (u, a_1, a_2, ..., a_J, v)$ can be expressed by the product,

$$P(\Omega_{vu}) = \frac{w_{a_1 u}}{s_u} \times \frac{w_{a_2 a_1}}{s_{a_1}} \times \cdots \times \frac{w_{v a_J}}{s_{a_J}}, \tag{7.9}$$

where w_{ij} is the weight of the edge from node j to node i and $s_j = \sum_{i \neq j} w_{ij}$ is the **strength** of node j, as defined in Chapter 4. The term $w_{a_1 u}/s_u$ is the probability that a random walker located at node u travels to node a_1, while $w_{a_2 a_1}/s_{a_1}$ is the probability of traveling from node a_1 to node a_2, and so on.

The formulation of search information given by Equation (7.9) enables a random walker to return to the node it just came from, whereas alternative formulations preclude walkers from doubling back (Sneppen et al., 2005). It is important that the edge weights are not remapped according to Equation (7.1) when computing $P(\Omega_{vu})$, since this would mean that the random walker is biased towards traversing edges with low connectivity values. In other words, edge weights should be remapped when computing shortest paths, but the original edge weights should be used when computing a shortest path's search information. Equation (7.9) applies to binary networks as well. For directed networks, we simply need to ensure that the outgoing edge strength is used. In general, the search information for a path between two nodes will be high when the intervening nodes along that path have high **degree**. This means that it is harder to find the shortest path under a diffusion model when that path comprises many high-degree nodes. Search information cannot be readily computed in networks with negatively weighted edges, owing to the difficulty in dealing with negative transition probabilities.

The second property considered in the analysis of Goñi and colleagues (2014) was path transitivity, which is the number of local detours along a path. Goñi and colleagues only consider local detours that traverse two edges when measuring path transitivity. To this end, the matching index is first computed for every possible pair of nodes comprising the path under consideration. The matching index between nodes i and j in an undirected network is given by,

$$m_{ij} = \frac{\sum_{k \neq i, j} (w_{ik} + w_{jk}) \mathbf{1}_{w_{ik}} \mathbf{1}_{w_{jk}}}{\sum_{k \neq j} w_{ik} + \sum_{k \neq i} w_{jk}}, \tag{7.10}$$

where $\mathbf{1}_{w_{ij}} = 1$ if $w_{ij} > 0$, otherwise $\mathbf{1}_{w_{ij}} = 0$, and w_{ij} is the weight of the edge between nodes i and j. The node indexed by k is the intermediate node along a detour between nodes i and j. For a detour via node k to exist, there must be an edge between nodes i and k as well as nodes j and k, in which case the weight of those two edges is contributed to the sum in the numerator of Equation (7.10). If one or both of those edges is absent, a detour cannot exist and thus $\mathbf{1}_{w_{ik}} \mathbf{1}_{w_{jk}} = 0$, ensuring zero weight is contributed to the sum in the numerator. The denominator is a normalizing constant that ensures the matching index is a value between zero and one. As with search information, the remapping given by Equation (7.1) should not be applied to the edge weights when computing the matching index.

The path transitivity of a path defined by the ordered sequence of nodes $\Omega_{vu} = (u, a_1, a_2, \ldots, v)$ is given by summing the matching index for every possible pair of nodes comprising the path,

$$M(\Omega_{vu}) = \frac{2}{|\Omega_{vu}||\Omega_{vu} - 1|} \sum_{i > j \in \Omega_{vu}} m_{ij}. \tag{7.11}$$

The notation $|x|$ denotes the cardinality of the set x, or the number of elements that it contains. As such, $|\Omega_{vu}||\Omega_{vu} - 1|/2$ is the total number of pairs of nodes comprising Ω_{vu}. Figure 7.8 provides an example illustrating the difference in search information and path transitivity.

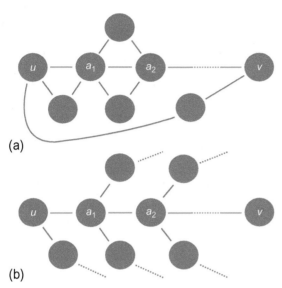

FIGURE 7.8 Search information and path transitivity. Search information and path transitivity measure the ease with which a diffusion process can find the shortest paths in a network. The shortest path between nodes u and v is shown in green and has the same path length and hop count in **(a)** and **(b)**. Each of the nodes comprising the shortest path provides a branch point at which information can be diverted. This is analogous to the flow of a fluid in a network of pipes, where the fluid disperses along all branch points rather than following a single path. In **(a)**, any information that is diverted along a branch point eventually returns to the underlying path at an upstream node, and thus each diversion can be considered a local detour. In **(b)**, each branch point does not return to the underlying path, at least not within two hops, and thus any information diverted along these branch points is unlikely to reach the desired destination. Accordingly, the shortest path in **(a)** has high transitivity and low search information, whereas the path in **(b)** has low transitivity and high search information. In other words, the path in **(a)** is more "accessible" in the absence of global knowledge of the network. If our random walker, Bob, begins his journey at node u, he will find node v with probability 1 in the case of **(a)**. In the case of **(b)** the probability that Bob finds node v is substantially smaller.

Goñi and colleagues (2014) computed the search information and path transitivity for each pair of nodes in human structural connectivity networks mapped with diffusion imaging. Functional brain networks were then mapped in the same individuals using resting-state functional MRI. The authors tested whether the search information and path transitivity of the shortest path between a pair of regions in the structural network were predictors of the functional connectivity between that node pair. In this way, a strong functional connection was taken as an index of efficient diffusion between the corresponding pairs of regions.

In general, the presence of a direct structural connection is the most accurate predictor of functional connectivity, and the strength of both structural and functional connectivity decline as a function of the physical separation (measured by Euclidean distance) between regions (Goñi et al., 2014; Honey et al., 2007). However, two brain regions can still show strongly correlated activity in the absence of a direct structural connection (Chapters 2 and 10), presumably reflecting polysynaptic interactions. In these cases, Goñi et al., 2014 found that strong functional connectivity was not consistently observed between pairs of regions interconnected by shortest paths exceeding a hop count of two. Crucially, they found that in most cases, the shortest path's search information and transitivity were better predictors of functional connectivity strength compared to the path's hop count and Euclidean distance, and compared to predictions based on a biophysical neural mass model of brain dynamics (Deco et al., 2008). These findings are shown in Figure 7.9 and suggest that direct anatomical connections and two-hop anatomical paths with low search information and/or high transitivity can be used to accurately predict the presence of strong functional connections.

In the case of a direct connection, the search information reduces to a function of the degree of the pair of nodes linked by that connection. This implies functional connectivity is strong between pairs of low degree nodes that are linked by a direct structural connection. This is consistent with the intuition that any information generated by a node of degree one has no choice but to flow to the sole region to which it is connected. More generally, the findings of Goñi and colleagues (2014) suggest that neural signaling is not exclusively determined by the shortest paths in brain networks (if at all), especially if those paths are polysynaptic and traverse multiple hops. If neural information flow was exclusively constrained to these paths, then measures such as search information and path transitivity should not be able to predict functional connectivity—information should be routed with perfect fidelity along the shortest path. The strong relationship identified between functional connectivity, search information and path transitivity suggests that information does not necessarily flow along the shortest paths of nervous systems.

7.3.2   Measures of Diffusion Efficiency

Thus far, we have considered the difficulty with which a diffusion process can find the shortest paths in a network. Here, we turn our attention to quantifying the efficiency with which information can be communicated using a diffusion process.

The measures considered in this section can be contextualized in terms of our random walker Bob, introduced at the start of this chapter. We will compute the probability of Bob serendipitously taking the shortest hop path from his hotel

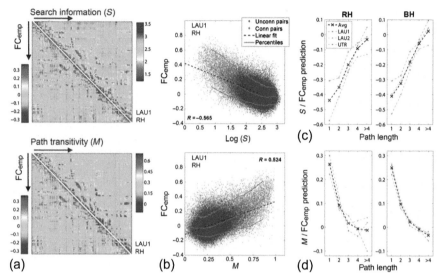

FIGURE 7.9 Search information and path transitivity as predictors of functional connectivity. The search information and path transitivity of the shortest paths in human brain structural connectivity networks mapped with diffusion MRI were used to predict functional connectivity strength, as measured with resting-state functional MRI. **(a)** The upper triangular part of each matrix shows the search information (upper matrix) and path transitivity (lower matrix) corresponding to the shortest path for each node pair in the right hemisphere (RH). The lower triangle of each matrix shows the corresponding functional connectivity strength for each node pair, measured using Pearson correlation (FCemp). The matrices have been scaled to ±3 standard deviations. The variation in functional connectivity strength across all node pairs is well predicted by the shortest path's search information and transitivity. **(b)** Functional connectivity plotted as a function of search information (upper scatter plot) and path transitivity (lower scatter plot). Each data point represents a unique node pair located in the RH. Node pairs with a direct structural connection are colored red, while all other connections are colored blue. **(c-d)** Node pairs were stratified into bins based on the number of hops (steps) comprising their shortest path in the structural network. For each bin, the extent to which functional connectivity was correlated with search information **(c)** and path transitivity **(d)** is shown for three different data sets (LAU1, LAU2, and UTR) and for RH node pairs as well as all node pairs in both hemispheres (BH). The accuracy with which search information and path transitivity predict functional connectivity rapidly decreases with increasing number of hops comprising the shortest paths. *Image reproduced from Goñi and colleagues (2014), with permission.*

to a specified destination (Eiffel Tower), as well as the average number of hops that Bob traverses to reach this destination. The former is called the *shortest path probability* and the latter is known as the *mean first passage time* (Wang and Pei, 2008). These measures reveal the efficiency with which information can be communicated between a pair of nodes using a diffusion process. Following the work of Goñi and colleagues (2013), we will also compute the number of times that Bob needs to commence a *new* random walk from his hotel to ensure that he reaches the Eiffel Tower on at least one occasion by the shortest hop path. This is called the *diffusion efficiency*.

Suppose Bob commences a new random walk from his hotel on each day of his holiday. How many days must Bob stay in Paris for him to encounter a day where he walks to the Eiffel Tower via the shortest path? This is known as the *resource efficiency* of a diffusion process, and quantifies the amount of resources required (e.g., the number of messages that need to be sent) to ensure at least one of them reaches the desired destination using the shortest hop path.

To derive these measures of diffusion efficiency, we make use of the matrix $U = WS^{-1}$, where S is the diagonal matrix,

$$S = \begin{bmatrix} s_1 & 0 & 0 \\ 0 & \ddots & \vdots \\ 0 & \cdots & s_N \end{bmatrix}, \tag{7.12}$$

s_i is the strength of the ith node, and W is our adjacency matrix (weighted or binary). It can be seen that $U_{ij} = w_{ij}/s_j$ is the probability that a random walker located at node j steps to node i. For those readers familiar with Markov chains, U is the transition matrix of the Markov chain with states corresponding to the network nodes. Furthermore, the probabilities used to compute search information in Equation (7.9) are elements of U. For example, the probability w_{a_1u}/s_u appearing in Equation (7.9) is given by U_{a_1u}. The matrix U is also related to the **Laplacian matrix** (Box 7.2).

We can quantify the probability of a random walker traversing the least number of hops, H, from node i to node j using a measure called the shortest path probability. The shortest path probability is derived by setting to zero all the elements of the jth row of U, which effectively means that a walker cannot enter node j from any node. If we let U_j denote the matrix U with all elements in the jth row set to zero, the shortest path probability from node i to node j is then,

$$\pi_{ij} = 1 - \sum_{n=1}^{N} \left[U_j^H \right]_{ni}. \tag{7.13}$$

The matrix U_j^H expresses the probabilities of a random walker traveling from one node to another in exactly H steps, with the option of traveling to node j removed. Summing the (n, i) the elements of U_j^H overall nodes $n = 1, ..., N$ gives the probability that the walker is at any node other than node j after H steps, and thus the complement of this probability gives the probability that the walker is at j in H steps. Following Goñi and colleagues (2013), the shortest path probability of the whole network is given by the average $\sum_{i \neq j} \pi_{ij}/N(N-1)$.

BOX 7.2 THE GRAPH LAPLACIAN

The Laplacian of a graph, or Laplacian matrix, is a transformation of a network's connectivity matrix. The eigenvalues of a network's Laplacian matrix provide clues about its connected components, **community** structure, **motif** distribution, bipartiteness, and other topological properties. Networks can also be classified according to differences in the distribution of the **eigenvalues** of their Laplacian matrix (Banerjee and Jost, 2009; de Lange et al., 2014).

The Laplacian matrix is given by,

$$\Lambda = S - W = \begin{cases} s_i, & i=j \\ -w_{ij}, & i \neq j \end{cases}, \qquad (7.2.1)$$

where S is the diagonal matrix of node degrees (or strengths), as specified in Equation (7.12), and W is the adjacency matrix. In other words, to generate Λ, we reverse the polarity of the off-diagonal elements of the adjacency matrix and set the diagonal equal to the node degrees. This formulation holds for both weighted and binary networks, although some care is required when dealing with weighted networks. While most texts denote the Laplacian matrix with L, we use Λ here to avoid confusion with the characteristic path length. The Laplacian matrix can also be normalized such that,

$$\Lambda' = 1 - WS^{-1} = \begin{cases} 1, & i=j \\ -w_{ij}/s_j, & i \neq j \end{cases}. \qquad (7.2.2)$$

The eigenvalues of the normalized Laplacian are between zero and two (Chung, 1997), which enables comparison of eigenvalues between networks of different size (an overview of eigenvalues and **eigenvectors** is provided in Box 5.1). The normalized Laplacian can be understood with respect to the transition probabilities governing a random walker stepping across the network. Specifically, the term w_{ij}/s_j is the probability of a random walker stepping to node i from node j. Recall that to derive the measures of efficiency under a diffusion model of information flow, we define the matrix $U = WS^{-1}$ to denote these transition probabilities. Accordingly, we can write the normalized Laplacian as $\Lambda' = 1 - U$.

In some texts, it is assumed that w_{ij} is the weight of the edge from node i to node j, and thus the normalized Laplacian is more commonly written as $\Lambda' = 1 - S^{-1}W$, with w_{ij}/s_i giving the probability of a random walker stepping to node j

from node i. Fortunately, the eigenvalues of the normalized Laplacian are the same for both conventions. The symmetric normalized Laplacian matrix is given by $\Lambda'_{sym} = 1 - S^{-1/2}WS^{-1/2}$. It shares many properties with the normalized Laplacian but has the advantage of being a symmetric matrix. For example, the matrix $S^{-1/2}WS^{-1/2}$ can be used to determine the communicability of weighted networks (Section 7.3.3; Crofts and Higham, 2009).

The eigenvalues of a network's normalized Laplacian matrix reveal several properties about its topological organization. The smallest eigenvalue is always zero and the number of eigenvalues equal to zero gives the number of connected components (Chung, 1997). The smallest eigenvalues of the normalized Laplacian and their corresponding eigenvectors can be used to partition the network into **modules** (Shi and Malik, 2000; see also Newman, 2013). Based on the sign of each eigenvector element, the set of nodes can be partitioned into two modules; that is, positive elements indicate the nodes in one module, and negative ones the other module. The eigenvectors with the smallest eigenvalues delineate the strongest modules, since small eigenvalues are associated with long diffusion times that occur when the number of connections within modules is greater than the number of connections between them (de Lange et al., 2014). In this way, the smallest nonzero eigenvalue reveals the best possible cut of the network into two modules.

The largest eigenvalue of a network's normalized Laplacian indicates the extent to which it is a bipartite graph. A graph is bipartite if its nodes can be partitioned into two mutually exclusive groups in such a way that nodes within the same group are not connected. Partitioning the nodes based on the polarity of the eigenvector associated with the largest eigenvalue yields the most bipartite division of the graph (Bauer and Jost, 2012). The normalized Laplacian also provides clues about a network's motif distribution. Peaks in the distribution of the eigenvalues of the normalized Laplacian indicate multiple instances of a particular node or path motif. A peak at the eigenvalue of 1 indicates the presence of multiple nodes with a similar connectivity profile (Banerjee and Jost, 2008). However, not all motif types can be mapped to a specific eigenvalue. Figure 7.2.1a shows which eigenvalues of the normalized Laplacian matrix are most informative with respect to modular structure, motifs, and bipartite organization.

Continued

BOX 7.2 THE GRAPH LAPLACIAN—CONT'D

de Lange et al. (2014) investigated the normalized Laplacian of structural brain networks mapped for the macaque, cat, and *C. elegans*. In all species, the smallest nonzero eigenvalue of the normalized Laplacian matrix was significantly lower than appropriately matched random networks, indicating a high level of modular structure in the brain networks. As shown in Figure 7.2.1b, the distribution of eigenvalues in all species was also found to exhibit a distinct peak at approximately one, whereas the eigenvalues were more uniformly distributed for random networks. This indicates that the brain networks include certain motifs that recur with a frequency that is greater than chance. The largest eigenvalue in the empirical networks was not significantly greater than that found in the random networks, suggesting that neural networks are not bipartite. In fact, the largest eigenvalue of the Laplacian of the cat brain network was significantly smaller than expected. de Lange and colleagues suggest that this may indicate an abundance of path motifs with an odd number of nodes in the cat cortex.

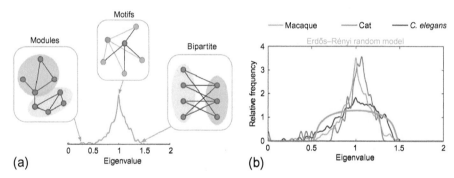

FIGURE 7.2.1 The graph Laplacian and network topology. **(a)** Modular structure is informed by the smallest eigenvalues. The polarity of the eigenvectors associated with the smallest eigenvalues can be used to partition the nodes in mutually exclusive groups. A peak at the eigenvalue of 1 indicates the presence of multiple nodes with a similar connectivity profile, giving insight into network motifs (see Chapter 8). In this case, the green and blue nodes are both connected to the same set of nodes. The largest eigenvalue reflects the extent to which the network is a bipartite graph. **(b)** The distribution of eigenvalues of the Laplacian matrix for brain networks mapped in the macaque, cat, and *C. elegans* (blue lines), compared with appropriately matched random networks (green line). The brain networks show increased modular structure and a distinct peak at the eigenvalue of 1, but they are not more bipartite than the random networks. *Image reproduced from de Lange et al. (2014), with permission.*

Another measure of communication efficiency based on the diffusion model is the *average* number of hops that a random walker must traverse to reach node *j* from node *i*; the so-called mean first passage time. If X_{ij} denotes the number of hops that a random walker takes to walk from node *i* to node *j*, it follows that (Wang and Pei, 2008; Zhou, 2003),

$$\langle X_{ij} \rangle = \sum_{t=0}^{\infty} t P(X_{ij} = t) = \sum_{t=0}^{\infty} P(X_{ij} > t)$$

$$= \sum_{t=0}^{\infty} \sum_{n=1}^{N} \left[U_j^t \right]_{ni} = \sum_{n=1}^{N} \left[(1 - U_j)^{-1} \right]_{ni}. \tag{7.14}$$

The first line of Equation (7.14) follows from the fact that the expectation of a discrete random variable is equal to the sum of its complementary cumulative distribution function (Chapter 4). The upper terminal of the sum is infinity because the number of hops comprising a walk is unlimited. To understand the second line, it is important to note that $P(X_{ij} > t) = \Sigma_{n=1}^{N} \left[U_j^t \right]_{ni}$. This relation reveals that the probability of a random walker requiring more than t hops to arrive at node j is given by the sum of the probabilities of the walker still being at one of the nodes other than node j after exactly t hops. It may seem strange that the sum includes node j, but remember that the jth row of U is set to zero, which ensures that $[U_j^t]_{ji}$ contributes zero to the sum.

Goñi and colleagues (2013) average the reciprocal of the mean first passage time over all node pairs to yield a summary measure called diffusion efficiency,

$$E_{\text{diff}} = \frac{1}{N(N-1)} \sum_{i \neq j} \frac{1}{\langle X_{ij} \rangle}, \tag{7.15}$$

Whereas E_{glob} (Equation 7.5) quantifies communication efficiency according to a model in which information is routed via the shortest paths in a network, E_{diff} is the analogous measure when the flow of information is better described by a diffusion process.

Diffusion may require more resources to communicate information than shortest path routing. Information routed via the shortest paths in a network is certain to reach the desired destination unless a disruption occurs. In contrast, under a diffusion process, many random walkers (messages) may need to be sent from node i to ensure at least one of them reaches node j using the shortest hop path; that is, in the shortest possible time. Rebroadcasting the same signal consumes metabolic resources, and thus it is important to quantify the number of random walkers, r, that must be sent to ensure that the probability of at least one of them conveying the signal using the fewest number of hops, H, is η. By rearranging Equation (17) appearing in Wang and Pei (2008), it can be shown that,

$$r_{ij}(\eta) = \frac{\log(1-\eta)}{\log \left(\sum_{n=1}^{N} \left[U_j^H \right]_{ni} \right)}, \tag{7.16}$$

where $r_{ij}(\eta) = 1$ for the special case $U_{ji} = 1$. When $U_{ji} = 1$, node j is the only neighbor of node i, meaning only one walker is ever required, irrespective of η, because only one path exits to node j from node i. Let's now consider the general case where $U_{ji} < 1$. If we initiate r random walkers from node i, the probability of *all* those walkers requiring more than H hops to reach node j is $\left(\sum_{n=1}^{N} \left[U_j^H \right]_{ni} \right)^r$. This follows from the assumption that the walkers are identical and independently distributed. The probability of at least one of the r walkers arriving at node j in H hops is therefore the complement of this probability; that is, $\eta = 1 - \left(\sum_{n=1}^{N} \left[U_j^H \right]_{ni} \right)^r$. Taking the logarithm of both sides of this equality and rearranging gives Equation (7.16).

To quantify the resources required to sustain diffusion-based communication in a network, the reciprocal of $r_{ij}(\eta)$ can be averaged over all node pairs to yield a summary measure called *resource efficiency*,

$$E_{\text{res}} = \frac{1}{N(N-1)} \sum_{i \neq j} \frac{1}{r_{ij}(\eta)}. \tag{7.17}$$

Goñi and colleagues (2013) applied the concept of a network morphospace to understand the trade-off between shortest path routing and diffusion-based measures of communication efficiency in various network topologies. Shortest path routing efficiency was measured using E_{glob} (Equation 7.5), and diffusion efficiency was measured using E_{diff} (Equation 7.15). A morphospace in this context defines the set of combinations of E_{diff} and E_{glob} that can be actually realized in terms of a particular network topology. For example, a network that achieves arbitrarily high values of E_{diff} and E_{glob} may be unattainable, and thus the boundary of the morphospace cannot extend to encompass these values. For a review of network morphospaces and their application to brain networks, see Avena-Koenigsberger et al. (2015).

Goñi and colleagues (2013) explore the boundaries of the efficiency morphospace by evolving synthetic network topologies that were carefully optimized to maximize or minimize E_{diff} and/or E_{glob}. This gives four possible optimization combinations, yielding four fronts in the morphospace (Figure 7.10a). Modular networks display higher E_{glob} than E_{diff} (Front 1), which can be attributed to the difficulty a random walker has in escaping from the module in which it originates. The random walker is likely to walk to other nodes within its module, but the likelihood it walks to another module is lower. This may be desirable for containing the spread of pathology to a single module within a brain network (Fornito et al., 2015; see also Chapter 9). In contrast, small-world networks with a highly connected core structure display higher E_{diff} than E_{glob} (Front 3), since the core structure facilitates efficient diffusion to all "corners" of the network, after which local connections guide information to desired nodes in the vicinity of a hub node. This has been described as a zoom-out/zoom-in mechanism, in which information zooms out from a local node to reach the **rich-club** core (see Chapter 6), and then zooms in on a target node after exiting the rich club (Boguñá et al., 2009; see also Section 7.4.1). Brain networks are both modular and display rich-club organization. It is therefore likely that brain networks do not reside at an extreme point of the morphospace, having evolved to achieve a balance between efficient diffusion and shortest path routing. Star-like topologies display high E_{diff} and high E_{glob}

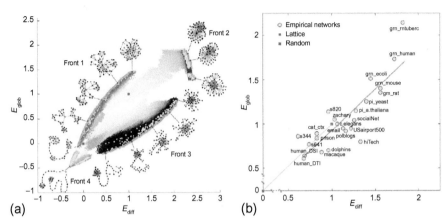

FIGURE 7.10 Using a network morphospace to characterize the efficiency of diffusion and shortest path routing. **(a)** Synthetic network topologies were evolved to maximize or minimize E_{diff} (diffusion efficiency) and E_{glob} (global efficiency of shortest path routing). The coordinate of each network in the efficiency morphospace is determined by its normalized value of E_{diff} and E_{glob}. The morphospace can be seen to evolve along four distinct fronts. Front 1 (high E_{glob}, low E_{diff}) comprises modular networks. Front 2 (high E_{glob}, high E_{diff}) is dominated by star-like networks. Front 3 (low E_{glob}, high E_{diff}) mainly comprises small-world networks with a highly connected core structure. Chain-like networks underpin Front 4 (low E_{glob}, low E_{diff}). Blue and red squares indicate the reference points of regular lattices and randomized networks, respectively. Green points indicate the initial seed population from which the synthetic networks were evolved. Gray circles indicate evolving networks over epochs, with darker shades of gray indicating networks encountered in later epochs. Orange points show networks residing along one of the four fronts. Black asterisks denote positions of the representative networks shown in insets. **(b)** Several empirical networks positioned in the efficiency morphospace. Blue and red squares indicate lattice and random reference points, respectively, linked by the green reference line. The gray line represents the linear regression across all 23 empirical networks. *Image reproduced from Goñi and colleagues (2013), with permission.*

(Front 2), but their high communication efficiency comes at the cost of low resource efficiency and vulnerability to attack. Figure 7.10b shows several empirical networks positioned within the efficiency morphospace. It can be seen that most of the morphospace is empty, suggesting that empirical networks are tightly constrained to reside along a linear axis that represents a balance between E_{diff} and E_{glob}.

It is important to note that the diffusion-based models of neural information flow we have considered in this chapter are static. Measures such as the diffusion and resource efficiency do not reveal anything about the characteristics of a diffusion process as a function of time. Diffusion is, however, an inherently dynamic process, with information diffusing to new elements of the network at different time epochs based on a diffusivity constant that controls propagation speeds (Raj et al., 2012). Alternative models of diffusion characterize communication processes in a network as they unfold over time. These are considered in Box 7.3.

BOX 7.3 SIMULATING DIFFUSION-LIKE SPREADING DYNAMICS

Mišić and colleagues (2015) simulated diffusion-like spreading processes in structural human brain networks mapped using diffusion MRI and tractography. Their simulation captures the dynamics of diffusion, whereas the measures considered in this chapter are static characterizations of a diffusion process. Edge weights were determined by streamline counts. Their simulation is based on a so-called linear threshold model of spreading dynamics (Granovetter, 1978) and synchronously evolves in time at discrete time epochs. At each time point, t, the state of the ith node, $x_i(t)$, is either active, $x_i(t) = 1$, or inactive, $x_i(t) = 0$. Once a node becomes active, it remains active forever. At time $t = 0$, all nodes are in the inactive state, except for a subset of seed nodes at which an initial perturbation is introduced. At times $t = 1, 2, 3, \ldots$ the state of the ith node is updated according to the following rule,

$$x_i(t+1) = \begin{cases} 1, & \theta k_i < \sum_{j \in N_i} x_j(t), \\ 0, & \text{otherwise,} \end{cases} \quad (7.3.1)$$

where N_i is the set of nodes that are neighbors of node i, k_i is the degree of node i and θ is a parameter giving the minimum proportion of a node's neighbors that are required to be active at time t in order for it to adopt an active state at time $t + 1$. When $\theta = 0$, we have a pure diffusion process, where the active state propagates according to a breadth-first search (Box 6.3) and develops along the shortest paths of a binary network. In contrast, propagation may deviate from shortest paths when $\theta > 0$. Mišić and colleagues base their simulations on the highest value of θ that ensures all nodes in the network eventually adopt the active state. They also consider an asynchronous update model where states are no longer updated at discrete times, but rather after a finite conduction delay, which is estimated as the ratio of the connection length to the streamline count. In this way, spread of the active state becomes fastest along short, well-myelinated axonal fiber bundles.

The dynamic spread of the active state across the network can potentially model the flow of neural information, prion-like spread of a pathological mechanism, or neuronal avalanches (Beggs and Plenz, 2003). Mišić and colleagues do not specifically model any of these spreading phenomena, instead choosing to keep their model conceptual. Their focus is on evaluating the speed at which the active state is adopted at each node. A node's adoption time is the period of time that elapses before it adopts the active state. Figure 7.3.1a shows snapshots at various time epochs of the spread of the active state, which in this example was initiated at a single seed node in left frontal cortex. The simulation was repeated for different seed nodes to yield an adoption time matrix. Figure 7.3.1b shows the adoption time matrix ordered according to hemispheres, anatomical communities and putative resting-state networks. Each row represents a particular seed region and columns represent target nodes. It can be seen that adoption times are fastest among nodes in the same anatomical community as the seed node, as expected by the difficulty a diffusion process has in escaping from its local module. The adoption time matrices are also highly symmetric, indicating that adoption times between pairs of nodes are not particularly sensitive to the choice of seed and target. As shown in Figure 7.3.1c, adoption times are also related to resting-state networks. In particular, propagation is significantly slower between sensory resting-state networks, but faster within associative networks.

Mišić and colleagues (2015) also quantified the time elapsed before an edge is used in communicating the active state. As shown in Figure 7.3.1d, edges interconnecting medial cortical regions facilitate early spreading, whereas lateral connections are on average utilized at a later time, after the network core has adopted the active state. Interestingly, these medial cortical regions overlap, to a certain extent, with several of the brain's rich-club hubs (Chapter 6). Finally, Figure 7.3.1e shows that the number of times an edge is utilized during the spreading process is positively associated with edge **betweenness** centrality. This suggests that shortest paths play an important role in spreading the active state across the network, but spreading can additionally occur along alternative pathways according to a diffusion process. The authors also simulate perturbations that are simultaneously initiated from two distinct nodes, where the perturbations can either cooperate or compete with each other. Simulation of other diffusion models on empirically-derived structural connectivity networks in humans has shown promise in predicting the spatial pattern of atrophy in neurodegenerative disease (Raj et al., 2012).

BOX 7.3 SIMULATING DIFFUSION-LIKE SPREADING DYNAMICS—CONT'D

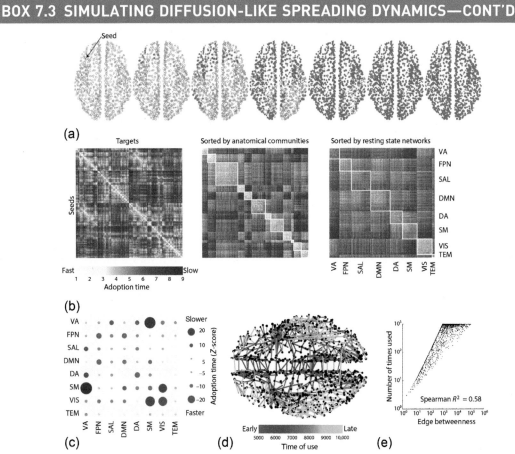

FIGURE 7.3.1 Simulation of diffusion-like dynamics in human brain networks. **(a)** A perturbation was initiated at a seed node located in the left frontal cortex. This perturbation spreads according to a simple diffusion model via the structural connections of a human brain network mapped with diffusion imaging techniques. The nodes to which the perturbation has spread at various snapshots in time are shown in red. The spreading process is initiated at a single node and completes when the perturbation spreads to all nodes. **(b)** The adoption time matrix, with the rows/columns ordered according to hemispheres (leftmost matrix) as well as communities identified in the structural (central) and functional brain network (rightmost). Element (i, j) of the adoption time matrix stores the time that elapses until a perturbation initiated at node i spreads to node j. The squares shown along the matrix diagonal delineate network communities. It can be seen that adoption times are fastest for the nodes of the anatomical community in which the perturbation was initiated. **(c)** Adoption times between human resting-state functional connectivity networks are expressed as Z-scores relative to a null distribution in which the assignment of nodes is permuted. Blue circles indicate that the adoption times are slower than the null distribution, while red circles indicate the adoption times are faster. **(d)** The color of an edge can be modulated by the time elapsed before that edge is used in propagating the perturbation. **(e)** The number of times edges are used to facilitate the spreading process is positively correlated with edge betweenness centrality. VA, ventral attention; FPN, front-parietal; SAL, salience; DMN, default mode; DA, dorsal attention; SM, somato-motor; VIS, visual; TEM, temporal. *Image reproduced from Mišić and colleagues (2015), with permission.*

7.3.3 Communicability

Communicability (Estrada and Hatano, 2008) is a popular measure of network integration that is consistent with the diffusion model of information flow. It accounts for the contribution of all possible *walks* between a pair of nodes. If A is a binary adjacency matrix, the communicability between nodes i and j is given by,

$$Com_{ij} = \sum_{n=0}^{\infty} \frac{[A^n]_{ij}}{n!} = [e^A]_{ij}, \tag{7.18}$$

where e^A denotes the matrix exponential of the matrix A, which is the matrix analogy of the ordinary exponential function. This formulation relies on the fact that A^n characterizes all walks in the network that traverse exactly n hops. More specifically, $[A^n]_{ij}$ is the total number of walks between nodes i and j comprising n edges. Communicability is the weighted sum of the total number of walks between a node pair. Walks of length n are weighted by the factor $1/n!$. Longer walks contribute less to the sum, while direct connections contribute the most. A summary measure of communicability can be computed by averaging Com_{ij} over all node pairs, as long as the network is not fragmented. Alternatively, the matrix Com_{ij} can be used to define a new network, whose nodes coincide with the nodes of the original network, but whose edges are now weighted and given by Equation (7.18). This is called the *communicability network* (Crofts et al., 2011).

The computation of communicability is best demonstrated with a simple example. Consider a binary, undirected network comprising three nodes, where an edge is found between nodes 1 and 2, as well as between nodes 1 and 3. We have an open triangle motif in other words. The adjacency matrix of this network raised to $n = 1, 2, 3, 4$ is as follows,

$$A = \begin{bmatrix} 0 & 1 & 1 \\ 1 & 0 & 0 \\ 1 & 0 & 0 \end{bmatrix}, \quad A^2 = \begin{bmatrix} 2 & 0 & 0 \\ 0 & 1 & 1 \\ 0 & 1 & 1 \end{bmatrix}, \quad A^3 = \begin{bmatrix} 0 & 2 & 2 \\ 2 & 0 & 0 \\ 2 & 0 & 0 \end{bmatrix}, \quad A^4 = \begin{bmatrix} 4 & 0 & 0 \\ 0 & 2 & 2 \\ 0 & 2 & 2 \end{bmatrix}. \tag{7.19}$$

It can be easily seen that when the matrices are raised to an odd power, no walks exist between nodes 2 and 3, since $A_{23}^n = 0$ when $n = 1, 3, 5, \ldots$. This node pair can only be connected with a walk that traverses an even number of hops, and thus $A_{23}^n > 0$ if and only if n is even. For example, $A_{23}^4 = 2$, meaning there are two walks between nodes 2 and 3, each of which comprise exactly four hops. These two walks are given by 2-1-2-1-3 and 2-1-3-1-3. Excluding the $n = 0$ term, the first four terms of the sum in Equation (7.18) for the communicability between nodes 2 and 3 are thus, $Com_{23} = 0/1 + 1/2 + 0/6 + 2/24 \approx 0.583$, which is quite close to the value of 0.589 that is obtained when summing across all powers. This suggests that the contribution of walks that traverse back and forth along the same edge more than four times is minimal in this example due to the weighting provided by the $1/n!$ factor in Equation (7.18).

Communicability can be generalized to weighted networks by introducing a normalization factor to regulate the undue influence of nodes with high strength (Crofts and Higham, 2009). This generalization involves replacing the binary adjacency matrix A in Equation (7.18) with $S^{-1/2} W S^{-1/2}$, where W now denotes our weighted connectivity matrix and S is the diagonal matrix previously defined in Equation (7.12). However, the interpretation of weighted communicability requires care, since we no longer count the number of walks between a node pair. The matrix $S^{-1/2} W S^{-1/2}$ is called the reduced adjacency matrix. The element $\left[\left(S^{-1/2} W S^{-1/2}\right)^n\right]_{ij}$ is related to the probability of a random walker reaching node i from node j in n steps. Another way to understand the matrix $S^{-1/2} W S^{-1/2}$ is in terms of a normalization of the connectivity weights, such that w_{ij} becomes $w_{ij}/\left(\sqrt{s_i}\sqrt{s_j}\right)$. Crofts and Higham (2009) suggest that this normalization suppresses the undue influence of high strength nodes. Equation (7.18) can be used without modification to compute the communicability of directed networks.

Communicability reflects a network's capacity for parallel information transfer under a diffusion model of information flow. It is particularly sensitive to the presence of lesions in human brain networks. Croft and colleagues (2011) were able to separate stroke patients from healthy controls based on a weighted communicability network derived from brain networks mapped with diffusion MRI and tractography. They differentiated patients from controls using communicability values computed in either the lesion-affected hemisphere or the contralesional hemisphere, despite the lack of pathology in the latter. Similar techniques have been applied to identify widespread reductions in communicability in patients with multiple sclerosis, with the extent of impaired communicability being associated with motor symptoms (Li et al., 2013).

Andreotti et al. (2014) conducted a comprehensive study evaluating the accuracy of communicability to detect the effects of lesions. Their extensive analyses are based on simulated pathology and demonstrate that communicability is sensitive to a wide variety of lesions. Moreover, these authors contend that the communicability network might provide a more accurate characterization of a network's rich club compared to the conventional approach for defining this core of hub regions (Chapter 6), because the edge weights in this network implicitly capture the extent of integrative processing and communication under a diffusion model of information flow.

7.4 NAVIGATION AND OTHER MODELS OF NEURAL COMMUNICATION

We consider navigation in the final section of this chapter. Navigation and shortest path routing are similar in that both models involve use of a single path. Unlike shortest path routing, navigation is performed "on the fly" without a global map of the network. The downside of navigation is that reaching a desired destination is not guaranteed. However, as we will see in this section, some types of networks can be efficiently navigated, in the sense that a desired destination can be reached with the help of a locally greedy navigation scheme.

Let's begin by revisiting our tourist Jane, who was introduced at the start of this chapter. Recall that Jane faced the predicament of walking to the Eiffel Tower from her hotel room without a map or knowledge of the streets of Paris. Since Jane could see the Eiffel Tower from afar, one option available to her was to walk along the road that was most closely oriented in the direction of the Eiffel Tower. In this way, Jane relied on the local information available to her at each point (node) of her journey and effectively generated a path on the fly. This is a simple example of a locally greedy navigation strategy. The term "locally greedy" refers to the fact that each decision is made to maximize immediate benefit; in Jane's case, the immediate benefit is a reduction of the distance to the Eiffel Tower. Here, we consider the kinds of networks that can be efficiently navigated using greedy schemes, and whether navigation is a feasible model of information flow in brain networks.

7.4.1 Navigation of Small-World Networks

Milgram's (1967) small-world experiment is best known for identifying the concept of "six degrees of separation" in social networks, but perhaps the most striking conclusion of this study is that people are able to collectively identify the *pathways* underlying these "six degrees" without any global knowledge of the social network. Milgram asked the participants in his experiment to forward a letter to a designated target person living in Boston, MA, USA. The participants were told the name, address, occupation, and some personal information about the target person. The catch was that they were only allowed to forward the letter to a single acquaintance who they knew on a first-name basis, with the goal of reaching the target person in the least number of steps. They were not allowed to send the letter directly to the target person. Remarkably, about a third of the letters reached the target person and they did so in an average of about six steps (Easley and Kleinberg, 2010). Milgram's experiment is an example of the efficiency of a locally greedy navigation strategy. In particular, it demonstrates that social networks can be navigated using only local knowledge about who, among a participant's acquaintances, is "closest" to the target person. Follow-up studies suggest that participants in Milgram's experiment forwarded the letter to their acquaintance who most closely matched the target person in terms of geographical closeness, occupation, and ethnicity (Killworth and Bernard, 1978).

What kind of greedy routing strategies might be used for efficient navigation in neural systems? Most of the social attributes used in Milgram's experiment are clearly irrelevant to biological networks. However, geography—in particular, spatial embedding and distance constraints—is an important exception, being of paramount importance in nervous systems (Chapter 8). A distance-based greedy navigation scheme for biological networks might involve simply forwarding information to the next node that is closest in distance to the desired target node. Crucially, this greedy scheme only requires each node to be able to determine which of its neighbors is closest in distance to the target node. In fact, this scheme is sometimes used to communicate data in wireless telecommunications networks, where it is known as position-based, or geographic navigation (Stojmenovic, 2002). This kind of geographic scheme is an attractive navigational model for biological networks because it is decentralized and does not require complete global knowledge of the network topology. For example, with navigation, if node B is not a neighbor of node A, then node A does not need to know anything about the location of node B or how close it is to the desired target, whereas this information is required by shortest path routing. As with shortest path routing, navigation implies the existence of a desired target, whereas diffusion is not specific to a target.

Not all network topologies can be successfully navigated with geographic navigation schemes. Unsuccessful navigation attempts can cause information to circulate in loops or reach a dead-end, signified by cases in which the current node is closer in distance to the target than all other neighboring nodes. Boguñá and colleagues (2009) and Boguñá and Krioukov (2012) investigated network topologies that can be efficiently navigated using distance-based greedy navigation, where information is forwarded to the next node that is closest in distance to the desired target node. Of course, as mentioned above, this paradigm assumes that neurons know the desired target neuron with which they want to communicate, which may not be a realistic assumption. Nonetheless, it is still interesting to consider whether some network topologies are easier than others to navigate using locally greedy strategies. In particular, Boguñá and colleagues demonstrate that navigability is greatest in **scale-free** networks where clustering is strongest among nodes that are spatially proximate. These navigable networks form a subset of small-world networks, which describe many natural and engineered networks. Spatial proximity is conveniently understood with respect to the Euclidean distance between the physical locations of spatially embedded nodes of networks like the brain; however, proximity can also be defined with respect to "hidden" metrics that are independent of physical distances (Boguñá et al., 2009). For example, **geodesic** distances defined on the two-dimensional cortical surface might serve as a hidden metric.

Boguñá and colleagues (2009) analyze a spatially embedded **generative model** defined by two parameters: (i) γ, the exponent governing the power-law degree distribution (see Chapter 4); and, (ii) α, a parameter controlling the preference for edges to be drawn between spatially proximate nodes. As α is increased, edges are more likely to be drawn between proximate nodes, leading to stronger clustering in the network. They evaluated the navigability of their simulated networks based on the probability that distance-based greedy navigation can successfully navigate between randomly chosen pairs of nodes. For a given level of clustering, as determined by α, a critical value $\gamma_c(\alpha)$ can be identified such that when $\gamma > \gamma_c(\alpha)$, the success probability decreases as the network size is increased, whereas it remains constant when $\gamma < \gamma_c(\alpha)$. This critical value facilitates the delineation of navigable and non-navigable regions in the two-dimensional parameter space of γ versus α. It remains to be determined whether brain networks reside in the navigable region.

The two key ingredients promoting network navigability are the presence of spatially distributed hubs, together with strong clustering among spatially proximate nodes. Hub-to-hub connections facilitate long-distance communication between broadly defined locales, while strong local clustering enables the navigational process to "home in" on the specific target node within the locale. This organizational motif is consistent with the existence of hubs and rich clubs in brain networks (Chapters 4 and 6) and the presence of strongly interconnected modules of neural elements that are spatially close to each other (Chapter 9). Given such a topology, navigability may be construed as comprising two phases: coarse-grained navigation via hubs, followed by a fine-grained search using strongly clustered local communities. Figure 7.11a shows example paths navigated using distance-based greedy navigation in simulated networks with different combinations of α and γ. It can be seen that the structure of these paths in navigable networks (i.e., when $\gamma < \gamma_c(\alpha)$) is such that they: (1) first navigate a relatively short distance to a hub node in the network core; (2) navigate to another hub node via the core, resulting in a dramatic reduction in the distance to the destination; and, (3) exit the core towards a low degree node that is either the destination or densely clustered with the destination. As shown in Figure 7.11b, this routing structure resembles the "feeder to rich club to feeder" pathway motif that is overrepresented in brain networks (de Reus and van den Heuvel, 2013b; Chapter 6). A similar pattern has also been identified for the macaque connectome (Harriger et al., 2012) and for *C. elegans* (Towlson et al., 2013).

The Kleinberg model (Kleinberg, 2000; Easley and Kleinberg, 2010) considers greedy navigation on a simpler network model embedded on an $N \times N$ grid, where each grid point defines a node. Each node forms a connection to all other nodes that lie within the radius of a fixed number of edges. This ensures strong local clustering. Each node additionally forms a fixed number of random connections. The probability that node i forms a random connection to node j is proportional to $d(i,j)^{-q}$, where $d(i,j)$ is the number of edges separating nodes i

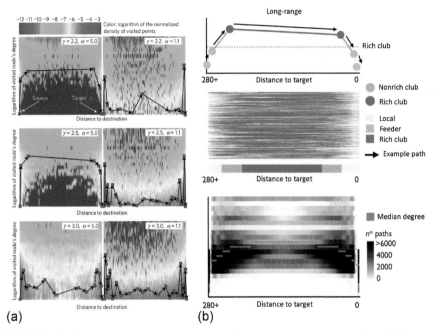

FIGURE 7.11 Structure of paths generated using distance-based greedy navigation. **(a)** The solid black lines represent sample paths that begin and end at low degree nodes, respectively shown in the bottom left and right of each diagram. The red squares represent nodes. Spatially embedded, scale-free networks were generated with various levels of clustering among spatially proximate nodes, as determined by the parameter α. The parameter γ is the exponent in the power-law **degree distribution** and governs the number of hubs in the network. The sample path shown for the network that can be navigated using the fewest hops ($\gamma = 2.2$, $\alpha = 5.0$; top-left) features navigation of a long hub-to-hub connection, followed by a fine-grained "homing in" on the target in its strongly clustered local area. When the level of clustering is insufficient (e.g., top-right; $\gamma = 2.2$, $\alpha = 1.1$), hub nodes cannot efficiently navigate to low degree target nodes (zoom-in mechanism) and source nodes cannot efficiently navigate to the network core (zoom-out mechanism). When the number of hubs is insufficient (e.g., bottom-left; $\gamma = 3$, $\alpha = 5$), the core does not span enough of the network to ensure all broad locales comprise a hub. **(b)** Shortest paths in structural brain networks also follow a zoom-out and zoom-in structure. A typical path zooms out to the rich club via a feeder link, utilizes the rich club to reach the locale of the desired destination and exits the rich club via another feeder link. The vertical axis of the upper and lower panel represents the node degree. For the middle panel, each row represents a particular path. *Image reproduced (a) from Boguñá and colleagues (2009) and (b) from van den Heuvel et al. (2012), with permission.*

and j, and q is a parameter determining the extent to which the random connections are biased towards linking spatially proximate nodes. Kleinberg shows that $q \approx 2$ is optimal for efficient navigation of the network. At this critical value, the random connections are uniformly distributed across concentric "scales of resolution," delineated by circles of radius d, $2d$, $3d$, ... drawn around a given node. Kleinberg argues that this ensures greedy navigation strategies can be "funneled" towards a finer scale at each step, eventually reaching the target. If $q > 2$, the random connections are mostly short-range and do not provide the long-distance ties characteristic of a small-world network; whereas if $q < 2$, they are essentially too random to facilitate efficient navigation. The trade-off between navigation efficiency and network cost has also been studied in the context of game theory (Gulyás et al., 2015).

7.4.2   Internet and Computer Analogies

Shortest path routing, diffusion, and navigation are models that describe the paths along which information flows in a brain network. But understanding these paths reveals little about the mechanisms and dynamics of network communication. Is neural communication analogous to the Internet, where data is broken up into small packets that individually navigate to their desired destination; or is it more like a traditional telephony network, where circuit switching is used to reserve a dedicated path each time a call is initiated? In the final section of this chapter, we consider the broader aspects of neural communication with the help of some popular technological metaphors.

The Internet, computers, and telecommunications networks can potentially serve as useful analogies for neural communication, as long as these analogies do not fuel elaborate fictions (O'Reilly, 2006). As described by Graham and Rockmore (2011), history is littered with numerous technological analogies for the brain, often corresponding to whatever is cutting-edge at the time. The water gardens at Versailles inspired Descartes to imagine the brain as an intricate waterworks controlled by a master valve, the pineal gland. Later, Leibniz saw the brain as behaving like a mill, while Norbert Weiner adopted a cybernetic perspective of neural communication and mental illness (Weiner, 1948). Today, in the age of computers and the Internet, it is no surprise that theories of how information is communicated in nervous systems have been articulated in the terms of these pervasive technologies.

The most basic measure of information flow—bandwidth—comes from the analysis of computer networks. Bandwidth expresses the volume of information communicated in a system per unit of time (e.g., bits per second or mutual information rate; Antonopoulos et al., 2015). To make the notion of neural information flow more concrete, let's assume for the moment that an axon's bandwidth is in the order of 100 bits/s (Strong et al., 1998). Based on this estimate, the amount of information that can be theoretically communicated between different elements of the human brain in a given second is about 1 TB (10^{11} neurons $\times$ 100 bits/s $\approx$ 1 TB), which is equivalent to the bandwidth of the US Internet backbone in 2002 (Laughlin and Sejnowski, 2003).

However, it is likely that metabolic demands associated with generating an action potential render this theoretical limit an overestimation. When energy demands are taken into account, it appears that the brain's actual bandwidth is less than a single bit per neuron per second on average (Lennie, 2003). Compared to a computer, this is remarkably slow. To overcome this low bandwidth, nervous systems are likely to have evolved efficient ways to dynamically communicate information between different neural elements, as well as clever representational codes to encode neural signals and govern which set of neurons communicate concurrently at any given time (e.g., sparse coding; Olshausen and Field, 2004). All neurons cannot communicate at the same time, since this would result in network congestion, and the rate of energetic expenditure required to support such massive signaling would greatly outpace the rate of supply.

Graham and Rockmore (2011) argue that the brain should be considered in terms of the Internet metaphor, thereby placing greater emphasis on the nature of the transmission of information across nervous systems, rather than computations performed by individual neurons, as would be emphasized by the computer metaphor. On the Internet, information is typically transmitted using a protocol called packet switching. Messages are broken up into discrete blocks known as packets, each of which is stamped with the address of a desired destination. This enables each packet to traverse the individually most efficient route to its destination, which increases communication efficiency through a process called statistical multiplexing. Multiplexing in this context refers to the fact that multiple pairs of communicating nodes can share a common connection by utilizing it at different times. Moreover, packets can be buffered at nodes until congestion on an outgoing connection has dissipated, after which forwarding of the packet occurs in a decentralized manner based on the packet's address and the most efficient path at the time. Packets are recompiled into the original message when they all reach their destination.

A potential downfall of the packet switching metaphor is lack of evidence for buffering in nervous systems as well as the assumption that neural signaling can be decomposed into discrete, neatly defined packets. It is also unclear how the address of a packet might be encoded in a spike train. It has been suggested that addresses could be encoded in spike timing, with the data modulating the spike rate (Graham, 2014). On the other hand, one of the appealing parallels between packet switching and neural signaling is that packets can be dynamically rerouted to bypass damaged or heavily congested connections on the Internet via a mechanism called deflection routing (Zalesky et al., 2007). This might be considered analogous to the changes in structural and functional brain connectivity that occur following focal damage such as stroke (Crofts et al., 2011; Fornito et al., 2015).

Mišić and colleagues (2014) simulated a variant of packet switching operating in a network defined by the structural connectivity of the macaque cortex. Instead of breaking up messages into smaller packets, they communicated messages as a whole to their desired destination. This is known as message, or burst, switching (Zalesky, 2009). It reduces the data overheads associated with stamping individual packets with an address, but often does not yield the same level of statistical multiplexing as packet switching. Mišić and colleagues assumed messages arrive at nodes according to a Poisson process, meaning that arrivals are random and the delay between arrivals is an exponentially distributed random variable. Messages are thought to represent "signal units," although the precise correspondence between signal units and neural signaling characteristics is unclear. Messages are communicated to randomly chosen destinations according to a diffusion process. Whenever a message arrives at a node, it remains there for a random "service time," after which it is forwarded to the next node. Service times could model axonal conduction delays, for example. Messages are buffered (queued) if they arrive at a node that is already processing another message, and they are lost if they arrive at a node with a buffer that is full. This model is known in the telecommunications literature as a Markovian queuing network. It is a commonly studied model of Internet traffic and other data networks.

Mišić and colleagues (2014) treat the topology of the macaque connectome, as collated from published tract tracing studies using the CoCoMac database, as the connections in a Markovian queuing network model. Compared to various reference topologies, message switching in the macaque network was characterized by higher message loss rates, faster transit times, and lower throughput, suggesting that neural connectivity may be optimized for speed rather than fidelity. Interestingly, the rich club of the macaque network carried the bulk of the messages between most node pairs, with connections that feed into the rich club showing the next highest utilization. We have already seen that hub-to-hub connections comprising the rich club are one of the two key ingredients mandated by greedy navigation, and thus it is not surprising that these connections display the highest utilization and throughput in a queuing network. These connections may thus be elements of nervous systems that are under a high signaling load, a result consistent with evidence for their high metabolic requirements (Collin et al., 2014; Fulcher and Fornito, 2016; Chapter 8). Figure 7.12 shows various node-specific measures that reveal the heterogeneity in the extent to which different nodes comprising the macaque connectome are utilized to perform message switching.

Although many of the elements underlying queuing network models cannot be readily mapped to specific neural signaling characteristics, the work of Mišić and colleagues demonstrates that these models can provide a valuable level

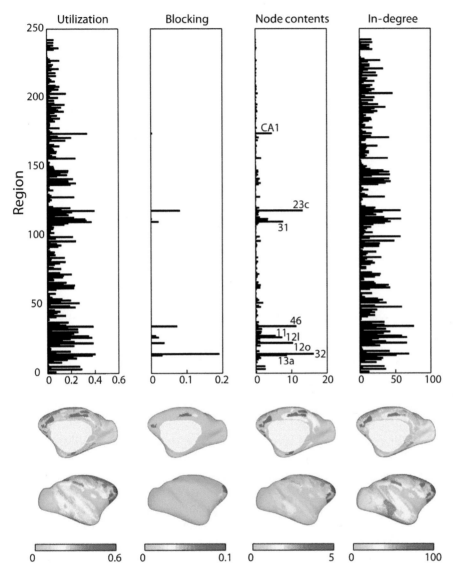

FIGURE 7.12 Markovian queuing network simulated on the macaque connectome. The transmission of messages between pairs of regions was constrained to the connections of the 242-node macaque connectome, as collated from published tract tracing studies using the CoCoMac database. Messages are processed at each node for a random "service" time, after which they are forwarded to subsequent nodes for processing. Messages are buffered (queued) if they arrive at a node that is already processing another message, and they are lost if they arrive at a node with a buffer that is full. *Utilization* reflects the proportion of time a node is busy processing a message, *blocking* is the probability of a message being lost at a node (which occurs when messages arrive at a node with a full buffer) and *node contents* is the average queue length at each node. Cortical surfaces show the anatomical distribution of regional variations of each measure. High-degree nodes are generally the most utilized and have the longest queue lengths. *Image reproduced from Mišić and colleagues (2014), with permission.*

of abstraction that emphasizes global patterns of communication over the microscopic nuances of neuronal signaling.

7.5 SUMMARY

Three models of neural information flow were considered in this chapter: shortest path routing, diffusion, and navigation. The efficiency with which information can be communicated under these models differs markedly and can be estimated with network simulations or the analysis of network models of the brain. We introduced three tourists—Alice, Bob, and Jane—to serve as metaphors for these three models of neural information flow. Having a map at her disposal, Alice epitomized shortest path routing. Jane, on the other hand, was a navigator and used local information to navigate to her destination without the help of a map. Finally, Bob was a random walker, who diffused through the network in a random manner, without a map and without a particular destination in mind.

Of these three models of neural information flow, shortest path routing is the most studied. It is fundamental to the concept of the brain as a small-world network and underpins the idea of a characteristic path length. Whenever the efficiency of brain networks is discussed in the literature, it is usually with respect to a network measure quantifying the efficiency of shortest path routing in the connectome (e.g., global, local, and regional efficiency). Dijkstra's algorithm can be used to compute the shortest paths in both directed and undirected networks, as long as the edge weights are non-negative.

Computing shortest paths in networks with negatively weighted edges is challenging. Most investigators eliminate negatively weighted edges by considering their absolute values, or by adding the absolute value of the most negative edge weight to all other edge weights. Examples were presented in this chapter showing that these edge weight transformations do not necessarily yield physiologically meaningful shortest paths. More generally, we saw that shortest paths are not amenable to straightforward interpretation in correlation-based functional brain networks.

Shortest path routing is contingent on centralized control of the network. This is an unlikely mode of communication in neural systems. Diffusion and navigation are two models of information flow that are not centralized, and are thus attractive models for the brain. A diffusion process implies that the flow of information is dispersive, propagating simultaneously along multiple "fronts," without being directed towards a single destination. The analogy of fluid flow in a network of pipes was used to explain diffusion. Communicability is the simplest measure quantifying the efficiency with which information can be communicated under a diffusion model. This chapter reviewed the utility of diffusion models in improving the accuracy with which functional connectivity can be predicted from structural connectivity in human brain networks, and introduced the concept of a morphospace to elucidate the key topological attributes conducive to efficient diffusion in a network. The downsides of diffusion as a model of neural information flow are that it implies communication is not directed to a particular target and allows information to flow through the same connection on multiple occasions in an inefficient back and forth manner.

Scale-free networks that display strong clustering among spatially proximate nodes can be efficiently navigated using greedy navigation schemes. One such scheme is where each node simply forwards information to its neighbor that is closest in distance to the desired destination. In this way, each node only needs to know the location of its neighbors relative to the destination. This is less knowledge than required for shortest path routing, but may still be an unrealistic amount of knowledge for any single neuron or population of neurons to possess. The advantages of navigation as a model for neural communication are its decentralized nature and the fact that it performs particularly well in spatially embedded networks that contain hubs, of which nervous systems are prime examples. We saw that navigation generates "feeder to rich club to feeder" type pathway motifs. In this way, hub-to-hub connections facilitate long-distance communication between broadly defined locales, while strong local clustering enables the navigational process to "home in" on the specific target node within the locale.

To conclude this chapter we considered models of neural communication that have been inspired by the Internet and telecommunications protocols. We considered a Markovian queuing network as a model of neural communication and emphasized the abstract nature of these kinds of models. In particular, these models are based on discrete messages being queued at network nodes, which is difficult to reconcile with our current understanding of neural signaling.

Future work in this area will benefit from both modeling and empirical studies seeking to test whether shortest path routing, diffusion, or navigation best embody the flow of information in brain networks. The dynamics of information flow—which node pairs communicate at which time—is also poorly understood and requires study. The reader may have noticed that most of the summary measures of communication efficiency considered in this chapter represent averages over all pairs of nodes. However, it is likely that some node pairs do not interact at all, or only during short intervals. Understanding communication dynamics will help in refining these summary measures.

CHAPTER 8

Motifs, Small Worlds, and Network Economy

The diversity of topological properties that can be computed for all nervous systems, at scales ranging from individual **nodes** through **subgraphs** to entire networks, raises the question: do these properties arise from simpler network wiring rules? In particular, do certain, recurring connectivity **motifs** act as basic building blocks for the emergence of more complex topological properties?

Connectivity motifs in networks can be analyzed at the level of node pairs, triplets, or any other higher-order combination. They can also be studied at the level of **paths**, allowing us to understand how different types of **edges** are arranged with respect to each other. In this chapter, we consider approaches for characterizing different kinds of network motifs and show how such analyses help us to understand structural constraints on brain function. We then discuss the **clustering coefficient**, an index of the frequency of a specific class of three-node motif that, together with low average **path length**, is a defining characteristic of **small-world** organization. We review how small-world properties provide a topological substrate for functional segregation and integration in the brain, thus giving rise to complex dynamics (Sporns et al., 2000; Tononi et al., 1994). In the final section, we investigate how network wiring **costs** (Box 8.1) play an important role in shaping small-world and other topological properties of brain networks.

8.1 NETWORK MOTIFS

A motif is a **subgraph** with a particular topological configuration. Node motifs are defined at the level of connected subgraphs of nodes. Path motifs are defined at the level of sequences of edges. The aim of motif analysis is to determine which specific node or path motifs recur in a network with a frequency that is significantly greater than chance. Any motifs that meet this criterion are thought to represent important building blocks of the network.

Fundamentals of Brain Network Analysis. http://dx.doi.org/10.1016/B978-0-12-407908-3.00008-X

BOX 8.1 WIRING COST

Brain networks are embedded in physical spaces such as the intracranial volume (for mammals). Connections between neural elements traverse this space. The physical cost of connecting a spatially embedded network is often referred to as its wiring cost. In brain networks at a cellular scale, the wiring cost is proportional to the number of axonal projections, and the length and cross-sectional area of each axon. Since measures of axonal cross-sectional area are difficult to estimate in large-scale brain networks, most connectomic studies generally approximate wiring cost using **connection density**, connection distances, or a combination of both.

Connection density is the number of edges that are present in a network relative to the total possible number of edges (Chapter 3). The accuracy with which this quantity is measured depends on the fidelity of the method used to map neuronal connectivity. Methods that reconstruct many spurious connections (false positives) or fail to map actual connections (false negatives) will limit the accuracy of any connection density estimate (see Chapter 2).

To estimate connection distances, we would ideally trace the trajectory of the axonal fibers linking each pair of nodes and compute the length of that trajectory. For example, for a structural network constructed with axonal **tract tracing**, we can identify the fiber tract linking two regions, draw a line following the trajectory of the tract, and estimate the distance of the axonal projection (Rubinov et al., 2015). In other datasets, such as those acquired with in vivo neuroimaging, such detailed anatomical measurements are often not possible and the Euclidean distance between pairs of network nodes is commonly used instead. The Euclidean distance is the shortest, straight-line distance between two points. This measure will provide a more accurate estimate of actual connection distances in nervous systems with a simple geometry, such as the neuronal network of *C. elegans*. In more complex systems such as the highly-folded mammalian cortex, Euclidean distance can underestimate the true connection distance between node pairs. The problem arises because, unlike the trajectory of axonal fibers, Euclidean distances are not constrained by anatomy. For example, connections between the left and right cingulate gyrus must travel from the gray matter, down into the white matter, through the corpus callosum and back up the other side. The shortest straight line (Euclidean) distance between these areas intersects the midline cerebrospinal

fluid. The Euclidean distance is thus a conservative and sometimes inaccurate index of the true wiring cost of a nervous system.

In general, wiring costs can be estimated with less ambiguity in structural compared to **functional connectivity** networks. In **structural connectivity** networks, the edges represent physical **links** between neural elements, and their wiring length corresponds to a sensible notion of wiring cost. In functional networks, two regions may show strong functional connectivity in the absence of a structural connection due to polysynaptic pathways (Honey et al., 2009; Vincent et al., 2007). Because we cannot be sure that two functionally connected regions share a direct structural link, estimates of connection distances in functional networks are an indirect approximation of network wiring costs.

If brain network wiring costs are related to the number and length of connections, we should expect **hubs** to be particularly costly elements of nervous systems. Consistent with this hypothesis, studies of human, rat, mouse, and *C. elegans* **connectomes** have revealed that hub nodes, and particularly the **rich-club** links between them, account for a disproportionate fraction of network wiring costs (Collin et al., 2014; Fulcher and Fornito, 2016; Towlson et al., 2013; van den Heuvel et al., 2012, 2015; Figure 8.1.1). Topologically central areas with either high **betweenness** centrality or high **degree** in human structural connectivity networks also have high rates of metabolic activity (Bullmore and Sporns, 2012; Collin et al., 2014; Figure 8.1.1). Similar findings have been reported in analyses of functional connectivity networks (Liang et al., 2013; Tomasi et al., 2013; Figure 1.11). Energetic demands dominate the genetic signature of hub connectivity: connected pairs of hubs show more tightly coupled gene expression than other pairs of brain regions, and this finding is primarily driven by coordinated expression of genes regulating oxidative metabolism (Fulcher and Fornito, 2016). Together, these results may partly explain findings that hub areas of the brain are particularly vulnerable to disease (Buckner et al., 2005, 2009; Crossley et al., 2014), as the high energetic demand of these regions may increase their sensitivity to any disruptions of metabolic supply caused by damage or dysfunction (Bullmore and Sporns, 2012). In cases where a particular neural insult triggers compensatory activation of other areas, the activity level of hubs may be pushed beyond their high basal rate, causing a functional overload and subsequent degeneration (de Haan et al., 2012; Fornito et al., 2015; Stam, 2014).

BOX 8.1 WIRING COST—CONT'D

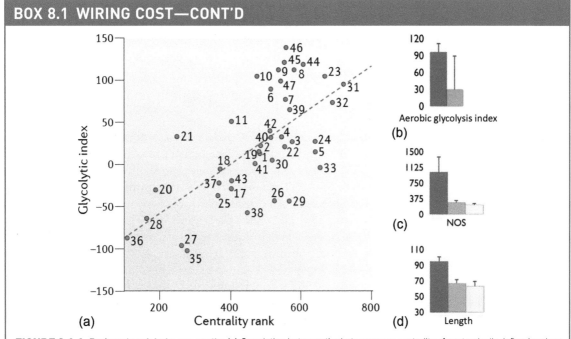

FIGURE 8.1.1 Brain network hubs are costly. **(a)** Correlation between the betweenness centrality of anatomically defined regions of the human cortex, as measured with **diffusion MRI**, and an index of regional aerobic glycolysis quantified with **PET** (Vaishnavi et al., 2010). Regions with higher **centrality** have higher levels of aerobic glycolysis. **(b)** A separate analysis found that network hubs belonging to the human brain's rich club (red) show higher average levels of aerobic glycolysis than other areas (gray). **(c)** Rich-club connections between hub nodes are particularly costly. This graph shows the average connction weight, measured as the number of streamlines (NOS) reconstructed with diffusion MRI, that link hub areas belonging to the rich club of the human brain (red), compared to the NOS reconstructed for feeder (orange) and local (yellow) connections. **(d)** Shows the same comparison, but for a measure of connection length. Rich-club connections between hub areas extend over longer distances, suggesting that they are more costly. *(a) Reproduced from Bullmore and Sporns (2012) and (b–d) from Collin et al. (2014) with permission.*

8.1.1 Node Motifs

Consider a pair of network nodes. How many different ways can these two nodes be connected? In a binary, undirected network, the answer is one: there can be only one connection between a pair of nodes; otherwise the pair is not connected. In a binary, directed network, we can define three possible wiring configurations: (1) node A projects to B; (2) node B projects to A; or (3) nodes A and B are reciprocally connected.

Now consider a triad of nodes. In an undirected network, these nodes can be connected in two ways: they form either an open triangle or a closed triangle. In an open triangle, one node is linked to two other nodes that do not directly connect to each other. In a closed triangle, each node in the triad is linked to the other two. In a directed network, there are 13 possible ways of connecting three nodes. Each of these configurations represents a candidate network motif.

Directed networks give rise to a richer diversity of motifs than undirected networks. For this reason, node motifs are more commonly analyzed in **directed graphs**. The full range of three-node motifs that are possible in a directed graph, and the distinction between open and closed triangle motifs, is shown in Figure 8.1.

Formally, a node motif may be defined as a **node-connected** subgraph that consists of M nodes linked by at least $M-1$ edges. For each subgraph size, M, there is a limited number of possible wiring configurations. These configurations are referred to as motif *classes*. The number of possible motif classes increases with M. For example, for $M=3$ in a directed network there are 13 classes, for $M=4$ there are 199 classes, $M=5$ has 9,364 classes and $M=7$ has 880,471,142 classes (Harary and Palmer, 1973). When counting distinct motifs, we do not separately count isomorphisms of the same graph. For instance, an open triangle with edges $A \rightarrow B$ and $A \rightarrow C$ is considered to belong to the same motif class as the open triangle with edges $B \rightarrow C$ and $B \rightarrow A$. It is rare for analyses of large networks to consider values of $M > 5$, since the number of possible subgraphs increases exponentially with M. A review of algorithms for finding motifs of different sizes is provided by Wong et al. (2012).

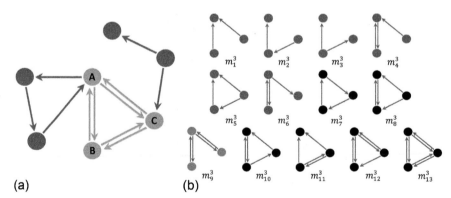

(a)   (b)

FIGURE 8.1 Three-node motifs in directed networks. **(a)** An example directed network. We select a subgraph comprising three nodes (red) and consider the possible ways in which they can be connected, excluding isomorphisms. **(b)** The 13 different ways to connect three nodes in a directed network. Motifs with gray-colored nodes correspond to the different types of open-triangle motifs that can be realized in a directed network. Motifs with black-colored nodes correspond to the different types of closed-triangle motifs that can be realized in a directed network. Blue-colored nodes highlight the m_9^3 apical open-triangle motif, which occurs with a greater frequency in brain networks compared to random and lattice-based reference networks. Note that an undirected network has only two classes of three-node motif: an open triangle and closed triangle. *Adapted from Sporns and Kotter (2004).*

By distinguishing between incoming and outgoing edges, directed networks give rise to a larger number of possible motif classes than undirected networks. The number of possible motif classes can be even higher in graphs with heterogeneous edges, such as those that distinguish between inhibitory and excitatory interactions. These kinds of motifs have been studied in metabolic and gene expression networks, but have not traditionally been considered in neuroscience due to the limited data available for distinguishing between different types of neuronal connections. Interested readers are referred to Alon (2007), Shoval and Alon (2010), and Shen-Orr et al. (2002) for examples of motif analyses in graphs with heterogeneous edges.

The variable topological configurations of different motif classes imply that each class serves a distinct functional role. Let us denote each motif class as m_i^M, where the superscript M refers to the number of nodes in the motif and the subscript i is the specific motif class. Consider the motif m_5^3 in Figure 8.1b, where node C sends an outgoing link to both A and B, and B also sends an outgoing edge to A. If we assume that all links in this class are excitatory, this configuration is consistent with a classic feed-forward loop. Feed-forward loops are commonly found in gene regulatory networks, where a general transcription factor (e.g., node C) and a specific transcription factor (node B) jointly regulate an effector operon (node A), while the general factor also regulates the specific factor (Alon, 2007; Shen-Orr et al., 2002). Contrast this motif with m_9^3, where node A is reciprocally connected to nodes B and C, which are otherwise unconnected with each other. In this motif, all communication between nodes B and C must flow through node A.

A motif analysis typically proceeds as follows:

1. Identify all possible subgraphs of size M within a network.
2. Count the number of instances of each motif class to estimate a *motif spectrum* for the network.
3. Repeat this process many times in an ensemble of suitably matched randomized and/or **lattice**-like benchmark networks (Chapter 10), to generate an empirical null distribution of frequencies for each motif class.
4. For each motif class, compute a *p*-value as the fraction of times the frequency count estimated in the benchmark data is higher than the count obtained in the empirical data.

If the *p*-value calculated in step 4 is low (e.g., <0.05), we conclude that the motif class recurs in our empirical network with a probability greater than expected under the null hypothesis, thus suggesting that the motif class is an important building block for the topological organization of the system.

Sporns and Kötter (2004) performed the first detailed motif analysis of brain networks, focusing on the chemical synapse network of *C. elegans*, interregional connectomes of the macaque visual system and whole cortex, and the interregional cortical connectome of the cat. For each of these directed networks, motif frequency spectra were analyzed at values of M ranging between 2 and 5. At each level of M, a distinction was drawn between structural and functional motifs. A *structural* motif is the actual connection pattern that defines a motif class. *Functional* motifs are defined with respect to a given structural motif, and represent the family of motifs that can be realized by eliminating different combinations of edges from the structural motif. In other words, functional motifs are subgraphs of structural motifs that are obtained by removing specific edges. We can think of functional motifs as subgraphs of a structural motif that may be selectively activated at any given time.

Any particular structural motif can give rise to a number of different functional motifs. For example, Figure 8.1b shows the full set of structural motifs for $M = 3$. Consider the m_{10}^3 motif. If we remove the edge running from node B to C, we get the m_4^3 motif. If we remove the edge running from node A to B, we get the m_5^3 motif and so on. Therefore, m_4^3 and m_5^3 are functional motifs of the structural motif m_{10}^3.

In brain networks, we can think of structural motifs as representing the anatomical building blocks of the network, and functional motifs as representing the diversity of interactions that can be realized on those structural foundations. Functional motifs in this context should not be confused with a direct index of functional connectivity or brain activity; rather, functional motifs are purely a topological property of the network. Some structural motifs will have a larger number of functional motifs. For example, all possible motif classes at $M = 3$ are subgraphs (and thus functional motifs) of the m_{13}^3 motif. In contrast, the m_4^3 has only two functional motifs: m_1^3 and m_2^3. As a corollary, the recurrence of certain kinds of structural motifs in brain networks may allow for greater diversity of functional motifs. Motif analysis can thus offer insight into the topological foundations of the brain's functional repertoire.

In their analysis, Sporns and Kötter (2004) found that brain networks across all the species and scales analyzed displayed the maximum diversity of functional motifs; that is, at least one instance of all possible functional motifs was found in all networks and at each level of M. In contrast, the diversity of structural motifs was lower and, for the macaque, was significantly less than the diversity observed in corresponding randomized networks. This result suggests that brain networks are configured to support high functional diversity through a small number of structural motif classes. Indeed, among the possible motifs of directed edges between a subgraph of three nodes ($M = 3$), the only structural motif that was significantly overrepresented in cat and macaque connectomes was an open triangle with an apical node reciprocally connected to two basal nodes that are not connected to each other (that is, the motif denoted m_9^3; see Figure 8.1b). At $M = 4$, only five classes occurred with greater frequency in the macaque visual system when compared to both random and lattice-based reference networks. Nine motifs showed similar evidence for over-representation in the macaque cortex, and five for the cat, out of a total of 199 possible four-node motifs. Thus, only a small fraction of the total possible number of structural motifs showed statistical evidence of overrepresentation in these brain networks.

Sporns and Kötter (2004) used a genetic algorithm to generate synthetic networks that were optimized for functional motif number and diversity. They found that these networks displayed structural motif spectra that were very similar to the empirical brain networks. For example, the m_9^3 apical open-triangle motif that was overrepresented in the macaque and cat networks was also the only motif overrepresented at $M = 3$ in these synthetic networks. In contrast, the motif spectra of networks optimized for *structural* motif number and diversity did not resemble the brain spectra. Collectively, these findings suggest that brain networks are topologically configured to realize a large number of functional states (as indexed by the number and diversity of functional motifs) from only a small set of structural building blocks (structural motifs). Subsequent work has shown that it is possible to generate synthetic networks with very similar motif spectra to brain networks using simple wiring rules that either favor hierarchical, clustered connectivity, or penalize the formation of long-range connections (Song et al., 2014; Sporns, 2006). These findings are consistent with the hypothesis that a pressure to minimize network wiring costs constrains the topological organization of the brain. We return to this issue in Section 8.3.

Harriger et al. (2012) replicated the statistical overrepresentation of the m_9^3 motif in a later analysis of a more fine-grained cortical connectome for the macaque, comprising 242 regions and 4090 connections, as collated from published tract-tracing studies. The authors also found evidence for an increased frequency of two other motifs: m_4^3 and m_6^3 (Figure 8.2a and b). An important property that is common to these three motifs is that they are all open triangles, meaning that two nodes connect to a third, *apex* node, without connecting to each other. In the brain, hub nodes are overrepresented as the apex node in these three-node motifs (Harriger et al., 2012; Honey et al., 2007; Sporns et al., 2007; Figures 5.6 and 8.3c). This overrepresentation can be measured using the apex ratio, a node-wise index that quantifies the proportion of open three-node motifs (e.g., $m_1^3, m_4^3, m_6^3,$ and m_9^3 and so on) for which that node occupies the apex position. A high apex ratio for hub nodes is consistent with a role in integrating otherwise segregated network elements (Gollo et al., 2015). Extending this idea, Harriger et al. (2012) moved beyond simple three-node motifs to examine the size of star-like structures centered on hub and nonhub regions. Star-like structures comprise a central node that is directly linked to other nodes that, in turn, are disconnected from

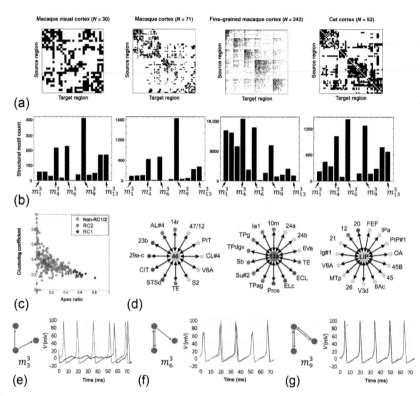

FIGURE 8.2 Structural and functional properties of three-node motif classes in brain networks. **(a) Adjacency matrices** for four different brain networks and **(b)** their corresponding motif spectra at $M = 3$. The frequency of the m_9^3 motif is high across all four brain networks. Analysis of the fine-grained macaque cortical network (middle right) additionally revealed a high frequency of the m_4^3 and m_6^3 motifs. **(c)** Nodes with a high apex ratio in the fine-grained macaque cortical network tend to have a low clustering coefficient (Section 8.2.1). This result is expected, since a node can only have a high apex ratio if it is involved in many open-triangle motifs, whereas high clustering implies a high frequency of closed-triangle motifs (Figure 8.1). Hub nodes that belong to the rich club (red) of the macaque brain have the highest apex ratio. Shown here are data for rich clubs defined at two levels of degree: RC1 is the most densely connected club and RC2 is a more liberal definition. **(d)** Star-like motifs centered on hub areas 46, 13a, and LIP of the macaque cortex. All surrounding regions are reciprocally connected to the center node, but are otherwise unconnected to each other. Colors represent the affiliation of each node to a different **module**. **(e–g)** Model-based estimates of activity time courses for simulated Hodgkin-Huxley excitatory neurons connected according to the m_3^3 **(e)**, m_6^3 **(f)**, or m_9^3 **(g)** apical motifs. Time courses for the two nonapical nodes in each motif are shown. **(e)** The m_3^3 motif supports a common driving input from the apex as the cause for zero-lag synchronization in the nonapical (basal) nodes, but this motif lacks reciprocal connectivity. Zero-lag synchrony within this motif is only achieved under specific conditions, such as when coupling strength is strong or there is negligible variation in conduction delays along the edges that link the apex and nonapical nodes. **(f, g)** Stronger and more robust zero-lag synchronization is achieved when the apex node has at least one reciprocal connection with at least one of the nonapical nodes, as in the m_6^3 and m_9^3 motifs. This result suggests that enhanced synchronization of activity in otherwise unconnected nodes is putatively a functional advantage for apical motifs that have reciprocal connectivity between the apex and nonapical nodes. *(a), (b), and (e–g) Adapted from Gollo et al. (2014) and (c) and (d) from Harriger et al. (2012) with permission.*

each other (much like the hub-and-spoke organization of a bicycle wheel; see also Figure 5.1). The average size of star-like motifs that were centered on high-degree hubs comprising the rich club of the macaque cortex (Chapter 6) was significantly larger than the size of star motifs centered on regions outside the rich club. Figure 8.2d shows examples of these star motifs for three hub regions: area 46 in dorsal prefrontal cortex, area 13a in orbitofrontal cortex and lateral inferior parietal cortex (LIP). Each of the outer nodes in these star motifs belongs to a distinct module, consistent with a role for the hubs in integrating diverse neural systems.

Apex nodes in open three-node motifs are thought to act as dynamic relays that promote zero-lag synchronization between the outer nodes (Vicente et al., 2008). Instantaneous synchronization of neuronal signals (zero-lag synchrony) has been found in multiple neural systems and is thought to underlie several important processes, such as the promotion of reliable neuronal signaling, integrated perception and behavior, and the temporal alignment of spikes to enable spike-timing dependent plasticity (Fries et al., 2007; Roelfsema et al., 1997; Singer, 1999; Varela et al., 2001). Zero-lag synchrony can be observed between distant anatomical locations, a phenomenon that seems to contradict the physical limits imposed by axonal conduction delays, which should cause lags of several tens of milliseconds. Zero-lag synchronization must therefore be established by a mechanism that can compensate for such delays. In this context, apex nodes are thought to exert a common, driving influence that, under certain conditions, is sufficient to induce zero-lag synchronization in the outer nodes (Vicente et al., 2008). However, computational modeling indicates that more robust and stable zero-lag synchrony occurs if the apex node has at least one reciprocal connection with one of the other vertices, as in the m_6^3 and m_9^3 motifs (Gollo et al., 2014; Figure 8.2e–g). This work also found that the addition of an edge that closes the triangle by directly linking the two outer nodes prevents the formation of zero-lag synchrony. Thus, the specific overrepresentation of open-triangle motifs with reciprocal connections in brain networks may promote rapid and stable synchronization across anatomically distributed neural elements.

8.1.2 Path Motifs

Node motifs represent the possible configurations of pairwise links between specific subsets of nodes. We can also define motifs at the level of paths of edges in a graph. In this analysis, each path consists of an ordered sequence of edges, and the edges are categorized into different classes according to some topological criterion.

The first path motif analysis of a brain network was performed by van den Heuvel et al. (2012). In an analysis of human structural connectivity networks constructed with diffusion MRI, these authors first categorized edges based on their relation to hubs comprising the brain's rich club. Edges were thus classified as rich (R) if they directly linked two hub nodes; feeder (F), if they linked a hub with a nonhub area; and local (L), if they linked two nonhub nodes that were not part of the rich club. Based on this categorization of edges, all shortest paths (Chapter 7) between pairs of brain regions were classified according to the order in which these edge types were traversed. For example, if a shortest path between two nodes comprised three edges, and these edges moved from a feeder to rich to local link, the path motif was classified as F—R—L. van den Heuvel and colleagues found that path motifs involving rich links were statistically overrepresented in the brain, indicating that the bulk of shortest paths in the brain pass through connections between hub areas. In particular, the most common path motif followed a L—F—R—F—L trajectory, such that a path started in the periphery, linked to a hub via a feeder edge, transferred between hubs through a rich link and moved out of the rich club via another feeder edge before linking to the periphery again (Figure 8.3a). Similar findings have been reported in the 279-neuron network of *C. elegans* (Towlson et al., 2013), in which the rich club constitutes 11 command interneurons with known functional roles in coordinated movement and adaptive behavior.

Directed networks allow us to infer directions of information flow along path motifs. More specifically, we can distinguish between feed-in edges, which project into the rich-club, and feed-out edges, which project out of the rich club. We should thus expect that most shortest paths in the brain will follow a directed route such as $L \rightarrow F_{in} \rightarrow R \rightarrow F_{out} \rightarrow L$, where F_{in} and F_{out} denote feed-in and feed-out edges, respectively (Figure 8.3b). We can also consider how these paths link nodes in different modules. For example, we might expect that the source and target nodes of the L—F—R—F—L motif are likely to be in distinct modules, and that the primary link between modules is through the rich club. We can thus develop a more fine-grained understanding of different shortest paths in the brain by considering the directionality of connectivity and the modular organization of the network. Analyses of such path motifs in the cat and macaque connectomes have confirmed that the majority of shortest paths linking nodes in different modules pass through rich links (de Reus and van den Heuvel, 2013b; Harriger et al., 2012; Figure 8.3c and d).

Note that the motif classes considered here—both at the level of nodes and paths—are only a small fraction of the diverse array of different types of motifs that can be defined in brain networks. As already mentioned, the distinction between inhibitory and excitatory interactions allows us to define a variety of feed-forward, feed-back, and other regulatory loops (Alon, 2007; Shoval and Alon, 2010). Different motifs can also be defined based on the distribution of weights amongst edges (Onnela et al., 2005). Alternatively, we can define motifs by categorizing nodes and edges on topological or anatomical properties other than those related to rich club and modular organization, such as whether links are between cortical and subcortical areas, whether they are ipsilateral or contralateral, and so on. Others have proposed using the motif spectra of individual

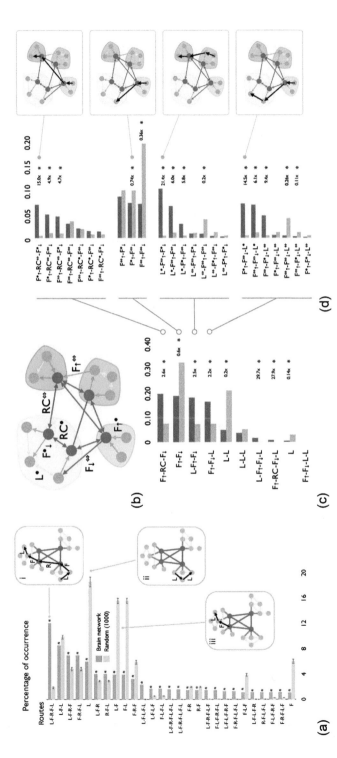

FIGURE 8.3 Path motifs in brain networks. **(a)** The path motif spectrum for undirected structural connectivity networks generated with human diffusion MRI. Edges are categorized as either local (L), feeder (F), or rich (R), and different ordered sets of these edges on the shortest paths between nodes represent different classes of path motifs. Plotted here are the frequencies of some key motifs in the empirical brain networks and matched randomized networks. Insets show a schematic for the route followed by some of the key paths. The L-F-R-F-L motif is the most common. **(b)** Schematic showing how additional path motif classes can be characterized by considering the directionality of connectivity and the modular organization of the network. Local edges (L) are shown in yellow, feeder (F) in orange, and rich links (R) in red. Colors depict different topological modules. In directed networks, a distinction can be drawn between feed-in edges (F ↑) and feed-out edges (F ↓). We can also categorize rich or feeder edges as linking nodes that are in the same module (intramodular, denoted with •) or as linking nodes that are in different modules (intermodular, denoted with ⇔). Thus, for example, $F_{\downarrow}^{\Leftrightarrow}$ denotes an intermodular feed-out edge. **(c)** Directed path motif spectrum in the cat cortical connectome. Pink bars represent empirical motif frequencies, gray bars represent the corresponding values in appropriately randomized control networks. The most common motif follows a feed-in—rich—feed-out (i.e., $F_{\uparrow} - RC - F_{\downarrow}$) trajectory, consistent with most shortest paths traversing the rich club (Chapter 6). Extended motifs incorporating local connections could not be properly quantified in this low-resolution network, since 97% of intermodular paths involved just two or three links and none comprised five edges. **(d)** Additional subclasses of directed motifs that distinguish between intramodular and intermodular links. Intermodular communication paths in the cat brain either pass through rich links, or through a feed-out link. *(a)* Reproduced from van den Heuvel et al. (2012) and *(b)* from de Reus and van den Heuvel (2013b) with permission.

nodes across different levels of M to define measures of node centrality (Wang et al., 2014b). We are only limited by our imagination and the connectivity data available when characterizing motif spectra in brain networks.

8.2   CLUSTERING, DEGENERACY, AND SMALL WORLDS

Analysis of three-node motifs suggests that open-triangle structures, in which two nodes link to a third apical node but are otherwise unconnected, are overrepresented in brain networks. This is not to say that the apical motif characterizes all triadic interactions in the brain. In fact, the frequency of closed-triangle motifs is directly related to the clustering of a network, such that a network with a high number of closed triangles will necessarily have high clustering (Figure 8.1). As discussed in Chapter 1, clustering is one of the defining characteristics of small-world networks, and it is also a prominent feature of nervous systems. In fact, it was first noted in the seminal analysis of *C. elegans* by White et al. (1986), whose qualitative descriptions predated subsequent attempts to quantify the clustering of neuronal connectivity with **graph theory**. In their words:

> One of the striking features of the connectivity diagrams is the high incidence of triangular connections linking three classes ... A typical neuron in *C. elegans* is accessible (i.e., adjacent) to a fairly limited subset of the total complement of neurons but is fairly highly locally connected within this subset ... Thus, if a neuron has synaptic contacts with two partners, these two partners must be neighbors to the neuron and therefore are likely to be neighbours themselves. It is therefore quite probable, given the high level of local connectivity, that there will be a synaptic contact between them, which will close the triangle. The abundance of triangular connections in the nervous system of *C. elegans* may thus be simply a consequence of the high levels of connectivity that are present within neighbourhoods.
>
> **White et al. (1986, p. 47)**

It should therefore come as little surprise that the clustering coefficient, a measure of triangular motif frequency, was later shown by Watts and Strogatz (1998) to be quantitatively greater in *C. elegans* than in a comparable random graph. However, clustering of neuronal connectivity, which implies a high frequency of closed-triangle motifs, appears to contradict the results of the motif analyses considered in the previous section, which have generally found evidence of an overrepresentation of open-triangle motifs in brain networks. One reason for this apparent discrepancy is that the motif analysis of Sporns and Kötter (2004) declared motifs as significantly overrepresented if their frequency was higher relative to comparable randomized *and* lattice-like null networks (for details on how these null networks are generated, see Chapter 10). The lattice benchmarks will, by construction, have near-maximal clustering, which makes it more difficult to find evidence for overrepresentation of closed-triangle motifs. The Watts-

Strogatz analysis implies that closed triangles are indeed overrepresented in the brain if we only use random networks as a benchmark for comparison. In this section, we examine different ways of measuring clustering in networks and consider what clustered connectivity means for brain organization.

8.2.1 The Clustering Coefficient

Clustering can be measured either at the level of an individual node or across the entire network. At the node level, we are interested in the relationships between the pairs of **neighbors** of any given node i. A triad or triangle is formed when node i is connected to any two neighbors, h and j. If h and j are also directly connected to each other, the triangle is closed. If h and j are unconnected, the triangle is open.

We can compute the clustering coefficient of node i in a binary, undirected network by counting the number of pairs of node i's neighbors that are connected with each other. To ensure the clustering coefficient is between zero and one, we normalize this number by the total number of pairs of neighbors that node i has, irrespective of whether they are connected or not. We thus have,

$$Cl(i) = \frac{\text{Number of pairs of } i\text{'s neighbors that are connected}}{\text{Number of pairs of } i\text{'s neighbors}}. \qquad (8.1)$$

We can think of the clustering coefficient as the probability of finding a connection between any two neighbors of node i. This is equivalent to computing the proportion of closed triangles that are attached to node i, relative to the total number of closed triangles that are possible between i's neighbors,

$$Cl(i) = \frac{2t_i}{k_i(k_i - 1)}, \qquad (8.2)$$

where k_i is the degree of node i and t_i is the number of closed triangles attached to i. We double the numerator in Equation (8.2) because this definition is for an undirected network.

To quantify the clustering of the entire network, we can simply average our nodewise clustering values to obtain the network-wide clustering coefficient,

$$Cl = \frac{1}{N}\sum_{i \in N} Cl(i) = \frac{1}{N}\sum_{i \in N} \frac{2t_i}{k_i(k_i - 1)}. \qquad (8.3)$$

Both $Cl(i)$ and Cl take values between zero and one, where a value of zero indicates a complete absence of clustering (i.e., no closed triangles) and a value of one indicates full connectivity between a node's neighbours or, for the global coefficient, a fully connected graph. Note that the clustering coefficient will be zero for any node with degree less than two. This is because a node must have at least two neighbors for it to be part of a triangle. If a network is fragmented, isolated nodes will contribute a value of zero to the overall mean clustering coefficient of the network, thereby reducing the global average. In this case, the calculation can be restricted to the largest **connected component** of the network (Kaiser, 2008).

The clustering coefficient uses a nodewise normalization procedure. If a node has low degree, the denominator in Equation (8.2) will be small, and the clustering coefficient may be inflated simply because the node has few connections. An alternative method is to use a network-wide normalization. This results in a measure called *transitivity*, which is defined as

$$T = \frac{\sum_{i \in N} 2t_i}{\sum_{i \in N} k_i(k_i - 1)}. \tag{8.4}$$

Transitivity estimates the probability that any two nodes connected to a third are also connected to each other, whereas the clustering coefficient estimates the proportion of closed triangles attached to a node, separately for each node (Equation 8.2). These node-specific probabilities are then averaged to estimate the global clustering coefficient (Equation 8.3). This average is not the same as the probability estimate given by transitivity. Why? Some nodes may have a higher degree than others, and thus the number of potential closed triangles that can be formed around these nodes is larger. Taking the average across these nodes, as per Equation (8.3), does not weight these higher degree nodes accordingly. All nodes are weighted equally, and thus Equation (8.3) tells us the average proportion of closed triangles around a node. This is not necessarily the probability of finding a closed triangle across the entire network.

The clustering coefficient can be generalized to weighted networks. Several definitions have been proposed, each of which varies in the way it normalizes the edge weights of the network. Here, we focus on the definition proposed by Onnela et al. (2005) because it robustly characterizes edge weight variations in different kinds of triangles. Interested readers are directed to an article by Saramäki et al. (2007) for a more systematic comparison of different weighted clustering coefficient measures.

The weighted definition of clustering proposed by Onnela et al. (2005) is based on the concept of subgraph intensity. The intensity of a subgraph comprising a triangle attached to node *i* is defined as the geometric mean of its edge weights. The geometric mean is the *n*-th root product of *n* numbers. When calculating the clustering coefficient, we only consider triplets of nodes, so $n = 3$. We can thus define the weighted clustering coefficient as

$$Cl^w(i) = \frac{2}{k_i(k_i - 1)} \sum_{j, h} \left(\hat{w}_{ij} \hat{w}_{jh} \hat{w}_{hi} \right)^{1/3}, \tag{8.5}$$

where the weights have been scaled relative to the maximum edge weight in the network, such that $\hat{w}_{ij} \leftarrow w_{ij}/\max(w)$. Values of this weighted coefficient vary between zero and one. A value of zero indicates an open triangle. A value of one indicates that all of node i's neighbors form closed triangles, and that the weights on those edges are all equal to the maximum weight found in the network. We can thus think of Equation (8.5) as a measure of the average weighted intensity of all triangles attached to a given node, relative to the maximum edge weight of the network. Importantly, $Cl^w(i) = Cl(i)$ when the network is binary. The 2 in the numerator of Equation (8.5) assumes that the summation only counts each triangle once.

The picture is more complicated in directed networks as we can define different kinds of triangles by considering the directions of the edges within the triad. Generalizations of the clustering coefficient that account for this heterogeneity have been proposed (Fagiolo, 2007), but have rarely been used in the study of brain networks (for an exception, see Rubinov et al., 2015). In directed networks, analysis of motif spectra can provide a more comprehensive understanding of the relative frequency of distinct three-node motifs.

In **connectomics**, the clustering coefficient is often interpreted as a measure of functional segregation or specialization. A few caveats to this kind of interpretation are worth noting. First, the clustering coefficient is purely a topological measure and does not directly index network dynamics. Second, the clustering coefficient does not directly index segregation. Segregation implies some topological or functional separation between subsets of network nodes. The clustering coefficient characterizes the properties of triadic subgraphs in the network, but does not measure anything about how different subgraphs relate to each other. The clustering coefficient is more accurately interpreted as an index of the topological capacity of a network for integration within three-node motifs. This capacity may facilitate functional specialization, to the extent that specialization is supported by tight integration within small subgraphs. Functional specialization implies segregation, but the clustering coefficient does not directly measure segregation.

We must also consider the spatial scale of functional specialization when interpreting the clustering coefficient. It is commonly assumed that a functionally specialized unit in the brain corresponds to a spatially contiguous collection of neurons, such as a cortical column, hypercolumn, or cytoarchitectonic region. Under this assumption, functional specialization is a spatially localized property of a specific brain region. The clustering coefficient has sometimes been referred to as an index of the "local" information-processing capacity of a network. However, when considering a spatially embedded network such as the brain, we must be careful to distinguish *spatial* proximity from *topological* proximity. None of the definitions presented in this section for measuring clustering use spatial information. This means that "locality" is defined purely by **topology**, and neighbors are classed as such just because they share a direct connection and not because they are physically close to each other. As a result, two topological neighbors may be located at opposite anatomical ends of the brain, and a topological neighborhood can comprise directly connected areas that are separated by large anatomical distances. It is therefore possible for the clustering coefficient to be defined for subgraphs of nodes that are distributed throughout the brain. Integration within these distributed subgraphs may support functional specialization at a systems-level, but the clustering coefficient in and of itself does not tell us anything about how these specialized processes are anatomically localized.

8.2.2   Redundancy, Degeneracy, and Structural Equivalence

The clustering coefficient can also be thought of as a very simple measure of the topological redundancy of a network. **Redundancy** is the repetition of identical elements of a system. In contrast, **degeneracy** is the capacity of structurally different system elements to perform the same function (Tononi et al., 1999). Redundancy is common in engineering, where repetition of identical elements ensures that failure of one node does not compromise the operation of the entire system. For example, a computing infrastructure can have multiple back-up systems to protect against data loss. Degeneracy is common in biological systems, where resources are scarce and repetition of identical elements is costly (Kitano, 2004). A cost-effective solution is to allow structurally different elements to perform overlapping functions. This organization affords both flexibility and robustness to perturbation—the functional diversity of different elements increases the system's repertoire (flexibility), while the capacity to assume overlapping functions promotes resilience to the failure of any individual element.

It is easy to see why degeneracy is an attractive property for the brain. Limits on metabolic resources prohibit the large-scale replication of individual neural elements (i.e., redundancy), since repetition comes at the expense of diversity and thus flexibility and adaptability. With degeneracy, individual system elements can assume the functions of other units, while still retaining a capacity to operate independently. It is precisely this combination of shared and unique contributions to the outputs of a system that characterizes degeneracy. A certain degree of redundancy is necessary to support degeneracy. However, a completely redundant system will not be degenerate, because it does not allow for independent contributions from individual elements (Tononi et al., 1999). In other words, a degenerate system will only show partial redundancy. An example of degeneracy in the brain can be seen in two anatomically and functionally dissociable neural systems that have been implicated in reading (Seghier et al., 2008). Damage to one system on its own is insufficient to cause a reading impairment because the intact system is able to assume the functions of its lesioned counterpart; however, deficits will arise if both systems are affected (Friston and Price, 2011). Neural degeneracy can be expressed in different ways, as described in detail by Noppeney et al. (2004).

We can understand how the clustering coefficient relates to redundancy and degeneracy by comparing an open and closed-triangle motif in an **undirected**

graph. In the open triangle, dysfunction of the apex node will fragment the triad, since there is no direct connection between the remaining two nodes. In contrast, removal of any single node in the closed-triangle motif will not disrupt communication between the remaining elements. In this case, redundancy in the neighborhood of the nodes comprising the triad allows the motif to continue to function as a network. It is for this reason that measures of clustering or local **efficiency** are sometimes also referred to as measures of local fault-tolerance, and why clustered networks are resilient to random attack (Latora and Marchiori, 2003; see also Chapter 7). We should therefore expect that a neural system characterized by high clustering should also show redundancy. Hub nodes with low clustering, which link otherwise unconnected areas, may play a pivotal role in degeneracy by enabling communication between different functional systems. As a consequence, dysfunction of hub nodes with low clustering will have a more profound impact on brain network integrity, and compromise the system's capacity for compensation (Fornito et al., 2015).

We can extend this logic beyond subgraphs of three nodes to consider similarities in the extended neighborhoods of pairs of nodes. In this case, we are examining a property known as **structural equivalence**—the extent of overlap in the connection profiles of pairs of nodes. To quantify this equivalence, we could calculate the correlation between two rows (or columns) i and j of a binary, undirected adjacency matrix. A high positive correlation would suggest that nodes i and j have very similar neighborhoods—they connect to the same nodes. In the case of identical neighborhoods, the two nodes are completely redundant. Short of this limiting case, topological overlap may promote degeneracy.

An alternative method for assessing topological overlap is to compute the number of neighbors that nodes i and j share in common. To account for variations in node degree, this value can be normalized by the total number of unique nodes in the neighborhoods of nodes i and j, yielding the *normalized matching index* (de Reus and van den Heuvel, 2013b; Zamora-López et al., 2010),

$$nmi(i, j) = \frac{|\Gamma(i) \cap \Gamma(j)|}{|\Gamma(i) \cup \Gamma(j)|}, \tag{8.6}$$

where $\Gamma(i) = \{j : A_{ij} = 1\}$ is the neighborhood of node i. This quantity, which is also called the Jaccard index (Jaccard, 1912), will equal one when nodes i and j connect to exactly the same neighbors and zero when they connect to completely nonoverlapping sets of nodes. For directed networks, Equation (8.6) quantifies the overlap with respect to neighbors that are directed to nodes i and j (i.e., incoming connections). To evaluate overlap with respect to the edges outgoing from nodes i and j, we redefine the neighborhood of node i such that $\Gamma(i) = \{j : A_{ji} = 1\}$.

Figure 8.4 shows a simple example of how the normalized matching index is computed. In an analysis of the cat cortical connectome constructed from a

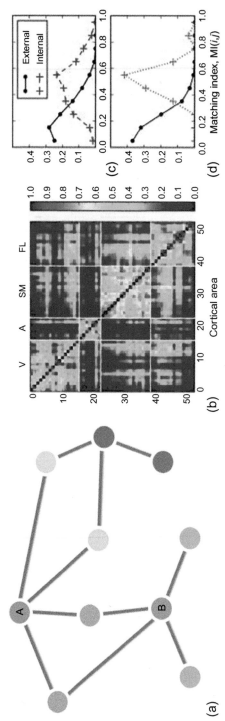

FIGURE 8.4 Topological overlap and the modular organization of brain networks. **(a)** An example network showing the calculation of the normalized matching index. Take two index nodes A and B (red). The neighborhood of A is shown in green, the neighborhood of B is shown in blue, and nodes that belong to both neighborhoods are shown in purple. Purple nodes thus represent the intersection of the neighborhoods of A and B, so that we have $|\Gamma(A) \cap \Gamma(B)| = 2$. The union of the two neighborhoods is the number of purple, blue, and green nodes in this graph, which is $|\Gamma(A) \cup \Gamma(B)| = 6$. The normalized matching index is thus $nmi(A, B) = 2/6 = 0.33$. **(b)** A **matrix** of normalized matching index values estimated for every pair of 53 regions of the cat cortex. Ordering this matrix to emphasize its organization into visual (V), auditory (A), sensorimotor (SM), and frontolimbic (FL) modules shows that topological similarity is higher within modules than between modules. **(c)** Distributions of normalized matching index values for pairs of regions in the same module (dashed lines) and pairs of regions in different modules (solid lines). Many of the high values for intermodule pairs are contributed by hub nodes comprising the rich club of the network, which were found to have high topological overlap. This is demonstrated in **(d)**, which shows the distributions for pairs of nodes belonging to different modules after exclusion of hubs (solid line). The tail of the distribution decays faster than in **(c)**. The dashed line is the distribution of matching values for hubs in the topological rich club. The values are as high as those observed for intramodular pairs, suggesting that the rich club shows a degree of topological overlap that is commensurate with pairs of nodes belonging to well-defined functional systems. **(b–d)** *Reproduced from Zamora-López et al. (2010) with permission.*

synthesis of published tract-tracing studies, Zamora-López et al. (2010) found that pairwise measures of the normalized matching index, *nmi*, cluster into modules that correspond closely with known functional systems of this network (Figure 8.4a). They also found that the average nmi of node pairs within the same module was higher than the average *nmi* of node pairs in different modules (Figure 8.4b), suggesting that topological overlap, a marker of redundancy and degeneracy, is high amongst pairs of nodes within functionally specialized neural systems. Interestingly, hub nodes belonging to the rich club of the network also had very high topological overlap, despite belonging to distinct modules (Figure 8.4c and d). This finding suggests that connectivity between hub areas may support degeneracy *between* neural systems.

An alternative measure of topological overlap is the *cosine similarity* (Salton, 1989),

$$cs_{ij} = \frac{\sum_h A_{ih} A_{jh}}{\sqrt{\sum_h A_{ih}^2} \sqrt{\sum_h A_{jh}^2}}. \tag{8.7}$$

In a binary, undirected network, the numerator of Equation (8.7) counts the number of neighbors shared by *i* and *j*, and the denominator is the geometric mean of the respective sums of their squared degrees. In a directed network, Equation (8.7) measures the overlap of the incoming connections of nodes *i* and *j*. The name cosine similarity comes from the fact that this expression can be rewritten in geometric terms as the cosine angle between the vectors of connections attached to nodes *i* and *j*. As with the normalized matching index, values of cosine similarity vary from zero, indicating no common neighbors, to one, indicating identical neighborhoods. In directed networks, we can compute cosine similarity separately for incoming and outgoing links.

Song and colleagues (2014) used this measure to examine the pairwise similarity of projection targets for 29 regions of the macaque cortical connectome, as mapped with viral tract tracing. Node pairs with high similarity clustered together into known functional subdivisions of the brain, suggesting that functionally similar areas have high topological similarity (Figure 8.5a). Similar findings were obtained in an analysis of the mouse connectome (Figure 8.5b). In the macaque, the authors found that the probability of connection between two regions depended in a similar way on the topological similarity of those regions (Figures 8.5c and d) and their physical separation (Figure 8.5e). In particular, regions were more likely to be connected if they were more topologically similar and located closer together in anatomical space. One possible reason for this dependence is that cortical regions are more likely to form connections with anatomically proximal areas, which in turn increases the likelihood that two nearby areas will connect to similar regions (including each other). Thus, nearby regions

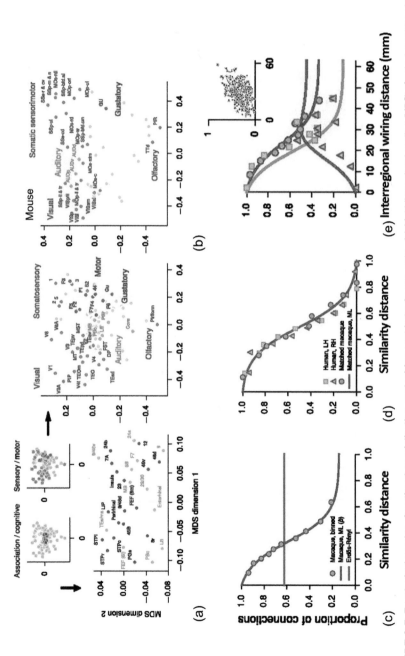

FIGURE 8.5 Topological overlap, functional specialization, and wiring cost in brain networks. **(a)** Multidimensional scaling (MDS) of pairwise cosine similarity distance values in the 29-region cortical connectome of the macaque. Topological similarity was computed as $1 - CS_{ij}$, so higher values reflect less similarity. The similarity values cluster nodes into functionally specialized systems. In this projection, regions are located closer together in Euclidean space if they have a similar profile of cosine similarity distances to other areas. The axes correspond to the first two dimensions identified by the MDS analysis. The MDS projections for the association/cognitive (gray, bottom) and sensory/motor systems (color, right) are magnified to show further detail. **(b)** Similar results were obtained in an analysis of a mouse connectome constructed by Zingg et al. (2014). **(c)** The dependence of connection density on topological similarity distance. Each dot represents the number of present connections divided by the total possible number of connections in different bins of similarity distance. The plot shows that two regions are more often connected when they have similar neighbors (i.e., they have high topological similarity). **(d)** The same relationship between connection density and topological similarity is observed in the undirected structural connectivity network of the human, as constructed with diffusion MRI. "Matched macaque" refers to a version of the macaque connectome that has been symmetrized and thresholded to match the human network. LH and RH represent left and right hemispheres, respectively. **(e)** The dependence of connection density on interregional wiring distance. Pairs of regions are more often connected when they are spatially close. The form of the relationship is very similar to **(c)** and **(d)**. Green represents reciprocal connections and orange represents unidirectional connections. Inset shows the association between pairwise similarity distance (vertical axis) and interregional wiring distance (horizontal axis). Regions that are further apart have less topological similarity. *Images reproduced from Song et al. (2014) with permission.*

will have a higher probability of connection and higher topological similarity. In this manner the cost of forming long-range projections interacts with the spatial location of different brain regions to produce clustering, functional specialization, redundancy, and degeneracy.

We have only considered two measures of topological overlap, but others are available (Dice, 1945). Many of these can be generalized to examine similarities in higher-order, so-called m-step, neighborhoods rather than just the direct neighbors of each node (Yip and Horvath, 2007). In this analysis, we set $m = 1$ to investigate whether nodes i and j have the same neighbors, $m = 2$ to determine whether the neighbors of i and j are themselves similar, and so on. In general, these measures quantify the redundancy of connections between pairs of nodes (for an alternative measure of topological redundancy, see Albert et al., 2011). However, this redundancy is seldom complete, since it is rare for two nodes in a brain network to have identical neighborhoods. Given that partial redundancy is a marker of degeneracy, we can think of measures of topological overlap as providing an indirect index of a network's capacity for degeneracy.

We can distinguish more precisely between degeneracy and redundancy if we consider how different elements of a neural network relate to a given output. In a framework proposed by Tononi et al. (1999), this distinction is made by examining how the outputs of a neural system are affected by perturbation of different system elements. In this context, we can think of the nervous system as a network comprising nodes connected by edges and the output as any generic output of this system, ranging from a simple physiological or motor response to a complex cognitive operation.

The analysis proceeds as follows. We first map a neural system, X, and its interconnections. This map will generally correspond to the adjacency matrix of our brain network. We then define how the elements of X relate to a specific output, O. For example, in Figure 8.6a, X is a graph of nodes connected by edges and O is a separate sheet of units defining the output of interest. Specific elements of X impact units of O in a precisely defined way, as indicated by the edges running from nodes in X to units of O. Once these mappings are defined, we then perturb every possible subsystem of X and quantify the impact of this perturbation on O (Figure 8.6b). These subsystems are defined as different partitions or subgraphs of X. Perturbation of a partition of X will affect O if, and only if, that partition contributes to the output in some way. If we repeat this process across all partitions, we can assess whether any two or more partitions make overlapping contributions to the output.

Formally, we can quantify the association between X and O using mutual information, a general measure of the association between two variables that is sensitive to both linear and nonlinear interactions (Box 8.2). We can define many different subsystems j of X at various scales, comprising $n = 1 \ldots N$ nodes. At each scale n, we compute the average mutual information between the output

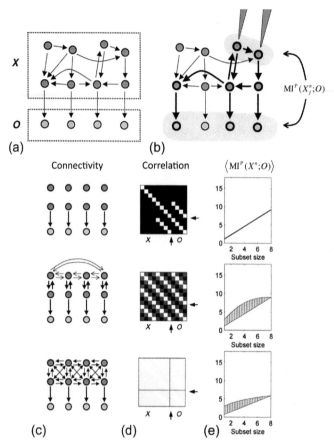

FIGURE 8.6 Quantifying neural degeneracy and redundancy. **(a)** To accurately distinguish between degeneracy and redundancy in a neural system, we first define the intrinsic connectivity of this system, X, and how its nodes influence a generic output, O. In this representation, O is modeled as a sheet of distinct units receiving inputs from specific nodes of X. **(b)** A subset of X (gray shading) is perturbed by injecting uncorrelated noise (gray arrows, top right). The perturbation propagates along the connections of X (thick edges) and changes the activity of other elements, as well as units of O. The process is repeated for all possible j partitions of X at each scale n, and the mutual information between each subset X_j^n and O under perturbation, $MI^P(X_j^n; O)$, is computed. **(c)** Model network architectures that vary in terms of redundancy and degeneracy. Top row shows a model neural system that lacks intrinsic connectivity, and in which all system elements make independent contributions to O. Middle row shows a system with high degeneracy that comprises four strongly interconnected modules (each comprising two nodes) that are sparsely connected with each other. This topology allows for both independent and overlapping contributions to O. Bottom row shows a system that is densely interconnected and thus highly redundant. **(d)** Correlation matrices for each system depicted in (a), showing the correlations that are intrinsic to X and O, as well as the correlation between them. The arrows delineate the output portion of the matrices. Light colors indicate higher correlations. **(e)** The average mutual information under perturbation between O and all subsets of X at each subset size, n, denoted $\langle MI^P(X^n; O)\rangle$. The shaded area shows the extent to which this quantity exceeds a linear increase with increasing subset size. The independent network (top) shows a linear increase of this quantity with subset size. The degenerate network (middle) shows a greater increase in $\langle MI^P(X^n; O)\rangle$ with subset size, indicating that different subsets make overlapping contributions to O, consistent with neural degeneracy. The redundant network (bottom) also shows a greater increase in $\langle MI^P(X^n; O)\rangle$ than the independent system (top), but this quantity is lower for larger subsets of X when compared with the degenerate system (middle). This is because the quantity $\langle MI^P(X^n; O)\rangle$ in a redundant system will be highest for small partitions of X. *Reproduced from Tononi et al. (1999) with permission. Copyright (1999) National Academy of Sciences, U.S.A.*

and all possible subsets of nodes X_j^n. We then quantify how this average deviates from the mutual information of O and the entire system X to obtain a measure of the degeneracy of X with respect to O.

$$D(X; O) = \sum_{n=1}^{N} \left[\left\langle MI^P \left(X_j^n; O \right) \right\rangle - \left(\frac{n}{N} \right) MI^P(X; O) \right], \qquad (8.8)$$

where $MI^P(X_j^n; O)$ is the mutual information, under perturbation, between the subset X_j^n and the output O; and $\left(\frac{n}{N} \right) MI^P(X; O)$ is the mutual information, under perturbation, between the entire system X and O, scaled at each level of n. The triangular brackets denote the mean mutual information across all possible partitions of X into subsystems of size n nodes. If the contributions of different partitions of X to the output O are completely independent, the contribution that the entire system makes to the output will be equal to the sum of the contributions of each partition. In other words, the mean quantity $\langle MI^P(X_j^n; O) \rangle$ will be the same as the mutual information between O and the entire system at that scale, $\left(\frac{n}{N} \right) MI^P(X; O)$, and $D(X; O)$ will equal zero. If subsets of X make overlapping contributions to the output, $\langle MI^P(X_j^n; O) \rangle$ will be larger than $\left(\frac{n}{N} \right) MI^P(X; O)$, and $D(X; O)$ will be greater than zero (see also Figure 8.6c–e).

We can define the redundancy of X with respect to O by focusing on subsets of X of size $n = 1$,

$$R(X; O) = \sum_{j=1}^{N} \left[MI^P \left(X_j^1; O \right) \right] - MI^P(X; O). \qquad (8.9)$$

By this definition, redundancy is high if the sum of the mutual information, under perturbation, between O and each individual node in X is higher than the mutual information between the entire system and the output. In this case, each of the elements contributes similar information to the output, meaning that $MI^P(X; O)$ will be less than the sum of the individual $MI^P(X_j^1; O)$ values. The quantity $R(X; O)$ is zero if all nodes contribute independently to the output.

Degeneracy and redundancy are related as

$$D(X; O) = \sum_{n=1}^{N} \left[\left(\frac{n}{N} \right) R(X; O) - \left\langle R \left(X_j^n; O \right) \right\rangle \right]. \qquad (8.10)$$

According to this equation, degeneracy is high when the average redundancy of small subsets of X is lower than expected from a linear increase of redundancy with increasing subset size. In other words, degeneracy is high when the redundancy of small subsets of X is low and the redundancy of the entire system is high. In this case, we have partial redundancy for large subsets of X but not individual elements, allowing for the possibility of both overlapping and independent contributions to the output (Figure 8.6c–e).

Tononi, Sporns, and Edelman (Sporns et al., 2000; Tononi et al., 1999) have used computational models to show that network topologies optimized for degeneracy have a modular, small-world topology and complex dynamics (Box 8.2). In this work, the links between elements of X and O can be precisely defined, and different partitions of X can be perturbed by injecting random noise into specific system elements (Figure 8.6b). Computing these measures for empirical data is challenging. The large number of possible subgroupings of elements in a neural network makes an exhaustive search over all possible perturbations intractable for anything but small systems (although analytic approximations are available; see Tononi et al., 1999). Furthermore, the connectivity between elements of X and output units of O must be known a priori. This requires a precise knowledge of how specific network elements relate to different functions. Such a mapping may be possible in simple model organisms such as C. elegans but can be difficult to discern in more complex nervous systems.

A more practical approach to characterizing neural degeneracy has been proposed by Noppeney et al. (2004). Principally developed for the study of human clinical populations, they suggest that functional imaging can first be used to identify the candidate neural substrate of a particular behavior or output. Second, the effects of perturbing each activated region of this system must be characterized. The perturbation could be studied by, for example, identifying patients with suitably localized lesions or through the use of **transcranial magnetic stimulation** to temporarily interrupt the function of each implicated brain area. Degeneracy is evident when the perturbation of a specific system element does not impair the output behavior. As a final step, Noppeney and colleagues suggest measuring brain activation as participants perform the target behavior under perturbation of each element to examine whether there are any alternative or latent degenerate systems that can support behavior. Such systems are revealed by the activation of additional regions that are not seen under normal performance conditions. If such a system is discovered, the process should be repeated again for the new system.

Noppeney and colleagues (2004) review examples where this strategy has provided insights into various types of neural degeneracy. This approach is nonetheless difficult to implement in practice, as it requires either access to patients with a sufficient diversity of lesions to cover the system of interest, or the capacity to reversibly induce lesions in all relevant brain regions. This can be a problem for noninvasive transcranial stimulation techniques, which can only stimulate superficial cortical areas, and with limited spatial precision. Techniques such as optogentics, when applied to animal models, may offer a powerful alternative for studying degenerate neural systems, but they have not yet been used for this purpose.

8.2.3 Small Worlds

In Chapter 1, we discussed the seminal work of Watts and Strogatz (1998) in defining a new class of networks with topological properties that are neither completely random nor completely regular. These so-called small-world networks show high levels of clustering, much like a regular lattice, combined with low average path length, much like a random network.

The small-world idea was inspired by studies of social networks. Social networks tend to be clustered, since there is a high probability that any two friends of an individual person will also be friends themselves. This tendency suggests that social networks are like lattices. However, the pioneering work of Stanley Milgram showed that messages are able to navigate through social networks, often extending across large geographical distances, via only a short number of steps—the famous "six degrees of separation" rule (Milgram, 1967). Such a low average path length is suggestive of a more random topology. Small-world networks are interposed between regular and random graphs, showing both high clustering and low average path length. As we have seen, clustered connectivity supports degeneracy and triangular integration within subgraphs of three nodes, and may facilitate functional specialization. Low average path length facilitates efficient integration across the network (Chapter 7). Thus, small-world properties provide a topological substrate for simultaneous specialization and integration of function (see also Box 8.2).

In a formal analysis, we want to be more rigorous than simply concluding that clustering is "high" and path length is "low" in a network. We typically want to quantify which specific values of clustering should be considered high and which values of path length should be considered low. A simple solution to this problem is to compare the values of clustering and path length in our empirical network to comparable values computed in appropriately randomized control networks. We can thus define the normalized clustering coefficient of a network as

$$\gamma = \frac{Cl}{\langle Cl_{\mathrm{rand}} \rangle},\tag{8.11}$$

where $\langle Cl_{\mathrm{rand}} \rangle$ is the average clustering coefficient computed in an ensemble of randomized surrogate networks (Chapter 10). If $\gamma = 1$, the empirical network has the same clustering as a random network. Values exceeding one suggest greater-than-random clustering and values below one suggest less clustering than a random network. We can compute a normalized measure of path length in the same way,

$$\lambda = \frac{L}{\langle L_{\mathrm{rand}} \rangle},\tag{8.12}$$

where L is the average shortest path length between nodes in the network (Chapter 7). Once again, values greater than one suggest a longer average path length than random and values less than one suggest shorter average path length.

In a small-world network, we expect that $\lambda \sim 1$, suggesting a comparable average path length to a randomized network, and that $\gamma > 1$, suggesting greater clustering than random. Watts and Strogatz (1998) provided the first evidence that nervous systems satisfy these conditions and thus show evidence of small-world organization. In their analysis of the neuronal network of C. *elegans*, they estimated an average path length of 2.65 and clustering coefficient of 0.28, compared to values of 2.25 and 0.05 in **Erdös-Rényi graphs** of the same size and connection density.

Humphries et al. (2006) proposed a single scalar index to quantify the "small-worldness" of a network. This index is a ratio of the normalized clustering and path length measures,

$$\sigma = \frac{\gamma}{\lambda}.$$
(8.13)

It should be evident that if a network shows small-world properties, such that $\lambda \sim 1$ and $\gamma > 1$, then the scalar index σ will exceed one. As such, values of σ that are greater than one are often used as a simple indicator of small-world organization.

The metrics λ, γ, and σ have been applied to characterize diverse neural systems, and small-world organization has been a robust finding (Hilgetag et al., 2000; Micheloyannis et al., 2006a,b; Salvador et al., 2005; Stam et al., 2007; Yu et al., 2008). Intuitively, we can understand the benefits that such a topological organization provides: high clustering enables tight integration and specialization within specific subsystems, while low path length allows for the efficient integration of these specialized processes. Indeed, it has been shown that a small-world topology supports the emergence of complex neural dynamics—dynamics characterized by the simultaneous presence of functional segregation and integration (Sporns and Zwi, 2004; Sporns et al., 2000; Box 8.2). Other studies have shown that small-world networks facilitate synchronizability and information propagation (Barahona and Pecora, 2002; Hong et al., 2002), enhance computational power (Lago-Fernández et al., 2000), and can provide many other functional benefits (reviewed by Bassett and Bullmore, 2006).

Alternative measures of small-world organization have been proposed. Telesford and colleagues (2011b) argue that appropriately matched lattice networks provide a more suitable **null model** to normalize clustering coefficients than randomized networks. Their reasoning is that the two ends of the topological spectrum ranging from lattice to random organization provide inherent reference points, and thus clustering is best normalized to maximally clustered networks (i.e., lattice networks), just as path length is best normalized to networks with minimal path length (i.e., random networks). This proposal deviates from the logic of the Humphries index of small-worldness, where both path length and clustering are normalized with respect to random networks. Telesford et al. (2011b) thus propose an alternative index to quantify small-worldness,

$$\omega = \frac{\langle L_{\text{rand}} \rangle}{L} - \frac{Cl}{\langle Cl_{\text{latt}} \rangle}, \tag{8.14}$$

where L and Cl are the average path length and average clustering coefficient, respectively, of the empirical network; $\langle L_{\text{rand}} \rangle$ is the path length averaged across an ensemble of matched random networks; and $\langle Cl_{\text{latt}} \rangle$ is the clustering coefficient averaged over an ensemble of matched lattice networks. The ω index ranges between -1 and 1. Values close to zero are indicative of small-world networks, positive values suggest more random characteristics, and negative values indicate a lattice-like structure. Methods for generating lattice-like graphs matched to an empirical network are discussed in Chapter 10.

BOX 8.2 NEURAL COMPLEXITY AND SMALL WORLDS

Recall from Chapter 1 that two fundamental principles of brain function are segregation and integration. Tononi et al. (1994, 1998) have argued that the simultaneous presence of these two characteristics underlies the complexity of brain dynamics, and have proposed an information theoretic framework to quantify such complexity.

Consider a neural system X comprising N nodes. Functional segregation implies that all nodes will show divergent patterns of activity and will therefore be statistically independent. Interactions between these nodes (i.e., integration) will introduce some degree of statistical dependence, or conversely, a deviation from statistical independence. We can measure these deviations from independence using two information theoretic quantities: (1) entropy, which is a general measure of uncertainty or variance in a system; and (2) mutual information, which is a measure of statistical dependence that is sensitive to linear and nonlinear interactions.

For simplicity, assume that X can take on a number of discrete states, $m = 1 \ldots M$. Let p_m denote the probability of each state, the sum of which equals 1. The entropy of X is given by

$$H(X) = -\sum_{m=1}^{M} p_m \log_2(p_m). \tag{8.1.1}$$

The value $H(X)$ will be large if the system has many, equally likely states, reflecting high uncertainty about X. Conversely,

$H(X)$ is zero if the system has a single state with $p_m = 1$ (i.e., there is no uncertainty).

If the nodes of X are completely segregated, their states will be statistically independent. Integration amongst nodes will be associated with a deviation from independence: that is, the states of one subset of nodes are informative with respect to the other. We can use mutual information to quantify these deviations from independence. Let X_j^n denote the j th subset of X comprising n nodes and $X - X_j^n$ denote its complement (i.e., all nodes that are not part of X_j^n). Tononi et al. (1994) define the mutual information between these two groups of nodes as

$$MI\left(X_j^n; X - X_j^n\right) = H\left(X_j^n\right) + H\left(X - X_j^n\right) - H\left(X_j^n; X - X_j^n\right). \tag{8.1.2}$$

If the two subsets X_j^n and $X - X_j^n$ are statistically independent, the sum of their individual entropies, $H(X_j^n)$ and $H(X - X_j^n)$, respectively, will equal the joint entropy of the two subsets X_j^n and $X - X_j^n$, denoted $H\left(X_j^n; X - X_j^n\right)$, and $MI\left(X_j^n; X - X_j^n\right)$ will be zero. Conversely, if X_j^n and $X - X_j^n$ show some degree of dependence (i.e., integration), then the sum of their individual entropies will be greater than their joint entropy and $MI\left(X_j^n; X - X_j^n\right)$ will be greater than zero. In general, $MI\left(X_j^n; X - X_j^n\right)$ will be high when two conditions are met: (1) the individual entropies of X_j^n and $X - X_j^n$ are high, reflecting

Continued

BOX 8.2 NEURAL COMPLEXITY AND SMALL WORLDS—CONT'D

a high variance in their potential states; and (2) the states of X_j^n and $X - X_j^n$ are statistically dependent (i.e., knowing something about the states of one subset is informative about the other).

Now imagine computing $MI\left(X_j^n; X - X_j^n\right)$ for every possible bipartition of X. Figure 8.2.1 illustrates this procedure for a hypothetical network. We start with $n = 1$, and compute the mutual information between each individual node and the rest of the network. We then move to $n = 2$ and compute the mutual information between subsets of two nodes and all others. We continue this process for all $n = 1, \ldots, N/2$ subset sizes. The complexity of the neural system X can be defined as the average mutual information for all possible bipartitions of the network,

$$C_N(X) = \sum_{n=1}^{N/2} \left\langle MI\left(X_j^n; X - X_j^n\right)\right\rangle, \qquad (8.1.3)$$

where $\langle\rangle$ denotes the ensemble average over all bipartitions of a given subset size, n. Remember that the quantity $MI\left(X_j^n; X - X_j^n\right)$ will be high if each subset has many different states, and these states are informative with respect to the rest of the system. The first condition implies functional segregation or specialization of individual elements, since only a small number of states is possible if all elements have identical function. The second condition implies integration, since knowing something about the states of one set of nodes tells us something about others. Thus, $C_N(X)$ will be

high when the system X shows both functional segregation and integration (Figure 8.2.1; Tononi et al., 1994).

Computational modeling has shown that network topologies associated with maximal $C_N(X)$ demonstrate the low average path length and high clustering that is characteristic of small-world systems (Sporns et al., 2000). This work also found that the cortical networks of the cat and macaque show near optimal complexity, defined using a simple model of dynamics simulated on empirically derived structural connectivity matrices. In particular, random rewiring of these structural networks almost always led to a decrease, rather than increase, of their complexity. This link between small-worldness and complexity is not surprising, since a small-world network should, by definition, support both locally clustered and globally integrated dynamics. However, this work demonstrates how a complex topological structure involving an admixture of random and regular properties also gives rise to complex dynamics that are characterized by simultaneously segregated and integrated function.

In practice, computing $C_N(X)$ in empirical data is complicated by the large number of possible bipartitions of X across all scales. Analytic approximations can be used (Avena-Koenigsberger et al., 2014; Barnett et al., 2009, 2011; Sporns et al., 2000; Tononi et al., 1994), but they often assume stationary, multivariate Gaussian and linear dynamics which may not be realistic given the known **nonstationarity** of neural activity (Hutchison et al., 2013; Zalesky et al., 2014; see also Box 10.3).

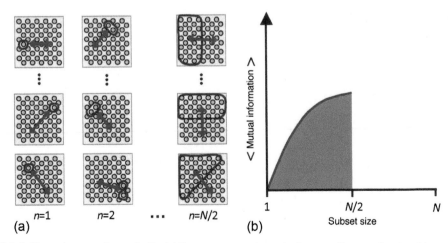

FIGURE 8.2.1 Measuring neural complexity. (a) To measure neural complexity, we split a neural system into every possible bipartition across all subset sizes $n = 1, \ldots, N/2$. Here, we start with $n = 1$ and compute the mutual information between each individual node and the rest of the system. We then move to $n = 2$, and compute the mutual information between subsets of two nodes and all remaining network elements. We then repeat this process until we sample all possible subsets up to size $N/2$. **(b)** If we plot the average mutual information of all possible bipartitions across all subset sizes, n, the area under the curve (red) corresponds to the complexity of the system. *Image adapted from Tononi et al. (1998) with permission.*

In clinical applications, the trade-off between clustering and path length has been used to categorize disorders of the brain as being associated with either a more regular or more random topology (Sporns, 2011b; Stam, 2014). In this context, disorders in which either λ or γ are decreased relative to healthy controls are thought to be associated with a randomization of network connectivity. Conversely, disorders characterized by increased λ or γ are thought to have a more regular or lattice-like topology. Healthy brains inhabit the small-world regime between these extremes, while also possessing many other complex topological properties, such as a heterogeneous **degree distribution** and a hierarchical modular organization, which are not necessarily aspects of the class of small-world networks in general (Figure 8.7). For example, the original model of small-world networks proposed by Watts and Strogatz (1998) had a homogeneous degree distribution and lacked modular organization.

In practice, we should be cautious about drawing strong inferences about brain organization from global summary measures such as γ and σ. As global summaries, they can obscure more meaningful variations occurring at individual nodes or subgraphs, and it can often be unclear whether changes in global topological measures are caused by either a focal disruption that affects other network elements, or a more global dysfunction. For this reason, analyses of global measures should be supplemented by a consideration of nodal, connectional, and subgraph properties to understand the key causes of variations in brain network topology.

The small-world organization of nervous systems has not gone unchallenged. Analysis of tract-tracing data describing axonal connectivity between 29 anatomically defined regions of the macaque cortex has identified a network connected at a density of 66%, due to the discovery of a large number of axonal connections that had previously not been discovered (Markov et al., 2013a,b). At this density and resolution scale, no small-world properties for the binary topology of the network are detected (Markov et al., 2013c). As discussed in Chapters 2 and 3, high-density networks arise when a low-resolution regional parcellation is combined with a high-resolution tract-tracing method, since there is a high probability of finding at least one axon connecting any pair of large areas. Many binary topological properties are difficult to compute in dense networks, since most nodes are interconnected. Importantly, a binary analysis will ignore variations in edge weights, which vary over five orders of magnitude in the macaque data. Notably, an analysis of a mesoscale mouse connectome that shows comparable variation in edge weights did find evidence of small-world organization by conventional criteria ($\sigma > 1$) when using weighted measures of clustering and path length (Rubinov et al., 2015). The findings were consistent across a range of connection densities up to and including the limit (53%) imposed by a probabilistic threshold for false positive connections.

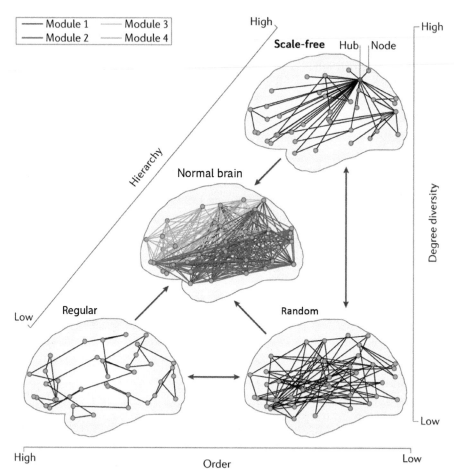

FIGURE 8.7 Three dimensions of the global topological organization of brain networks. Brain networks have been categorized along three broad topological dimensions, shown by the horizontal, vertical, and diagonal axes. The horizontal dimension of order describes whether the topology is either regular, like a lattice, or completely random. The vertical dimension of degree diversity characterizes the degree distribution of the network, with a random, Erdös-Rényi-like topology at one extreme and a hub-dominated **scale-free** topology (Chapter 4) at the other. Healthy brains show an organization that is intermediate between these extremes. The diagonal dimension of **hierarchy** represents the relationship between degree and order; highly ordered lattices have low degree and are relatively nonhierarchical, whereas scale-free networks are highly hierarchical because many of the hubs have low clustering. Degree and clustering are inversely related in hierarchical networks (Ravasz and Barabasi, 2003). In brain networks, this hierarchical relationship is consistent with the observation that high degree hubs are often also the apex nodes in open triangle motifs (Figure 8.2c and d). Another aspect of the hierarchical organization of brain networks is the concept of **hierarchical modularity**, which is discussed in Chapter 9. *Reproduced from Stam (2014) with permission.*

More generally, it is clear that the brain is not strictly configured to match the small-world topology originally envisaged by Watts and Strogatz (1998); that is, of a lattice with a small number of random projections. Indeed, the simple **generative model** they used in their analysis does not reproduce many other important topological features of brain networks, such as the existence of hubs and modules. Rather, it seems likely that the small-world properties of high clustering and short path length often found in nervous systems are a consequence of the hierarchical, modular organization of these networks (Chapter 9).

8.3 NETWORK ECONOMY

As we have seen, the high clustering and short average path length that is characteristic of small-world organization provides a topological foundation for complex brain function, which is characterized by the simultaneous presence of segregated and integrated dynamics (Box 8.2). A simple way to maximize clustering and minimize path length is to add more connections to the network. As we add connections, we close more and more triangles, thereby increasing the clustering coefficient. Each new connection also provides a direct link between two nodes, thereby reducing average path length.

Ramón y Cajal (1995) noted over a century ago that limitations on metabolic and material resources prevent the continual addition of new connections to nervous systems. It has since become clear that brains are disproportionately hungry for energy: in humans, the brain accounts for approximately 20% of energy consumption despite making up just 2% of body weight (Clark and Sokoloff, 1999). A large proportion of this metabolic budget is consumed by the costs (in molecular units of adenosine triphosphate) associated with active transport of ions across neuronal membranes to maintain nonequilibrium resting potentials; and this component of metabolic cost will evidently scale with increasing surface area of cellular membrane in the nervous system (Attwell and Laughlin, 2001; Lennie, 2003). It thus seems reasonable to assume that there is a strong pressure to conserve metabolic and other limited physiological resources.

In the context of neuronal connectivity, we can think of this pressure as a drive to minimize network wiring costs, where wiring refers to the axonal, dendritic, and synaptic connections between neural elements and cost refers to the energy and biological material required to form and sustain this wiring. Each connection in a network will be associated with a particular cost that generally scales in proportion to its distance, volume, and activity: longer connections, and those with greater cross-sectional area, are more costly because they occupy more physical space, require greater material resources, and consume more energy (Box 8.1). Moreover, axonal projections that mediate a higher frequency or volume of spiking signals will require more energy to sustain their higher rates of activity by frequent, energy-consuming repolarization of membranes after spike depolarization.

These considerations lend weight to Ramón y Cajal's laws of conservation of space and material (see Chapter 1). Short distance connections will conserve space and materials. Therefore, they will have low wiring cost and consume fewer metabolic resources. However, cost minimization of spatially embedded networks tends to result in a lattice-like topology, which is incompatible with efficient transfer of information across the network as a whole (Chapter 7). In other words, networks that strictly conserve material and space will likely pay a price in terms of conservation of time: it will take longer to communicate an electrophysiological signal between nodes separated by the longer path lengths that are characteristic of lattices. Precisely how nervous systems negotiate this competition between conservation laws, or more generally the trade-offs between biological cost and topological value, is at the heart of brain network **economy**: the careful management of finite biological resources in the service of robust, adaptive, and efficient network performance (Bullmore and Sporns, 2012). In this section, we consider how economical trade-offs can be measured in connectomes using graph theoretic techniques. We note at the outset that brain network costs have typically been approximated quite crudely, for example, using the Euclidean distance between nodes as a proxy for the (generally longer) distance of the axonal projections that often follow a curvilinear trajectory through anatomical space (Box 8.1).

8.3.1   Wiring Cost Optimization in Nervous Systems

Conservation of space and material dictates that most connections in brain networks will be short-range. In support of this hypothesis, Figure 8.8a shows the fiber length distributions of connections in the macaque cortical connectome and the *C. elegans* nervous system (Kaiser and Hilgetag, 2006). We can see that most connections are indeed short, and that the probability of finding links that span very long distances is low. In fact, several studies suggest that the probability of finding a connection between any pair of neural elements decays roughly exponentially with increasing spatial separation, a trend that has been called an exponential distance rule for neuronal connectivity (Ahn et al., 2006; Ercsey-Ravasz et al., 2013; Fulcher and Fornito, 2016; Markov et al., 2013c). Similar trends have been noted in human structural and functional connectivity networks constructed with MRI (Achard et al., 2006; Alexander-Bloch et al., 2013c; Hagmann et al., 2007).

While such data suggest that the wiring cost of brain networks is generally low, they also highlight the existence of some nodes and edges that are more expensive than a simple penalty function of distance would generally allow (the nodes in the tails of the distributions in Figure 8.8a). This raises the question:

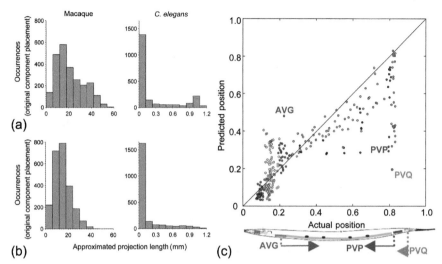

FIGURE 8.8 Wiring cost minimization in brain networks. **(a)** Connection length distributions for a macaque connectome comprising 95 cortical areas, as constructed from the collation of published tract-tracing studies (left), and for a network of 277 neurons of the *C. elegans* nervous system (right). **(b)** Connection length distributions for the macaque (left) and *C. elegans* (right) connectomes after component placement optimization. The probability of finding long-range connections is lower in these optimized networks, suggesting that the empirical networks are not configured solely for minimal wiring cost. **(c)** The actual spatial locations of neurons in the *C. elegans* nervous system relative to their positions predicted by a generative model that minimizes wiring costs. Positions have been normalized with respect to the body of the worm such that a position of zero is in the head and a position of one is in the tail. Colors denote different ganglia. The diagonal line represents a perfect fit between model and data. Not all points fall along this line, indicating that wiring cost minimization alone cannot account for the spatial locations of neurons in this network. Three classes of pioneer neurons, AVG, PVP and PVQ, are labeled and arrows at the bottom indicate the direction of neurite outgrowth during development. The positions of these pioneer neurons are poorly predicted by the model because they have long axonal processes that span the length of the worm. *(a) and (b) Reproduced from Kaiser and Hilgetag (2006), and (c) from Chen et al. (2006), Copyright (2006) National Academy of Sciences, U.S.A., with permission.*

is the wiring cost of brain networks as low as possible? Or is it possible to con-figure the network in an alternative way that leads to further savings in wiring cost? One way of addressing this question is from the perspective of component placement optimization. Specifically, we can ask whether it is possible to per-mute the spatial locations of nodes, while keeping their patterns of connections with other nodes constant, so that the total wiring length of the network is reduced. If we are unable to find such an alternative, we can conclude that wir-ing costs are minimized relative to the topological complexity of the brain. A series of investigations across the *C. elegans*, fruit fly, rat, cat, and monkey ner-vous systems has marshaled evidence to suggest that component placement is indeed near-optimal for minimizing wiring cost (Cherniak, 1994; Chklovskii et al., 2002; Klyachko and Stevens, 2003; Rivera-Alba et al., 2011). In other words, the spatial locations of individual neurons, neuronal populations, and large-scale cortical areas are positioned to keep wiring costs as low as possible, given the connection topology of the network.

Other work has shown that cost minimization can explain numerous other properties of neural organization, including the geometry and branching patterns of neuronal arbors (Cherniak et al., 1999; Chklovskii, 2004), the emergence and spatial arrangement of functionally specialized patches of cortex within an area, such as topographic maps and ocular dominance columns (Chklovskii and Koulakov, 2004; Mitchison, 1991), the fraction of neuropil volume occupied by axons and dendrites (Chklovskii et al., 2002), and possibly cortical folding in gyrencephalic species (Scannell, 1997; Van Essen, 1997). Accordingly, simple generative models of brain networks (Chapter 10), in which graphs are grown by adding edges according to a distance-dependent rule (usually an exponential penalty on the likelihood of forming long-range connections), are able to reproduce many global and local properties of brain networks, such as their clustering, mean path length, **modularity**, and degree distributions (Ercsey-Ravasz et al., 2013; Henderson and Robinson, 2013; Kaiser and Hilgetag, 2004; see also Samu et al., 2014).

Despite these findings, a simple cost minimization principle is insufficient to account for all features of brain organization (Chen et al., 2006; Costa et al., 2007; Klimm et al., 2014; Samu et al., 2014). If nervous systems were configured strictly to minimize wiring costs, they would have a lattice-like topology in which each neural element was only connected to spatially **adjacent** nodes. In fact, in the macaque and *C. elegans*, the probability of finding long-range projections is low, but these projections exist nonetheless (Figure 8.8a). For example, more than 10% of the projections in a macroscopic macaque connectome link regions separated by a Euclidean distance of at least 40 mm, which is more than half of the maximum possible distance in this species (Kaiser and Hilgetag, 2006). In *C. elegans*, some connections extend across nearly the entire length of the organism. Kaiser and Hilgetag (2006) explored the effects of these links in nematode and macaque data for which connectivity had been mapped more comprehensively than in the connetcomes used in earlier analyses. They found that alternative spatial arrangements of nodes with lower wiring cost could indeed be realized for both species when using a component placement optimization algorithm (Figure 8.8b; see also Raj and Chen, 2011). Similar findings were reported in an independent analysis of *C. elegans* that tried to predict the spatial position of each neuron using a wiring cost minimization model (Chen et al., 2006): despite a generally good agreement between model and data, the positions of specific, so-called pioneer neurons with very long processes that span the length of the worm were not predicted by the model (Figure 8.8c).

8.3.2   Cost-Efficiency Trade-Offs

One interesting finding in the analysis of Kaiser and Hilgetag (2006) was that the **characteristic path length** of the macaque and *C. elegans* networks was much lower than comparable minimally-wired benchmark networks. In these benchmarks, the number and spatial locations of nodes, and the number of edges, were the same as the empirical networks, but only short distance connections were formed. In fact, the characteristic path length of the macaque and *C. elegans* networks was comparable to alternative benchmark networks that were optimally wired to minimize path length (Figure 1.7d). This result suggests that the pressure to minimize wiring costs must be balanced with the functional requirement of maintaining an integrated topology, as indexed by a short average path length. Recall that Ramón y Cajal's (1995) cost conservation principle states that nervous systems are configured to conserve time, space, and material. Conservation of space and material is achieved through minimization of wiring costs. Conservation of time can be achieved by a short average path length between nodes, on the reasonable assumptions that (1) a shorter path length means fewer synaptic junctions must be negotiated by signal traffic; and (2) the time taken for a signal to pass across a synaptic junction is long compared to the time constants of axonal signal propagation. As we saw in Chapter 5, a short average path length can still promote rapid information transfer, even if neural signals are not strictly routed along shortest paths (see also Mišić et al., 2015).

The mean shortest path length of a network is inversely related to its global efficiency, such that networks with lower average path length have a more integrated and globally efficient topology (Chapter 7). There is an inherent trade-off between cost and efficiency in networks (Figure 8.9a). In principle, it is possible to maximize the efficiency of a network by forming a connection between every possible pair of nodes. In a fully connected graph, all nodes are directly connected with each other, and the average path length is equal to one, which is the minimum possible. Global efficiency is thus maximal. However, as we have seen, metabolic and spatial constraints limit the connection density of nervous systems.

Small-world networks offer an effective solution to the trade-off between wiring cost and topological efficiency. Latora and Marchiori (2003) used a spatially embedded variation of the classic **Watts-Strogatz model** for small-world networks (Chapters 1 and 10) to show that these networks offer high global efficiency for low connection cost. Specifically, they found that the addition of just a few long-range connections to a lattice-like topology, though individually costly, dramatically increased network efficiency while keeping overall network costs low (Figure 8.9b). They then showed that the connectomes of the macaque, cat, and *C. elegans* each shared the characteristic pattern of high global and local efficiency for low connection cost (Figure 8.9c): evidence of so-called economic small-world behavior.

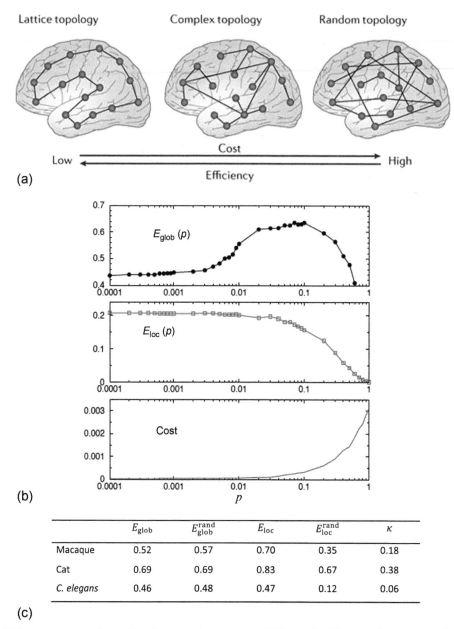

FIGURE 8.9 The economic small-world topology of nervous systems. **(a)** Schematic of the cost-efficiency trade-off in brain network wiring. A lattice-like topology (left) has minimal wiring cost, since only short-range connections between spatially adjacent areas are formed. The efficiency of this network is also low. A random topology (random) has a short average path length and very high efficiency, but the presence of many long-range connections results in a high wiring cost. Networks with complex, small-world-like topologies, such as nervous systems, make many short-range connections, but also possess a small number of long-range projections which dramatically increase efficiency for relatively low wiring cost. **(b)** The global efficiency (top), local efficiency (middle), and wiring cost (bottom) of a spatially embedded generative model for small-world topology. In this model, each node has a fixed spatial position around the perimeter of a circle and the

Achard and Bullmore (2007) proposed a simple measure for capturing the trade-off between efficiency and cost in correlation-based functional connectivity networks measured with **functional MRI**. In correlation-based networks, it is common to set a threshold to distinguish real from spurious connections (Chapter 11). This threshold defines the connection density of the network. Figure 8.10 shows how global efficiency and wiring costs scale as a function of connection density in human functional connectivity networks. Global efficiency shows a monotonic rise with increasing connection density (Figure 8.10a). If we approximate the wiring cost of the network as the total Euclidean distance of supra-threshold edges at each connection density (Box 8.1), we find that this cost increases linearly (Figure 8.10b), which is a slower rate of acceleration than global efficiency at lower connection densities. Achard and Bullmore suggested that a simple index of the cost-efficiency of a network is given by normalizing both efficiency and cost so that they are defined in the unit interval and then estimating the difference between normalized efficiency and normalized cost at a given connection density. If we plot the behavior of this metric as a function of density we obtain an inverted U-shaped curve (Figure 8.10c) with two important properties: (1) all values are positive, suggesting that brain networks are economical or cost-efficient (i.e., the cost as a proportion of maximum cost never exceeds the efficiency as a proportion of maximum efficiency); and (2) the curve has a clear peak, suggesting that there is a specific connection density at which the networks attain maximum cost-efficiency. Achard and Bullmore found that the height of this peak was lower in older compared to younger adult volunteers, suggesting that brain network cost-efficiency deteriorates with aging.

A subsequent resting-state functional MRI study of a small sample of healthy twins found that individual differences in peak cost-efficiency of brain functional networks are highly heritable, with approximately 60% of the variance in global cost-efficiency being attributable to additive genetic effects (Fornito et al., 2011b). For some specific regions, particularly in the prefrontal cortex,

---

wiring cost is computed as the summed Euclidean distance of the edges. Starting with a lattice-like configuration, edges are gradually rewired at random with a given probability, p, until the network starts to resemble a random graph at $p \sim 1$. For low values of p we see that wiring cost is low, local efficiency is high, but global efficiency is low. For $0.01 < p < 0.1$, we see an economic small-world regime in which both global and local efficiency are high while wiring cost is low. **(c)** The global efficiency (E_{glob}), local efficiency (E_{loc}), and connection density (κ) of a 69-region cortical connectome of the macaque, a 55-region cortical connectome of the cat, and a 282-neuron connectome of the *C. elegans* nervous system. Corresponding values in randomized benchmark networks are also presented. Each network has a global efficiency close to the random network, higher local efficiency, and a relatively low connection density. Formal definitions of E_{glob} and E_{loc} are presented in Chapter 7. **(a)** *Reproduced from Bullmore and Sporns (2012) and* **(b)** *and* **(c)** *reproduced from Latora and Marchiori (2003) with permission.*

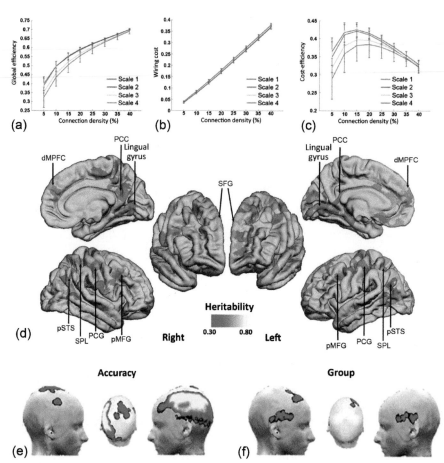

FIGURE 8.10 Cost-efficiency trade-offs in human brain networks. **(a)** The global efficiency of correlation-based functional connectivity networks measured with human functional MRI shows a monotonic increase with increasing connection density. Regional time series were filtered using a discrete **wavelet** transform and the different lines represent data at different wavelet scales, corresponding to the frequency intervals 0.18-0.35 Hz (scale 1), 0.09-0.18 Hz (scale 2), 0.04-0.09 Hz (scale 3), and 0.02-0.04 Hz (scale 4). **(b)** Network wiring cost, estimated as the total Euclidean distance between connected pairs of brain regions, increases linearly with connection density. **(c)** Global efficiency increases with connection density more rapidly than wiring cost at low densities. If we compute an index of cost-efficiency as the difference between efficiency and cost at each connection density, we obtain an inverted U-shaped curve as shown here. The height of the peak and the connection density at which it occurs vary between people. **(d)** These variations in network cost-efficiency are strongly heritable. Shown here are cortical areas where regional peak cost-efficiency was significantly heritable. Abbreviations correspond to region names, as explained in Fornito et al. (2011b). **(e)** Individual differences in the cost-efficiency of brain functional connectivity networks measured with **MEG** during a working memory task correlate with task performance. Shown here are scalp locations where network nodal cost-efficiency measured for functional networks in the β frequency band (15-30 Hz) correlates with performance accuracy. **(f)** Scalp locations where nodal cost-efficiency is reduced in patients with schizophrenia compared to controls. *(a–d) Reproduced from Fornito et al. (2011b) and (e) and (f) from Bassett et al. (2009) with permission.*

up to 80% of the variance was heritable (Figure 8.10d). An independent study found that individual differences in peak network cost-efficiency correlate with working memory performance, and that this measure is reduced in patients with schizophrenia (Bassett et al., 2009; Figure 8.10e and f). Collectively, these results suggest that cost-efficient organization is a heritable and functionally important aspect of brain organization.

In the twin study, reliable genetic influences were not found for measures of network cost or efficiency alone (Fornito et al., 2011b). This result suggests that it is the trade-off between these two properties that is specifically under tight genetic control, and is consistent with the hypothesis that natural selection may favor networks that can achieve high efficiency for low wiring cost (although see Smit et al., 2008; van den Heuvel et al., 2013b). The strong genetic effects observed in regions of association cortex (Figure 8.10d), which correspond to known connectivity hubs of the brain, suggest that these areas may be particularly important in supporting the cost-efficient integration of distinct brain regions. This conclusion aligns with evidence that most of the shortest paths between neural elements pass through rich-club links between hubs (Figure 8.3), and a report that the transcriptional activity of these areas is tightly coordinated, particularly with respect to the expression of genes regulating energy metabolism (Fulcher and Fornito, 2016). Hubs are particularly costly aspects of brain networks (Box 8.1). Strong genetic control over the connection topology of hub nodes may thus play a critical role in promoting network economy.

Further evidence supporting the importance of cost-efficiency trade-offs in brain network topology has come from generative modeling. Chen et al. (2013) generated synthetic networks that were topologically configured to optimize the trade-off between network wiring cost and average path length by minimizing the objective function,

$$F = (1 - \alpha)\mathcal{L}' + \alpha D', \tag{8.15}$$

where $\mathcal{L}'$ is a normalized sum of the shortest path lengths between all pairs of nodes in the network and D' is a normalized index of the total wiring cost of the network, estimated as the summed Euclidean distance between all node pairs (Box 8.1). The tunable parameter α varies between zero and one and is used to differentially weight the optimization towards minimizing wiring cost or minimizing path length. When $\alpha = 0$, the optimization will only minimize network path length and when $\alpha = 1$ it will only minimize wiring cost. The authors found that, for a broad range of of α, minimization of Equation (8.15) generated networks that had many properties in common with the neuronal network of *C. elegans* and the cortical connectome of the macaque. These properties included the formation of clustered connectivity and hub regions, the spatial position of the hubs, the modular organization of the networks, and the placement of a significant proportion of individual connections (up to 60% for the macaque and 30% for *C. elegans*). However, empirical properties such as the degree distribution and the proportion of long-range connections could not be reproduced, suggesting that the trade-off between cost and efficiency may not be the only factor influencing brain network topology.

The inability of a simple cost-efficiency model to replicate all key features of brain networks is consistent with other evidence that connectomes, particularly those characterized at the macroscale, are not configured to optimize topological efficiency, regardless of whether efficiency is measured using traditional path-length based metrics or diffusion-based measures (Avena-Koenigsberger et al., 2014; Goñi et al., 2013; Samu et al., 2014; see Chapter 7). Together, these findings suggest that efficiency may represent just one of several attributes that collectively endow nervous systems with sufficient topological complexity to adapt to their surroundings. Achieving the required level of complexity for low wiring cost is the essence of network economy.

In general, generative models of network organization that form connections according to an economical trade-off between cost and some other adaptive topological property are better able to reproduce a broader range of empirical properties of neural networks when compared to models based on cost minimization alone. For example, Vértes et al. (2012) showed that an economical trade-off between distance and clustering more accurately reproduced the topological properties of human functional MRI networks than a simple spatial distance penalty or the purely topological model of **preferential attachment**. Nicosia et al. (2013) demonstrated that an economical trade-off between distance and nodal degree more accurately reproduced the development of the *C. elegans* connectome through various larval stages than generative models that optimized only spatial or topological criteria. These and other generative models for networks are discussed in more detail in Chapter 10.

8.3.3   Rentian Scaling

An elegant framework for analyzing the trade-off between the topological complexity of a network and its wiring cost comes from the optimization of high performance computer chips, such as very-large-scale integrated (VLSI) circuits. These circuits can be modeled as graphs where nodes represent circuit elements such as logic gates and edges represent the interconnecting wires. In VLSI circuits, the number of nodes within any given partition or subgraph of the system can be related to the number of edges at the boundary of that partition (i.e., the number of edges that link nodes in the partition to nodes outside the partition) as a simple power law of the form

$$E_{\text{B}} = tn^{\rho}, \tag{8.16}$$

which is also called Rent's rule after the IBM employee who first discovered it (Landman and Russo, 1971). In Equation (8.16), E_B is the number of edges over the boundary of the partition or subgraph, n is the number of nodes in the subgraph, t is called the Rent coefficient and corresponds to the average node degree in the subgraph, and ρ is called the Rent exponent. The Rent exponent determines the slope of the power-law relationship and is regarded as a measure of topological complexity (Christie and Stroobandt, 2000). Values of ρ can range between zero and one. Higher values indicate a more rapid acceleration in the number of boundary edges, E_B, with partition size, n. Recall from Chapter 4 that power-law distributions follow a straight line when plotted on logarithmic axes. Thus, if a system shows Rentian scaling, a plot of $\log E_B$ against $\log n$ will approximate a straight line. The parameter t can be estimated from the intercept of this line and the exponent ρ is the slope (for a discussion of issues associated with fitting power laws to empirical data, see Chapter 4).

Rentian scaling can be measured in terms of either the topology or the physical layout of the network. The topological Rent exponent, ρ_T, is defined by partitioning the network into subgroups of connected units without regard for their spatial location (Figure 8.11a-c). The physical Rent exponent, ρ, is defined by partitioning the network into spatially contiguous units (Figure 8.11d). In one approach to partitioning, we start with a box covering the entire network and recursively partition the box into halves, quarters, and so on until each individual node is in a unique partition (Bassett et al., 2010). The partitions can be defined in either topological space (to estimate the topological Rent exponent, ρ_T) or physical space (to estimate the physical Rent exponent, ρ), and each partition is drawn to minimize the number of edges between partitions. For each partition size, we count the number of nodes, n in that partition, and number of edges, E_B, crossing the boundary of that partition (Figure 8.11e). The result is a set of values for E_B estimated at each partition size, n. We can then plot these vectors on logarithmic axes to compute the Rent exponent (Figure 8.11f-j).

What does Rentian scaling have to do with the brain? We know that computer chips with a higher topological Rent exponent, ρ_T, have a more complex wiring pattern and higher dimensional topology, resulting in greater logical capacity (Bassett et al., 2010). The topological complexity of a network is given by its topological dimension, $\mathcal{D}_T$, which is related to the Rent exponent as (Bassett et al., 2010; Ozaktas, 1992),

$$\rho_T \geq 1 - \frac{1}{\mathcal{D}_T}. \tag{8.17}$$

If the topological dimension of a network is greater than its Euclidean dimension, $\mathcal{D}_E$, its connection topology is more complex than a $\mathcal{D}_E$-dimensional

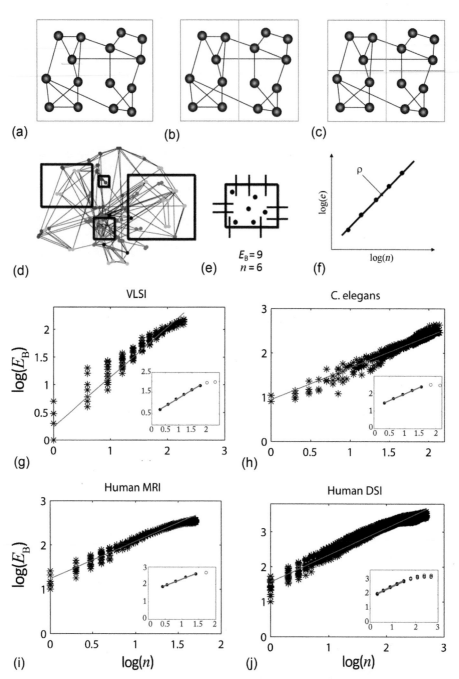

FIGURE 8.11 Rentian scaling in the brain and computer chips. To analyze topological Rentian scaling, we can start with a box that covers the entire network **(a)** and then recursively partition the network into subgraphs by dividing the box into halves **(b)**, then quarters **(c)**, and so on. At each iteration, the partition is made such that the partition boundaries cut through the minimum possible number of edges. In the example shown here, the partitions are made on a purely topological basis and spatial relations between nodes are not considered.

lattice. Note that the Euclidean dimension is simply the number of spatial dimensions that the network occupies. For example, the human brain is physically embedded in three spatial dimensions, so $\mathcal{D}_E = 3$. If the neuronal connectivity of the human brain is arranged as a lattice, then $\mathcal{D}_T = \mathcal{D}_E = 3$. If the brain has a more complex topology than a lattice, then the topological dimension, $\mathcal{D}_T$, will be greater than the Euclidean dimension, $\mathcal{D}_E$. In this way, $\mathcal{D}_T$ can be used to index the complexity of brain network topology. Rearranging Equation (8.17), a bound for the topological dimension is given by $\mathcal{D}_T = 1/(1 - \rho_T)$.

A central problem in computer chip design is how to construct a circuit with high topological complexity for low wiring cost. Topological complexity provides functional advantages such as higher information-processing capacity, but commercial imperatives and restrictions on space place pressure on circuit designers to minimize wiring costs. Brain networks face similar competitive pressures (Bullmore and Sporns, 2012; Fornito et al., 2011b). When $\mathcal{D}_T > \mathcal{D}_E$, we have a network with complex topology, but we know that wiring costs will not be absolutely minimal because the network cannot be embedded as a lattice. However, we can ask whether a brain network shows near-minimal wiring cost relative to its level of topological complexity by considering the *physical* Rent exponent of the connectome, denoted here simply as ρ. More specifically, we can compare the value that we estimate for this exponent in our brain network to the theoretical minimum Rent exponent that can be realized for a network with the same topological dimension. This theoretical minimum is given by (Verplaetse et al., 2001)

$$p_{min} = \max\left(1 - \frac{1}{\mathcal{D}_E}, \rho_T\right), \tag{8.18}$$

(d) To investigate the physical Rentian scaling of a network, we partition the physically embedded graph (here we see a human brain network projected into anatomical space) by drawing a large number of randomly sized boxes around spatially adjacent collections of nodes. (e) For both topological and physical analysis, we count, for each partition, the number of nodes within that partition, n, and the number of edges crossing the partition boundary, E_B. (f) Either the topological or physical Rent exponents can be estimated by plotting these two variables on logarithmic axes. A straight line is indicative of Rentian scaling. The remaining figure panels present the empirical relationships between $\log E_B$ and $\log n$, defining physical Rentian scaling in a VLSI circuit comprising 440 logic gates (g), the neuronal network of *C. elegans* (h), a human **structural covariance** network comprising 104 cortical and subcortical regions, as measured with structural MRI (i), and a 1000-region human structural connectivity network constructed with a variant of diffusion MRI called diffusion spectrum imaging (DSI) (j). Insets show the corresponding relationships for topological Rentian scaling. The scaling relationship breaks down at high n due to boundary effects. Black stars and circles correspond to data points drawn from different network partitions. The best-fitting straight line is in red. *Images reproduced from Bassett et al. (2010), with permission.*

where $1 - 1/\mathcal{D}_E$ is the physical Rent exponent for a lattice with the same Euclidean dimension as the nervous system under investigation and $\rho_T = 1 - 1/D_T$ is the topological Rent exponent for a minimally wired network with a topological dimension equal to $\mathcal{D}_T$. Equation (8.18) thus indicates that a network with a topology that is more complex than a lattice will have a minimum physical Rent exponent that is equal to the topological Rent exponent of the network; that is, $\rho_{min} = \rho_T$. The parameter ρ_{min} thus corresponds to the physical Rent exponent of a network with optimally placed components for a given topological dimension.

To summarize, if a given nervous system has a higher topological dimension than the three dimensions of space in which it is embedded, the minimum Rent exponent $\rho_{min} = \rho_T$. If the physical Rent exponent ρ that we estimate by spatial partitioning of the connectome is close to ρ_{min} (which, in this case has been defined by topological partitioning of the network), we can conclude that the connectome has near minimal wiring cost relative to its high-dimensional topology.

Bassett et al. (2010) conducted a detailed investigation of topological and physical Rentian scaling in brain networks. They report topological dimensions of 4.42, 4.54, and 4.12, respectively, for the neuronal network of *C. elegans*, a human structural connectivity network measured with diffusion MRI, and a human structural covariance network measured with T1-weighted MRI (Figure 8.11g–j). Each of these values is higher than the Euclidean dimension of these networks (which is equal to 3 in each case), suggesting that the networks are indeed topologically complex. The fractional values of these estimates of $\mathcal{D}_T$ point to a **fractal** (self-similar) organization, which the authors argued was consistent with their observation that the networks could be decomposed into hierarchical modules (i.e., modules within modules) across several levels of topological resolution (see Chapter 9). Importantly, Bassett and colleagues found that the physical Rent exponent was close to the topological Rent exponent (corresponding to ρ_{min}) in all cases, indicating that the actual wiring costs of these networks is near-minimal relative to their level of topological complexity. Applying an edge-rewiring algorithm to the empirical networks confirmed that alternative network configurations with lower cost were indeed possible, but that these cost savings came at the expense of reduced topological complexity. Rentian scaling has also been observed in functional connectivity networks constructed from spike correlations across hundreds of neurons in the primary visual cortex of the mouse (Sadovsky and MacLean, 2014). Collectively, these findings support the hypothesis that brain networks are economical, having near minimal wiring cost given their level of topological complexity.

8.4 SUMMARY

We have seen how networks can be decomposed into basic building blocks, called motifs, at the level of both subgraphs of nodes and paths of distinct classes of edges. Node motifs that recur in brain networks with a frequency that is greater than chance consist of open triangles with reciprocal connections (Harriger et al., 2012; Sporns and Kotter, 2004). In these motifs, a hub node is often located at the apex position that links two otherwise unconnected nodes (Sporns et al., 2007). The position of the hub in this motif facilitates rapid and stable dynamical synchronization of its neighbors (Gollo et al., 2014). Path motifs have been studied in relation to the rich club and modular organization of brain networks. These analyses indicate that one of the most common motifs is one in which communication paths between distal network elements pass through rich-club connections.

The clustering coefficient quantifies a specific type of motif in which all three nodes of a triad are connected. Brain networks show higher clustering than corresponding random graphs. Together with a short average path length between nodes, this high clustering is the defining feature of small-world systems.

Small-world networks are thought to provide a topological substrate for the simultaneous integration, specialization and segregation of brain function. They also support functional degeneracy, and thus robustness to perturbation. Importantly, small-world networks embedded in physical space also economically support efficient communication for relatively low wiring cost. A recurrent theme throughout this book has been that wiring cost minimization is an important principle for brain network organization, but wiring cost minimization alone seems insufficient to account for all topological characteristics of nervous systems. Rather, nervous systems are subject to the competitive selection pressures of wiring cost minimization and topological complexity. In optimizing this trade-off, neural systems are highly economical, such that that their wiring cost is near minimal given their level of topological complexity.

CHAPTER 9

Modularity

The **nodes** of many real-world networks aggregate into densely connected sub-groups called **modules** or **communities**. Nodes within these modules are more strongly connected with each other than with other parts of the network (Figure 9.1a). For example, in the global air transportation network, airports in countries that are in the same geographic area, or which share close geopolitical ties, are linked by more direct flights than airports that do not share these two factors in common (Guimerà et al., 2005; Figure 1.4). In social networks, friendship ties also tend to be dense within certain groups of people, such as when high school students are more often friends with peers in the same year level than with students in other year levels (Moody, 2001; Newman, 2003a). In scientific citation networks, where journals are nodes and citations to papers in other journals are **edges**, citations are more likely to occur between journals in related fields than those in different disciplines (Rosvall and Bergstrom, 2008). Scientific citation networks thus form communities that closely match the traditional boundaries between disciplines.

Another important property of real-world networks is that their modules are often organized hierarchically, such that they contain modules within modules and so on, over several topological scales of resolution (Figure 9.1b). This Russian doll-like organization is consistent with a **fractal** community structure and is called **hierarchical modularity** (Bassett et al., 2010; Meunier et al., 2011). Extending our high school example, suppose that we first find that people within the same year level are more likely to be friends with each other. We also find that, within each year level, people cluster into friendship groups based on gender. Within gender groups, we then find that students aggregate into groups who like basketball, tennis, athletics, and so on. The friendship groups defined according to year level, gender, and sporting interest define a modular organization that spans three hierarchical levels.

The brain also operates as a hierarchical, modular system. Considered at the level of the entire body, the brain may be viewed as a single module that interacts with other organs such as the heart, kidney, lungs, and so on. If we narrow our focus to consider the brain alone, we can discern coarse divisions such as the forebrain, cerebellum, and brainstem. Focusing on the forebrain, we can delineate subdivisions at progressively finer scales, moving from hemispheres

Fundamentals of Brain Network Analysis. http://dx.doi.org/10.1016/B978-0-12-407908-3.00009-1

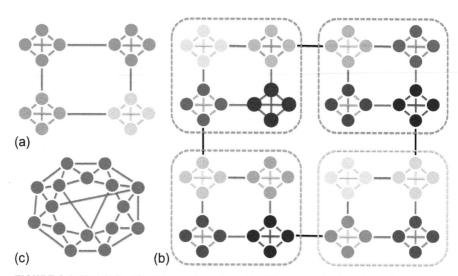

(a)

(c) (b)

FIGURE 9.1 Modularity, hierarchy, and small-worlds. **(a)** An example of a network that can be decomposed into four modules, represented by different colors. Nodes within each module are strongly interconnected (colored edges) and there is sparse connectivity between modules (gray edges). **(b)** Example of a hierarchical and modular network. At the coarsest resolution, the nodes can be grouped into four modules (blue, red, purple, and green boxes). Each of these modules can be subdivided into four additional modules (different shades of blue, red, purple, and green). This organization is consistent with a hierarchical organization in which modules contain nested submodules. Intermodular connections at the coarsest resolution are shown in black, and then in gray at the next resolution scale. **(c)** An example small-world network, based on the model originally described by Watts and Strogatz (1998). Each node is connected to its four nearest **neighbors** yielding high clustering. A few long-range **links** act as topological shortcuts that reduce the **characteristic path length** of the network. This network has no discernible modules.

to lobes, lobes to cytoarchitectonic areas and subcortical nuclei, and so on until we reach the level of individual cells. Indeed, we can press even further to study the modular organization of molecular interactions within cells, the genetic interactions that influence the structure of specific proteins, and so on. At each resolution scale, we can partition the network into modules comprising nodes that preferentially interact with each other.

Modular networks generally have **small-world** properties (Chapter 8). Strong within-module connectivity results in a high **clustering coefficient**, whereas a small number of intermodular links is sufficient to maintain a low characteristic path length of the network. However, not all small-world networks are modular. In fact, the original Watts and Strogatz (1998) small-world network model distinctly lacks modular structure (Figure 9.1c). Therefore, it is possible that many of the functional benefits that have been associated with small-worldness and which are thought to underlie brain function, such as high dynamical complexity (Sporns et al., 2000; see Chapter 8), may result from the modular **topology** of neural connectivity.

Modularity affords many other benefits. In a prescient essay, Nobel laureate Herbert A. Simon argued that modularity and hierarchical organization are essential ingredients of evolvability, flexibility, adaptability, and complexity (Simon, 1962). To illustrate this point, he compared two hypothetical watch-makers, Hora and Tempus, both of whom make watches consisting of 1000 parts each. Tempus assembles his watch so that each part is dependent on another. If construction of the watch is interrupted by, for example, a phone call, Tempus must reassemble his watch from the beginning. In contrast, Hora assembles his watch using a piecemeal approach, in which he constructs subassemblies consisting of ten parts each, which he then combines into larger subassemblies. Each time Hora is interrupted, all subassemblies completed to that point remain intact and he need only reassemble the subassembly that he is currently working on.

Simon's parable effectively highlights several important features of hierarchical modular architectures; in particular, that they can adapt and evolve to changing environmental circumstances. This is because new modules can be added to the system without drastically altering the existing elements. Modeling studies have shown that modular architectures naturally arise in systems that must optimize performance under changing environmental circumstances (Kashtan and Alon, 2005). In these situations, distinct and specialized modules rapidly evolve in order to achieve different subgoals of a larger evolutionary task. Much like the contrast between Tempus and Hora, each module can act in parallel and nearly independently to work toward these subgoals, and the success or failure of each module does not have a major impact on other modules. This level of independence between subsystems reduces the potential for dysfunction or damage to propagate throughout the system and cause catastrophic failures (Buldyrev et al., 2010). In other words, perturbations to a modular system will have a limited capacity to impact other system elements (Brummitt and D'Souza, 2012; Variano and Lipson, 2004). In brain networks, this means that a focal pathology affecting a tightly interconnected module will be less likely to spread and affect other areas (Fornito et al., 2015). Consistent with this view, modeling studies have shown that a breakdown of modularity can lead to a propensity for hyper-synchronized, seizure-like dynamics (Kaiser et al., 2007a). Other work indicates that hierarchical modularity enhances the variety, complexity, and stability of network activation patterns (Robinson et al., 2009; Shanahan, 2010; Variano and Lipson, 2004), and is vital for the emergence of self-organized critical dynamics (Box 4.1; Rubinov et al., 2011).

The strong connectivity of nodes within a module suggests that they share common functions. Weak connectivity between nodes in different modules suggests that these functions are segregated from each other. It is therefore plausible that a modular architecture can support the twin pillars of brain function: functional segregation and integration (Tononi et al., 1994).

There is a close correspondence between topological characterizations of the modular architecture of brain networks and long-standing descriptions of the modularity of mental function. Indeed, the idea that cognitive and other psychological abilities can be subdivided into component functions is ancient. Perception, memory, and action were first explicitly recognized as somewhat separable components of conscious experience by Aristotle. This basic idea re-emerged in the early nineteenth century in the form of phrenology. Proposed by Franz Joseph Gall (1758-1828), the phrenological principle was that mental function could be decomposed into multiple component functions or faculties, and that each faculty could be discretely localized to an area of the brain. Classical phrenology, as practiced by Gall and other early enthusiasts, assumed that marked expression of a faculty would be associated with gross enlargement of the cortical area where it was localized, and that local cortical enlargement could be diagnosed by feeling for corresponding bumps in the skull around that location. Although phrenology has since been discredited, the basic principle that psychological faculties can be localized to specific brain regions has endured. Indeed, there are many important examples of the successful localization of component functions, dating back to Broca's demonstration that language production was specifically impaired by destruction of the left inferior frontal gyrus (Broca, 1861).

More recently, the principles of phrenology have been evaluated and reformulated in the context of contemporary cognitive neuroscience. As first described in detail by Fodor (1983), cognitive modules correspond to innate, domain-specific, hard-wired, autonomous, and informationally encapsulated psychological faculties that are implemented by a discrete, specialist, and localized neural architecture. A prototypical example is the visual motion module. The faculty for automatically detecting motion in the visual scene emerges early in development, is localized to area V5 on the dorsolateral surface of occipital cortex, and can be specifically impaired in patients with a focal lesion to this area, causing the clinical syndrome of akinetopsia (Zeki, 2001).

In this context, it seems reasonable to expect that cognitive modules might correspond to network topological modules, and there is already good evidence to suggest this is the case. It has been recognized since some of the first graph theoretical studies of **tract-tracing** data (Hilgetag et al., 2000; Hilgetag and Kaiser, 2004; Young, 1992; Young et al., 1995) that cortical areas known to be specialized for sensory and motor functions tend to form densely connected clusters. This result has been corroborated and quantified more formally by many studies showing that topological modules tend to be spatially localized, and that major topological modules typically include cortical areas that are known to be specialized for visual, auditory, and motor functions (Crossley et al., 2013; de Reus and van den Heuvel, 2013b; Rubinov et al., 2015; Zamora-López et al., 2010).

Other cognitive functions, such as working memory, attention, and planning, are not informationally encapsulated or automatic, and have not been localized to a discrete cortical area. Such "higher-order" or effortful cognitive processes seem more likely to depend on a global neuronal workspace than segregated modular function (Figure 1.10; Baars, 1989; Dehaene et al., 1998). Translating this to the language of brain graphs, we might expect that topological measures of network integration would be more closely linked to "higher-order" task performance than topological measures of network segregation, such as network modularity. Once again there is some evidence that this is the case. Lower characteristic path length and higher topological **efficiency** of human brain networks is associated with higher IQ (Li et al., 2009; van den Heuvel et al., 2009). Lower **path length**, higher topological efficiency, and a disruption of modular **community** structure has also been reported in human functional networks during the performance of challenging cognitive tasks (Fornito et al., 2011a; Kitzbichler et al., 2011). Therefore, while there is some correspondence between the topological modules of brain networks and the Fodorian concept of cognitive modules, the formation of workspace networks for higher-order processing may depend on a breakdown of traditional modules and enhanced intermodular communication.

In the rest of this chapter, we focus on methods for defining modules based on network topology. The development and validation of algorithms for characterizing the modular structure of networks is a large and intensive field of investigation, with many alternative methods available. Here, we focus on those most commonly used in neuroscience, or those that usefully highlight important concepts in the analysis of modular systems. For further details, interested readers are referred to thorough reviews of the topic by Fortunato (2010) and Schaeffer (2007).

9.1 DEFINING MODULES

The accurate decomposition of a network into modules with high intrinsic connectivity and weaker extrinsic connectivity is an example of a more general class of problems concerned with data clustering. These problems have a long history in statistics and data mining. The basic idea is to reduce a large set of observations across different measures into a smaller subset of informative clusters by exploiting redundancies and associations in the data. In many analyses, we often do not know how many clusters to extract, or where the divisions between clusters should be drawn, so we require a data-driven method for finding the "natural" divisions in the data. Techniques such as cluster analysis, multidimensional scaling, and principal component analysis are examples of different solutions to this problem (Gan et al., 2007; Jain et al., 1999).

In general, data clustering methods can be categorized into two broad classes: agglomerative and divisive. Agglomerative methods start with individual data points and group these together into larger clusters. Divisive methods start with all nodes in a single cluster and attempt to find divisions that delineate cohesive subsets of observations. In this section, we consider key examples of both approaches, and examine how they have informed the evolution of community detection algorithms developed specifically for the topological analysis of complex networks. We will use the terms 'module' and 'community' interchangeably throughout our discussion.

9.1.1 Agglomerative and Divisive Clustering

Hierarchical clustering is a popular agglomerative method. It starts with each data point in a separate cluster and aggregates points into larger sets of similar observations. A critical step is to define a measure of similarity between observations. In network analysis, the data points correspond to nodes and the observations to a topological property that is used to index relationships between pairs of nodes. In principle, we could use the direct connectivity weight between two nodes as a measure of their similarity, but it is also possible to use some of the more general measures of **structural equivalence** discussed in Chapter 8, such as the normalized matching index and cosine similarity. Once an appropriate metric is chosen, the similarity between every pair of nodes is computed to generate an $N \times N$ similarity **matrix**.

The values of the similarity matrix are used to group nodes into clusters, such that the similarity *within* clusters is high and the similarity *between* clusters is low. Hierarchical clustering typically does this using a greedy approach. Specifically, we start with each node in its own cluster. We then join pairs of nodes with the highest similarity to form clusters of size two. At the next iteration, we join pairs of clusters with the highest similarity. Because our attention is now trained on clusters of nodes, but our similarity index is only defined for pairs of individual nodes, we must aggregate similarity measures according to a specific rule. For a given pair of clusters of size n_1 and n_2, it is common to take either the maximum similarity value observed across all $n_1 n_2$ pairs of nodes (called single-linkage clustering), the minimum value (complete-linkage clustering), or the average value across all pairs (average-linkage clustering). One of these cluster-level similarity measures is then used to successively merge nodes into larger sets until all nodes are in a single cluster. The results are presented in the form of a tree-like structure called a **dendrogram**, which shows the nested, hierarchical relations between clusters. We can cut this tree at distinct levels to obtain a partition at a particular topological scale of the system.

Figure 9.2 presents the results of a hierarchical clustering analysis applied to a human **structural connectivity** matrix constructed with **diffusion MRI** (Moreno-Dominguez et al., 2014). Individual voxels at the interface between gray and white matter were seeded for diffusion **tractography** and treated as nodes. Hierarchical clustering with spatial constraints was then used to group these voxels into anatomically contiguous clusters of voxels at increasingly coarse resolution scales. Cutting the dendrogram at a particular level yields a partition or community structure at a specific resolution scale.

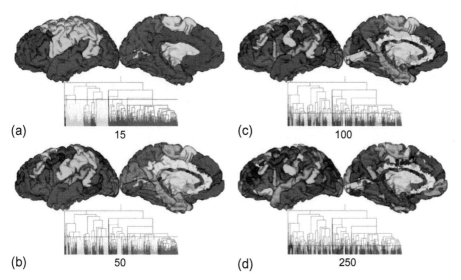

FIGURE 9.2 Hierarchical clustering of human structural connectivity networks. Diffusion MRI tractography was performed using each voxel positioned at the interface between gray and white matter as a separate seed. A spatially constrained hierarchical clustering algorithm was used to merge voxels into progressively larger clusters of topologically similar and anatomically contiguous voxels. Topological similarity was measured based on each node's profile of connectivity to the rest of the brain, and quantified using a variant of the correlation coefficient. The result is a parcellation of the cortex into putative areas with a common connectional fingerprint (see Chapter 2) at different levels of spatial resolution. Here we see the parcellations that are obtained when cutting the dendrogram to yield 15 **(a)**, 50 **(b)**, 100 **(c)**, and 250 **(d)** clusters, or putative modules. The corresponding dendrogram is displayed below each anatomical map. At the lowest level, each voxel is treated as a separate cluster. As we move up the **hierarchy**, the clusters merge into larger and larger groupings. The red horizontal line on each dendrogram indicates the level at which the tree has been cut to generate the anatomical display. *Figure reproduced from Moreno-Dominguez et al. (2014), with permission.*

Despite its popularity and flexibility, hierarchical clustering can be an inaccurate method for delineating modules in networks with known community structure. In particular, Newman and Girvan (2004) consider how hierarchical clustering is biased toward grouping together the core nodes of a community at the expense of peripheral nodes. As shown in Figure 9.3, core nodes have a high degree of similarity and are grouped together early in the agglomeration procedure. Peripheral nodes have only weak similarity to other nodes, causing them to be overlooked by the clustering procedure.

Girvan and Newman (2002) proposed a divisive method for decomposing a network into hierarchical modules, as an alternative to agglomerative clustering. The method defines communities based on the ordered removal of edges according to their **betweenness** centrality. The algorithm proceeds as follows:

1. Calculate the betweenness centrality of each edge in the network.
2. Remove the edge with the highest score. If two or more edges have the same score, remove one chosen at random, or remove all at the same time.
3. Recalculate the betweenness centrality of all remaining edges.
4. Repeat steps 2 and 3 until no edges remain.

Removing edges in this way causes the graph to fragment, and the **connected components** that remain represent putative modules of the network. As the algorithm progresses and more edges are removed, we obtain smaller and smaller modules.

Why should removal of edges with high betweenness centrality identify the natural modules of a network? Recall from Chapter 5 that the betweenness centrality of a node is estimated as the fraction of all shortest **paths** in the network that pass through that node. It follows that nodes lying on a large fraction of these shortest paths represent putative bottlenecks of traffic flow, under the assumption that information travels along shortest paths (see also Chapter 7). We can extend this logic to individual edges, and define edges with high betweenness centrality as those that lie on a large fraction of the shortest paths between all nodes in the network. In a modular network, any message that travels from one community to another must pass through one of a sparse set of intermodular links, meaning that these links will generally have high betweenness centrality (Figure 9.3). Targeted removal of edges with high betweenness centrality should therefore reveal natural divisions between network modules. As with hierarchical clustering, the result of the procedure is a dendrogram that reveals the nested community structure of the graph. Each level of the dendrogram corresponds to a point at which the removal of an edge fragments one of the graph's components.

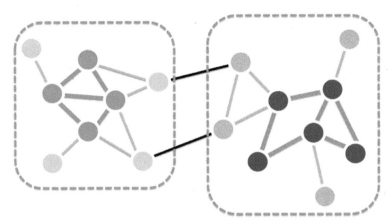

FIGURE 9.3 Agglomerative and divisive clustering. Shown here is a network that can be split into two modules, red and blue. The rectangular boxes delineate the natural borders of the modules. Agglomerative methods do well at clustering together core members of modules with strong pairwise similarity (bold node and edge colors), but often fail to group together peripheral community members with only weak similarity to the rest of the group (pale node and edge colors). Instead, if clusters are defined by removing links in decreasing order of betweenness centrality, the first edges that are removed are the intermodular links (black lines). In the example shown here, the removal of these links will fragment the graph into two components (one comprising blue nodes and the other comprising red nodes) that correspond to the intuitive modules of the network. This figure also shows why intermodular links tend to have high betweenness centrality—any shortest path linking a node in the blue module to one in the red module must pass between one of the intermodular edges. *Figure adapted from an example provided by Newman and Girvan (2004), with permission.*

An advantage of the Girvan-Newman algorithm for modular decomposition is that, unlike hierarchical clustering, it does not require us to define a measure of pairwise similarity between nodes; we only need the connectivity weight assigned to each edge. We are also not required to use a specific rule for aggregating this similarity measure across clusters of nodes. Instead, the edge betweenness centrality can be computed using standard graph theoretic methods, and it is used to distinguish between communities consistently at each resolution scale. The approach is also flexible, in that edge betweenness centrality can be substituted by any alternative edge-based measure of **centrality** that is relevant for community detection. Newman and Girvan (2004) report that edge betweenness centrality gives the most accurate results in many practical circumstances.

9.1.2 Quantifying Modularity

An advantage of the hierarchical analyses we have considered thus far is that they offer a full representation of the nested relationships between nodes, thus uncovering the fractal, Russian doll-like organization of modules at different resolution scales. Another advantage is that they do not require prior specification of the exact number of clusters that will be extracted from the data. By comparison, other popular clustering methods, such as k-means clustering (MacQueen, 1967), require this number to be specified a priori, and it is often unclear what specific number should be chosen in most brain network analyses.

The disadvantage of hierarchical analyses is that we have no way of determining whether the clustering solution obtained at one level is more valid or reliable than another. For example, Figure 9.2 shows the results obtained at different levels of a hierarchical clustering of structural connectivity data, but is the clustering of voxels into 15 regions better than the clustering into 100 regions? Or is the best solution at the top of the hierarchy, implying that all voxels should be grouped into a single cluster? Is the degree of clustering in any given partition greater than expected by chance? To answer these questions, we need a measure of the quality of a partition at each scale of the clustering solution.

Newman and Girvan (2004) propose two definitive criteria for a high quality partition. First, such a partition will cluster nodes into highly cohesive modules. In other words, nodes within the same module will be strongly connected with each other. This criterion is consistent with the very definition of a network module as a strongly interconnected subset of nodes. Second, the degree of intramodule connectivity in a good partition will be greater than expected by chance, as defined by a network in which edges are placed between nodes at random. To see why this second criterion is required, consider a partition in which all nodes are assigned to the same module. The fraction of edges that link nodes in the same module will be one, since all nodes are in the same module. So, if we rely only on the first criterion—the degree of intramodular connectivity—the trivial solution in which all nodes are placed into a single module will always be identified as the highest quality partition, despite the fact that it reveals nothing about the community structure of the network. This is where the second criterion helps. If we place edges at random in a network comprising just one module, the fraction of intramodular edges will also equal one, since there is nowhere else for these edges to go. In this case, the difference between the observed and chance-expected intramodule connectivity is equal to zero, a result that accords with our intuition that a partition with only one module is of low quality.

To incorporate these two criteria into a formal measure of partition quality—a so-called modularity index—we first count the number of edges between nodes that are in the same community. For a binary, undirected network this quantity is given by

$$\frac{1}{2}\sum_{ij}A_{ij}\delta(m_i, m_j),\qquad(9.1)$$

where $\delta(m_i, m_j)$ is called the indicator or Kronecker delta function and equals one if nodes i and j belong to the same module (i.e., $m_i = m_j$) and zero otherwise. This function ensures that we only count edges between nodes within the same module.

To quantify whether our modularity index is greater than expected by chance, we must specify a **null model** for intramodule connectivity. This null model provides an estimate of the number of edges that can be expected to link nodes within the same community by chance. In completely random, **Erdös-Rényi graphs**, the probability of finding an edge between any two nodes is equal to the **connection density** of the graph. However, in most analyses of real-world networks, we require that our null model preserves the empirical **degree distribution** of the observed network (Chapter 10). In this case, the probability of finding an edge between a pair of nodes by chance is no longer the connection density. Instead, this probability will depend on the **degrees** of the respective nodes, since it is more likely that edges are found between a pair of high-degree nodes than between a pair of low-degree nodes.

In modularity analysis, the so-called **configuration model** is used to define a null model that has the same **degree sequence** as the empirical network. The configuration model can be used to define the average of a given network property across the ensemble of all possible graphs that have a fixed degree sequence, but which are random in all other aspects (Bender and Canfield, 1978; Molloy and Reed, 1995; Newman et al., 2001). This model yields null networks that are similar to the null networks generated by the Maslov-Sneppen rewiring algorithm (Maslov and Sneppen, 2002), which is typically used as a null model in most network analyses (Chapter 10). The main difference is that Maslov-Sneppen rewiring does not generate self-loops or multi edges (multiple links between a pair of nodes), whereas the configuration model allows for them. The advantage of the configuration model is faster calculation of expected values under the null hypothesis.

The configuration model can be thought of in terms of cutting in half each edge in the network. Each half edge is called a stub, and there are $2E$ stubs in the network. Consider an edge with one stub attached to node i. The chance that the other stub of this edge is one of the k_j stubs attached to node j is $k_j/2E$, if connections are made at random and we also count the stub under consideration (otherwise, the denominator would be $2E-1$). This is because there is a total of k_j stubs attached to node j, out of a total possible $2E$ stubs. Given that node i is attached to k_i stubs, the total number of edges between nodes i and j that is expected by chance is $e_{ij} = k_i k_j/2E$. The number of edges that are expected to fall by chance between nodes within the same module is then

$$\frac{1}{2}\sum_{ij} e_{ij}\delta(m_i, m_j). \qquad (9.2)$$

Given Equations (9.1) and (9.2), we can define an index of modularity as the difference between the degree of intramodule connectivity observed in our network and that which is expected by chance,

$$Q = \frac{1}{2E} \sum_{ij} \left(A_{ij} - e_{ij} \right) \delta(m_i, m_j),$$

(9.3)

where E is the number of edges in the network. The normalization uses $2E$ because the sum double counts edges in an undirected network. Values of Q span the range between -1 and 1. When Q is positive, the network has higher intramodule connectivity than expected by chance; when Q is equal to zero, the degree of intramodule connectivity does not differ between the data and null model; when Q is negative, the network lacks modular structure.

We can compute Q for each level of the dendrogram obtained by a hierarchical decomposition of a network to obtain a measure of the quality of the partition at each scale. Local peaks in Q across scales can be used to identify levels of the dendrogram where the partition warrants further consideration (Newman and Girvan, 2004).

Equation (9.3) can be easily extended to the analysis of weighted networks by substituting the appropriate terms (Newman, 2004a),

$$Q = \frac{1}{2W} \sum_{ij} \left(w_{ij} - e_{ij}^w \right) \delta(m_i, m_j),$$

(9.4)

where $W = \frac{1}{2} \sum_{ij} w_{ij}$ is the total weight of the unique edges of the network, and w_{ij} is the weight of the edge linking nodes i and j. The connectivity weight on edges linking nodes in the same community that is expected by chance is $e_{ij}^w = s_i s_j / 2W$, where s_i and s_j are the respective **strengths** of nodes i and j (Chapter 4).

Some weighted brain networks, such as those generated using correlation-based analyses of **functional connectivity**, contain signed edge weights. Intuitively, we expect that nodes within a module should show strong positive connectivity. For example, if all such nodes show coherent dynamics, their activity time courses will be positively correlated. In contrast, negative edge weights imply an antagonistic or anticorrelated interaction between nodes, suggesting that these nodes should be placed in different modules. Therefore, a good partition will be one in which the positively weighted edges are within modules and the negatively weighted edges are between modules. To quantify the quality of such a partition, Rubinov and Sporns (2011) proposed the following definition,

$$Q^* = \frac{1}{2W^+} \sum_{ij} \left(w_{ij}^+ - e_{ij}^+ \right) \delta(m_i, m_j) - \frac{1}{2W^+ + 2W^-} \sum_{ij} \left(w_{ij}^- - e_{ij}^- \right) \delta(m_i, m_j),$$

(9.5)

where the superscripts + and − denote the relevant values computed over only the positive and negative edge weights, respectively. The modularity index $Q*$ is thus defined by subtracting the weighted Q index estimated over negatively weighted edges from the weighted Q index estimated over positively weighted edges. This subtraction accords with our intuition that positive weights between nodes in the same module increase modularity whereas negative edge weights within a module decrease modularity. Note that the absolute values of the negative edge weights are used when computing Equation (9.5).

An important characteristic of Equation (9.5) is the disproportionate scaling that is applied to the positive and negative modularity indices. The factor $1/2W^+$ scales the sum over positive weights only by the total positive weight of the graph, whereas the term $1/(2W^+ + 2W^-)$ scales the sum over negative weights by the total positive and negative weight of the network. This ensures that, negative edge weights make a smaller contribution to $Q*$. In particular, an increase in total positive weight will reduce the contribution of negative weights to $Q*$, whereas an increase in negative weight will not affect the contribution of positive weights to $Q*$. In fact, in a network with an equal proportion of positive and negative weight, the contribution of positive edge weights to $Q*$ will be twice as large as the contribution of negative weights. Rubinov and Sporns (2011) argue that this disproportional scaling captures the intuition that a strong positive edge weight explicitly tells us that two nodes belong in the same community, whereas a strong negative edge contributes to the community assignment problem only indirectly by telling us which nodes should *not* be placed in the same community. Generalizations of Q for the analysis of signed networks that do not assume disproportionate effects of positive and negative edge weights have also been proposed (Gómez et al., 2009; Kaplan and Forrest, 2008; Traag and Bruggeman, 2009).

For directed networks, the modularity index has been defined as (Leicht and Newman, 2008)

$$Q = \frac{1}{E}\sum_{ij}\left[A_{ij} - \frac{k_i^{in}k_j^{out}}{E}\right]\delta(m_i, m_j),\qquad (9.6)$$

where A_{ij} is the directed edge running from node j to i. The null model thus defines the likelihood that one of the outgoing edges of j will also be an incoming edge of i, which is given by $k_i^{in}k_j^{out}/E$. We use E rather than $2E$ in Equation (9.6) because edges are not counted twice in the analysis of a directed network—the upper and lower triangles of the **adjacency matrix** contribute unique information.

When analyzing directed networks, it is important to use a method for community detection that appropriately accounts for the directionality of the edges. For example, using Girvan and Newman's (2002) divisive approach, the betweenness centrality of each edge should be estimated using the directed shortest paths of the network (Chapters 5 and 7). Equation (9.6) can be easily generalized to weighted, directed networks by using the appropriate weighted analogs of the relevant quantities, as per Equations (9.4) and (9.5). Kim et al. (2010) discuss limitations of Equation (9.6) and propose an alternative definition of modularity for directed networks (see also Rosvall and Bergstrom, 2008).

9.1.3   Maximizing Modularity

The modularity index Q can be used to quantify the quality of a partition that has been defined using any method for data clustering or community detection. In this context, Q is applied posthoc, once communities have been algorithmically defined according to criteria that may not necessarily be optimal for yielding a network partition with the highest possible Q. For example, with agglomerative hierarchical clustering, clusters are formed according to a measure of pairwise node similarity. In Girvan and Newman's (2002) divisive algorithm, communities are defined by the removal of edges with high betweenness centrality. Neither of these approaches uses criteria for defining communities that are directly related to the quantities used in the modularity index, Q. If our goal is to find a partition with maximum modularity, it seems sensible to incorporate the modularity index into the community detection procedure itself. In this section, we consider some popular approaches to modularity maximization. An overview of alternative methods for community detection that do not rely on the optimization of modularity is provided in Box 9.1.

To find a partition with maximum modularity, we would ideally identify all possible partitions of a network and select the one with the highest Q. Unfortunately, such an exhaustive search is only feasible for very small networks, since the number of possible partitions grows faster than exponentially with network size (Brandes et al., 2006; Fortunato, 2010). We must therefore rely on heuristic algorithms to find a partition with near-maximal Q. For example, Newman (2004b) proposed a greedy agglomerative and hierarchical algorithm in which each node starts in a separate community. Communities are then merged repeatedly into larger and larger pairs, choosing at each step the amalgamation that results in the largest increase or smallest decrease (if increases are no longer possible) of Q. The best partition is the level of the dendrogram that is associated with the highest Q. Newman later proposed a spectral approach in which communities are identified using eigendecomposition (Box 5.1; Leicht and Newman, 2008; Newman, 2006). Others have used optimization algorithms such as **simulated annealing** (Guimerà et al., 2004).

Blondel et al. (2008) proposed one particularly popular agglomerative method, called the Louvain algorithm, for finding partitions with high modularity.

BOX 9.1 ALTERNATIVES TO MODULARITY

Maximization of network modularity, as defined by the Q index, is not the only method for uncovering the community structure of a network. One alternative to the modularity index, called surprise, considers both the number of edges *and* the number of nodes that fall within each community. Specifically, this measure of surprise quantifies the degree to which the observed distribution of nodes and edges within communities departs from a null model given by the cumulative hypergeometric distribution (Aldecoa and Marín, 2011, 2013). This method has been shown to overcome the resolution limit of modularity maximization (Box 9.2), but can result in modules comprising a single node.

Another class of methods identifies modules based on hypothesized patterns of information flow on the network, such as those based on models of random **walks** (Chapter 7) and other dynamical processes (Arenas et al., 2006; Boccaletti et al., 2007; Delvenne and Yaliraki, 2010; Reichardt and Bornholdt, 2006; Rosvall and Bergstrom, 2008). One popular example of such an approach, called InfoMap, describes the random walk with a binary code (Rosvall and Bergstrom, 2008). The code is defined in relation to the amount of time the walker spends visiting each node. The high interconnectivity of nodes within a module can ensnare a **random walker**, meaning that it spends more time visiting those nodes and less time transitioning between communities. This can lead to regularities in the code that is used to describe the walk. These regularities can be used as a basis for compression, much like when compressing a computer file. A good partition is thus one that minimizes the description length of the code, and the quality of the partition is measured by a function that considers the movement of the walker within and between modules. This approach has shown very high accuracy in evaluations on synthetic benchmark networks (Lancichinetti and Fortunato, 2009) and has been used to identify communities in human brain functional networks (Power et al., 2013).

It is also possible to characterize random walks over different temporal scales to gain insight into hierarchical modularity (Delvenne and Yaliraki, 2010), since walks examined over longer time scales will be more sensitive to identifying larger communities (Betzel et al., 2014). The advantage of such flow-based methods is that they define communities in relation to a model of the dynamics of information flow on the network, but their validity depends on the accuracy of the model in describing the actual dynamics occurring on the network (see Chapters 5 and 7).

Another important class of methods for community detection uses stochastic **block models** (see also Chapter 6). In this work, a **generative model** of the community structure of the network is fitted to the data, wherein each module is modeled as a block of nodes within which all nodes have the same probability of connection with other nodes. The advantage of this approach is that the problem of community detection is addressed from a statistical standpoint, allowing estimation of parameters and their errors. A disadvantage is that some parameters, such as the number of modules, must be specified a priori. In most applications, this number is not known, so model fits must be compared under multiple parameter values to identify the optimum model. Interested readers are referred to the text by Doreian et al. (2005) and the overview by Snijders and Nowicki (1997). For an application to the analysis of the community structure of the *C. elegans* neuronal network, see Pavlovic et al. (2014). An overview of flow-based algorithms, block models, and other approaches to community detection is provided by Fortunato (2010).

A schematic of how this algorithm proceeds is presented in Figure 9.4. The method consists of five major steps:

1. Start with all nodes in a distinct module (Figure 9.4a).
2. Choose a node at random, determine the change in Q that would result from merging that node with each of the existing modules, and implement the merge that yields the largest gain in Q.
3. Repeat step 2 until no further gains in Q can be achieved, each time selecting a new node at random. Any individual node may be considered more than once in this process (Figure 9.4b).

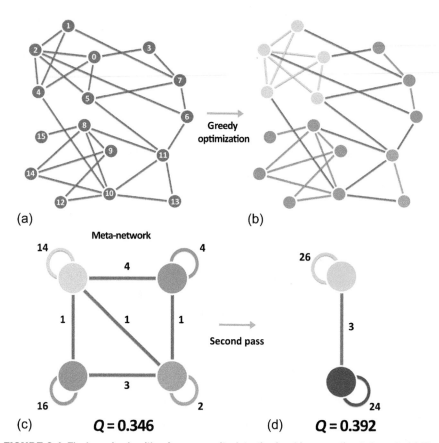

FIGURE 9.4 The Louvain algorithm for community detection in a binary, undirected graph. **(a)** The algorithm starts by placing each node in a separate module. **(b)** In random order, the algorithm sweeps across nodes and merges them with the module that produces the largest positive gain in Q. This greedy optimization phase proceeds until no more gains in Q are possible, resulting in an initial assignment of nodes into modules. Here, the optimization phase has identified four modules, represented by different colors. **(c)** The nodes within a community are then merged into super nodes to generate a meta-network. The connectivity within each module is represented as a self-loop. The weight of the self-loop is twice the number of edges that link the nodes within that module, since each edge within a module is counted twice. The weight of the edge between any pair of super nodes is the number of intermodular edges. Construction of the meta-network signals the end of the first pass of the algorithm. The Q score associated with this partition is depicted beneath the graph. **(d)** In a second pass, the meta-network shown in panel **(c)** is subjected to another round of greedy optimization, resulting in the meta-network shown here. This time, the light blue super node comes from the merger of the green and dark blue super nodes in panel **(c)**, and the dark purple super node comes from the merger of the red and light purple super nodes in panel **(c)**. The algorithm proceeds until no mergers that result in a positive gain in Q are possible. Here, we see that the partition of the network into two modules is associated with a higher Q score than the partition into four modules depicted in panel **(c)**. *Figure adapted from the example provided by Blondel et al. (2008), with permission.*

4. Merge nodes within a module into a single super node and generate a meta-network, such that edges between super nodes are weighted according to the edges that link their constituent nodes, and the connectivity of nodes within any single super node is represented by a self-loop (Figure 9.4c).

5. Repeat steps 2-4 until no further gains in Q are possible (Figure 9.4d).

We thus see that the algorithm involves two principal phases that are iterated repeatedly: steps 2 and 3, which comprise a greedy optimization phase that merges nodes to maximize Q; and step 4, which generates a meta-network of super nodes based on the communities identified in steps 2 and 3. At each successive iteration, the greedy optimization is applied to the meta-network generated in the previous iteration. Importantly, since nodes are selected for consideration at random in the greedy optimization phase, the results of the Louvain algorithm can depend on the order in which these nodes are chosen. One solution to this problem is to run the algorithm many times and generate a consensus partition across all runs. Methods for constructing such a consensus partition are discussed in Section 9.3.3.

An important difference between the Louvain algorithm and Newman's greedy approach is that the latter will merge nodes to produce the smallest decrease in Q if no positive gains in modularity are possible. In contrast, the Louvain algorithm does not merge nodes if there is no positive gain. This rule yields a natural point of termination of the Louvain algorithm, since the partition obtained at the final iteration is the one associated with the maximal Q. In contrast, Newman's approach will progress until all nodes are placed into a single community.

The Louvain algorithm is very fast, making it applicable to large networks (Blondel et al., 2008). Evaluations with respect to synthetic benchmark networks with known community structure have also found it to be more accurate than many other methods (Lancichinetti and Fortunato, 2009). There is thus good reason to think that the partition obtained by the final pass of the algorithm is close to the one associated with maximum modularity (some caveats to this interpretation are discussed in Box 9.2). Another advantage is that the algorithm can be used to investigate hierarchical organization, since optimization of modularity is performed at different scales: first, at the level of individual nodes in the initial iteration, then at the level of meta-nodes in the second iteration, and so on. This property means that the Louvain method is less susceptible to the well-known resolution limit of other methods for optimizing modularity (Box 9.2; Fortunato and Barthelemy, 2007; Lancichinetti and Fortunato, 2009).

An example application of the Louvain algorithm to characterize the hierarchical modular organization of brain networks is provided by a study of Meunier et al. (2009b). These authors examined human resting-state **functional MRI**

BOX 9.2 THE LIMITS OF MODULARITY

Most community detection algorithms try to find a partition of a network that maximizes the modularity index, Q. This approach has proven powerful, popular, and useful, but it does have some limitations. One important limitation, called the resolution limit, biases the algorithms against detecting small communities (Fortunato and Barthelemy, 2007). The problem arises because such algorithms will generally merge two subsets of nodes, A and B, into a single module when that merger is associated with a positive gain in Q. A positive gain occurs when the number of edges linking A and B is greater than the number of edges that is expected to fall between them by chance, which is given by $e_{AB} = K_A K_B / 2E$, where K_A and K_B correspond to the total degrees of nodes in modules A and B, respectively (Good et al., 2010). The precise value of e_{AB} will be small when the network is sparse (i.e., when K_A and K_B are small) and/or when the network is large (i.e., when E is large). In fact, if the network is sufficiently sparse or sufficiently large, e_{AB} can fall below one, meaning that less than one edge is expected to fall between modules by chance. As such, a pair of nodes or modules in an empirical network that is linked by the weakest possible connection (a single edge) will be merged because this link is considered "surprising" relative to the null model. This behavior counters our intuition that weakly connected modules should remain separate. The same logic extends to weighted networks in which edge weights are independent of network size (see Good et al., 2010, for a discussion). An example of the problem posed by the resolution limit is presented in Figure 9.2.1a.

The resolution limit arises because the null model used in the definition of modularity assumes that each node has an equal probability of connecting to every other node. This assumption may not hold in many real-world systems, particularly very large networks where any individual node may only have a limited "horizon" within which it is able to interact with other nodes (Fortunato, 2010). This problem may be particularly salient for spatially embedded networks in which each link carries a **cost**, and in which there is a physical or economical limit on the ability of a node to connect with another node that is separated by a long distance. Nervous systems are prime examples of such networks (Chapter 8).

One solution to the resolution limit is to use an alternative null model that is constrained to consider a more limited "horizon" of connectivity. For example, it is possible to define a null model that only considers a topologically local **subgraph** comprised of nodes in the same module and other, connected modules (Muff et al., 2005). For spatially embedded networks, a null model has been proposed that considers the physical distances of connections (Expert et al., 2011). This model may be suitable for brain networks, where very simple spatial constraints can generate graphs with a biologically realistic community structure (Henderson and Robinson, 2011, 2013, 2014; Rubinov et al., 2015). Using a spatially informed null model can help determine whether a given network shows modular organization beyond the effects of space alone (see also Chapter 10). Null models that are suitable for community detection in correlation-based networks have also been proposed (Bazzi et al., 2014; MacMahon and Garlaschelli, 2015).

Some algorithms overcome the resolution limit of modularity by incorporating a tunable resolution parameter that determines the topological scale at which communities are detected (Reichardt and Bornholdt, 2006). A typical approach involves multiplying the null model by a factor, γ. Large values of γ bias the algorithm toward detecting smaller communities, since fewer connections will be deemed "surprising" relative to chance expectations. Some of these approaches can successfully overcome the resolution limit (Traag et al., 2011), but the selection of a specific value for the resolution parameter can be arbitrary. Tests of these multi-resolution algorithms against synthetic, benchmark networks with known community structure indicate that they show two alternative biases: a tendency to merge small communities at low resolution and to split large communities at high resolution (Lancichinetti and Fortunato, 2011).

Good and colleagues (2010) have identified two additional limitations of modularity. First, the maximum modularity score for a given network, Q_{max}, increases with the number of nodes and modules in a network. The precise form of this relationship depends on the topology of the network, but the reason is intuitive: as the number of modules in a network increases, there is a smaller chance of an edge falling within a given community under the null model because there are so many other possible modules to which the edge could connect. In this case, a high modularity score suggests a large deviation between the data and null model, but not necessarily that the observed network is strongly modular. A practical consequence of this behavior is that Q cannot be used to compare the modularity of networks with different numbers of nodes and modules.

The second limitation of modularity identified by Good et al. (2010) is that there is generally a very large number of alternative partitions that are possible in a network, each with a modularity score that differs only slightly from Q_{max}. In other words, there is no clear global maximum of Q. Instead, the "landscape" of Q values for a given network is irregular and contains many local maxima (Figure 9.2.1b). This so-called degeneracy problem means that many different solutions, each of comparable quality, are possible. (Note that we use the term degeneracy here in a different sense to the formal definition of neural **degeneracy** considered in Chapter 8.) Consensus-based clustering methods that aggregate results across many iterations of a partitioning algorithm can be used to identify communities that are robust to this algorithmic degeneracy (see Section 9.3.3).

BOX 9.2   THE LIMITS OF MODULARITY—CONT'D

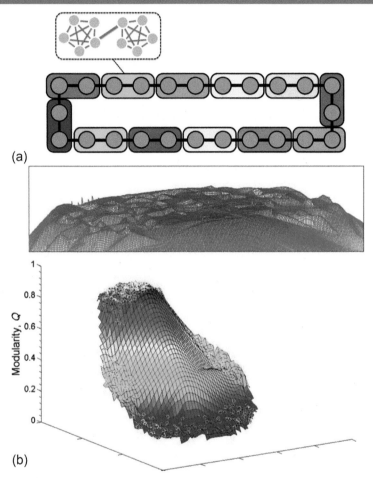

(a)

(b)

FIGURE 9.2.1 The resolution limit and degeneracy of modularity maximization. **(a)** A ring network used to illustrate the limits of modularity. The network comprises 24 maximally connected cliques comprising 5 nodes each. The connectivity of two cliques is illustrated in the dotted box. Each node within a clique is connected to every other node, and different cliques are only linked by a single edge. The full network is shown below, with each clique represented as a single, super node (gray circles, black outlines). The intuitive partition of such a network should assign each clique to a distinct module. This partition has optimal modularity when the network is small. However, when the network is sufficiently large, as in the 24-clique case here, the optimum partition ($Q = 0.871$) identifies communities in which pairs of **adjacent** cliques are merged together (colored tiles). The expected partition, in which each clique is assigned to a distinct community, is associated with slightly lower modularity ($Q = 0.867$). This inability to identify small communities in large networks is due to the resolution limit of modularity. **(b)** The modularity landscape of the network depicted in panel **(a)**. Plotted here are the Q values obtained from 997 runs of a modularity maximization algorithm. An irregular plateau of high modularity scores is apparent and shown in more detail in the upper panel. The large number of peaks in this plateau suggests that many alternative partitions of comparable quality are possible, even in this simple ring-like network. *Images adapted and reproduced from Good et al. (2010), with permission.*

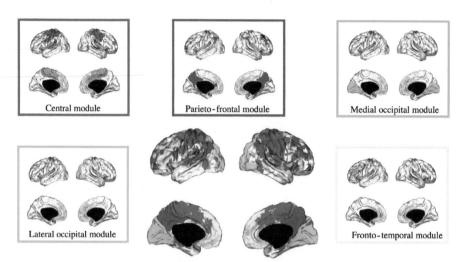

FIGURE 9.5 Hierarchical modularity of human functional connectivity networks. Shown here are the results of a hierarchical decomposition of correlation-based functional MRI networks comprising over 1800 voxel-level nodes. The assignment of these nodes into distinct communities based on the partition with the highest Q, as determined by the Louvain algorithm, is shown in the central panel. At this level, 8 large modules comprising at least 10 nodes were identified. Each module is represented by a different color. Boxes show the subdivision of five of these modules into submodules at the next level in the hierarchy. The colors of the boxes correspond to the module at the highest level of the hierarchy, as shown in the center panel. For example, the central, sensorimotor module depicted in red at the highest level of the hierarchy could be further subdivided into medial and lateral submodules (red box). *Figure reproduced from Meunier et al. (2009b), with permission.*

networks consisting of over 1800 nodes, defined at the level of coarse, 4 mm^3 voxels. The analysis revealed three hierarchical levels. The highest level of the hierarchy comprised 8 modules with at least 10 nodes each, and was associated with a maximum modularity score $Q = 0.6$ (Figure 9.5). By comparison, an average $Q = 0.3$ was obtained for an ensemble of randomized networks with the same number of nodes, edges, and degree distribution, as generated with the Maslov-Sneppen rewiring procedure (see Chapter 10). Figure 9.5 shows the results obtained at two levels of the hierarchy. Some modules obtained at the highest level, such as the lateral occipital module, could not be further subdivided. Other modules, such as the frontotemporal module, had many submodules that were often determined by spatial location. For example, the central module identified at the highest level of the hierarchy could be further subdivided into medial and lateral aspects.

Bassett and colleagues (2010) used a different approach to study the hierarchical modularity of nervous and other systems. They first applied the Louvain algorithm to a very large-scale integrated (VLSI) computer chip; the neuronal network of *C. elegans*; and two human structural connectivity networks, one constructed with diffusion MRI and the other constructed using interregional covariations in gray matter volume, as measured with structural MRI. For each network, they focused only on the highest-level partition generated by the Louvain method (i.e., the partition with the highest modularity). They calculated the modularity index, Q, separately for each community in the partition and compared this score to that obtained in an ensemble of appropriately randomized, surrogate networks (Chapter 10). They then applied the algorithm again only to subgraphs of nodes and edges corresponding to individual modules that were associated with a Q score that was significantly higher than the surrogate graphs. This procedure was repeated iteratively until no further modules with a statistically significant Q score could be identified. In this way, the authors were able to examine modules and their nested submodules, while only focusing on modules at each hierarchical level that were not expected to arise by chance. This analysis identified significant modular organization over four hierarchical levels in the VLSI chip and *C. elegans*, three levels in the human diffusion MRI network, and two levels in the human **structural covariance** network. This commonality of hierarchical modular organization across these diverse systems suggested that it was a general characteristic of spatially embedded networks that are subject to constraints on wiring cost (see also Chapter 8). These findings are summarized in Figure 9.6.

The analyses of Meunier et al. (2009b) and Bassett et al. (2010) both compared the modularity index obtained in their brain networks to the same quantity estimated in an ensemble of randomized surrogate graphs. Why is this comparison necessary given that the modularity index (Equation 9.3) already takes into account the null model given by e_{ij}? Remember that e_{ij} describes an *average* expectation of connectivity between nodes, taken over all possible configurations of random graphs with the same degree sequence as the empirical network. Any individual instantiation of a null network drawn from this model will naturally fluctuate around the average expectation. These fluctuations can result in a high modularity score even in Erdös-Rényi graphs with no intrinsic modular structure (Guimerà et al., 2004). To see how this can occur, consider an Erdös-Rényi graph comprising 100 nodes. This graph can be partitioned into 10 modules comprising 10 nodes each in $\sim 10^{85}$ different ways, which is more than the estimated number of stars in the universe. The vast majority of these partitions will be "typical", in the sense that the number of intramodular links per module is quite close to the expected value of $pn(n-1)/2$, where $n = 10$ is the number of nodes per module and p is the connection density. However, since there are so many partitions, it is inevitable that

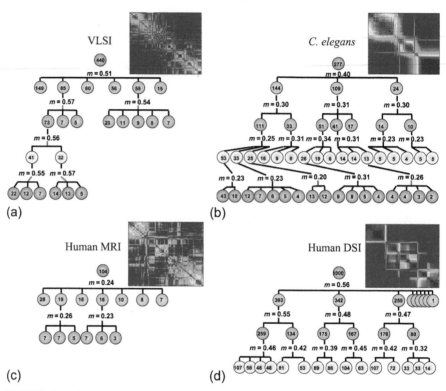

FIGURE 9.6 Hierarchical organization of brain structural connectivity networks. **(a)** For comparison, this panel shows the hierarchical modularity of a very large-scale integrated (VLSI) computer chip. The top level of the tree shows the number of nodes in the network with its modularity, denoted here as m, underneath. Each subsequent level shows the number of nodes in each module that was identified by a partition of the level above. At each level, nodes with a modularity score significantly greater than chance were further partitioned into submodules. For example, here the initial network of 440 nodes was partitioned into six modules (blue circles), two of which showed statistically significant modularity, with scores of 0.57 and 0.54. These two modules were then partitioned into submodules (cyan circles). At this level, only one significant community with modularity equal to 0.56 was identified. This community was then further subdivided, and so on, to reveal four hierarchical levels of the network. Inset is a representation of the connectivity matrix that has been organized and reordered to emphasize hierarchical modularity. **(b)** The same analysis applied to the neuronal network of *C. elegans* identified four hierarchical levels of significant modularity. **(c,d)** Analysis of human structural covariance networks measured with structural MRI **(c)** and human structural connectivity networks measured with a variation of diffusion MRI called diffusion spectrum imaging (DSI) **(d)** revealed two and three hierarchical levels of significant modularity, respectively. *Figure reproduced from Bassett et al. (2010), with permission.*

a highly *atypical* partition will be found where the number of intramodular links is much higher than expected. Community detection algorithms will do their best to identify this highly atypical partition, and yield what appears to be a high Q-value. In reality though, this Q-value will be quite *typical* among the set of *atypical* partitions that a modularity algorithm will identify if this process is repeated for another instantiation of an Erdös-Rényi graph. If we blindly accept the Q score from this analysis, without benchmarking against appropriate surrogate networks, we will conclude that such a graph shows modular organization.

Given that sampling variations can lead to high modularity scores even in non-modular networks, it is important to compare the score obtained in an empirical network to the same quantity estimated in a suitable ensemble of randomized networks. Generating a sufficient number of randomized graphs allows us to construct a null distribution that can be used to estimate the statistical significance of Q. The analysis of Bassett et al. (2010) shows that we can also use these surrogates to evaluate the statistical significance of individual modules (Figure 9.6). Other methods for statistical inference at the level of single modules have been proposed (Lancichinetti et al., 2010). Methods for generating appropriately randomized networks and null models are discussed in Chapter 10.

9.2 NODE ROLES

Mapping the community structure of a network allows us to characterize the topological roles of individual nodes in facilitating communication within and between modules. We can think of nodes that are highly connected to other nodes within the same module as facilitating functional specialization—they support communication within tightly linked and functionally related communities. Nodes with many connections to other modules play an important role in functional integration—they act as bridges that enable communication between otherwise segregated modules. Understanding patterns of intramodule and intermodule connectivity can therefore uncover the topological foundations of functional segregation and integration in brain networks.

9.2.1 Cartographic Classification of Nodes

Guimerà and Amaral (2005) proposed two measures for characterizing how a node's connectivity is distributed within and between modules. The parameter space defined by these two measures provides a cartographic representation or map that facilitates the classification of each individual node according to its functional role in the network. The first measure, called the within-module degree z-score, is estimated for a node, i, as

$$z_i = \frac{k_i(m_i) - \bar{k}(m_i)}{\sigma_{k(m_i)}}, \tag{9.7}$$

where $k_i(m_i)$ is the within-module degree (i.e., number of connections linking node i to other nodes in the same module, m_i), $\bar{k}(m_i)$ is the mean within-module degree of nodes in the same module as node i, and $\sigma_{k(m_i)}$ is the standard deviation of $k_i(m_i)$ values across all nodes in module m_i. A node with a within-module degree z-score equal to zero has an average level of intramodule connectivity relative to other nodes in the same module. A positive z-score indicates above-average intramodule connectivity and a negative score indicates below-average intramodule connectivity. Note that for Equation (9.7) to be interpreted as a z-score, the distribution of intramodule degree within a community must be Gaussian.

The second measure proposed by Guimerà and Amaral (2005) is called the participation coefficient, P. This measure quantifies how a node's links are distributed across different modules. It is estimated as the fraction of node i's connectivity that is attributable to each module,

$$P_i = 1 - \sum_{m=1}^{M} \left(\frac{k_i(m)}{k_i} \right)^2, \tag{9.8}$$

where M is the number of modules in the network and $k_i(m)$ is the number of links between node i and other nodes in module m. The participation coefficient is close to one when the edges of a node are distributed uniformly across modules and it is zero when all the links of a node are contained within its own module.

The participation coefficient is influenced by: (1) how evenly the connections of a node are spread across the modules of a network; and (2) how many modules these connections are spread across. The importance of the second property is shown by the following example. Consider a network with three modules. Node A has three links with nodes in module 1, three links with nodes in module 2, and no connections to the third module. Node B has two links to each of the three modules. Nodes A and B both have a degree of six and their links are distributed equally across the modules to which they connect. That is, modules 1 and 2 each receive 1/2 of the links attached to node A and modules 1, 2, and 3 each receive 1/3 of the links attached to node B. However, the links of node B are distributed across more modules, and so this node has a higher participation coefficient ($P = 0.5$ vs. $P = 0.67$ for node A). This result concurs with the intuition that nodes connected to more communities play a more integrative role and should therefore have higher participation. However, the dependence of P on M means that we cannot compare participation coefficients estimated in networks that comprise a different number of modules.

An alternative measure to the participation coefficient, called the diversity coefficient, is based on the normalized Shannon entropy (Rubinov and Sporns, 2011),

$$h_i = -\frac{1}{\log M} \sum_{m=1}^{M} p_i(m) \log p_i(m), \tag{9.9}$$

where $p_i = k_i(m)/k_i$. As with the participation coefficient, a value of h_i that is close to one indicates that node i has an even distribution of links across modules whereas a value of zero indicates that all of node i's links are contained within its own module. Generalizations of this measure to deal with signed and weighted networks are provided by Rubinov and Sporns (2011).

Guimerà and Amaral (2005) have used variations in z and P to define seven distinct, universal topological roles for network nodes. These roles can be visualized using a two-dimensional plot, in which variations in z are displayed on the vertical axis and variations in P on the horizontal axis (Figure 9.7a). According to this scheme, we first classify nodes as either module **hubs** or nonhubs based on their within-module degree, z. Nodes with $z > 2.5$ are deemed module hubs and nodes with $z < 2.5$ are nonhubs. We can then achieve a more fine-grained classification within these two categories by considering variations of the participation coefficient, P. Specifically, Guimerá and Amaral define four distinct roles for nonhub nodes:

- *Ultra-peripheral*, where nearly all the connections of a node are contained within the same module ($P \leq 0.05$; Figure 9.7a, R1);
- *Peripheral*, where most of a node's connections are in the same module ($0.05 < P \leq 0.62$; Figure 9.7a, R2);
- *Nonhub connector*, where nodes contain many links with nodes in other modules ($0.62 < P \leq 0.80$; Figure 9.7a, R3); and
- *Nonhub kinless*, where most of a node's links are distributed outside of its module ($P > 0.80$; Figure 9.7a, R4).

Using the same logic, we can define three distinct types of module hubs:

- *Provincial hubs*, which contain most of their connections within their own module ($P \leq 0.30$; Figure 9.7a, R5);
- *Connector hubs*, which have many links with other modules ($0.30 < P \leq 0.75$; Figure 9.7a, R6); and
- *Kinless hubs*, which have their links distributed uniformly across all other modules ($P > 0.75$; Figure 9.7a, R7).

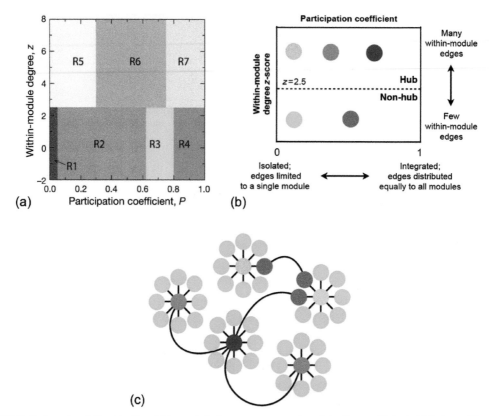

(a)

(b)

(c)

FIGURE 9.7 Nodes roles defined by the within-module degree z-score and participation coefficient. **(a)** Cartographic scheme for defining node roles according to variations in z and P. Each colored area defines a parameter regime associated with one of seven distinct topological roles. R1, ultra-peripheral; R2, peripheral; R3, nonhub connector; R4, nonhub kinless; R5, provincial hub; R6, connector hub; and R7, kinless hub. **(b)** A simplified scheme that focuses on the node role distinctions usually considered in the analysis of brain networks. **(c)** An example network used to demonstrate the connectivity profile of nodes with different topological roles, as indicated in panel **(b)**. *Reproduced **(a)** from Guimerà and Amaral (2005) and **(b-c)** from Power et al. (2013), with permission.*

Guimerà and Amaral (2005) computed z and P for over 26,000 nodes belonging to diverse classes of networks, including the *C. elegans* proteome, the global air transportation network, academic collaboration networks, and model networks such as Edös-Rényi and Barabási-Albert graphs (see Chapter 10). The authors found that the seven key topological roles approximated basins of attraction within the $z - P$ parameter space, such that each basin represented a parameter regime associated with a high density of nodes. Notably, the authors found that nonhub kinless nodes were present in model graphs but were not seen in real-world systems, suggesting that they are very rare in empirical networks. It is also worth noting that the assignment of a kinless hub to a community is largely a matter of chance, because it shows no strong preference for connectivity with any specific module. The role of these nodes may be better characterized using methods for overlapping community detection, which allow a node to belong to more than one module (Box 9.3). With such an overlapping or fuzzy decomposition, measures have been defined that allow the identification of bridge hubs—topologically central nodes that belong to many modules (Nepusz et al., 2008).

BOX 9.3 OVERLAPPING MODULES

Traditional approaches to community detection decompose networks into nonoverlapping modules, such that any individual node is assigned to one and only one module. This property disagrees with our intuition about the community structure of many real-world networks, where we expect that nodes should belong to more than one group or category. For example, an agent in a social network may belong to a group of friends that is associated through a sporting club, a group defined by shared music interests, and so on. The brain is no exception to this rule. In particular, it is known that certain neuronal populations, such as those in polymodal association cortices, participate in multiple functional systems (Buckner and Krienen, 2013; Mesulam, 1998).

Several methods have been proposed for decomposing networks into communities with overlapping structure, within which any individual node can belong to more than one community. In this case, we no longer attempt to find a partition of the network into nonoverlapping communities. Rather, the goal is to identify an appropriate *cover*: a set of clusters (communities) defined such that each node is assigned to one or more clusters, and in which no cluster is a subset of any other (Shen et al., 2009). Some methods for overlapping community detection focus on the analysis of fully connected subgraphs, called cliques, of different size in the network (Palla et al., 2005; Shen et al., 2009). Others use flow-based methods based on the characterization of random walks (Esquivel and Rosvall, 2011) or other dynamical processes (Gregory, 2010; Xie et al., 2011), while others have modified traditional, nonoverlapping community detection methods (Gregory, 2007; Nicosia et al., 2009). A systematic evaluation of these and other methods is provided by Xie et al. (2013).

One approach determines community assignment at the level of edges rather than nodes. In an application to brain networks, de Reus and colleagues (2014) started by building a **line graph** from human structural connectivity networks constructed from diffusion MRI data. A line graph is the reversal of a typical brain graph, in which the nodes of the original graph become the edges of the line graph and the original edges become the new nodes (Figure 9.3.1a). Applying a traditional, nonoverlapping community detection algorithm to these line graphs, de Reus and colleagues identified a set of distinct edge-centric communities in which each edge was assigned to a different community, but nodes could belong to more than one edge-based community (Figure 9.3.1b-c). With this approach, the authors found that more than half of the 68 brain regions studied participated in more than one community. Areas of medial parietal and prefrontal cortex showed a high degree of community overlap, respectively participating in seven and eight of the total of eleven communities identified (Figure 9.3.1c). The authors also found that high-degree hub nodes corresponding to the brain's **rich club** were found across all communities and participated in more communities, on average, than other nodes. This result is consistent with the hypothesized role of rich-club connectivity in integrating diverse neural systems (see also Chapter 6). More formal measures for quantifying the extent to which an individual node participates in multiple communities have been proposed (Nepusz et al., 2008).

This edge-centric approach is useful for characterizing how edges cluster together, but it can result in communities with a star-like configuration that converges on a single hub node (e.g., the community labeled "right PC" in Figure 9.3.1b). In such communities, the majority of nodes do not directly interact with each other. This property counters our intuition that a module should comprise a cluster of tightly linked nodes. Alternative methods for edge-based community detection can get around this limitation (Ahn et al., 2010).

Continued

BOX 9.3   OVERLAPPING MODULES—CONT'D

More generally, evaluations of algorithms for overlapping community detection using synthetic benchmarks with a known community structure have shown that there is considerable room for improvement. For example, one comparison of fourteen different methods that included a range of clique-based, edge-centric, and flow-based approaches, found that the algorithms were rarely able to recover the true community structure with 100% accuracy, and that the relative performance of each method strongly depended on the topological properties of the network. That is, methods that performed well on networks with certain topological characteristics performed poorly on other networks (Xie et al., 2013).

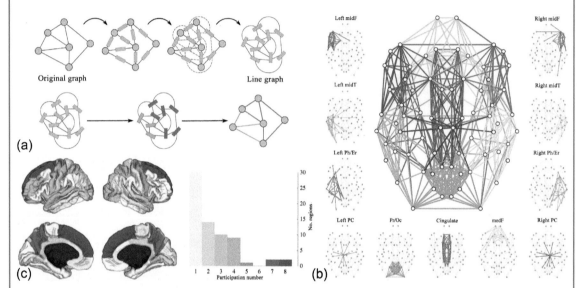

FIGURE 9.3.1 Link communities in human structural connectivity networks. **(a)** Top row presents a schematic of how a line graph is generated from a brain graph. In the original network (far left), brain regions are nodes and connections are edges. To generate a line graph, we first identify the edges and turn them into nodes, represented here as gray rectangles (middle left). We then draw edges between the rectangles if two edges of the original graph link to a common node (dotted lines, middle right). In the final line graph, the nodes of the original graph are represented as edges, and the edges of the original graph are represented as nodes (far right). Bottom row shows the application of a conventional community detection algorithm to the line graph (left). Nodes of the line graph (corresponding to edges of the original graph) are assigned to specific, nonoverlapping communities (middle). If we map these memberships back to the original graph, we see that the structural connections of the network have been assigned to specific communities, and the nodes (i.e., brain regions) can participate in multiple communities, depending on the module affiliations of their incident edges (right). **(b)** Application of this approach to a 68-region human structural connectivity network measured with diffusion MRI identified 11 link-based communities. The middle shows their spatial distribution in a whole-brain graph. The smaller graphs show each community individually. Note that some communities, such as the left and right PC communities, have a star-like configuration, in which a central node links to other nodes that are not directly connected to each other. This configuration violates traditional notions of a community in which nodes within a community are interconnected with each other. **(c)** Distribution of the participation number across nodes. The participation number counts the number of communities in which any given node participates. Over 50% of regions participate in multiple communities. *Images reproduced from de Reus et al. (2014), with permission.*

The within-module degree z-score and participation coefficient can be easily generalized to directed and weighted networks. In directed networks, either measure can be calculated separately using the **in-degree** and **out-degree** by substituting either k_{in} or k_{out} for k_i in Equations (9.7) and (9.8). Weighted versions of these measures can be estimated by using node strength rather than degree in the calculations.

9.2.2   Node Roles in Brain Networks

When network modules are identified in terms of a traditional, nonoverlapping, or mutually exclusive community structure, the $z - P$ classification of node roles offers a useful method for characterizing the way in which different nodes support specialization and integration within a network. However, there are some caveats when applying this analysis to the brain. One concerns the effect of network size. The graphs studied by Guimerà and Amaral (2005) were large, often comprising thousands of nodes. Large networks yield a wider range of degrees and a greater dynamic range of values for measures such as z and P. In smaller networks, such as those typically studied in neuroscience, these measures may have a restricted range, making it difficult to draw fine-grained distinctions between all seven topological roles proposed by Guimerá and Amaral. To get around this problem, some authors have altered the thresholds of z and P that are used to define node roles (Fornito et al., 2012a; Meunier et al., 2009a; Figure 9.7b,c). For example, Meunier et al. (2009a) identified the module hubs of human resting-state functional connectivity networks as nodes with $z > 1$, and delineated connector nodes from provincial nodes at $P = 0.05$. At this cutoff, a provincial node was one with no intermodular links whereas a connector node had one or more intermodular links (Figure 9.8a and b). Other work by the same group suggests that the use of higher resolution parcellations with a larger number of nodes results in a wider range of z and P values, and allows for an assignment of node roles that is more consistent with Guimerá and Amaral's original scheme (Meunier et al., 2009b; Figure 9.8c and d).

The analysis of correlation-based networks requires special consideration. Power et al. (2013) have shown that node strength correlates negatively with the participation coefficient in correlation-based networks, whereas strength and participation often correlate positively in noncorrelation networks. This negative correlation arises from the transitivity of the correlation coefficient, which means that if a node i correlates with another node j, and j correlates with node h, then i is likely to correlate with h. One practical consequence of this property is that nodes with high within-module degree—which represent module hubs in Guimerà and Amaral's (2005) scheme—will tend to be provincial. In other words, we will rarely find connector hubs, since a node cannot

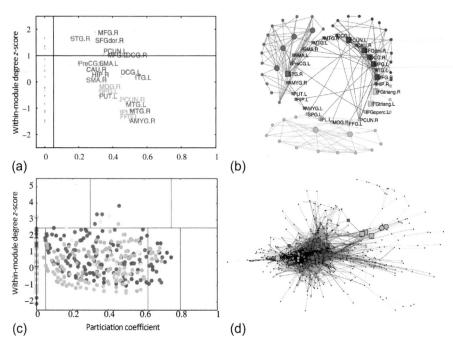

FIGURE 9.8 Network size and node role analysis. **(a)** Node roles identified for human functional connectivity networks measured with resting-state functional MRI. The network comprised 90 anatomically defined regions. In this case, the authors identified module hubs as nodes with $z > 1$ and delineated connector nodes from provincial nodes at $P = 0.05$ (black lines). Colors correspond to different modules and labels correspond to the names of the anatomical regions. **(b)** Topological projection of the nodes considered in panel **(a)**. Nodes with intermodular links (denoted by squares) are positioned around the perimeter of a central circle and provincial nodes (denoted by circles) are located on the periphery. **(c)** Node role analysis of human resting-state functional connectivity networks comprising over 1800 nodes. We see a much wider range of z and P values, allowing for more fine-grained distinctions between node roles. Black lines delineate the seven node roles originally defined by Guimerà and Amaral (2005; see Figure 9.7a). Colors correspond to different modules. **(d)** Topological projection of the nodes considered in panel **(c)**, generated using a force-directed algorithm (Chapter 3). *Reproduced **(a,b)** from Meunier et al. (2009a) and **(c,d)** from Meunier et al. (2009b), with permission.*

simultaneously show a strong correlation with two or more modules that are weakly correlated with each other. This constraint is not present in networks that are not based on correlations, such as those constructed with viral tract tracing or diffusion MRI. In these networks, any individual node is free to connect with one or more modules, irrespective of how those modules interact with each other. The transitivity of correlations also means that the strength or degree of a node in a correlation network will be be associated with the size of the module to which that node belongs, since a node embedded within a larger module will be correlated with more nodes, on average (see also Box 5.2).

These considerations mean that caution must be exercised when interpreting degree or strength-based measures estimated from such graphs. Power et al. (2013) advocated focus on the participation coefficient alone as a less biased measure for identifying hubs in these systems, based on the rationale that this measure adequately captures the intuition that network hubs play an important role in integrating diverse functional systems. In high-resolution voxel-based networks, Power and colleagues also proposed a measure of local community density to identify putative connector hubs. This measure is estimated at each voxel as the number of communities assigned to other voxels within the local ($\sim$5 mm radius) anatomical vicinity. Areas of high community density are thus in close physical proximity to multiple communities and represent likely convergence zones for different functional systems (Figure 9.9a). An alternative way to identify such areas is to use methods that allow for overlapping community structure (Box 9.3).

Warren and colleagues (2014) have provided experimental support for the hypothesis that nodes with a high-participation coefficient play a role in human correlation-based functional connectivity networks that is distinct from nodes with high degree (Figure 9.9a and b). Specifically, they found that patients with lesions to brain areas associated with a high-participation coefficient showed widespread impairment across a broad range of cognitive domains, consistent with the proposed role of connector hubs in integrating information across different modules. In comparison, patients with lesions to nodes with high degree showed more circumscribed deficits that affected specific cognitive domains (Figure 9.9c).

An important caveat against relying solely on the participation coefficient to characterize brain network hubs is demonstrated by the following example. Consider two nodes, A and B, within a larger network that has been partitioned into four modules. Node A is connected to a single node residing within each of the four modules. Its degree is thus four. Node B is connected to 25 distinct nodes that reside in each of the four modules. Its degree is one hundred. The participation coefficient for both these nodes will be equal to 0.75, even though the high degree of node B suggests that it plays a more important role in integrating the communities. Considering the participation coefficient alone will miss this important distinction. It is therefore useful to examine variations of P in relation to node degree or other measures of topological centrality (Chapter 5) to understand the precise role played by each node in the network.

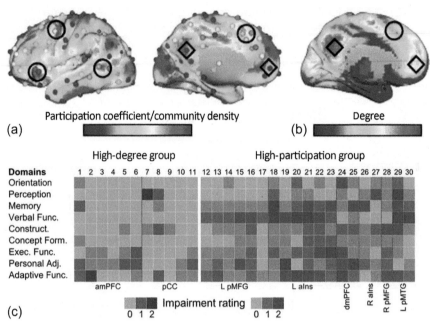

FIGURE 9.9 Community participation, community density, node degree, and cognitive performance. **(a)** Cortical surface map showing nodal variations in the participation coefficient as colored spheres, and voxelwise variations in community density as colors on the surface. The measures were computed from correlation-based functional connectivity networks constructed from human resting-state functional MRI data. **(b)** Variations in functional connectivity degree. Brain regions with high participation and community density (black circles) can be distinguished from areas with high functional connectivity degree but low community participation (black diamonds). **(c)** Summary of the neuropsychological impairment arising from lesions to areas with either high-participation (red) or high-degree (blue) nodes. Each row of the matrix corresponds to a different cognitive domain. Each column corresponds to a different patient. Patients are grouped according to whether they have a lesion in a region with high degree or a region with high participation. Colors indicate the magnitude of the impairment on a 3-point scale (0 is no impairment; 1 is moderate impairment; 2 is severe impairment). The deficits are more consistent and widespread in patients with lesions to areas with high community participation. *Images reproduced from Warren et al. (2014), with permission.*

9.3   COMPARING AND AGGREGATING NETWORK PARTITIONS

An important aspect of modularity analysis is the comparison of different network partitions. For example, we may want to quantify the extent to which a given community detection algorithm correctly recovers the known community structure of a benchmark network, to quantify the similarity of two partitions generated by the application of different community detection algorithms to the same network, or to examine the consistency of community structure across a population of networks.

A naïve approach to achieve these goals is to compare the modularity index, Q, across partitions and conclude that partitions with similar Q have a comparable community structure. This result may tell us that two networks have a similar degree of modularity, but it does not tell us anything about the composition of the modules, since two partitions with very different community structures can have similar values of Q (Box 9.2). To compare the actual community structure of two or more partitions, we want to understand whether the nodes have been assigned to the same or different modules. In other words, we want to index the similarity or overlap of the partitions at the nodal level.

Understanding the consistency of different partitions is also useful in generating an average or consensus representation of community structure across a population of networks. Such representations are particularly useful when comparing one group of networks with another, such as when contrasting a patient with a control group. In this section, we consider methods for comparing different partitions, and for generating consensus representations of community structure across a set of partitions.

9.3.1 Comparing Two Partitions

Suppose that we apply a community detection algorithm to the brain networks of two different people. Each network consists of N nodes. It is possible that our community detection algorithm will identify an optimal partition for each network that is distinct in terms of the composition of the modules and the number of modules. Let X be the partition of the first brain into M_X modules and let Y be the partition of the second brain into M_Y modules. The community assignments of X can be represented as a community affiliation vector of N elements, with each node assigned a value ranging between 1 and M_X that corresponds to its unique community label. Similarly, the community assignments of the partition Y are represented as an affiliation vector with values ranging between 1 and M_Y.

The community affiliation vectors of X and Y can be used to quantify the overlap between the partitions. Intuitively, we could just count the number of nodes that are assigned to the same community in the two partitions. However, the labels used in the two affiliation vectors may not correspond with each other. For example, community "1" in partition X may be more similar to community "3" in partition Y than community "1". Aligning these labels is deceptively challenging—at what point should we conclude that two communities in different partitions correspond to the same subset of nodes? We could choose some cutoff based on similarity, such as matching communities with at least

Table 9.1 Generic Confusion Matrix for Comparing Two Network Partitions

	y_1	y_2	$\cdots$	y_{M_Y}	Totals
x_1	n_{11}	n_{12}	$\cdots$	n_{1M_Y}	a_1
x_2	n_{21}	n_{22}	$\cdots$	n_{2M_Y}	a_2
$\vdots$	$\vdots$	$\vdots$	$\ddots$	$\vdots$	$\vdots$
x_{M_X}	n_{M_X1}	n_{M_X2}	$\cdots$	$n_{M_XM_Y}$	a_{M_X}
Totals	b_1	b_2	$\cdots$	b_{M_Y}	$\sum_{ij} n_{ij} = N$

50% of nodes in common, but this cutoff is arbitrary. Moreover, if partition X has more communities than Y, how do we align the "excess" modules?

We can avoid making such arbitrary choices by constructing an $M_X \times M_Y$ contingency table, also called a confusion matrix, which comprehensively characterizes the overlap between two partitions. The generic form of the confusion matrix is shown in Table 9.1 (Vinh et al., 2010). The modules of one partition are listed as separate rows and the modules of the other partition are listed as separate columns. Each cell counts the number of nodes that fall into every possible pairwise combination of modules across the two partitions. For example, the cell (x_1, y_2) counts the number of nodes that are assigned to module x_1 in the partition X and module y_2 in the partition Y.

One type of measure for quantifying partition similarity is based on counting different types of node pairs. More specifically, we can define four numbers that count the number of node pairs falling into one of the following categories: c_{11}, which is the number of node pairs that are assigned to the same module in both partitions X and Y; c_{00}, which is the number of node pairs that are assigned to different modules in both X and Y; c_{10}, which is the number of node pairs that are in the same module in X but different modules in Y; and c_{01}, which is the number of node pairs that are in the same module in Y but different modules in X. Note that when we say a node pair is assigned to the same module in X and Y, we do not mean that the partition label is necessarily the same. For example, nodes i and j might both reside within module "4" of partition X and within module "7" of partition Y. The critical property that we consider here is whether a pair of nodes resides in the same module within each partition, irrespective of the other nodes in that module.

The counts c_{11} and c_{00} are measures of the agreement between X and Y and the counts c_{10} and c_{01} are measures of the disagreement between the partitions. Given these four numbers, an intuitive measure of the overlap between X and Y is the number of node pairs showing agreement between the two partitions relative to the total number of possible node pairs. This measure, called the *Rand index* (Rand, 1971), is formally defined as

$$RI(X, Y) = \frac{c_{11} + c_{00}}{c_{11} + c_{00} + c_{10} + c_{01}}. \tag{9.10}$$

An advantage of this measure is that it avoids any assumptions about how to align the community labels of X and Y. Instead, overlap is quantified according to the consistency with which different node pairs are grouped together across the two partitions. Other pair counting measures are reviewed by Hubert and Arabie (1985) and Meilă (2007). Meilă also reviews alternative measures of partition overlap based on finding the best match for each module in X to one of the modules in Y.

An alternative set of measures for quantifying partition overlap is based on information theory. These measures are popular in network science and several different metrics are available. Interested readers are referred to overviews by Vinh et al. (2010), Meilă (2007), and Fortunato (2010). Here, we focus on two key measures that are commonly used to assess partition overlap in community detection.

In general, information theoretic measures of overlap estimate the mutual information between partitions X and Y. In doing so, we ask the following question: does knowing the community label of a node in partition Y tell us anything about the community label in partition X? More precisely, mutual information quantifies the degree to which knowledge of the community label of a node in partition Y reduces our uncertainty about that node's community label in partition X, and is defined as

$$MI(X, Y) = H(X) - H(X|Y). \tag{9.11}$$

The term $H(X)$ is the Shannon entropy of partition X, or the uncertainty associated with a community label in that partition. The term $H(X|Y)$ is the conditional entropy of X given Y and corresponds to the amount of information needed to predict partition X given that partition Y is known in full.

Using Table 9.1, the Shannon entropy of partition X can be estimated as

$$H(X) = -\sum_{i=1}^{M_X} p(x_i) \log p(x_i), \tag{9.12}$$

where $p(x_i) = a_i/N$ and is the probability that any one of the N nodes in the network will be assigned to the community x_i. From Table 9.1, we see that a_i is the number of nodes in community x_i of partition X.

The conditional entropy of X given Y is estimated as

$$H(X|Y) = -\sum_{i=1}^{M_X} \sum_{j=1}^{M_Y} p(x_i, y_j) \log p(x_i|y_j), \tag{9.13}$$

where $p(x_i, y_j) = n_{ij}/N$ is the joint probability of randomly selecting a node that belongs to modules x_i and y_j, and $p(x_i|y_j) = n_{ij}/b_j$ is the conditional probability that a node belongs to module x_i in partition X, given that it is in module y_j in partition Y. From Table 9.1, we see that b_j is the number of nodes in community y_j of partition Y.

Problems with the mutual information defined in Equation (9.11) can arise when one partition completely determines the other. Karrer and coworkers (2008) provide a useful example. Consider the following three partitions of a hypothetical network comprising seven nodes:

$$
\begin{aligned}
X &= \{\{1, 2, 3\}, \{4, 5, 6, 7\}\}; \\
Y &= \{\{1, 2, 3\}, \{4, 5\}, \{6, 7\}\}; \\
Z &= \{\{1, 2, 3\}, \{4\}, \{5\}, \{6, 7\}\}.
\end{aligned}
\tag{9.14}
$$

The second and third communities in partition Y are subsets of the second community in X. Similarly, the second, third, and fourth communities in partition Z are subsets of the second community in X. As a result, the conditional entropies $H(X|Y)$ and $H(X|Z)$ are both equal to zero despite Y and Z having a different community structure. We thus find that $MI(X,Y)$ and $MI(X, Z)$ both reduce to $H(X)$, such that $H(X)=MI(X,Y)=MI(X,Z)=0.68$, despite Y and Z having a different community structure. This equivalence occurs because we can deduce the communities in X based on knowledge of either Y or Z, given an appropriate mapping of community labels. We can get around this problem by defining a normalized mutual information measure with reference to the cells in Table 9.1 (Danon et al., 2005; Fred and Jain, 2003; Kuncheva and Hadjitodorov, 2004),

$$
MI'(X, Y) = \frac{-2\sum_{i=1}^{M_X}\sum_{j=1}^{M_Y} n_{ij} \log\left(\frac{n_{ij}N}{a_i b_j}\right)}{\sum_{i=1}^{M_X} a_i \log\left(\frac{a_i}{N}\right) + \sum_{j=1}^{M_Y} b_j \log\left(\frac{b_j}{N}\right)}.
\tag{9.15}
$$

This normalized measure can be expressed equivalently as

$$
MI'(X, Y) = \frac{2MI(X, Y)}{H(X) + H(Y)}.
\tag{9.16}
$$

The normalization accounts for the entropies of both X and Y, and ensures that values of $MI'(X,Y)$ span the unit interval. Returning to the example given in Equation (9.14), partitions Y and Z have different community structures, meaning that their respective entropies will not be the same. When we estimate their normalized mutual information with partition X, we incorporate their respective entropies into our calculations, and obtain two different values: $MI'(X, Y)=0.775$ and $MI'(X, Z)=0.697$. This result is consistent with our expectation that the different community structures of the partitions Y and Z should cause them to have a different degree of overlap with partition X.

The similarity measure $MI'(X,Y)$ will equal one if partitions X and Y are identical and will equal zero if they are completely independent. Variations of $MI'(X,Y)$ within these bounds are relative, and are not a direct measure of the "distance" between two partitions, in a metric sense. An alternative information theoretic measure that offers a more explicit metric of partition distance is called the variation of information (Meilă, 2007), and is defined as

$$VI(X,Y) = H(X) + H(Y) - 2MI(X,Y), \qquad (9.17)$$

or equivalently as $VI(X,Y) = H(X|Y) + H(Y|X)$. In other words, the variation of information is the sum of the information needed to describe X given Y, and the information needed to describe Y given X. As a measure of distance between partitions, low values indicate greater similarity. Specifically, a value of zero indicates that partitions X and Y are identical, whereas a maximal value of $\log N$ is obtained when the community assignments are as far apart as possible, which occurs when one partition places all nodes in a single community and the other partition assigns each node to a distinct community. The variation of information can be normalized by its maximum $\log N$ so that its values span the unit interval (Karrer et al., 2008; Rubinov and Sporns, 2011).

9.3.2 Comparing Populations of Partitions

It is often useful to compare more than two partitions, particularly when we want to understand differences in the community structure of two populations of networks. Suppose that we measure brain network connectivity in a group of patients and a group of healthy controls. How can we characterize differences in the community structure of the two sets of networks? At a global level, we could test for mean differences in the value of the modularity index, Q. However, as we have already noted, this analysis will only tell us about differences in the modularity of the networks and does not tell us anything about differences in the composition of the modules across the two groups.

One solution to this problem was proposed by Alexander-Bloch and colleagues (2012). These authors consider comparing modular decompositions between two groups of individuals, with the null hypothesis of equality in community assignments between groups. They start by estimating the normalized mutual information of partition similarity, as defined in Equation (9.15), for every pair of participants. They reason that a genuine difference between the two groups in community structure will result in a higher average pairwise similarity within groups compared to between groups. As such, the average within-group similarity observed in the data can be evaluated for statistical significance with respect to an empirical null distribution generated by computing the same quantity after permutation of group labels (**permutation testing** is discussed in more detail in Chapter 11).

Alexander-Bloch and colleagues (2012) suggested a similar approach to examine group differences in community assignments at the nodal level. Specifically, they defined, for each individual network and each node i, a binary vector comprising $N-1$ values, where each element j equals one if node j resides in the same community as i and zero otherwise. The correlation coefficient was then used to quantify the similarity of this vector for every pair of individual networks, separately for each node. Once again, we expect that a group difference will be apparent when the average correlation for a given node is higher within groups than between groups. The significance of any group differences can be tested separately for each node by comparing the average within-group correlation to an empirical null distribution of the same quantity generated under permutation of group membership.

Alexander-Bloch and colleagues (2012) used this approach to examine the community structure of resting-state functional connectivity networks, measured with functional MRI, in patients with childhood-onset schizophrenia and healthy controls. They found that patients showed reduced modularity (Figure 9.10a) but did not differ from controls with respect to the average number of modules identified across participants (Figure 9.10b). The normalized mutual information, $MI'(X, Y)$, was then computed for all possible pairs of individuals in the same clinical group. The within-group average of these pairwise measures was significantly greater than expected by chance (Figure 9.10c), suggesting that the global reduction of modularity found in patients was accompanied by a difference in the composition of the modules. Analysis at the nodal level suggested that the group differences in community structure were driven by changes in the module affiliation of regions in insula and sensorimotor cortex, as well as subcortical areas (Figure 9.10d).

9.3.3 Consensus Clustering

Some analyses of community structure may result in a large number of partitions. In some cases, we may obtain multiple partitions for the same network by running our community detection algorithm many times. Recall from Section 9.1.3 that many such algorithms use heuristics that can result in a different community structure from run to run. Related to this problem is the degeneracy of most modularity maximization algorithms, which means that a large number of alternative partitions of a network may be just as good as each other (Box 9.2; Good et al., 2010).

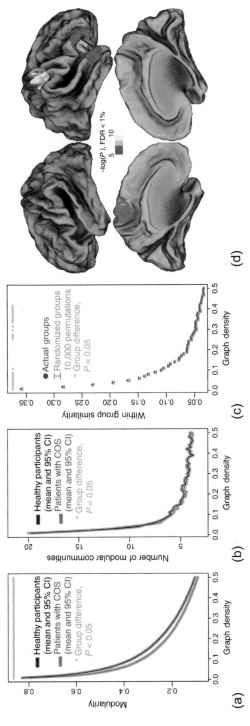

FIGURE 9.10 Differences in the community structure of brain networks in people with childhood-onset schizophrenia and healthy volunteers. **(a)** Group mean modularity scores, Q, in functional connectivity networks measured with resting-state functional MRI. The networks comprised 278 nodes defined using a random parcellation (see Chapter 2). The horizontal axis displays results obtained at different connection densities, generated after applying a range of thresholds to these correlation-based networks (see Chapter 11). Patients showed significant reductions in modularity across the full range of thresholds examined (green asterisks). **(b)** No significant group differences were observed in the average number of modules detected across participants. **(c)** The mean within-group partition similarity, measured with normalized mutual information, was significantly greater than expected by chance across several thresholds, suggesting a group difference in community structure. **(d)** Map of nodes showing group differences in community affiliation. Differences were localized to limbic, sensorimotor, and subcortical regions. *Images reproduced from Alexander-Bloch et al. (2012), with permission.*

Examining the consistency of each node's module affiliation across multiple partitions can reveal the core members of putative network modules. In general, we expect that core members of modules will be consistently assigned to the same community across partitions. We can exploit this consistency to generate a consensus partition that provides a representative summary of the community structure of the full set of individual partitions. Such a consensus partition allows us to identify a single, representative community structure for a network that is robust to the algorithmic degeneracy of community detection algorithms (Box 9.2). Consensus clustering is also useful when analyzing partitions obtained from a population of networks, such as when comparing a patient and a control group. In this case, we can generate a summary representation of the community structure of each group.

One way to examine the consistency and variability of module affiliation across a set of partitions is to construct an $N \times N$ coclassification matrix (Fornito et al., 2012a; Lancichinetti and Fortunato, 2012; Rubinov and Sporns, 2011; Sales-Pardo et al., 2007), D, which is sometimes also called a module allegiance matrix (Bassett et al., 2013a, 2015). Each element D_{ij} of this matrix represents the number of times nodes i and j have been classified in the same module across different runs of the algorithm. We can then analyze the community structure of this matrix to identify consensus modules. In other words, we perform community detection in two stages. In the first stage, we generate the partitions that will be used to construct the coclassification matrix. In the second stage, we run community detection on the coclassification matrix itself. Two nodes that are frequently coclassified across partitions in the first stage will have a high edge weight in the coclassification matrix, and are therefore more likely to be assigned to the same module by the second stage decomposition. The final result will be a consensus partition that is representative of the community structure of the individual partitions.

This consensus clustering approach has been used in studies of human functional connectivity networks to characterize degeneracy at the level of individual graphs and to generate a consensus partition across a sample of participants (Bassett et al., 2015; Dwyer et al., 2014; Fornito et al., 2012a; Rubinov and Sporns, 2011; van den Heuvel et al., 2008a). In one study, Dwyer et al. (2014) first used consensus clustering to generate a robust partition for each individual participant. They then used a similar method to generate representative community structures for different groups of participants. Specifically, they examined functional connectivity between 73 different regions-of-interest representing nodes of two large-scale networks, a cognitive control network commonly activated during the performance of attentionally demanding tasks, and the so-called default mode network, which characteristically shows reduced activation during such tasks (Buckner et al., 2008; Raichle et al., 2001; Shulman et al., 1997). These networks were of particular interest because they often interact antagonistically or competitively (Fox et al., 2005; Kelly et al., 2008; Sonuga-Barke and Castellanos, 2007).

To generate a robust partition for each participant, Dwyer et al. (2014) applied the Louvain algorithm 1000 times to each network (e.g., Figure 9.11a and b). For each person, they constructed a coclassification matrix that represented the fraction of runs in which each pair of nodes was assigned to the same community. The Louvain algorithm was then applied to this coclassification matrix to generate a consensus partition for each person (Figure 9.11c). To obtain a group-based representation of community structure, a second coclassification matrix was constructed using the consensus partition of each individual. A high edge weight in this matrix indicated that two nodes were placed in the same module across a large proportion of participants (Figure 9.11d and e). In fact, two separate group-level coclassification matrices were constructed: one for a subset of adolescents who performed well on a cognitive control task and the other for a subset who performed poorly on the task. The Louvain algorithm was applied separately to each matrix to obtain a consensus partition for each group. The community structure of good performers was dominated by two large modules corresponding to the control and default mode systems. This structure was consistently observed during rest and task performance, and was associated with strong segregation between the two modules (Figure 9.12a). In contrast, the community structure of poor performers contained a third, intermediary module that was interposed between the cognitive control and default mode modules (Figure 9.12b).

Coclassification matrices can also be used to examine the topological roles of different nodes with respect to network community structure. Dwyer et al. (2014) computed the within-module strength and diversity coefficient (Equation 9.9) of each node using the group-level coclassification matrix. A node with a high diversity coefficient in this matrix will have an approximately equal probability of being classified into different modules across partitions. Conversely, a node with high within-module strength will show a consistent pattern of module affiliation across networks. Dwyer and colleagues found that the nodes comprising the third intermediary module in the consensus partition of poor performers had the highest diversity coefficient, suggesting that these nodes are not consistently assigned to any single module across participants (Figure 9.12a and b). This result indicates that poor performance was associated with a failure to recruit the intermediary nodes into either the control or default mode modules, consistent with the hypothesis that functional segregation between these major modules supports optimal allocation of attentional resources (Sonuga-Barke and Castellanos, 2007).

In a similar analysis of task-related functional connectivity in healthy adults performing a memory task, Fornito et al. (2012a) found that the emergence of a third intermediary module bridging the default mode and control systems during task performance facilitated *better* performance (Figure 9.12c and d). This result is consistent with prior evidence that the memory task coactivates

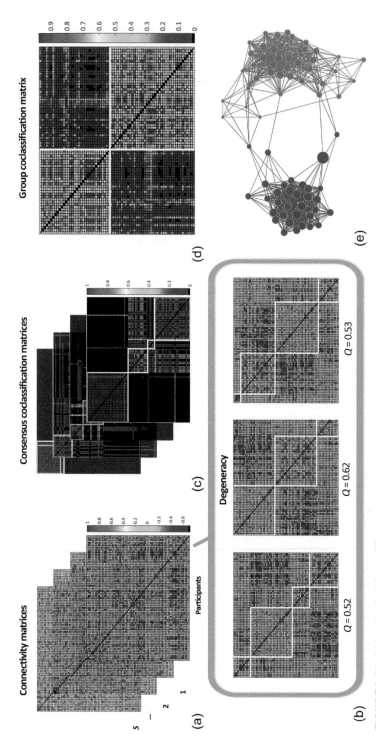

FIGURE 9.11 Addressing individual differences and the degeneracy of modularity maximization with consensus clustering. **(a)** We begin with a population of networks. Shown are human resting-state functional connectivity matrices measured with functional MRI, comprising 73 regions-of-interest sampling key areas of the default mode and cognitive control networks (for details, see Dwyer et al., 2014). We have a separate network for each of S participants. **(b)** For each individual, we run the community detection algorithm n times, yielding a set of n degenerate partitions. Here we see three example partitions, in which the network of a single person has been divided into four (left), three (middle), and four (right) modules, as delineated by the white boxes. Each partition is associated with a similar modularity score, Q. **(c)** For each person, we build a consensus coclassification matrix in which each element, D_{ji}, represents the fraction of the n partitions in which nodes i and j were coclassified into the same module. We then run community detection on this matrix to generate a consensus partition for each individual. In the example shown here, the matrix for the first participant is decomposed into four modules (white boxes). **(d)** Using the consensus community affiliations for each participant, we build a group-level coclassification matrix, in which the (i, j)th element represents the proportion of participants in which nodes i and j were coclassified into the same module, according to the consensus partition of each individual. Community detection is then performed on this matrix to yield a group-averaged representation of community structure. In this analysis, nodes that are frequently coclassified across individuals are likely to be assigned to the same module in the group representation. Here we see a clear division of the network into two large modules. **(e)** Force-directed projection of the group coclassification matrix showing a clear segregation between the two modules, which largely corresponds to the default mode (blue) and cognitive control (red) systems. Node size is proportional to node strength.

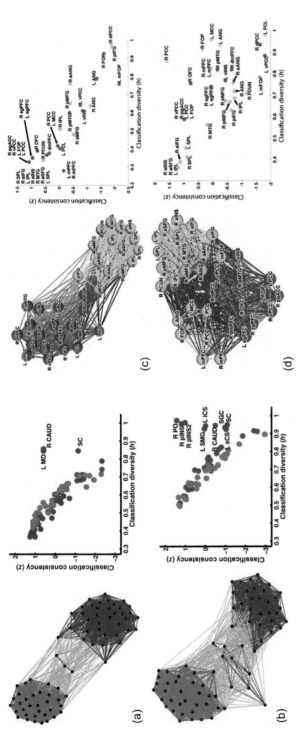

FIGURE 9.12 Characterizing modular architecture using consensus partitions. **(a)** An analysis of functional connectivity networks in healthy adolescents who showed good performance on a cognitive control task identified two large communities corresponding to a cognitive control network (red) and default mode network (blue). Left shows a force-directed projection of the group-level coclassification matrix, in which edges represent the likelihood that two nodes belong in the same module across participants. A strong segregation between the two large communities is evident. Right panel shows node roles, based on the within-module strength and diversity coefficient computed from the group coclassification matrix to yield measures of module classification consistency and diversity, respectively. These networks were mapped during performance of the cognitive task. **(b)** The same plots generated for a group of adolescents performing poorly on the same cognitive task. Here we see the emergence of a third, intermediary module interposed between the control and default mode systems (left panel). The classification diversity of the nodes in this third module was very high (right panel), indicating that they were not clearly integrated into one of the larger modules across participants. **(c)** The results of a similar analysis in adults performing a contextual recollection task. Analysis of task-unrelated functional connectivity, reflecting an index of spontaneous or intrinsic network organization, revealed two modules that corresponded to a frontoparietal control system (green) and a default mode module (purple). **(d)** Analysis of task-related interactions showed a reorganization of the network into three modules, such that the default mode system split into a core (purple) and intermediary module (cyan). Nodes in the intermediary module had higher classification diversity than other nodes, suggesting that they promoted functional integration. In this task context, greater integration between these systems was associated with better task performance. *Reproduced (a,b) from Dwyer et al. (2014) and (c,d) from Fornito et al. (2012a), with permission.*

regions of both networks (Simons et al., 2008), and highlights the context-dependent nature of interactions between large-scale functional communities in the brain: in certain contexts, strong segregation of communities supports better performance; in other contexts strong integration may be optimal, consistent with the emergence of a global workspace configuration (Figure 1.10).

The degeneracy of many community detection algorithms (Box 9.2) means that we could, in principle, iterate the consensus clustering procedure indefinitely. For example, after generating a consensus coclassification matrix from n iterations of the algorithm on a single dataset, we could run the algorithm another n times on the consensus matrix to generate a second consensus matrix, and so on, with no guarantee that the results will converge. To address this problem, Lancichinetti and Fortunato (2012) proposed a method for converging to a single final, consensus partition. There are five key steps to their approach:

1. Run community detection n times on the initial adjacency matrix to yield n partitions.
2. Compute the consensus coclassification matrix, D, where each element is the fraction of n times that a given pair of nodes has been assigned to the same module.
3. Apply a threshold, τ, to the matrix D, such that all elements D_{ij} with a value less than τ are set to zero.
4. Run community detection on the thresholded matrix n times to yield another n partitions.
5. Repeat steps 2-4 until the consensus matrix has a block-diagonal structure in which all edge weights equal one for node pairs in the same community and zero otherwise (i.e., all n partitions are identical).

The threshold τ ensures that the algorithm will converge to a final, consensus partition. Lancichinetti and Fortunato (2012) demonstrated that this consensus approach is able to more accurately recover the community structure of synthetic benchmark networks, when compared to simply choosing the partition with the highest modularity from multiple runs of the algorithm. They also found that around $n = 50$ runs of community detection were sufficient to yield stable results for their benchmark networks. The specific value of τ is an important choice. A high value of τ may lead to more rapid convergence but a less robust partition. Lancichinetti and Fortunato found that the optimal value of this parameter varied depending on the community detection method that was used. They found good results with the Louvain algorithm for values of τ less than about 0.40, but it is unclear how this choice depends on the precise topology of the network being analyzed.

9.4   DYNAMIC MODULARITY

We have so far considered methods for characterizing the static community structure of a network. However, the brain is not a static system. As discussed in Chapter 3, both structural and functional connectivity evolve dynamically on temporal scales ranging from milliseconds to years. In particular, we should expect functional connectivity networks to show pronounced changes in community structure over short timescales, as cell assemblies rapidly synchronize and desynchronize their activity in response to fluctuations in endogenous states and exogenous stimuli (Varela et al., 2001). Appropriate characterization of these dynamic changes requires an extension of traditional techniques for analyzing network modularity to adequately account for the temporal dimension.

9.4.1 Multilayer Modularity

A simple approach to examine how modules evolve over time is to split our data into distinct temporal windows, each of which represents network organization at a different point in time. We can then perform a static modularity analysis within each window separately, and compare results across windows. A limitation of this method is that it ignores dependencies between windows. These dependencies are likely to be an important factor in the temporal evolution of the modules of many real-world systems, since we can generally expect that the community structure of a network at one point in time will be related to its community structure in adjacent time points.

A more elegant solution to the analysis of dynamically evolving community structure is to incorporate dependencies across time into the community detection procedure. To this end, Mucha and colleagues (2010) suggested a method in which each window is treated as a different layer of a *multilayer network*, and the goal of the analysis is to find an optimum set of partitions across layers that maximizes a multilayer generalization of the modularity index, Q_{ML}. Here, we focus on the logic and application of this method to the analysis of dynamic brain networks. An alternative approach to dynamic community detection has been proposed by Palla and coworkers (2007). A thorough review of dynamic and other types of multilayer networks is provided by Kivelä and coworkers (2014).

To understand the multilayer community detection approach of Mucha and colleagues (2010), suppose that we have performed a **sliding window analysis** (Box 2.2) of functional connectivity data such that we have constructed a series of binary, undirected adjacency matrices that represent how the network evolves in consecutive time steps. We regard these matrices as different layers (also called slices) of a multilayer network, with each layer, l, representing a different point in time. At a given layer, l, the connectivity between nodes i and j is specified by the elements of the adjacency matrix, A_{ijl}. Between layers, the coupling of node j in layer l with itself in a different layer r is determined by the interlayer coupling parameter C_{jlr}. In networks with ordered layers, such as those evolving through time, interlayer coupling is generally specified only

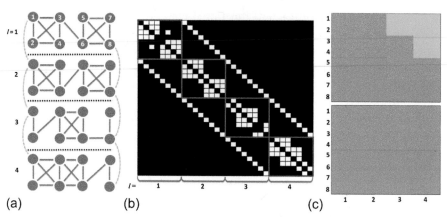

FIGURE 9.13 Quantifying dynamic modularity using multilayer networks. **(a)** Schematic structure for a simple binary network comprising eight nodes and four layers. Broken black lines delineate each layer. Within each layer, l, nodes are connected according to a particular topology. Each node is also coupled to itself in other layers (broken gray lines). In this example, only the interlayer links of two nodes (top left and bottom right) are shown, but interlayer links are defined for all nodes. The example here is for a single network examined at four different time points, but layers can represent any arbitrary set of suitable networks. For networks with ordered layers, such as temporal networks, it is common for a node only to be coupled to itself in adjacent layers, but it is also possible to define interlayer links for all pairs of layers. **(b)** We can represent this multilayer network with a single adjacency matrix. The red lines delineate the layers of the network. Nonzero elements outside the red boxes represent interlayer links between a node and itself in different layers. This matrix can be used as a basis for community detection using an algorithm that maximizes Q_{ML}. From this representation, we can see that the strength of the interlayer coupling parameter will determine the consistency of community assignments across layers. If the value of this parameter is large, nodes will be strongly coupled with themselves across layers and are therefore more likely to be assigned to the same module. **(c)** Module assignments generated for each layer using interlayer couplings $C_{jlr} = 0.1$ (top) and $C_{jlr} = 1$ (bottom). Each row is a node and each column is a layer. For both values of the interlayer coupling parameter, the first two layers have the same community structure. Panel **(a)** shows that there are clearly two modules in the first two layers, one comprising nodes 1, 2, 3, and 4 and the other comprising nodes 5, 6, 7, and 8. In layer 3, we see a major shift in network topology, such that nodes 3, 4, 5, and 6 become a maximally connected clique, and the remaining four nodes have relatively sparse connectivity. We should therefore expect that nodes 3, 4, 5, and 6 form their own module, nodes 1 and 2 form a second module, and nodes 7 and 8 form a third module. When $C_{jlr} = 0.1$, we are able to recover this intuitive partition, but when $C_{jlr} = 1$, the interlayer coupling is too strong and the community structure remains consistent across layers. A similar contrast is evident in layer 4.

for adjacent layers, but in principle we can specify interlayer links for all pairs of layers (Figure 9.13a). Note that distinct nodes are not coupled between layers—they are only coupled within layers according to a topology determined by each layer-specific adjacency matrix.

The layer-specific adjacency matrices can be combined to generate a single, multilayer matrix (Figure 9.13b). This matrix can be used as a basis for community detection, with the goal of optimizing a multilayer generalization of the modularity index that considers the connection topology *within* each network layer and the coupling *between* layers. Formally, this index is defined as

$$Q_{ML} = \frac{1}{2\mu} \sum_{ijlr} \left[(A_{ijl} - \gamma_l e_{ijl}) \delta_{lr} + \delta_{ij} C_{jlr} \right] \delta(m_{il}, m_{jr}), \qquad (9.18)$$

where $e_{ijl} = \dfrac{k_{il} k_{jl}}{2E_l}$, and k_{il} and k_{jl} are the degrees of nodes i and j, respectively, in layer l. Note that e_{ijl} is the same null model for intramodule connectivity that is specified in Equation (9.2), estimated separately for each layer l of the multilayer network.

To understand Equation (9.18), first consider the sum. In principle, this sum is computed for every pair of nodes, i and j, and for every pair of layers l and r. In practice, we only need to evaluate the summand when the indicator function $\delta(m_{il}, m_{jr})$ equals one, which is when node i in layer l is in the same community as node j in layer r. In all other cases $\delta(m_{il}, m_{jr})$ equals zero, which means that Q_{ML} also reduces to zero. This condition ensures that we only consider pairs of nodes that are in the same community, regardless of the layer to which they belong. When $l = r$, this condition is the same as the condition enforced by the indicator function in the definition of single-layer modularity given in Equation (9.3).

The term $(A_{ijl} - \gamma_l e_{ijl}) \delta_{lr}$ in Equation (9.18) considers the coupling *within* a layer and is similar to the quality function used to evaluate single-layer modularity (Equation 9.3), albeit with two important differences. First, the indicator function δ_{lr} equals one when $l = r$ and zero otherwise. This condition ensures that the difference between the observed and chance-expected intramodular connectivity, $(A_{ijl} - \gamma_l e_{ijl})$, is only calculated within a single layer. The second distinctive feature of the sum in Equation (9.18) is the inclusion of the resolution parameter, γ_l. This parameter can be used to tune the scale of the partition at layer l. A small value will reduce the influence of the null model, meaning that more connections are considered "surprising" with respect to the model, and that the final partition will comprise larger communities. A large value of γ will increase the influence of the null model and result in a partition with smaller communities. Setting γ to unity is equivalent to performing a traditional, single-scale modularity analysis. The parameter γ can be used to overcome the resolution limit of modularity maximization, and to investigate community structure across different topological scales (Box 9.2; see also Mucha et al., 2010).

The term $\delta_{ij} C_{jlr}$ in Equation (9.18) specifies the coupling *between* layers. The parameter C_{jlr} is the interlayer coupling between layers l and r for node j, and the indicator function δ_{ij} equals one if node $i=j$ and zero otherwise. This function ensures that we only consider how each node links to itself between layers and that we do not consider how one node couples to a different node across layers. The interlayer coupling parameter, C_{jlr}, is often set to either zero or an arbitrary value. Setting C_{jlr} to zero means that there is no coupling between layers. Each layer is thus independent, and maximizing Q_{ML} is equivalent to maximizing Q in each layer separately. When C_{jlr} is large and positive, nodes are strongly coupled between layers, and the optimization of Q_{ML} favors a consistent community structure across layers. A small, positive value of C_{jlr} allows for flexibility in community structure from layer to layer, and means that the community assignment of each node within a layer will be more strongly influenced by the connection topology of that layer (Figure 9.13b and c). When $l=r$, the value of C_{jlr} is zero to avoid a node connecting to itself (self-loop) within a given layer.

Another difference between Equations (9.3) and (9.18) is the term used to normalize the sum. In Equation (9.3), we normalize the sum by $2E$ since we count each edge twice in an undirected network. In Equation (9.18), we must consider both the number of edges in each layer and the number of edges between layers. Therefore, we have $2\mu = \sum_{jr} \kappa_{jr}$, where the multilayer degree, κ_{jr}, is the degree of node j in layer r, plus the sum of the interlayer coupling values that link node j to itself in all other layers; that is, $\kappa_{jl} = k_{jl} + c_{jl}$ and $c_{jl} = \sum_{r} C_{jlr}$. Note that in a single, undirected network, $2E = \sum_{i=1}^{N} k_i$. We can therefore think of μ as a multilayer generalization of E that counts links within and between layers.

To summarize, the multilayer modularity index, Q_{ML} (Equation 9.18), mandates that a good quality partition will show (1) a large difference between the null model and the observed degree of intramodule connectivity within each layer (consistent with the traditional definition of single-layer modularity); and (2) a consistent assignment of nodes into the same community across layers, conditioned on the precise value of the interlayer coupling parameter, C_{jlr}. In this context, C_{jlr} acts like a temporal resolution parameter that weights the importance of consistency across layers to the final Q_{ML} score (Figure 9.13c). Ideally, Q_{ML} should be optimized over different combinations of the temporal and topological resolution parameters C_{jlr} and γ. Mucha and colleagues (2010) also provide generalizations of Q_{ML} for directed, weighted, and signed networks.

It is possible to find a partition that maximizes the index Q_{ML} after adapting any one of the various community detection algorithms available to deal with multilayer networks. An adaptation of the Louvain algorithm has proven useful for this purpose (Bassett et al., 2011, 2015; Mucha et al., 2010). The output is a single community affiliation vector for all nodes across all layers, which can then be separated to examine how community structure evolves across layers

(Figure 9.13c). Details on the implementation of the algorithm for multilayer networks are provided by Jutla et al. (2011).

As with single-layer modularity (Section 9.1.2), it is important to evaluate multilayer community structure relative to an appropriate ensemble of randomized surrogate networks. Three categories of surrogate networks have been proposed for this purpose (Bassett et al., 2013a): a "connectional" surrogate, which is generated by randomizing the topology of each network layer independently; a "nodal" surrogate, which is generated by randomizing the links between layers; and a "temporal" surrogate, generated by preserving network topology but shuffling the order of the layers. Each of these surrogates can be used to test distinct hypotheses. The connectional surrogate allows us to test whether the network shows modular topology, the nodal surrogate allows us to investigate the consistency of a node's community structure across layers, and the temporal surrogate allows us to investigate whether community structure is stable or dynamic over time. Bassett et al. (2013a) give a detailed discussion of these and other possible null models. They also consider consensus clustering methods for dealing with the many degenerate solutions that can arise when maximizing Q_{ML} (see Section 9.3.3 and Box 9.2).

9.4.2 Dynamic Modularity of Brain Networks

Bassett and colleagues have used the multilayer modularity framework to examine how the community structure of human functional connectivity networks evolves over the course of a 42-day motor skills training program (Bassett et al., 2011, 2015). In one study (Bassett et al., 2015), participants were scanned with functional MRI at four different times: before training, early in training, midway through the training, and near the end of training. Within each scanning session, participants completed finger movement sequences for which they had received minimal, moderate, or extensive training. The authors first examined the consistency of community structure across participants, testing sessions, and training conditions, by constructing an $N \times N$ coclassification matrix, in which each element represented the probability of two nodes being classified in the same community. Analysis of this matrix identified two modules that were consistently defined across networks: one comprising primary visual cortex and the other consisting of primary and secondary sensorimotor areas (Figure 9.14a). Repeating the analysis separately for each scanning session revealed two important trends as training on the task progressed. First, the visual and sensorimotor modules became increasingly segregated over time (Figure 9.14b and c). Second, the integration of nodes outside the visual and sensorimotor modules, particularly cortical association and subcortical areas, was reduced over time (Figure 9.14b). Moreover, individuals who showed a

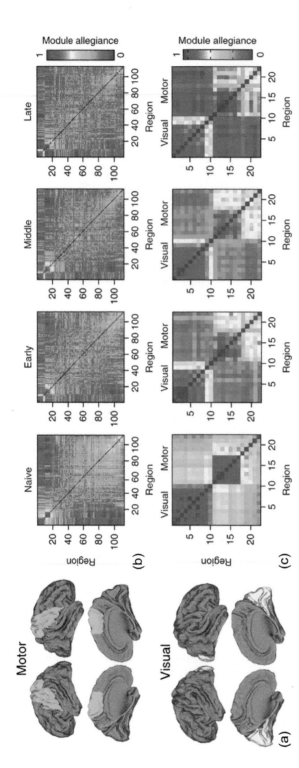

FIGURE 9.14 Dynamic modularity and motor learning. **(a)** A functional MRI study of human functional connectivity networks measured during a motor learning task identified two modules consistently across all of 20 participants, four different scanning sessions, and three different training manipulations. Shown here is the anatomical distribution of the two modules. One was localized to sensorimotor cortex (green) and the other to visual cortex (yellow). **(b)** Module allegiance matrices constructed separately for networks measured before task training had commenced (naïve), and then at early, middle, and later stages of the training program. Each element in the matrices represents the probability that two nodes have been coclassified into the same module. The matrices have been reordered to emphasize community structure. The visual and sensorimotor modules can be seen in the top left corner of the matrices. The allegiance of nodes outside these two modules decreases over time. **(c)** Zoomed in portion of the allegiance matrices presented in panel **(b)** to focus on the sensorimotor and visual modules. Allegiance between the two modules decreases over time, consistent with an increased segregation of these functional systems as training progresses. *Images reproduced from Bassett et al. (2015), with permission.*

larger reduction in the integration of these noncore nodes as a function of training showed a faster rate of learning on the task.

Together, these results suggest that early phases of training are associated with a high degree of integration across modules, as participants must process new and diverse types of information to adequately perform a novel task. As training improves proficiency, many of these processes can be automated and their corresponding neural substrates show greater specialization. This conclusion is supported by a separate study by the same group reporting that nodes are more likely to change their module affiliation early in training (Bassett et al., 2011). This apparent flexibility in community affiliation decreased as training progressed, and greater flexibility in one testing session predicted learning performance in the next testing session. Collectively, these findings suggest that novel tasks evoke a high degree of integration between different neural subsystems and that training induces a gradual segregation and specialization of nodes into distinct clusters, consistent with psychological characterizations of modules as supporting fast and automated processing (Fodor, 1983).

9.5 SUMMARY

Modularity is a ubiquitous property of many real-world networks. In nervous systems, strong connectivity within modules supports functional specialization whereas sparse links between modules support functional integration. Modules are organized hierarchically, such that any given module may contain submodules, and sub-submodules within submodules, and so on across several topological scales.

The degree to which a network can be decomposed into tightly interconnected modules can be measured with the modularity index, a simple measure that quantifies the difference between the intramodule connectivity of an observed network and a corresponding null model. Many methods are available for finding a partition of a network that is associated with near-maximal modularity. These methods are both powerful and flexible, but they often show a resolution limit, such that they are biased against finding small modules. They also suffer a problem of severe degeneracy, by which many alternative partitions can be associated with similar, near-maximal levels of modularity, suggesting that no single partition is best. Examining the consistency and variability of nodal module affiliation across a set of partitions is useful for dealing with the degeneracy problem, and for generating representative summaries of community structure across a population of networks. It is also important to evaluate the modularity of a network with respect to appropriate surrogate networks, as sampling fluctuations can cause high modularity scores even in nonmodular networks such as Erdös-Rényi graphs.

Given a decomposition of a network into modules, the topological roles of individual nodes can be characterized in terms of how well connected they are within their own module, and with other modules. Nodes with high intra-module connectivity act as provincial hubs and are thought to play an important role in functional specialization. Nodes with a relatively even distribution of connections across modules act as connector nodes that integrate disparate communities. Some nodes with a homogeneous distribution of links across modules are not easy to classify into a single module. Their role may be better characterized with community detection algorithms that allow for nodes to belong to more than one module.

The community structure of brain networks is not static, and evolves over both short and long timescales. Methods that incorporate temporal dependencies into the community detection procedure allow a principled analysis of the community structure of dynamic networks.

Modularity analysis is a diverse and rapidly evolving field, and further improvements in community detection algorithms will no doubt prove useful for neuroscience. A recurrent theme throughout this and other chapters of this book has been the importance of evaluating observed measures of network topology in comparison to appropriate null models. In the next chapter, we turn to an in-depth discussion of the rationale, construction, and application of these models to the analysis of brain networks.

CHAPTER 10

Null Models

Claiming that a brain network has a **clustering coefficient** of 0.4, for example, or an average **path length** of 4, is not particularly informative in the absence of any further information. Is an average path length of 4 unusually long or short, or is it typical? Since these and most other measures of network organization scale with the number of **nodes** and **edges** in a graph, the answer to this question depends on the size and **connection density**, as well as other basic properties of the underlying network. For small, densely connected networks, a path length of 4 may be considered unusually long, but for larger, more sparsely connected networks, it may be exceptionally short. To determine whether a path length of 4 is long or short, we can compare it to the path length in a random network that is matched for size, connection density, and other relevant properties. If the path length in such a random network is 8, for example, we can claim that the brain network has a relatively short path length; namely, half that of a random network.

What we have described above is a simple example of normalization, where a random network served as the benchmark, or **null model**. In **graph theory**, a null model is a model of a network that establishes expectations under the null hypothesis. In the path length example above, comparison of the observed path length in an empirical network to the path length of an appropriately matched random network allows us to determine whether the observed path length is longer or shorter than expected by chance. In other words, normalization with respect to a null model provides a benchmark to quantify the extent to which a network property deviates from what would be expected by chance alone. Many of the graph measures considered in this book require some form of normalization or benchmarking relative to a null model to enable meaningful conclusions to be drawn about network organization. Normalization is therefore an important step in making valid inference.

Normalization can also reveal differences *between* networks. To demonstrate this, Figure 10.1 shows the average path length and average clustering coefficient of each of 30 different networks generated using the **Watts-Strogatz model**. Recall from Chapter 1 that this model begins with a regular **lattice** and randomly rewires edges with a fixed probability. In Figure 10.1, the rewiring probability varies between 0.1 and 0.5 across the networks. As we will see

Fundamentals of Brain Network Analysis. http://dx.doi.org/10.1016/B978-0-12-407908-3.00010-8

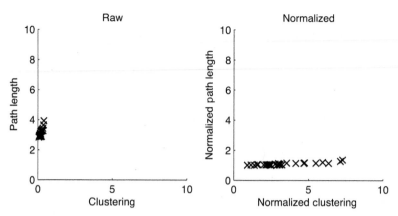

FIGURE 10.1 Normalization can reveal differences between networks. Each cross represents a network. Without normalization (left), the 30 networks appear quite similar in terms of their average clustering coefficient and path length. If anything, they appear to differ in path length, since the cluster of crosses is elongated along the vertical axis. However, normalization (right) reveals a strong difference in clustering among the 30 networks. Normalization was performed with Maslov-Sneppen rewiring (see Section 10.2.1). Each data point on the right was computed by dividing the corresponding data point on the left by the average value of the clustering coefficient (horizontal axis) or path length (vertical axis) in an ensemble of random networks.

later in this chapter, increasing the rewiring probability in the Watts-Strogatz model reduces the average clustering coefficient. Without normalization, the 30 networks appear quite similar in terms of both clustering and path length. It is only after we normalize these measures with respect to appropriate random networks that we see a strong difference in clustering among the 30 networks. This is because the extent of variation in clustering across the 30 networks is small relative to the extent of variation in path length. However, relative to appropriately matched random networks, the variation in clustering is considerable.

This chapter considers some of the different kinds of null networks that are available for normalization and the specific properties of a null network that should be matched in the analysis of brain networks. Ideally, the null network should be matched to our observed brain network for all uninteresting properties, leaving only the properties of interest to differ. Otherwise, a trivial, unmatched property can influence the measure we seek to normalize, and thus confound any subsequent analyses. For example, the path length of an empirical network may appear unusually low when it is normalized with respect to a null network that is not matched in terms of connection density, since path length is strongly influenced by density. In this case, we cannot be certain if the low path length in the empirical network is a distinctive property of that network, or merely a consequence of the low connection density of the network relative to the null.

Generating null networks that match all properties of an empirical network except for the one we seek to normalize is difficult in practice. It is therefore typical to match only basic properties, such as network size, connection density, and **degree distribution**. We will see that these kinds of null networks can be computed using either **generative models**, where a network is grown by adding connections to a set of nodes, or randomization strategies, where the observed network is randomly rewired according to a set of rules.

10.1 GENERATIVE NULL MODELS

The abstract nature of graphs allows us to grow networks in silico by adding nodes and edges according to specific rules. These rules represent a generative model of the network—a model of the processes thought to be important in shaping the **topology** of that network. We can compare the properties of the modeled network with those of an observed brain network. Successful replication by the model of the properties observed in the brain network suggests that the rules defining the model may play an important role in brain network development. In other words, they represent putative mechanisms involved in generating the observed data.

One of the simplest hypotheses that could explain the topology of an observed network is that it has arisen purely by chance. In this case, we could create a model network with the same number of nodes and edges, but this time with edges drawn between nodes completely at random (a so-called **Erdős-Rényi graph**, discussed in more detail in the following section). We would then measure key properties of this network, such as its degree distribution, clustering coefficient, path length, and so on, and compare these values to those computed in the empirical brain network. If the two sets of measures match, our analysis suggests that the brain is wired completely at random.

Another way of considering generative models is that they define expectations for network organization under a specific hypothesis. The hypothesis is determined by the rules of the generative model. In our concrete example above, the rule and corresponding hypothesis is random connectivity, although as we shall see, other rules are also possible. Viewing generative models from this perspective allows us to define two major applications in the analysis of brain networks. The first is to uncover critical constraints on brain network wiring. In this application, we aim to specify a model that will accurately reproduce the topological properties observed in the brain. Such an approach provides a powerful framework for testing competing hypotheses about the most important constraints on the development, aging, and disease-related degeneration of brain networks. For instance, if we find that the Erdős-Rényi model provides a poor fit to the wiring patterns seen in brain networks, we should think about other possible models with additional rules that better account for the data (see Box 10.1).

As mentioned above, the second application is to set expectations under the null hypothesis. Rather than attempting to model all relevant features of brain network organization, we may simply ask whether a particular property is expressed in the brain to a greater extent than expected by chance. In this case, we can generate a large ensemble of random networks to sample the distribution of network properties under the null hypothesis. If the value of a specific property in the observed network is highly unlikely under this null distribution, we can conclude that this property is due to a mechanism that is not accounted for by the null model, and thus we can reject the null hypothesis. These null networks can also be used to normalize observed measures, allowing comparison across different datasets. Critically, the validity of this approach depends on the appropriateness of the null model. In this section, we consider some of the canonical generative null models used in the analysis of complex networks.

BOX 10.1 GROWTH CONNECTOMICS: GENERATIVE MODELS FOR BRAIN NETWORKS

Generative models aim to build a network according to specific rules that are thought to be important in explaining a particular aspect of the system. Most generative models for brain networks keep node positions fixed, as determined by their actual anatomical location in the brain network. They then add edges according to some penalty placed on the formation of long-range connections or, in some cases, a trade-off between this **cost** and some other topological parameter (Vértes et al., 2012; Chen et al., 2013). The penalty is designed to capture the wiring cost of each edge (see also Chapter 8).

How can we specify a generative model according to such rules? In the simplest case, we can add a distance penalty to an Erdős-Rényi graph (Section 10.1.1). Instead of treating all pairs of nodes uniformly when deciding where to insert connections, we now favor adding connections between node pairs that are spatially proximal. In particular, the probability of adding a connection between nodes i and j is,

$$p_{ij} \propto e^{-\eta \mathcal{D}_{ij}}, \qquad (10.1.1)$$

where $\mathcal{D}_{ij}$ is the physical distance between nodes i and j, and η is a parameter to be fitted. As η is increased, the distance penalty becomes more severe and connections are more likely to be placed between pairs of nodes that are anatomically closer. When $\eta = 0$, we have the usual Erdős-Rényi model. To generate a network according to this model, we follow the same iterative process described in Section 10.1.1, where a new connection is added at each iteration based on p_{ij} (note that we never add more than one connection to the same pair of regions). We renormalize p_{ij} at each iteration according to $p_{ij} \leftarrow p_{ij}/\sum_{uv} p_{uv}$, where $u = 1, \ldots, N$ and $v = 1, \ldots, N$ index the set of all nodes in the network.

The distance penalty in this model is exponential, based on empirical evidence for a negative exponential decay in the probability of finding connections with increasing spatial distance (Fulcher and Fornito, 2016; Roberts et al., 2015). Indeed, Ercsey-Ravasz et al. (2013) found that a model wired according to such an exponential distance rule (EDR) was able to reproduce many key features of a dense, 29-region interareal directed macaque **connectome** reconstructed with axonal **tract-tracing**, including observed asymmetries in the proportion of unidirectional connections, its three-node **motif** frequency spectrum, its **eigenvalue** spectrum, and its core structure. Similar results have been reported for the analysis

BOX 10.1 GROWTH CONNECTOMICS: GENERATIVE MODELS FOR BRAIN NETWORKS—CONT'D

of human **structural connectivity** networks measured with **diffusion MRI** (Henderson and Robinson, 2013, 2014).

Song and colleagues (2014) proposed a more sophisticated variant of the EDR model that attempts to accurately capture axonal growth processes. The cranium is modeled as a spheroid and partitioned into an arbitrary set of N nodes. Axons are then randomly seeded throughout the spheroid and grown along a linear trajectory for a length that is sampled from an exponential distribution. The target of each axon is determined by the sum of distance-dependent attractive forces exerted by the centroids of all other nodes. To generate a weighted connectivity **matrix**, thousands of axons can be grown according to this rule and uniquely assigned to a pair of volumes. Each element of the model's directed connectivity matrix therefore enumerates the total number of axons initiated from one volume and landing in another. This model successfully reproduced several features of the 29-region macaque connectome (Markov et al., 2014), such as the distance-dependence of edge weights, in- and out-degree distributions, nodal clustering coefficient distribution, and its three-node motif frequency spectrum (Figure 10.1.1a–c). It was also successful in reproducing the motif frequency spectrum of a 49-region cortical mouse connectome (Zingg et al., 2014).

Incorporating knowledge of the timing of critical developmental events in brain networks can also facilitate generative modeling. Nicosia and colleagues (2013) observed that the relationship between the number of neurons, N, and the number of synapses, E, in C. elegans undergoes a phase transition around the time of hatching, with E growing roughly as N^2 before hatching and slowing to a linear increase after hatching. They found that this transition could be captured by a model that dynamically negotiates a trade-off between a penalty on forming long-range connections and the **preferential attachment** of new neurons to high-**degree** nodes. In this model, elongation of the worm with development increases the distance between neurons, which in turn increases the penalty on forming long-range synapses. The model also reproduced several topological properties of the empirical network, include nodal distributions for degree,

connection distance, and nodal **efficiency**, as well as the pattern of intraganglia and interganglia connections (Figure 10.1.1d–f). Models based on simpler, static distance-dependent and preferential attachment rules offered poorer fits to the data.

Vértes and colleagues (2012) found that simple distance-based or preferential attachment models were not sufficient to account for variations in key topological properties of human **functional MRI** networks. Instead, they found that a model incorporating a penalty on long-range connectivity and a bias towards forming clustered connections was best able to reproduce the degree distribution, clustering, and global efficiency of human **functional connectivity** networks (Figure 10.1.1g and h). Formally, this so-called *economical clustering model* is given by

$$p_{ij} \propto k_{ij}^{\gamma} \mathcal{D}_{ij}^{-\eta}, \qquad (10.1.2)$$

where k_{ij} now denotes the number of common neighboring nodes to which *both* nodes i and j are connected. This model favors new connections that are likely to result in triangles, but this favoritism is tempered by wiring length. When k_{ij} is large, nodes i and j share many common **neighbors**, and thus adding a connection between them is likely to complete the formation of many triangles. However, adding this connection is a tradeoff with the increase in wiring length. The choice of γ and η dictates the balance between this tradeoff.

Generating networks with a prescribed number of nodes and connection density is straightforward, but how should the model parameters γ and η be selected to match more complex topological properties of the empirical network? Vértes and colleagues (2012) used **simulated annealing** to optimize γ and η to generate networks most closely matched to functional brain networks in terms of specific topological properties. Simulated annealing is a heuristic approach that is more efficient than exhaustively searching across all parameter combinations. Collectively, this modeling work has underscored the important role of spatial constraints in determining brain network organization (Chapter 8), while also highlighting a role for additional constraints in shaping the topology of neural systems.

Continued

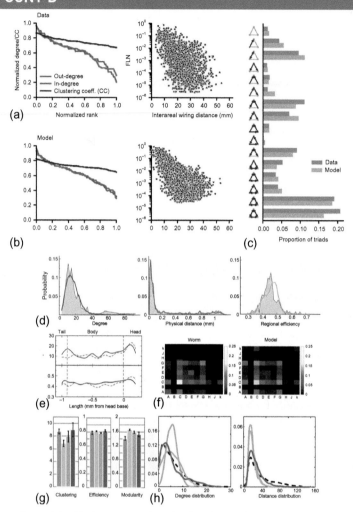

FIGURE 10.1.1 Generative models of brain networks. **(a and b)** An analysis of a 29-region cortical connectome of the macaque constructed with axonal tract-tracing. A model of axonal growth that uses an exponential decay rule for connectivity reproduces the nodal distributions of **out-degree**, **in-degree**, and clustering coefficient (left panel), as well as the dependence of connection weight (estimated as the fractional of labeled neurons, or FLN) on interregional spatial distance (right panel). The empirical data are shown in panel **(a)** and the model estimates in **(b)**. **(c)** The same model also reproduces the three-node motif spectrum of the macaque cortical network. **(d)** A preferential attachment model of the development of the *C. elegans* nervous system that incorporates a dynamic penalty on the formation of long-range connections can reproduce nodal distributions of degree (left), connection distance (middle), and nodal efficiency (right). The empirical data are shown in gray and the model as thick lines. **(e)** The same model predicted with reasonable accuracy the anatomical distribution of node degree (top) and node efficiency (bottom) across the length of the body of the nematode (black lines correspond to the data, red lines to the model). **(f)** The model also reproduces the pattern of connectivity between neurons in the same ganglia and between neurons in different ganglia (letters denote different ganglia). **(g)** A simple exponential decay rule (green) does not accurately reproduce relevant features of human functional connectivity networks measured with functional MRI (blue), nor does a preferential attachment model that incorporates a penalty on forming long-range connections (yellow). Instead, an economical clustering model (red; see Equation 10.1.2) is able to reproduce variations of mean clustering (left), global efficiency (middle), and **modularity** (right) in these data. **(h)** The economical clustering model provides a good fit to the degree distribution (left) and connection distance distribution (right) of the empirical data (dashed black line). *(a–c) Reproduced from Song et al. (2014), (d–f) from Nicosia et al. (2013), and (g and h) from Vértes et al. (2012) with permission.*

10.1.1   Erdős-Rényi Networks

The concept of randomness is central to any discussion of null networks. What do we really mean by *random* in this context? This question was first formalized by Erdős and Rényi (1959), who defined a generative model for the simplest random graph, the so-called Erdős-Rényi network. In this model, each node has a uniform probability of being connected. Here, we will consider a slight variation of the Erdős-Rényi model that was described by Gilbert (1959), as Gilbert's variation is tacitly assumed in most of the literature today. In Gilbert's variation, Erdős-Rényi networks are characterized by two parameters, the number of nodes in the network, N, and the probability, p, that a connection is found between an arbitrary pair of nodes. To construct an undirected Erdős-Rényi network with parameters (N, p), we create N nodes and then independently generate a random number between zero and one for each of the $m = N(N-1)/2$ possible pairs of nodes. If the random number that we generate for a given node pair is less than p, a connection is drawn between that pair of nodes, otherwise it remains disconnected. By construction, the number of connections in the network is binomially distributed, and is given by

$$P(\text{number of edges} = E) = \binom{m}{E} p^E (1-p)^{m-E}, \tag{10.1}$$

where $0 \le E \le m$ (see also Chapter 4).

Equivalently, we can understand Equation (10.1) to mean that all networks with N nodes and E connections have equal probability of being created . There are 2^m possible distinct networks that form the set of all Erdős-Rényi networks. This is an enormous set. For example, when $N = 10$ nodes, we have $m = 45$ possible pairs of nodes, and thus $2^m \approx 10^{13}$. For $N = 100$ nodes, the set of all networks is too large to enumerate. It is important to remember that this set will include unrealistic networks, such as the fully-connected network and the network with no connections at all. The probability p is essentially a weighting parameter between zero and one that governs connection density. As p is reduced below 0.5, networks with fewer and fewer connections are more likely, until we are guaranteed a network with no connections. As p is increased above 0.5, networks with more and more connections are likely, until we are guaranteed a fully-connected network. This demonstrates that the diversity in randomness is maximal for $p = 0.5$. For networks with very high or very low connection densities, the set of all possible random graphs becomes small and we therefore have less diversity.

Erdős and Rényi (1959) derived several topological properties of their model. Here we focus on degree distribution, average path length, and average clustering. As we saw in Chapter 4, the degree distribution of an Erdős-Rényi network is binomial, and thus,

$$P(\text{degree} = k) = \binom{N-1}{k} p^k (1-p)^{N-1-k}, \qquad (10.2)$$

where $0 \leq k \leq N-1$. The mean node degree is therefore $\langle k \rangle = (N-1)p$. For large networks with many nodes and a moderate connection density, the node degree can be well approximated with the Poisson distribution,

$$P(\text{degree} = k) \approx \frac{(Np)^k e^{-Np}}{k!}. \qquad (10.3)$$

For this reason, random networks are sometimes referred to as Poisson or binomial networks. Random networks are sometimes also referred to as Gaussian networks, since the binomial distribution can also be approximated by a Gaussian when N is large (at least 30) and p is not close to either zero or one. A general rule of thumb is that Np and $N(1-p)$ should exceed 5 for the Gaussian approximation to be accurate. It can be shown that the average path length in an Erdős-Rényi network is $L_{\text{rand}} = \log N / \log \langle k \rangle$ and the average clustering coefficient is $C_{\text{rand}} = \langle k \rangle / N$, where $\langle k \rangle$ denotes the mean node degree.

Ensembles of Erdős-Rényi networks can be used to normalize network measures observed in empirical networks. Erdős-Rényi networks can be matched to the size, connection density, and mean degree of an empirical brain network by setting the parameter p such that $p = \langle k \rangle / (N-1)$. Moreover, when normalizing the average path length or clustering coefficient, we do not need to generate multiple instances of the Erdős-Rényi graph; we can simply utilize the above closed-form expressions and avoid the computational expense associated with generating an ensemble.

The advantage of the Erdős-Rényi null model is that it is simple to generate, and analytic expressions exist for computing many commonly used topological metrics, such as clustering and path length. This ease of computation makes it particularly attractive for the analysis of large networks. However, as we saw in Chapters 1 and 4, the Erdős-Rényi network does not provide a very good model of many real-world systems, lacking basic properties such as a **small-world** organization and a heterogeneous degree distribution that allows for the presence of highly connected **hubs**. The last consideration is an important limitation, as many properties in a brain network may be driven by its degree distribution. In practice, the error in matching only the mean degree may be small, if not negligible and the Erdős-Rényi model can be generalized to match exact degree distributions (Molloy and Reed, 1995). However, alternative null models are often favored in the analysis of brain networks. Next, we will briefly describe two popular generative models that attempt to capture small-worldness and heterogeneous (power-law) degree distributions.

10.1.2 Watts-Strogatz Networks

Watts and Strogatz (1998) proposed a model to generate random networks with small-world organization. To sample from their generative model, we begin with N nodes and a desired mean node degree $\langle k \rangle$, which is assumed to be an even integer. The nodes are equidistantly positioned in two-dimensional space along a ring, such that each node has exactly two immediate neighbors (Figure 1.3a). We then add $N\langle k \rangle/2$ connections to construct a lattice network. In particular, for each node, connections are drawn between it and its $\langle k \rangle$ closest neighbors, $\langle k \rangle/2$ on each side. Only one connection is ever drawn between any pair of nodes. The final step is to randomly rewire each connection with probability p_{WS}. Rewiring is performed in such a way that avoids self-loops and connection duplication. A connection that is chosen for rewiring therefore has equal probability of being positioned between any of the distinct pairs of nodes that are not already connected.

When $p_{WS} = 0$, no rewiring is performed, yielding a lattice network. When $p_{WS} = 1$, all connections are rewired, completely randomizing the lattice structure, and leaving an Erdős-Rényi network. Setting the value of p_{WS} between these two extremes generates networks displaying small-world organization (Figure 10.2). This is consistent with random ($p_{WS} = 1$) and lattice ($p_{WS} = 0$) networks occupying the extremes of a topological spectrum whose middle ground is spanned by small-world networks. Rewiring a small proportion of the connections introduces topological shortcuts that intersect the ring along which the nodes are positioned. These shortcuts serve to drastically reduce the network's average path length, while causing only a modest decrease in the average clustering coefficient (Figure 1.3).

Both the Watts-Strogatz and Erdős-Rényi networks have a similar degree distribution, with a peak at $\langle k \rangle$ and an exponential decay at either side of this peak. As

$p_{\text{ws}} =$ 0   0.2   0.4   0.6   0.8   1

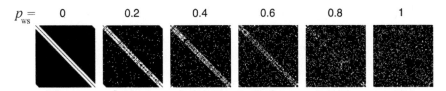

FIGURE 10.2 Adjacency matrices for the Watts-Strogatz model with different rewiring probabilities. The rewiring probability, p_{WS}, dictates the proportion of connections that are randomized. With no rewiring ($p_{WS} = 0$), we have a lattice network, which can be represented as an **adjacency matrix** with connections positioned close to the matrix diagonal. As p_{WS} is increased, the lattice structure is progressively diminished and replaced with connections that are far from the matrix diagonal. Small-world networks are produced due to the considerable reduction in path lengths afforded by this rewiring step. An Erdős-Rényi network results when $p_{WS} = 1$, in which case all connections are randomly rewired, eliminating all vestiges of the lattice structure.

a result, there is a low probability of finding nodes with degree values that are either much lower or much higher than $\langle k \rangle$. Null networks generated using the Watts-Strogatz model are therefore appropriate for making inference about the degree distribution of small-world networks. For example, we might want to ask whether an empirically measured brain network, which is known to have small-world organization, shows a degree distribution that differs from what we could expect under the null hypothesis of a small-world network with a Poisson degree distribution. However, since few real-world networks possess a homogeneous degree distribution of the type implied by the Erdős-Rényi or Watts-Storgatz models, we often want to generate null networks with a more realistic degree distribution. For this reason, while the Watts-Strogatz model provides important insights into how we can obtain a relatively complex topological organization from a simple, random rewiring procedure, it is seldom used as a null model for the analysis of brain networks.

10.1.3   Barabási-Albert Networks

One generative model that yields networks with more realistic, **scale-free** degree distributions was developed by Barabási and Albert (1999; see also Albert and Barabási, 2002). Their generative model encompasses two mechanisms that are important for scale-free networks: preferential attachment and network growth.

Unlike the Watts-Strogatz model, where the number and position of nodes is fixed from the outset, the Barabási-Albert model adds a new node at each iteration. Furthermore, preferential attachment requires that new nodes preferentially form connections with existing high-degree nodes. This preference results in the amplification of any initial random variation in node degrees. Node degrees therefore grow with iterations in a "rich get richer" manner. To generate Barabási-Albert networks, we begin with a few interconnected nodes. We then iteratively add further nodes, where each new node forms exactly v connections with existing nodes. Crucially, these v connections are not randomly assigned to existing nodes. Instead, the probability that a connection is assigned to an existing node is proportional to its current nodal degree. This results in the formation of hub nodes, since new nodes have a preference to make connections with high-degree nodes. New nodes are added until the network grows to a desired size.

Barabási-Albert networks have a connection density of $2v/(N-1)$ and a scale-free node degree distribution that follows a power law of the form $P(\text{degree} = k) \sim k^{-\gamma}$, where $\gamma = 3$ irrespective of the connection density (Chapter 4). Exact analytical expressions for the average path length and clustering coefficient are difficult to derive, although asymptotic results are available. Importantly, as the network size grows, the clustering coefficient approaches zero. Therefore, large Barabási-Albert networks are not small-world networks, since clustering is dependent on network size. Klemm and Eguíluz (2002) have advanced the preferential attachment algorithm to generate small-world networks with a scale-free node degree distribution. However, the degree distribution for brain networks may not conform to a power law (Figure 4.3), meaning that scale-free null models may not be the most appropriate for analysis of **connectomic** data. For this reason, null models based on the rewiring of connections in an empirical network are the most commonly used in neuroscience. These models are considered in the next section.

In summary, we have considered three models for generating networks that display a prescribed set of properties: Erdős-Rényi networks, the simplest of the three, allow us to formalize the concept of *random* networks. They can be generated to a prescribed size, connection density, and mean node degree. Watts-Strogatz networks can also be generated to satisfy these specifications, but additionally display small-world organization. Finally, Barabási-Albert networks display a scale-free degree distribution, but are not necessarily small-world networks. Figure 10.3 shows example adjacency matrices for these three generative models. We have only considered the most basic models here, as the concepts illustrated by these networks provide an important foundation for understanding some of the processes involved in the emergence of some key canonical properties of the brain. For example, we have seen that preferential attachment can lead to the appearance of hub nodes, while random rewiring of just a few connections in a lattice can produce small-world organization. More sophisticated generative null models are available for producing random networks with arbitrary degree distributions and other constraints (Gleeson, 2009; Newman, 2009; Wang et al., 2014a); **exponential random graph models** (ERGMs) are an important class of such models and can be used to flexibly model a wide variety of network properties (Box 10.2). A taxonomy of some of these different models is provided by Klimm et al. (2014).

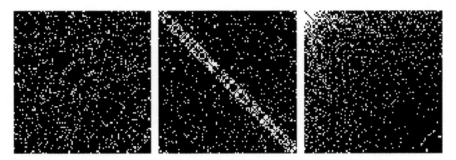

FIGURE 10.3 Adjacency matrices for different generative network models. These matrices were constructed for a network of 1000 nodes and a connection density of 10%, wired according to the Erdős-Rényi (left, $p = 0.5$), Watts-Strogatz (middle, $p_{WS} = 0.5$), and Barabási-Albert (right, $v = 50$) models. Black cells indicate the absence of a connection. The diagonal elements in the Watts-Strogatz network are remnants of a lattice structure that has been partially replaced with the addition of random connections during the rewiring step. The nodes that are located in the upper left corner of the degree-ordered Barabási-Albert matrix are hub nodes that have been generated by the preferential attachment mechanism.

BOX 10.2 EXPONENTIAL RANDOM GRAPH MODELS

Exponential random graph models (ERGMs), or p^* models as they are sometimes known, are a principled and very flexible network modeling tool that enables us, in theory at least, to generate networks displaying a diverse combination of topological and/or spatial properties. In fact, many of the models considered in this chapter can be parameterized in terms of ERGMs. These models have enjoyed widespread use in the social sciences, where networks can be used to model relationships between individuals (Robins et al., 2007). However, the technique has not gained substantial traction in the connectomics literature (although see Simpson et al. (2012) and Van Wijk et al. (2010) for exceptions). The limited use of the approach may be due to its high complexity and high computational burden, particularly in the case of large networks, among other factors.

ERGMs involve sampling networks from an exponential probability distribution. The sample space is typically taken to be the set of all $2^{N(N-1)/2}$ undirected networks that comprise N nodes. We denote this set as Q. If we let Y denote a random variable defined on this sample space, we can write,

$$P(Y=y)=\frac{e^{\sum_{i=1}^{K}\theta_i g_i(y)}}{\kappa(\theta)}, \quad y \in Q, \qquad (10.2.1)$$

where $\theta = (\theta_1 \dots \theta_K)$ are the model parameters to be fitted, $g_i(y)$ are real-valued functions defined over the sample space and $\kappa(\theta)$ is a normalizing constant that ensures $\sum_{y \in Q} P(Y=y) = 1$. Each $g_i(y)$ is chosen to measure a particular network property, such as the number of connections, triangles, stars, **cycles**, and so on. We fit the model parameters θ so that drawing samples from the distribution $P(Y=y)$ yields networks that are statistically matched to our empirical network in terms of the K properties we have specified using the statistics $g(y) = (g_1(y) \dots g_K(y))$. The model parameters control the relative importance of each network property. For example, suppose one of our properties is the number of open triangle motifs (i.e., disconnected pairs of nodes that have at least one shared neighbor; see Chapter 8). If our model parameter for this property is large and positive, we will generate networks comprising a greater proportion of open triangle motifs than expected under chance alone. If it is zero, we will generate networks that have the same proportion of open triangle motifs as expected under chance (assuming other model properties

do not indirectly affect the proportion of open triangle motifs), while if our model parameter is negative, we will generate networks with fewer open triangle motifs.

Generating samples from the exponential distribution associated with ERGMs relies on sophisticated numerical schemes, such as **Markov Chain Monte Carlo**. The basic idea is to draw samples, each of which is an adjacency matrix, that are closely matched to the properties stipulated by our network statistics and model parameters. Estimating $\kappa(\theta)$ and then fitting the model parameters θ can be computationally intensive. The interested reader is referred to Hunter et al. (2008) for details and a guide to the popular "ergm" software package for ERGMs. For a more technical treatment, see Geyer and Thompson (1992).

The ability to model any combination of network statistics is what makes ERGMs such a flexible network modeling tool. But how do we choose a suitable set of statistics? Should we focus on statistics that capture degrees, triangles, stars, cycles, or a combination thereof? One approach is to begin with a very comprehensive set of statistics and then use model selection strategies such as backward selection to eliminate statistics that do not significantly improve model fits. Using a sophisticated model selection approach, Simpson et al. (2012) conclude that the best fitting model for their human functional brain networks is,

$$P(Y=y) = e^{\theta_1 E(y) + \theta_2 C(y) + \theta_3 L(y)} / \kappa(\theta), \qquad (10.2.2)$$

where $E(y)$ is the total number of connections in y, $C(y)$ is the weighted sum of the number of closed triangle motifs, and $L(y)$ is the weighted sum of the number of open triangle motifs. This suggests that connection density and network transitivity (i.e., open vs. closed triangle motifs) have the most influential impact on the overall organization of functional brain networks. This is not surprising: we have already seen that connection density is one of the most basic properties that should be matched. Network transitivity implies that two nodes that are both connected to a common node are more likely to be functionally connected than expected by chance alone. In the case of the functional MRI networks examined in this study, this may be due to the nature of the statistical measures used to quantify functional connectivity (see Section 10.3 and Box 7.1).

Importantly, the basic concepts underlying these generative null models can be extended to test different generative models for brain networks. In this application, we grow networks according to simple, generative mechanisms, and examine whether the properties of the resulting graph match those of the brain. This rapidly developing area, sometimes called *growth connectomics* (Vértes and Bullmore, 2015), is yielding new insights into which constraints may be most important for governing the development of brain network connectivity (Box 10.1).

10.2 NULL NETWORKS FROM REWIRING CONNECTIONS

In this section, we consider null networks generated through an iterative process in which an observed network is rewired to generate a random topology, albeit with specific constraints. In this sense, these networks are not completely random, like the Erdos-Renyi model, but they are randomized with respect to an initial, empirically determined starting point. These rewiring algorithms often allow us to achieve a better match between the properties of an empirical and null network, and are thus very useful for normalization and statistical inference. The Maslov-Sneppen rewiring algorithm is the most well-known of these rewiring algorithms.

10.2.1 Maslov-Sneppen Rewiring

Maslov and Sneppen (2002) devised an algorithm to randomize protein interaction networks through an iterative rewiring process. The Maslov-Sneppen algorithm, as the approach is now often called, annuls all network properties except for network size, connection density, and degree distribution. The rewired network will have the same number of nodes, same number of edges, and same **degree sequence** as the original network, with all other properties being randomized. The algorithm was originally used to show that pairs of highly connected hub proteins interact less frequently than expected under the null hypothesis of interactions occurring as chance phenomena. Milo and colleagues (2002) used the algorithm soon after to identify network motifs that were present in genetic networks and food webs with a frequency greater than expected by chance alone. Due to its simplicity, the Maslov-Sneppen algorithm is now widely used to test hypotheses about many kinds of empirical networks.

The Maslov-Sneppen algorithm uses an iterative rewiring procedure. At each iteration, we randomly select two connections, (u_1, u_2) and (u_3, u_4). We then move (rewire) these **links** to introduce two new connections (u_1, u_4) and (u_2, u_3). If one or both of these new connections already exist, we abandon the iteration and randomly select another pair of connections. This rule ensures

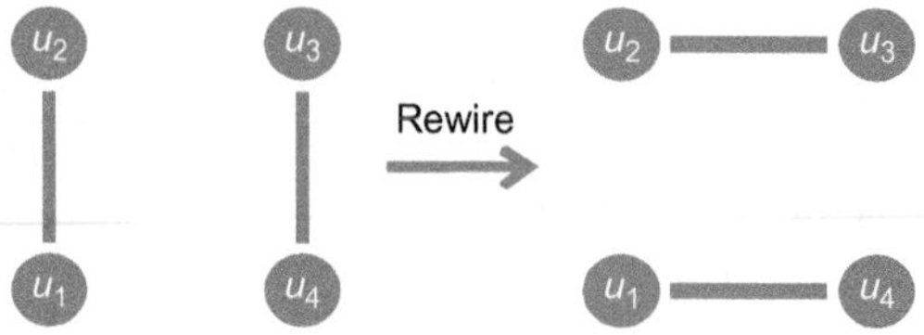

FIGURE 10.4 The Maslov-Sneppen rewiring algorithm. A single iteration of the algorithm is shown here. The four nodes on the left are rewired as shown on the right.

that multiple connections are not placed between the same pair of nodes. Repeated application of this rewiring step progressively randomizes the network topology. Since we do not add or remove nodes or edges, the rewired network will have the same size (number of nodes) and connection density as the original network. Moreover, the procedure ensures that the same number of edges remains attached to each node, thus preserving the degree distribution of the original network. The output of the algorithm is thus a randomized network that is matched to our empirical network in terms of size, connection density, and degree distribution. Figure 10.4 demonstrates a single iteration of the algorithm.

The Maslov-Sneppen algorithm is a computationally efficient approach, although we can never be sure how many rewiring steps are required. Empirical evidence suggests that the number of iterations should exceed at least 100 times the number of connections in the network (Milo et al., 2004). This can be computationally expensive for large networks. Viger and Latapy (2005) provide a theoretical analysis of the algorithm's complexity and discuss heuristics to improve running times for large networks.

The rewiring algorithm can be trivially extended to the case of directed networks to ensure both in- and out-degrees are preserved. In this case, we ensure that each connection switch is such that an outgoing edge at one node is always replaced with an outgoing edge from another node, thus preserving the in- and out-degrees. In other words, an outgoing edge is never replaced with an incoming edge.

We can generate an ensemble of randomized networks by repeatedly applying the Maslov-Sneppen algorithm, each time with a different initialization of the random number generator. We can then use this ensemble to approximate the mean and standard deviation of any given network measure under the null hypothesis of a randomized network. Repeating the randomization many times ensures that we will obtain a more stable estimate of the expectations under the null hypothesis. An ensemble of 100 or so randomized networks is usually sufficient to yield reasonable approximations for the mean and standard deviation of most network measures, although this may be contingent on the size of the network and complexity of the measure. If we wish to use the randomized networks to test for the statistical significance of a particular property, we need to generate a larger ensemble to sample the null distribution with sufficient precision to allow accurate statistical inference. A one-sided p-value can be computed as the percentile of an observed network measure in the null distribution. In this case, we ask the question: is the value of a given property computed in an empirical network significantly larger than expected by chance?

The mean of an ensemble of randomized networks is commonly used to normalize an observed measure of network topology. The normalization is computed as the ratio of the observed measure, M_{obs}, to the mean value of that property measured in the randomized network ensemble, $\langle M_{rand} \rangle$; that is, $M_{obs}/\langle M_{rand} \rangle$. This ratio is referred to as the *normalized* network measure and is nothing more than a dimensionless factor quantifying how much more or less a particular property is represented in an empirical network compared to an ensemble of null networks. Humphries' index of small-worldness (Humphries and Gurney, 2008) is a classic example of this kind of normalization (Chapter 8). The index is computed by normalizing the average path length and average clustering coefficient with respect to an ensemble of random networks generated with the Maslov-Sneppen algorithm. Evidence of small-world organization is suggested by a normalized clustering coefficient that is considerably larger than one and a normalized path length that is close to one (Watts and Strogatz, 1998).

10.2.2 Rewiring Connections in Weighted and Signed Networks

Rubinov and Sporns (2011) have generalized the Maslov-Sneppen algorithm to deal with weighted and signed networks, which is particularly important in analyses of correlation-based functional connectivity networks. It can also be useful in analyzing graphs that distinguish between different types of structural connections. For example, we could represent inhibitory connections with negative weights and excitatory connections with positive weights.

The Rubinov and Sporns algorithm comprises two steps. The first step involves an iterative connection-switching rule. At each iteration, we randomly select four nodes such that we have positive connections (u_1, u_2) and (u_3, u_4), and negative connections (u_1, u_3) and (u_2, u_4). We then reverse the signs of these four connections. This reversal ensures that the proportion of positive and negative connections associated with each node is preserved. Zero-weight edges (i.e., the absence of an edge) can either be designated to the set of negative connections, or they can be excluded from consideration so that the connection-switching rule only involves switches between strictly positive and strictly negative connections. While Rubinov and Sporns (2011) propose the latter approach in their original work, this might not achieve sufficient randomization when the proportion of positive and negative edges is highly skewed. This is because in the skewed case, the number of distinct sets of four nodes comprising two negative and two positive connections will be limited. For this reason, it may be advantageous to include edges with zero-weight (assuming there are many of them) in the smallest set of edges, either positive or negative. This ensures sufficient randomization can be achieved in cases where the proportions of positive and negatives edges are skewed, but this also implies that the total number of negative or positive edges emanating from a node is not preserved.

The second step involves randomizing the connection weights so that the positive and negative **strengths** of each node are approximately preserved. The node strength here is the sum of the weights across all connections associated with that node (see Chapter 4). For illustrative purposes, we assume connections are either all positive or all negative. If this is not the case, the weight rewiring procedure described next can be applied separately to the sets of positive and negative connections. Let $\hat{w}_{ij}$ denote the new weight that will be assigned to the connection between nodes i and j, and let w_{ij} denote the original weight for that connection. First, we begin by setting $\hat{w}_{ij} \leftarrow 0$ for all connections. Second, we compute the amount of surplus weight, e_{ij}, that can be added to each connection,

$$
e_{ij} = \left(s_i - \sum_h \hat{w}_{ih} \right) \left(s_j - \sum_h \hat{w}_{jh} \right),
\tag{10.4}
$$

where $s_i = \sum_h w_{ih}$ is the node strength. Third, we order the list of surplus weights from lowest to highest. Finally, we randomly choose a connection that has not yet been assigned a weight. Suppose (i, j) is that randomly chosen connection. We determine its rank, r, in the ordered list of surplus weights and set $\hat{w}_{ij} \leftarrow w^r$, where w^r is the rth largest weight that has not already been assigned. This process continues until all weights have been assigned. The surplus weights must be recomputed and reordered after each new assignment to ensure convergence to the original node strength values. Connections that have already been assigned a weight are excluded from the list of surplus weights.

A more sophisticated approach described by Serrano and colleagues (2006) randomizes weights in a way that preserves correlations between the strength and degree of each node. We know the *average* connection weight of the non-zero edges of a node with degree k and strength s is s/k, but this does not characterize fluctuations around this average. For example, a node with degree 3 and strength 6 can be realized with three connections, each with weight 2, or alternatively, one connection with weight 4 and two connections with weight 1. The generalization owing to Rubinov and Sporns (2011) does not preserve this distinction, whereas Serrano and coworkers' approach does.

It is important to note that the Rubinov and Sporns (2011) algorithm simultaneously randomizes both the brain network's topology (i.e., the location of positive and negative edges, as per step 1) and its connection weights (step 2). This may not be appropriate for analysis of all properties computed in weighted networks. For example, Alstott et al. (2014b) argue against this approach when analyzing weighted **rich-club** coefficients (Chapter 6). Specifically, they highlight how the selection of an appropriate null model can be used to separate the contributions of binary connection topology and weight distribution to specific properties computed for a network.

10.2.3   Lattice Null Networks

We know from the work of Watts and Strogatz (1998) that random graphs such as those studied by Erdős and Rényi are positioned at one end of a spectrum of possible network topologies. The other end of that spectrum is occupied by regular, lattice-like networks. Small-world networks, which include the brain as well as many other real-world systems, occupy the middle ground. Using the Maslov-Sneppen rewiring procedure will thus only provide a reference with respect to the random end of this spectrum. To obtain a more comprehensive view, it may be useful to generate lattice-like reference networks as well.

When embedded in Euclidean space, lattice networks appear as tiling patterns or grids, since most of their connections are between spatially proximal nodes (Figure 1.3). Lattice networks are highly clustered with long path lengths, whereas random networks have low clustering and short path lengths.

Telesford and colleagues (2011b) argue that appropriately matched lattice networks provide a more suitable null model to normalize clustering coefficients than Maslov-Sneppen randomized networks. Their reasoning is that the two ends of our topological spectrum provide inherent reference points, and thus clustering is best normalized to maximally clustered networks (i.e., lattice networks), just as path length is best normalized to networks with minimal path length (i.e., random networks). This proposal deviates from the logic of the Humphries index of small-worldness, where both path length and clustering are normalized with respect to random networks. Telesford and colleagues propose an alternative measure of small-worldness, in which clustering and path length are respectively normalized relative to lattice and random benchmarks (Equation 8.14).

Randomized networks with lattice structure can be generated with a slight modification of the Maslov-Sneppen rewiring algorithm. In particular, we randomize the graph using the same iterative rewiring process, however, we now abandon any iteration that does not bring nonzero elements (edges) of the connectivity matrix closer to the matrix diagonal. Note that the connectivity matrix

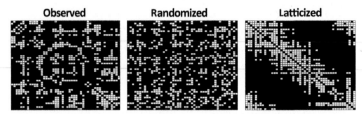

FIGURE 10.5 Rewiring the macaque connectome. Left panel shows the observed adjacency matrix for a 47-region connectome of the macaque, generated from the collation of a large number of published tract-tracing studies (Honey et al., 2007). Middle panel is the same matrix after randomization with the Maslov-Sneppen rewiring procedure. Right panel shows the outcome of rewiring the macaque network (left) to achieve a lattice-like configuration. Black cells indicate the absence of a connection. Most connections are positioned close to the matrix diagonal with lattice rewiring, whereas connections are randomly positioned with Maslov-Sneppen rewiring, subject to preservation of the node degree distribution.

of a lattice network can be reordered so that most nonzero elements are positioned close to the matrix diagonal, and thus connections only link nearby regions. Therefore, by selectively implementing only those rewiring steps that result in a connectivity matrix with more connections clustered around the matrix diagonal, we can generate random networks with lattice structure. Sporns and Zwi (2004) were the first to consider lattice-based reference networks. Another early example is provided by Sporns and Kötter (2004). Figure 10.5 shows examples of Maslov-Sneppen and lattice rewiring applied to a 47-region macaque connectome.

10.2.4   Spatial Embedding

It is often overlooked that nervous systems are spatially embedded within the limited volume of the organism to which they belong. The spatial locations of the nodes of a connectome are thus physically constrained by the geometry of the space in which they are embedded. In turn, these constraints impact network topology (Chapters 1 and 8), raising an important question: to what extent are the various nontrivial topological properties of brain networks, such as small-worldness, modularity, and rich-club organization, simply a consequence of spatial embedding? To address this question, we can generate randomized networks that also preserve the specific spatial characteristics of our empirical brain networks, thus allowing us to formally test whether a particular property is expressed in the brain to a greater (or lesser) extent than would be expected based on spatial embedding alone.

The Maslov-Sneppen algorithm can be modified straightforwardly to randomly rewire an empirical network in a manner that preserves basic spatial characteristics as well as the usual topological properties of degree distribution, connection density, and network size. Connection distance, or *wiring length*, is the most widely considered spatial characteristic. Wiring length is usually approximated as the Euclidean (straight line) distance between node pairs (Kaiser and Hilgetag, 2004). However, axons and fiber bundles can exhibit complex geometries, meaning that this Euclidean approximation is a lower bound for true wiring lengths (see Box 8.1). Better estimates of wiring length can be obtained by tracing the course of axons or bundles that connect different brain regions, which may be possible for tract-tracing studies, but is not possible for other datasets such as functional connectivity matrices.

Samu and colleagues (2014) describe a simple modification to the Maslov-Sneppen algorithm that preserves the total wiring length associated with each node. In this procedure, we retain information about the wiring length of each edge, and iterations of the rewiring algorithm are abandoned if the associated rewiring step results in a change in the total wiring length of any node. This generates null networks that are matched in terms of wiring length for each node. Algorithm running times may be slow because the likelihood of abandoning iterations is high. Consequently, many iterations are required to achieve a sufficient level of randomization. An alternative approach is to compute a minimum spanning tree (Box 6.1) to ensure that all nodes are connected, and then add connections in order of increasing distance until a prescribed connection density is achieved (Bassett et al., 2010). A drawback of this approach is that the degree distribution is not preserved.

Based on their spatially constrained null networks, Samu and colleagues (2014) report that most topological features of the human connectome, as reconstructed with diffusion MRI, can be explained, to a certain extent, by wiring constraints. In particular, they show that the efficiency, clustering, and small-worldness of the human brain as a point on a three-dimensional axis. On the same axis, they also show points for randomized versions of the same network with and without total wiring length preserved (Figure 10.6). Based on this figure, Samu and colleagues conclude that topological features that cannot be explained by the spatial embedding of the brain may be of particular functional relevance. We note however, that topological properties tightly coupled with spatial embedding in the brain are no less real, in the sense that they serve the same functional role as if generated with a purely topological rule independent of any geometry. In other words, a brain with modular organization will still show some degree of functional specialization, regardless of whether the **modules** were determined by spatial constraints or some other functional or biological imperative. Nonetheless, spatially constrained reference networks can be useful for understanding the degree to which spatial embedding impacts distinct properties of a connectome.

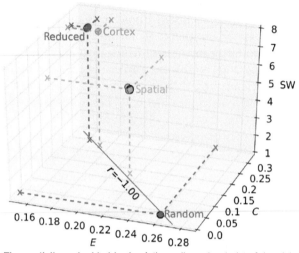

FIGURE 10.6 The spatially embedded brain. A three-dimensional plot of the global efficiency (*E*), clustering coefficient (*C*), and small-worldness (*SW*) of the 998-node human connectome generated with diffusion MRI by Hagmann et al. (2008; represented by the label "cortex", in green), and the same measures computed in a randomized versions of this network generated with either the Maslov-Sneppen algorithm (random; purple), or spatially embedded rewiring algorithms that preserve (spatial; yellow) and reduce (reduced; red) the total wiring length. The blue line indicates a negative association between clustering and efficiency, where *r* denotes the correlation coefficient. Randomized networks for which the total wiring length is reduced are closest to the human connectome in terms of efficiency, clustering, and small-worldness. This suggests that the connectome might favor topological segregation over integration, due to the suboptimal placement of long-range connections. Note however, that diffusion MRI tends to underestimate long-range connectivity in the brain (Chapter 2), and may thus over-emphasize the importance of short-range connectivity in connectome topology. *Reproduced from Samu et al. (2014) with permission.*

In summary, we have seen that the Maslov-Sneppen algorithm is a flexible approach, and its connection-switching rule forms the backbone of a variety of other methods for rewiring the connectome, which include random, lattice, and spatially embedded rewiring strategies. These rewiring strategies differ only in the criteria used to reject certain rewiring steps. The main disadvantage of rewiring is that iterations may need to be frequently abandoned if the goal is to preserve higher level connectome properties, such as spatial characteristics. This will increase algorithm running times and limits the extent of diversity that can be achieved in an ensemble of null networks. We can imagine the space of feasible null networks as a series of concentric circles, with the central point representing the topology of our empirical brain network. The outermost concentric circle encapsulates the diverse space of all possible random networks. The next concentric circle might represent the space of all random networks with matched network size and the next after that might represent the space of all random networks with matched size *and* connection density, and so on. As more features are matched, the diversity of feasible random networks decreases until we inevitably converge to our empirical network. In this sense, additional constraints can be added to the rewiring rules in order to test specific hypotheses about the impact of certain characteristics (e.g., spatial embedding) on network organization.

10.3 FUNCTIONAL CONNECTIVITY NETWORKS

The rewiring algorithms and generative models that we have considered thus far are general algorithms that operate on a network's adjacency matrix. In some cases, the analysis of functional connectivity networks can pose special problems that are not fully addressed by topological randomization. These problems arise because, in comparison to a structural network, the interpretation of an edge in a functional connectivity network can be somewhat ambiguous. In a structural network, connectivity is quantified as some property of the physical links between two nodes and thus aligns well with traditional concepts of network wiring. In contrast, edges in functional connectivity networks are often quantified using some measure of statistical dependence, such as a correlation coefficient. Edge weights are thus an abstract quantity for which a high value does not necessarily mean that there is a physical link between nodes. For example, analyses of human neuroimaging data have revealed that functional connectivity networks, as measured through functional MRI time series correlations, have a higher connection density than structural networks, as measured with diffusion MRI, because correlations are sensitive to both direct and indirect interactions between regions (Honey et al., 2009; Vincent et al., 2007; Zalesky et al., 2012b; see also Box 7.1). In other words, two regions i and j may show correlated activity because they are both connected to a third region h, despite sharing no direct structural link between them. As a result, the use of Pearson correlation as a measure of functional connectivity *in and of itself* gives rise to networks that are inherently more clustered than random networks. Therefore, topological randomization can exaggerate the extent to which functional brain networks display certain topological properties such as small-worldness (Zalesky et al., 2012b).

To see this bias, we can generate a set of N independent, random vectors and compute the Pearson correlation between every pair of vectors to yield a $N \times N$ symmetric connectivity matrix representing a hypothetical brain network. Intuition suggests that correlating random vectors that are completely independent of each other is very unlikely to yield meaningful topological structure. However, if we test this hypothesis using null networks generated with Maslov-Sneppen rewiring, we find that our hypothetical brain network is substantially more clustered than randomly rewired null networks (see Figure 10.7).

How can this be, given that we are randomly rewiring a network that is already supposedly random? The apparent paradox is resolved once we recognize that randomness is a relative concept. Our hypothetical brain network is not random in the Erdős-Rényi sense. It is therefore important to make a distinction between random correlation networks and Erdős-Rényi random networks. Mathematically, all correlation networks, random or otherwise, display a property called positive semidefiniteness. This property is a consequence of the transitive nature of Pearson correlation; that is, if region i is correlated with j, and j with h, then i and h must also be correlated. This effect manifests as greater clustering than would be expected in an equivalent Erdős-Rényi network. It is important to note that the partial correlation coefficient does not resolve this issue; in fact, partial correlations underestimate the extent of clustering that would be expected in an equivalent Erdős-Rényi network (Zalesky et al., 2012b).

In summary, using Maslov-Sneppen rewiring or Erdős-Rényi networks to normalize functional brain networks may not always be ideal, as certain measures of statistical dependence used to quantify functional connectivity can bias network organization in a way that cannot be removed by topological rewiring. Fortunately, the magnitude of this effect decreases as more time points are acquired: for 512 time points or more, the effect can be considered relatively

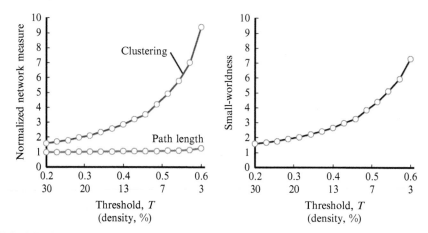

FIGURE 10.7 The topological structure of random correlation-based networks. The average path length and clustering coefficient (left), and Humphries' index of small-worldness (right), calculated for a network formed by computing the Pearson correlation between $N = 90$ independent, random vectors. Maslov-Sneppen rewiring was used to generate null networks for normalization. The horizontal axis shows the binarizing threshold, T, and the corresponding connection density. It is evident that correlating random vectors yields a network that is more clustered than randomly rewired networks, suggesting small-world organization. However, this small-world organization is due to the transitive nature of Pearson correlation. It can be factored out with the use of appropriate null networks. *Reproduced from Zalesky et al. (2012b) with permission.*

small in some circumstances (Zalesky et al., 2012b). In this regard, high-resolution physiological recording techniques will be relatively unaffected. However, this bias does pose a problem for lower-resolution techniques such as functional MRI, where the number of time points available for analysis is often less than 512.

In any case, methods are available for generating random correlation networks, thus providing an ensemble of reference graphs to complement those generated with more traditional topological rewiring methods. We discuss one such approach in the following section. These methods supplement a broader class of null models for functional connectivity networks in which randomization occurs at the level of the time series, rather than at the level of network topology. These methods thus randomize the data before the network has been constructed, and are better able to account for intrinsic characteristics of the measurement process (Box 10.3).

10.3.1   Null models for Correlation-Based Networks

Random correlation networks can be generated by sampling from a Wishart distribution or using the Hirschberger-Qi-Steuer (HQS) algorithm (Hirschberger et al., 2007). Although the HQS algorithm can be numerically unstable in some circumstances, it often yields a distribution of correlation values that is better matched to the distribution of actual correlation values of an empirical network, compared to sampling from the Wishart distribution. Matching the distribution of correlation values is important to ensure connection density is preserved.

The HQS algorithm generates random covariance matrices that are matched to the distributional properties of an observed covariance matrix. In our case, the observed covariance matrix is computed using our time series data. The algorithm matches the mean, e, and variance, v, of the off-diagonal elements (covariances) as well as the mean, $\hat{e}$, of the diagonal elements (variances) of the empirical covariance matrix. Higher order moments can also be matched, however, this can lead to numerical instability. The random covariance matrices that are generated can be transformed to random correlation matrices with variance normalization. The algorithm comprises the following steps:

1. $m \leftarrow \max\left(2, \lfloor \hat{e}^2 - e^2/v \rfloor\right)$
2. $\mu \leftarrow \sqrt{e/m}$
3. $\sigma^2 \leftarrow -\mu^2 + \sqrt{\mu^4 + v/m}$
4. $x_{i,j} \sim N\left(\mu, \sigma^2\right)$ $i = 1, \ldots, N; j = 1, \ldots, m$
5. $X = \left(x_{i,j}\right)$ $i = 1, \ldots, N; j = 1, \ldots, m$
6. $C = XX^{\mathrm{T}}$

BOX 10.3 TIME SERIES RANDOMIZATION AND DYNAMIC FUNCTIONAL CONNECTIVITY

Functional connectivity networks are constructed by computing some form of statistical dependency between the neural activity measured at different sites (Chapter 2). Rather than randomizing the topology of a functional connectivity network with the methods described in this chapter, it is sometimes advantageous to perform randomization at the level of the measured neural time series data. An ensemble of null networks can then be generated by computing the same form of statistical dependency across different instantiations of the randomized time series data.

Time series randomization is particularly useful when testing hypotheses about time-resolved, or *dynamic*, functional connectivity networks. Functional connectivity displays spontaneous fluctuations across a diverse range of timescales (Hutchison et al., 2013; Buzsáki and Draguhn, 2004). Much interest has recently arisen in understanding how these fluctuations relate to changes in the topological properties of brain networks over time (Bassett et al., 2011; Zalesky et al., 2014). In this case, the weight of each edge is no longer fixed, but now varies as a function of time. Sliding window correlation (Leonardi and Van De Ville, 2015; Zalesky and Breakspear, 2015) is the most popular approach for mapping dynamic functional connectivity, although temporal **independent component analysis** (Smith et al., 2012), model-based approaches (Lindquist et al., 2014), time-frequency coherence analysis (Chang and Glover, 2010), and change-point detection methods to identify stationary time segments (Cribben et al., 2012) have been used.

Once functional connectivity has been mapped using one of the above methods, we are interested in testing whether any temporal fluctuations in brain network topology are richer and more structured than would be expected if we were to apply the same time-resolved connectivity mapping methods to null data that preserves basic properties of the time series. The most commonly tested null hypothesis is that the connectivity dynamics are due to a linear, stationary system. Rejecting this null hypothesis allows us to conclude that the connectivity dynamics are nonlinear or **nonstationary** phenomena, although it is important to remember that physiological and other forms of nonneural noise can cause these phenomena to appear in neural time series (Chang and Glover, 2009). What do we mean here by nonstationary phenomena?

Stationarity is defined as a stochastic process whose mean, variance, and other moments do not change as a function of a given timescale. Perhaps the easiest way to understand the concept of stationarity is to understand the runs test (Bendat and Piersol, 1986), a very basic test that can be used to determine whether a time series is stationary or not with respect to a given timescale. The runs test begins by dividing the time series into contiguous intervals of equal length. The interval length must be commensurate with the timescale that interests us. If we are interested in testing stationarity at a timescale of days, for example, then the interval length should encompass several days, but not months or years. The mean of the time series is then computed in each interval. Finally, for each interval, we record whether the mean of the time series is above (A) or below (B) the mean value of the *entire* time series. Consider the sequence:

ABBAAABBABAABBABABAA.

In this example, we have a total of 13 runs, where a *run* is a segment of the sequence consisting of adjacent elements. To be clear, the first run is *A*, the second run is *BB*, the third run is *AAA*, and so on. For a stationary time series, we should not see long runs of *A*'s or *B*'s, since this would imply dependency between consecutive time intervals (i.e., a long term trend at the timescale of interest). In very basic terms, stationarity requires the absence of any long terms trends or drifts. Under the null hypothesis of stationarity, whether an *A* or *B* is found at a given time interval should thus be independent of all other time intervals. This null hypothesis can be evaluated using permutation testing or parametric approaches. In our example, the probability of observing 13 runs in a sequence comprising 20 elements is about 0.4 under the null hypothesis, thus we cannot reject the null hypothesis and we find no evidence of nonstationarity at the timescale considered.

Suppose we now consider a longer timescale. In particular, we double the interval length and in doing so record the following sequence:

AAAAABBBBB.

We now see a long run of *A*'s followed by a long run of *B*'s. Runs of this length would be highly unlikely if the time series was stationary. At this timescale, it turns out we can reject the null hypothesis of stationary for an α-significance of 0.05.

BOX 10.3 TIME SERIES RANDOMIZATION AND DYNAMIC FUNCTIONAL CONNECTIVITY—CONT'D

In the context of brain connectivity dynamics, nonstationary phenomena are thought to give rise to complex spatiotemporal network structures, such as repeating patterns of spontaneous synchronization between spatially distributed network elements. Trivial synchronization patterns can however emerge between linearly correlated, Gaussian noise signals. It is therefore important to test whether the observed spatiotemporal patterns are sufficiently long, repeat with sufficient frequency, involve coordination between sufficiently many network elements, or exceed some other meaningful discriminating statistic. This is analogous to the runs test, which uses sufficiently long runs of A's or B's as evidence against the null hypothesis of stationarity.

To this end, the usual approach is to generate random, or *surrogate*, time series data that are stationary by construction and possibly matched to some of the linear properties of our observed neural time series data. Dynamic functional connectivity is mapped using the surrogate time series data exactly as it was mapped in the neural time series data. This generates a set of null networks without the need for any of the topological randomization methods discussed in this chapter. A discriminating statistic can then be computed in the observed network and all the null networks. If the value of the statistic is significantly different between the observed network and the set of null networks, the null hypothesis is rejected and nonstationarity can be assumed.

Methods for generating surrogate time series data have been around for a long time (Schrieber and Schmitz,

2000). These methods have been traditionally used to test hypotheses about the neural time series data itself, but not the time-resolved measures of statistical dependency that are computed between them to estimate connectivity dynamics. Further work is needed to determine whether these old and trusted methods are therefore appropriate. Phase scrambling is perhaps the most widely used method for generating surrogate time series data (Prichard and Theiler, 1994). It involves taking the **Fourier** transform of each time series, adding a sequence of random phases, and then inverting the transform to yield randomized time series with preserved power spectra and **autocorrelation**. If the same sequence of random phases is added to each time series, the linear correlation between the time series is also preserved. This generates surrogate data that is consistent with the null hypothesis of a stationary, linearly correlated multivariate process. Similar resampling approaches can be performed in the **wavelet** domain (Breakspear et al., 2004; Van De Ville et al., 2007). While these approaches yield stationary time series data, they might not ensure that the time-resolved correlation between these time series are stationary. Chang and Glover (2010) suggest fitting stationary **vector autoregressive** (VAR) **models** to the neural time series data. VAR models are commonly used in econometrics and allow us to generate a set of stationary time series exhibiting linear dependencies that are matched to our observed data.

The algorithm returns C, which is a random covariance matrix of dimensions $N \times N$. To compute this covariance matrix, we generate samples from a Gaussian distribution (step 4), arrange these samples into a matrix (step 5), and then multiply this matrix by its transpose to yield a covariance matrix (step 6). Crucially, the number of Gaussian samples generated (step 1), and their mean (step 2), and variance (step 3) are tunable parameters that are chosen to ensure that the observed covariances are matched to the covariances returned by the algorithm.

Figure 10.8 shows the extent of small-world organization quantified by the HQS algorithm and Maslov-Sneppen rewiring in resting-state functional brain networks mapped in 10 healthy human volunteers. Although functional brain

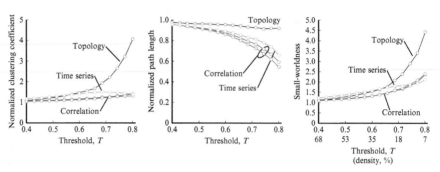

FIGURE 10.8 Time series randomization with the HQS algorithm. Average clustering coefficient (left), path length (middle), and Humphries' index of small-worldness (right) averaged across 10 human resting-state functional brain networks measured with functional MRI. The measures have been normalized with: Maslov-Sneppen rewiring (topology), random correlation matrices (correlation), and phase scrambling with independent random phases (time series; see Box 10.3). The red lines denote the HQS algorithm. The blue lines denote an alternative brute-force approach discussed in Zalesky et al. (2012b). The horizontal axis shows the binarizing threshold, T, and the corresponding connection density. Topological randomization (Maslov-Sneppen rewiring) exaggerates the extent of small-world organization. *Reproduced from Zalesky et al. (2012b) with permission.*

networks appear to be small-world networks under all normalization strategies, it can be seen that normalizing with respect to randomized networks generated with Maslov-Sneppen rewiring substantially exaggerates the extent of small-world organization.

10.4 SUMMARY

We have considered a wide range of network models in this chapter that can serve as reference networks or be used for normalization and other purposes related to statistical inference. In particular, we have considered generative network models, rewiring approaches, and randomization methods suited to correlation-based networks. Which one is the most appropriate for brain network analysis?

This is a difficult question to address. The choice of null model should be dictated to a large extent by the network measure that we seek to normalize and the connectivity measure that we use to construct our network. Maslov-Sneppen rewiring, or one of its generalizations, is a good first choice for most experiments. For particularly large networks, Maslov-Sneppen rewiring may become computationally burdensome, in which case Erdős-Rényi networks can be efficiently generated as a parametric alternative, although they do not allow matching of the full node degree distribution. The choice is less clear for functional brain networks, since these networks can exhibit topological properties that are owing to the very process of measuring functional connectivity. In these cases, alternative randomization strategies may be required to generate null networks that preserve the topological structure introduced by the connectivity measurement process.

A guiding principle in choosing an appropriate null model is that it should annul the particular network property that we seek to normalize or make inference about, but preserve all other features of our brain network, including its spatial embedding. In practice, however, we have seen that it is usually impossible to annul one specific property while preserving all others, since most properties are interrelated. Moreover, as we add more constraints to our null models, they begin to look more and more like our empirical data; in this sense, there is a gradual transition between a null model and generative model of the brain. A sensible first step for most purposes is to use a Maslov-Sneppen-style rewiring procedure to match empirical networks for the basic properties of size, density, and degree distribution. Any properties that differ between this null model and the empirical network should then be flagged for further investigation, using either mechanistically informed generative models (Box 10.1) or increasingly constrained null models (e.g., Samu et al., 2014; Klimm et al., 2014) to understand potential generative mechanisms.

CHAPTER 11

Statistical Connectomics

Performing statistical inference on brain networks is an important step in many applications of **connectomics**. Inference might involve comparing groups of patients and healthy controls to identify which of the hundreds (or thousands) of connections comprising the **connectome** have been impacted by disease, or correlating measures of brain connectivity with some index of cognitive performance. Statistical connectomics is concerned with the methodologies used to perform this kind of inference.

In this chapter, we examine two aspects of connectomics that are important to the statistical analysis of populations of brain networks. The first is thresholding. Brain connectivity data are noisy, and it is sometimes desirable to apply a threshold to the **adjacency matrix** to distinguish true connections from spurious ones. As we will see, there are many methods for applying such thresholds and the specific technique chosen can have a major impact on the conclusions drawn from any statistical analysis that follows.

The second part of this chapter covers statistical inference on brain networks. The statistical testing of connectomic data falls within three broad categories: global or omnibus testing, mass univariate testing, and multivariate approaches. In global testing, we investigate between-group differences or associations using one or more of the global topological measures considered in other chapters. The global approach is simple but lacks specificity: global measures often provide useful summaries of network properties, but they cannot tell us whether the effects are indeed distributed throughout the brain, or confined to a specific subset of **nodes** and/or **edges**.

Mass univariate or *connectome-wide* hypothesis testing involves testing the same hypothesis over many different elements of a brain network. Typically, this is done either at the level of each region, in which case we test a hypothesis about some node-level property of interest; or at the level of each edge in the connectivity **matrix**, in which case we seek to identify connection-specific effects. Unlike global testing, these analyses allow for the localization of effects to specific brain regions (nodes) or neural connections (edges). Mass univariate testing across a family of nodes or edges naturally gives rise to a **multiple comparisons problem**, which we will see is particularly sizeable, even for

383

coarse mappings of a connectome. The problem of multiple comparisons imposes a need for stricter significance thresholds when a family of independent hypotheses is tested simultaneously. We will describe various methods that can be used to correct for multiple comparisons when performing mass univariate testing with connectomic data. We will also see that multivariate methods such as machine learning have begun to gain traction in the field. These approaches seek to recognize patterns and high-level abstractions across multiple connections and nodes that discriminate between groups more accurately than any one measure alone.

How can statistical inference be performed on the topological measures considered in this book? This chapter seeks to address this question, focusing on global, connectome-wide, and multivariate analyses. The methods discussed in this chapter have emerged from the human connectomics literature, where access to large samples of data is readily available. Indeed, a primary goal of the large-scale Human Connectome Project is to understand how individual differences in brain connectivity relate to behavior (Van Essen et al., 2013). For this reason, the techniques presented in this chapter are largely focused on the human literature, although they are equally applicable to data acquired in other species.

11.1 MATRIX THRESHOLDING

As we saw in Chapter 3, it is common practice, particularly in the human connectomics literature, to apply a threshold to a connectivity matrix in order to remove noisy or spurious **links** and to emphasize key topological properties of the network. This step can complicate statistical analysis. For example, if we have two populations of networks—one from a patient group and one from a control group—which single threshold should we choose to make a fair comparison between them? This question is critical because many graph theoretical measures are dependent on the number of edges in the graph. We must therefore choose our threshold with caution. This section considers different approaches to threshold brain networks.

11.1.1 Global Thresholding

The simplest thresholding strategy is to apply a single, *global* threshold to all elements of the connectivity matrix. Elements below the threshold are set to zero. Surviving elements can either be set to one, resulting in a binary adjacency matrix, or they can retain their original connectivity weights if the goal is to then undertake an analysis of weighted networks. This threshold can be determined using one of two methods: weight-based or density-based thresholding.

Weight-based thresholding involves choosing a threshold value, τ, based on the weights in the connectivity matrix. We could choose a value for τ arbitrarily. For example, we could set $\tau = 0.25$ in a correlation-based network, meaning that any pair-wise correlations below this value will be set to zero. Alternatively, τ could be determined using statistical criteria, such as one based on statistical significance. For example, we could retain only correlation values within individual networks that satisfy an α-significance level.

One consequence of these weight-based approaches is that the **connection density** of each network may vary from network to network after the threshold has been applied. To illustrate this point, Figure 11.1a depicts **functional connectivity** data in a single patient with schizophrenia and a healthy control. In this case, the patient has, on average, lower connectivity compared to the control (Figure 11.1b). Assume that we now apply the same correlation threshold of $\tau = 0.20$ to both networks. The resulting adjacency matrix of the patient has a connection density of 53%, whereas the control matrix has a density of 75% (Figure 11.1c and d). This discrepancy arises because the higher average connectivity of the control network means that there are more connections satisfying the condition $w_{ij} > \tau$. Since most graph theoretic measures are sensitive to variations in the number of edges in a graph, any differences in density must be considered when comparing two or more networks.

Density-based thresholding explicitly addresses this problem. With this method, τ is allowed to vary from person to person to achieve a desired, fixed connection density κ. Consider our patient and control in Figure 11.1. Say that we threshold both networks to achieve a density of $\kappa = 0.20$ (i.e., 20% density). We need to set $\tau \approx 0.30$ for the patient and $\tau \approx 0.41$ for the control (Figure 11.1c and e). The value of τ for the control network is higher because the connectivity weights are, on average, higher in this network. Conversely, we use a lower value of τ when thresholding the patient network in order to achieve the same connection density as the control network. As a result, more low-value correlations, which may represent spurious connections, are included in the patient's thresholded network. One consequence of retaining a higher proportion of potentially spurious connectivity estimates is that the network **topology** will appear more random. Several brain disorders, including schizophrenia, have been shown to display a more random topology compared to controls, in the presence of mean functional connectivity reductions (Rubinov et al., 2009; Lynall et al., 2010; Stam, 2014). In such cases, it can be difficult to disentangle variations in connectivity weight from differences in network topology. The increased randomness in network topology seen in schizophrenia may therefore be a consequence of the density-based thresholding procedure itself; namely, the need to retain more low-weight edges in patient networks in order to achieve a particular connection density. These low-weight edges will

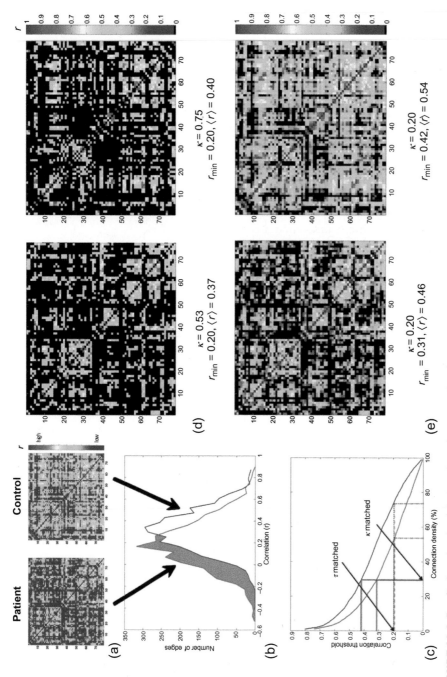

FIGURE 11.1 Density-based and weight-based matrix thresholding. Shown here is an example of the relative strengths and weakness of these two thresholding approaches. **(a)** Two individual functional connectivity matrices for 78-region networks estimated using **functional MRI**. One is from a patient with schizophrenia and the other is from a healthy control. The data are from the study reported by Fornito et al. (2011a). **(b)** The distribution of connectivity weights in each network. On average, connectivity is lower in the patient. **(c)** The relationship between τ and κ in these two networks. This plot illustrates how matching for τ results in networks with different connection densities while matching for κ requires us to apply a different weight threshold, τ, to both networks. **(d)** The thresholded adjacency matrices obtained for the patient and control after applying the same τ threshold. The minimum correlation value in the matrix, r_{min}, is the same, and the average correlation value, $\langle r \rangle$, across the two networks is comparable, but the connection density, κ, is different. **(e)** The thresholded adjacency matrices after applying the same κ threshold. The connection densities are the same, but the minimum and average correlations are different. *Figure adapted from Fornito et al. (2013) with permission.*

most likely be positioned randomly in the network. This problem affects both weighted and binary network analyses.

A possible alternative is to retain connections if they are statistically robust across the entire sample. For example, in a functional connectivity network, a one-sample t-test can be used to determine which correlations have a group mean that is significantly different from zero. This approach thus allows us to retain only those connections that are consistently identified across the sample, at some level of statistical confidence. In **diffusion MRI**, group-level thresholds have been shown to minimize the inclusion of false positive and false negative edges in the connectivity matrix (de Reus and van den Heuvel, 2013a). However, they also reduce inter-individual variability in network architecture, since the binary structure of the adjacency matrix is reduced to the connections identified with greatest consistency across the sample. As a result, any variations in network organization will be largely determined by differences in connectivity weight. Another limitation of group thresholds is that any reconstruction bias that consistently generates a false positive connection across the sample will also be declared significant. In other words, a one-sample t-test cannot suppress false positives that are consistent across a population.

The thresholding of connectivity matrices with signed edge weights (e.g., a correlation-based network) poses a special problem. In such cases, we must make a decision on how to treat negative weights. In general, a positive edge weight implies some degree of functional cooperation or integration between network nodes; a negative edge weight implies segregation or antagonism. If we wish to capture this qualitative distinction, we should threshold based on the absolute value of the correlation and then consider the polarity of the weights in subsequent analyses. This will ensure that both positive and negative edges are retained in the thresholded matrix. On the other hand, if we are only interested in positive connections, we can threshold the signed connectivity matrix (e.g., if our threshold is set to $\tau = 0.25$, any negative values will be set to zero).

The comparative strengths and limitations of weight-based and density-based thresholding highlight a conundrum. On the one hand, we know that topological properties vary as a function of the number of edges in a network, suggesting that we should use density-based thresholding. On the other hand, individual or group variations in connection density may themselves convey meaningful information, suggesting that any effect of these variations on network topology is real and should not be removed by density thresholding. This argument favors weight-based thresholding, but we know that comparing networks with different numbers of edges is problematic. The choice between the two methods depends on whether or not a difference in connection density is viewed as a confound. In some cases, it may be useful to understand how results vary across different thresholds selected using both approaches. Detailed discussions of these issues are provided by Ginestet et al. (2011), van Wijk et al. (2010), and Simpson et al. (2013).

11.1.2 Network Fragmentation

One limitation of using a global threshold is that it can cause networks to fragment at sufficiently stringent thresholds. Recall from Chapter 6 that a network is **node-connected** when all of its nodes can be linked to each other by a **path** of edges, thus forming a single **connected component**. **Fragmentation** occurs when the network splits into disconnected components.

In the absence of significant pathology, we expect nervous systems to be node-connected networks (Chapter 6). Thresholding can violate this basic property. As we increase our threshold and use more stringent criteria for retaining links, the connectivity matrix will become sparser and begin to fragment. This fragmentation will affect the measures that we compute on the networks. For example, traditional measures of **path length** cannot be computed between disconnected components (Chapter 7).

One solution to this problem is to constrain the range of τ and/or κ values, such that only node-connected networks are analysed. With this approach, we do not examine any thresholds that result in fragmentation of the network. However, in noisy datasets it is sometimes difficult to ensure connectedness for all individual networks. One potential solution is to first compute the **minimum spanning tree** (MST) of the network (Box 6.1), and then add connections in decreasing order of weight until the desired connection density is achieved. This method will ensure that a **path** can be found between all pairs of nodes and that the thresholded network will be node-connected.

11.1.3 Local Thresholding

Global thresholding is a simple and intuitive approach, but it may miss important structure in the data. For example, networks with fat-tailed weight distributions such as the brain (Chapter 4) may possess interesting features that span multiple scales. These features will be overlooked when a global threshold is applied, since connections to nodes with low average connectivity will be penalized (Serrano et al., 2009). Local thresholding methods aim to address this problem by computing thresholds determined locally at the level of each node, rather than at the level of the entire network.

One such approach is the so-called *disparity filter* (Serrano et al. 2009). The disparity filter operates on each node independently. For a given node, we first normalize the weights of each of its edges by the **strength** of that node (Chapter 4). In particular, we divide the weight of an edge by the strength (i.e., the sum of all edge weights) of the node under consideration to yield a node-wise normalized connectivity weight. The edges that account for a significant proportion of a given node's total connectivity with the rest of the network (i.e., its strength) are then identified with respect to a **null model** defining chance-expected fluctuations in regional weights. The null model for the disparity filter assumes that the node-wise normalized weights of a given node with binary **degree**, k, arise from random assignment from a uniform distribution. Since the node-wise normalized weights span the range [0, 1], the null model is given by distributing $k-1$ points with uniform probability in the unit interval, so that it is divided into k subintervals. The magnitude of these uniform subintervals is then compared to the node-wise normalized weights. Observed weights exceeding the chance-expected interval magnitudes for a given α-significance are declared statistically significant. Since weights are normalized on a node-wise basis, the new normalized edge values may be asymmetric, even in undirected networks. That is, the precise value of any particular normalized edge weight may vary depending on whether the strength of node i or j is used in the normalization. In such cases, the edge is retained if it meets the significance criterion for at least one of the two nodes to which it is attached.

An important property of the disparity filter is that it will only declare edges significant when there is considerable heterogeneity in the edge weights attached to a given node. To illustrate this point, take the following example. Consider a node in a correlation-based network with three edges, each of weight 0.70. The corresponding normalized weight of each edge will be $0.70/(3 \times 0.70) = 0.33$. If we then place $k-1 = 2$ points randomly on the unit interval, on average, the distance between consecutive points will be 0.33 which is identical to our observed normalized edge weights. In this case, none of the edges will be declared significant even though their weight is relatively high (on the scale of Pearson correlation). This example illustrates why the method is called a disparity filter—it is a method that is predominantly sensitive to large variations in edge weights at the local level of individual nodes. Nonparametric variants of this method have been proposed (Foti et al., 2011).

The disparity filter does not enforce a specific connection density on the data. Instead, the density is determined only by the proportion of edges that meet a given statistical significance. An alternative local thresholding method was proposed by Alexander-Bloch et al. (2010). Starting with the MST, they added edges incrementally based on the *k-nearest neighbor graph (k-NNG)*. The k-NNG comprises edges linking each node to its k nearest **neighbors**—the k nodes most strongly connected to the index node. With this method, k can be gradually increased until a desired connection density is achieved (Figure 11.2a). However, the method cannot guarantee a precise match to a given target density, since the number of edges added with a single increment of k may be larger than one. An analysis of resting-state functional MRI connectivity networks in children with childhood-onset schizophrenia and healthy controls

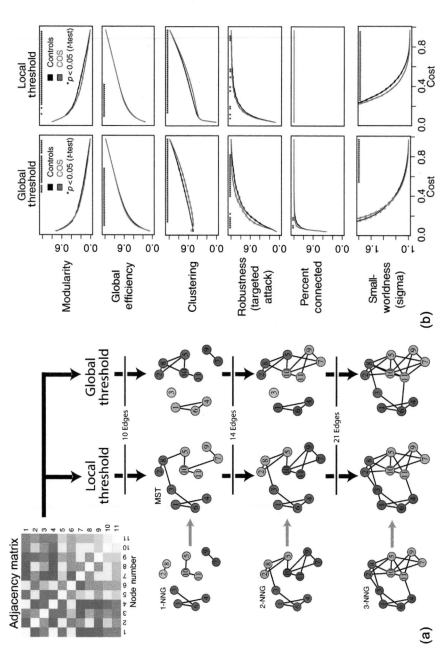

FIGURE 11.2 Comparison of global and local matrix thresholding. **(a)** An illustration of the difference between a local thresholding approach using k-nearest neighbor (k-NNG) graphs and global thresholding. The local threshold (left) first finds the MST of the graph, of which the 1-NNG is a subset. The parameter k is then increased to allow more edges into the network. For each k, we find the k-NNG graph, which includes each node's k most highly weighted connections. The global threshold (right) simply retains the most highly weighted edges in the graph at each threshold. **(b)** Different results can be obtained in comparisons of brain networks when these two thresholding methods are used. Shown here are differences in the topological properties of functional connectivity networks measured with functional MRI in people with childhood-onset schizophrenia (red) and healthy controls (black). The data are thresholded across a range of connection densities **(cost)**. As shown by the asterisks, which represent group differences significant at $p < .05$, uncorrected, the two thresholding methods vary in their sensitivity to group differences depending on the specific topological metric and connection density considered. *Images reproduced from Alexander-Bloch et al. (2010) with permission.*

found that the results of this method sometimes diverged with those obtained using a global thresholding approach, depending on the specific metric and threshold used (Figure 11.2b). The advantage of the k-NNG method is that any group differences in network measures cannot be attributed to network fragmentation, which is a problem when applying a global threshold. The disadvantage of this method is that the local thresholding process can, in and of itself, introduce nontrivial topological structure. For example, there will never be a node of degree less than k if the edges of the k-NNG are added to the thresholded network.

11.1.4   Normalization with Reference Networks

Global and local thresholding methods share the common goal of distinguishing real from spurious connections. These methods alter the network's connection density, or the range of connectivity weights in the adjacency matrix. We have already seen that most measures computed in a thresholded matrix depend on these basic properties. It is therefore important to interpret measures computed in thresholded networks with respect to the connection density that has been imposed by the thresholding process.

One potential approach to deal with the dependence of network measures on connection density is to normalize the measure with respect to an appropriate benchmark network, such as a **lattice** or randomized network matched to the empirical network for size, connection density, and **degree distribution** (see Chapter 10). These benchmarks set a baseline expectation of how a given network property, M, varies as a function of connection density in the absence of structured topology. However, detailed analysis of this method by van Wijk et al. (2010) has demonstrated that normalization with respect to either lattice-like or randomized networks does not completely remove the threshold dependence of commonly used topological measures such as the **characteristic path length** and **clustering coefficient**. Moreover, they found that the threshold-dependence varied according to the specific measure and benchmark network that was used.

One method that van Wijk et al. (2010) endorsed as a reasonable solution to the problem of threshold dependence was to normalize the observed measures relative to the most extreme values that these measures can assume, given the connection density. For **small-world** networks such as the brain, suitable bounds for this range are provided by the values obtained in matched lattice and randomized graphs. We can thus formally define a normalized measure for any network property as,

$$M' = \frac{M - M_{\text{rand}}}{M_{\text{lattice}} - M_{\text{rand}}}, \tag{11.1}$$

where the subscripts 'rand' and 'lattice' refer to the values obtained in the appropriately matched random and lattice networks, respectively (Chapter 10). More sophisticated methods for comparing networks with different connection densities, such as **exponential random graph models** (Box 10.2), hold promise but have not been extensively applied to the analysis of brain networks.

11.1.5   Integration and Area Under the Curve

It should now be clear that both local and global thresholding methods mandate selection of a particular threshold. The choice of threshold is rather arbitrary. Should we threshold to achieve a connection density of 5% or 10%, or should we only consider edges in a correlation-based network with a weight exceeding 0.5? To address the arbitrariness in this choice, we can repeat the desired statistical analysis across a range of thresholds, and examine the consistency of the findings. However, this method introduces a multiple comparisons problem, since we are computing a test statistic at many different thresholds. Generic methods for dealing with multiple comparisons are discussed in Section 11.2.

To sidestep the multiple comparisons problem, we can integrate a given network measure across the full range of thresholds analyzed. Analysis is then performed on the integrated measure, rather than on each of the many threshold-specific measures. The network measure of interest is integrated over a range of thresholds to yield the area under the curve (AUC), and then statistical testing is performed on the AUC (Bassett et al., 2006, 2012). The lower bound of this threshold range, κ_-, might be the point at which the network becomes node-connected, and the upper range, κ_+, might be the point at which the network is just short of being fully connected, or fails to show interesting topological structure, such as small-worldness (Lynall et al., 2010). With this approach, we start at κ_- and add edges, one at a time in decreasing order of weight (or some other criterion of priority) and compute the value of our topological metric, M_κ, at each connection density until we reach κ_+. This results in a curve that defines how M varies as a function of κ. We can then integrate these values of M_κ across all levels of κ to compute the area under this curve, and use this value in subsequent analyses. For example, we can imagine choosing a specific range of thresholds for the curves shown in Figure 11.2b, computing the area under those curves, and comparing the resulting values obtained for patients and controls.

This approach is similar in principle to the method for computing cost-integrated metrics favored by Ginestet et al. (2011). Specifically, these authors present theoretical arguments supporting inference on threshold-integrated metrics, showing that such measures are invariant to any monotonic remapping of the connectivity matrix. An example of a monotonic remapping is where the elements in one connectivity matrix are proportional to the corresponding elements in another connectivity matrix. Ginestet and colleagues prove that any threshold-integrated measure will be equal for these two matrices, which follows from the fact that a monotonic remapping does not affect the relative ranking of the elements of the adjacency matrix. Importantly, this approach results in a single value summarizing the behavior of any topological measure across a broad range of thresholds, thus reducing the multiple comparisons problem and simplifying analysis. The lower and upper bounds of the threshold range, κ_- and κ_+, should be chosen carefully: analyzing a wide range of thresholds will generally obscure effects that are present within a specific regime, whereas choosing a range that is too narrow may not adequately sample the data or miss effects that are apparent at other densities.

Drakesmith et al. (2015) have extended the AUC approach with the inclusion of cluster-based **permutation testing**, where a cluster in this context is defined as a contiguous range of thresholds at which the size of an effect exceeds a critical threshold. In an approach called multi-threshold permutation correction (MTPC), they perform a statistical test at each threshold independently. A null distribution is then generated for the test statistic computed for each threshold using permutation testing (Box 11.1). Drakesmith and colleagues argue that a genuine effect should exceed the null distribution over a contiguous range of thresholds. They therefore identify clusters along the threshold axis at which the observed test statistics exceed a critical test statistic cut-off value. The size of each cluster is then measured by computing the AUC over the *circumscribed* range of thresholds defining the cluster. Note that this is in contrast to the conventional approach of integrating the network measure of interest over the *entire* range to yield the AUC (Bassett et al., 2006). The process of identifying clusters is repeated after permuting the data to build a null distribution for the size of the largest cluster under the null hypothesis (see Box 11.1). This null distribution can then be used to assess the significance of the clusters measured in the observed data. Figure 11.3 presents results obtained with the MTPC method. The approach provides greater sensitivity to detect between-group differences in some network measures when compared to the conventional AUC approach, although the improvement is marginal for others. As with the AUC approach, the MTPC may also better control for the presence of false positive connections arising from an imperfect connectome reconstruction process. Considering a range of thresholds ensures inference is not dependent on a particular threshold that results in many false positive connections.

11.1.6 Multiresolution Thresholding

The thresholding approaches that we have considered in this section typically apply a hard threshold, such that all matrix elements with $w_{ij} \leq \tau$ are set to zero and all elements $w_{ij} > \tau$ are retained for further analysis. Since the choice of τ is arbitrary, this approach may create an artificial distinction between matrix elements that are useful and not useful. To address this problem, Lohse et al. (2014) investigated three alternative methods that offer insights into the

BOX 11.1 PERMUTATION TESTS

Permutation tests, also known as *randomization tests* and *exact tests*, are a versatile class of statistical significance tests in which the null distribution of a test statistic is estimated empirically by repeatedly permuting data points between groups, with each permutation providing a new sample from the null distribution. Permutation tests were first described in the 1930s by one of the founders of statistical science, Sir Ronald Fisher, and the Australian mathematician Edwin Pitman (Fisher, 1935; Pitman, 1937). The ubiquity of inexpensive and fast computers in the last few decades has made the application of permutation tests a natural choice for a wide range of problems in statistical inference where the true distribution of a test statistic is unknown.

Permutation tests can be most easily understood in terms of a *t*-test. Generalization to other test statistics is usually straightforward, but not always. Suppose we measure the clustering coefficient (Chapter 8), for example, in two groups comprising n_1 and n_2 individuals, and we seek to test whether the sample means are significantly different, given an α-significance level. This can be achieved with a two-sample *t*-test, in which case we assume the *t*-statistic can be parameterized under the null hypothesis by a theoretical distribution known as the Student's *t*-distribution. However, the theoretical null distribution is sometimes unknown and might provide a poor fit for small sample sizes even if it is known.

A permutation test uses the observed dataset itself to estimate the null distribution empirically. In particular, we pool together the clustering coefficients from both groups and then compute the *t*-statistic for every possible way of dividing these pooled values into two groups of sizes n_1 and n_2. There are $(n_1 + n_2)!/(n_1!n_2!)$ such permutations. For example, if x_1, x_2, x_3 denote the network clustering values in one group, and y_1, y_2, y_3 denote the clustering values in the other group, then one of the 20 possible permutations is x_1, y_2, x_3 for one group and y_1, x_2, y_3 for the other. We compute a *t*-statistic comparing these permuted groups and then repeat the process again many times. The permuted *t*-statistics that we obtain from this procedure define an exact null distribution under some basic assumptions.

A one-sided *p*-value is then given by the proportion of permuted *t*-statistics that are greater than or equal to the actual *t*-statistic. A two-sided *p*-value is given by the proportion of permuted *t*-statistics with absolute value greater than or equal to the absolute value of the actual *t*-statistic. When n_1 and/or n_2 are large, enumerating all possible permutations is intractable, and thus we can randomly sample a sufficient subset of permutations, usually between 5000 and 10,000, which is known as a Monte Carlo permutation test. Although we use parametric tests to quantify the difference in sample means, the *p*-values we compute with permutation tests are nonparametric. We can in principle use any measure to quantify the difference in sample means (e.g., percent difference), although there are good reasons to use a pivotal (normalized) quantity such as the *t*-statistic.

Permutation tests assume a technical condition known as **exchangeability**, which requires that the true null distribution is insensitive to permutations of the data, or equivalently, the data labels. For example, exchangeability is not satisfied in situations where the sizes and variances of the two groups are heterogeneous. Permutation tests have also been criticized for being computationally intensive, although nowadays this criticism is largely unfounded.

Permutation tests are a good way to account for dependencies between tests when performing mass univariate testing. Dependencies may be of a spatial or topological origin, or due to the transitive nature of the correlation coefficient (e.g., in functional connectivity networks). Under some basic assumptions, the (joint) null distribution estimated with a permutation test preserves any dependencies that are otherwise ignored by parametric approaches. To preserve dependencies, it is important to ensure that permutation is performed on all tests as a whole, rather than independently on each test.

A deeper treatment of permutation tests is provided by Good (1994) and Peasarin (2002). The work of Nichols and colleagues demonstrating the utility of permutation tests in traditional neuroimaging applications is also an invaluable resource (Nichols and Holmes, 2001; Nichols and Hayasaka, 2003).

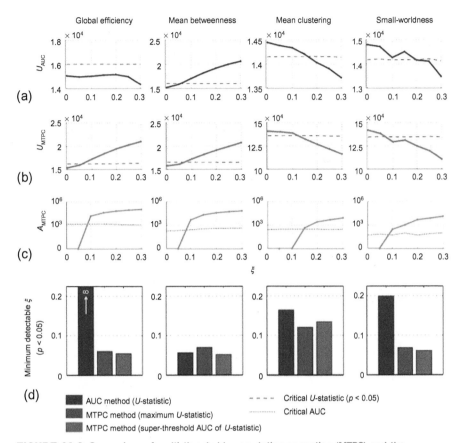

FIGURE 11.3 Comparison of multi-threshold permutation correction (MTPC) and the conventional area under curve (AUC) approach in detecting simulated between-group differences in four network measures. A difference between two populations of **structural connectivity** networks measured with diffusion MRI was simulated by removing interhemispheric streamlines from one of the populations. The proportion of streamlines removed was sampled from a half-normal distribution with standard deviation, ξ, and mean of zero. Increasing the standard deviation increased the magnitude of the between-group difference. The presence of a between-group difference in four network measures was then tested using the conventional AUC method **(a)**, MTPC based on the maximum test statistic observed across all thresholds considered **(b)**, and cluster-based MTPC **(c)**. The Mann-Whitney U-test was used in all cases. The vertical axis in panels **(a–c)** corresponds to the size of the test statistic. The minimum proportion of streamlines that needed to be removed for a significant difference to be declared (i.e., the smallest effect size required to reach significance) was then determined for the three methods **(d)**. The blue bar denotes AUC, brown denotes MTPC based on the maximum test statistic and blue is cluster-based MTPC. The MTPC methods offered improved sensitivity compared to the AUC method in most cases, although the improvement was marginal for the clustering coefficient and **betweenness**. *Images reproduced from Drakesmith et al. (2015) with permission.*

multiresolution structure of a brain network. One method involves first normalizing the weights of the adjacency matrix to the range [0, 1], and then raising the elements of the matrix to an arbitrary power r, such that $w_{ij} \rightarrow w_{ij}^r$. As r gets lower, the distribution of weights becomes more homogeneous until all elements have $w_{ij} = 1$ when $r = 0$. As $r \rightarrow \infty$, the distribution becomes more heterogeneous and the relative influence of strong connections is amplified. This method is suited to the analysis of weighted networks and is not a thresholding method, per se.

A second method investigated by Lohse et al. (2014) depends on the use of a multiresolution **module** decomposition of the network (see Chapter 9). With this approach, subsets of nodes within modules defined at particular scales of topological resolution are identified and their properties are investigated separately at each resolution. A third approach, termed *windowed thresholding*, involves independently assessing network properties within discrete windows of edge weights. A window is defined to span a limited range of edge weights. Edges with a weight that is spanned by the window are retained, while edges with a weight falling outside the window are set to zero. The length of the window dictates the number of edges that are retained and the position of the window determines the average weight of the retained edges. Examining network properties in different windows thus offers insight into the topological organization of connections within certain weight ranges. For example, we may wish to examine how the strongest connections are organized in the network, independently of the effects of weaker links. By construction, this method precludes an analysis of interactions between strong and weak links. Applying this method to functional connectivity networks measured with functional MRI in patients with schizophrenia and healthy controls, Lohse and colleagues only found between-group differences in windows with very low or high average percentile weight, but not for a broad range of intermediate threshold windows. This result suggests that topological disturbances in this group primarily affect the organization of very strong and very weak functional links.

11.2   STATISTICAL INFERENCE ON BRAIN NETWORKS

Connectivity weights and topological properties of brain networks can vary between different populations. Identifying differences in brain networks associated with brain disease, aging, and cognitive function (to name just a few) is an important application that requires statistical inference. In this section, we consider some of the main approaches to the statistical analysis of network measures.

11.2.1 Global Testing

Global, or omnibus testing, is the simplest kind of statistical inference that can be performed on brain networks. Global testing involves inference on one or more of the global topological measures considered in the earlier chapters, such as the average clustering coefficient, global **efficiency**, or Humphries' index of small-worldness (Humphries and Gurney, 2008). The exact network measure is not of concern to us here, since the basic approach is general. The goal is to determine whether one of these network parameters varies as a function of clinical diagnosis, age, cognitive ability, or some other characteristic of interest. Standard statistical tests can be employed to test for such associations, bearing in mind the usual caveats such as data normality, homogeneity of variance, and so on. Since it is often difficult to know the population distribution of most network measures a priori, the use of nonparametric statistics is recommended, as it offers a means for performing statistical analysis with a relaxed set of assumptions (Box 11.1).

In general, global testing is simple and offers insight into global network properties, but lacks specificity and may lack power when the effect of interest is limited to only a few connections or nodes. For example, if the clustering coefficient shows a difference at a single node in a network of hundreds of nodes, an omnibus test performed on the average clustering coefficient is unlikely to reveal a significant effect.

11.2.2 Mass Univariate Testing

Mass univariate testing is performed to localize effects to specific nodes or edges. It thus provides superior localizing power compared to global testing. Such an analysis allows us to move beyond global summary metrics to pinpoint which specific regions or connections might be driving the global alteration.

What exactly do we mean by mass univariate hypothesis testing? Mass univariate testing refers to the fact that a particular statistical test, such as a t-test, is performed independently across a large number of nodes or connections within a brain network. The *family* in the context of mass univariate testing defines the set of nodes or connections that we examine for an effect, which is often all of them. However, to reduce the number of multiple comparisons, it may be advantageous to constrain the family to a subset of connections or nodes based on prior beliefs.

When the family comprises nodes, we can perform inference on just about any node-specific measure, such as the node degree, local efficiency, or clustering coefficient. When the family comprises connections, we most often test for variations in connectivity strength (i.e., edge weights), or connection-specific topological properties, such as the edge betweenness centrality (Newman and Girvan, 2004). The particular measure is not of great concern to us here. The family might contain a mixture of nodes and connections, but this is rare.

Mass univariate testing yields a test statistic and corresponding p-value for each element in the family. Ginestet and Simmons (2011) refer to the family of test statistics as a statistical parametric network. The size of the family can range from hundreds to thousands, or even millions, depending on the size of the network. We denote the number of hypotheses in the family with J. We cannot simply reject the null hypothesis for all p-values that are less than a nominal significance of $\alpha = 0.05$, for example, since doing so can result in αJ false rejections, or *false positives*. This is an unacceptably large number of false positives. Usually, we must use a more stringent, or *corrected*, α-significance to ensure that the probability of one or more false positives across the entire family, known as the **familywise error rate** (FWER), is less than a prescribed level. This is known as the **multiple comparisons problem**.

There is a variety of old and trusted procedures for dealing with multiple comparisons. Bonferroni correction and control of the **false discovery rate** (FDR; Benjamini and Hochberg, 1995) are the most well-known. With the advance of statistical connectomics, more powerful correction procedures have been developed to exploit the spatial and topological properties of brain networks. These approaches can often provide greater statistical power than generic multiple comparison procedures such as the FDR. In other words, they provide us with greater power to detect true effects, while ensuring that the FWER is controlled at a prescribed level.

Most connectome-specific multiple comparison procedures have been designed for the case when the family comprises connections, since this is the most challenging case in terms of the scale of the multiple comparisons problem $(J \sim N^2)$. When the family comprises nodes, the multiple comparisons problem is substantially smaller $(J = N)$, and thus controlling the FDR may provide sufficient statistical power and is sometimes used in practice. The final goal of these methods is to allow the construction of a map depicting nodes and/or edges that show a particular effect of interest at a given level of statistical confidence. Interpretation of these maps can sometimes be complex, and is best done within an appropriate theoretical framework (Box 11.2).

11.2.3   Strong Control of Familywise Errors

The simplest and most conservative approach to the multiple comparisons problem is control of the FWER. This can be achieved with the classic Bonferonni procedure. To ensure a FWER no greater than a prescribed α, we reject the null hypothesis for all p-values satisfying,

$$p_j \leq \frac{\alpha}{J}, \quad j = 1, \ldots, J. \tag{11.2}$$

BOX 11.2 INTERPRETING CLINICAL DIFFERENCES IN BRAIN NETWORKS

Statistical connectomics is often used to characterize clinical differences between a group of patients with a given brain disorder and a healthy comparison group. For this purpose, the statistical methods discussed in this chapter offer a powerful framework for mapping abnormalities across the entire connectome. Once these maps are in hand, the major challenge is to interpret the findings in a way that clarifies disease mechanisms. Identifying a set of connections that is either increased or decreased in a given patient group may be useful, but the map will be most informative when it tells us something about the underlying disease process.

Interpretation of clinical differences in brain networks can be complicated by the complex pathological changes that often arise in brain disorders. For example, compared to healthy controls, patients with schizophrenia can show reduced structural connectivity, coupled with both increased and decreased functional connectivity (Skudlarski et al., 2010). Similarly, patients with multiple sclerosis, a disease that causes an inflammation and deterioration of structural connectivity, also show a complex pattern of increased and decreased functional connectivity (Hawellek et al., 2011). The simulation of large-scale brain dynamics with neural mass models (Deco et al., 2008) interconnected based on empirically derived structural connectivity networks suggests that distributed increases and decreases of functional connectivity can arise after the deletion of specific nodes (and their incident edges) of the structural network (Alstott et al., 2009; see also Figure 6.5). In each of these cases, a *reduction* in the integrity of the underlying structural network is associated with a complex combination of increases and decreases of functional connectivity.

We can reconcile these apparently contradictory findings by categorizing possible neural responses to insult into one of two broad classes: maladaptive and adaptive (Fornito et al., 2015). Figure 11.2.1 presents a schematic of how key examples of each class are expressed at the level of brain networks, and how they can be linked to underlying pathophysiological processes.

Maladaptive responses generally compound the effects of an initial insult. Three major examples are **diaschisis**, transneuronal degeneration, and dedifferentiation. Diaschisis was first described by von Monakow (1969) as the interruption in function of remote and otherwise intact areas that are connected to a lesioned site (Figure 11.2.1a). A classic example is crossed cerebellar diaschisis, where a unilateral lesion to the forebrain can result in a functional depression of the contralateral cerebellum (Gold and Lauritzen, 2002). These distal effects may arise from reduced afferent input to the intact area, or an impaired capacity for interregional synchronization following the lesion.

Transneuronal degeneration is the structural deterioration of these remote areas (Figure 11.2.1b). It may arise following prolonged exposure to aberrant signals emanating from the damaged site, which may lead to excitotoxic reactions in connected areas (Coan et al., 2014; Ross and Ebner, 1990). It may also be caused by deficits in axonal transport (Hirokawa et al., 2010) or the prion-like spread of pathological agents across synapses (Frost and Diamond, 2009).

Dedifferentiation refers to a breakdown of the brain's normally segregated and specialized processes (Li et al., 2001; Rajah, 2005; Figure 11.2.1c). For example, patients with schizophrenia show reduced activation of the canonical frontoparietal system during the performance of tests of executive function, coupled with increased activation in other areas (Minzenberg et al., 2009). Studies of functional brain network topology have also found evidence of reduced **modularity** in this patient group (Alexander-Bloch et al., 2010). This diffuse and desegregated pattern of activity may arise from aberrant plasticity, abnormal neurodevelopmental wiring, or altered modulatory influences from critical catecholamines such as dopamine, which may impact the fine-tuning and signal-to-noise ratio of neural information processing (Winterer and Weinberger, 2004).

Adaptive responses are changes in brain activity that are designed to maintain homeostasis and performance where possible. Three concepts that are important for these adaptive responses are compensation, neural reserve, and **degeneracy**. Compensation occurs when the activity of unaffected neural elements is increased to overcome deficits or dysfunction arising from an insult elsewhere in the brain (Figure 11.2.1d). For example, a unilateral stroke affecting motor cortex often results in increased activation of the motor area ipsilateral to the affected hand (Rehme et al., 2012). This recruitment of ipsilateral motor cortex has been shown to play a causal role in maintaining adequate motor performance (Johansen-Berg et al., 2002; O'Shea et al., 2007). Neural plasticity and cognitive flexibility facilitate the engagement of compensatory strategies.

Neural reserve is the amount of intact neural tissue within an affected system that is still able to support function. Reserve is implied if brain activity and performance are unaffected by an insult (Figure 11.2.1e). Degeneracy is the capacity of other neural elements to assume the function of the damaged system (Figure 11.2.1f; see also Chapter 8). The degeneracy of a nervous system depends on whether neurodevelopment and experience-dependent plasticity have sculpted network topology to enable distinct neural ensembles to support overlapping functions.

Continued

BOX 11.2 INTERPRETING CLINICAL DIFFERENCES IN BRAIN NETWORKS—CONT'D

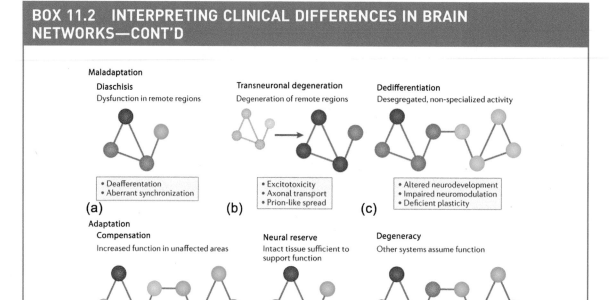

FIGURE 11.2.1 Maladaptive and adaptive responses to pathological perturbation of brain networks. This figure presents heuristic examples of how different neural responses to insult are expressed at the level of brain networks. Top row shows three examples of maladaptive responses. In each case, the behavioral output of the affected neural system will be impaired. Bottom row shows three examples of adaptive responses. In each case, the output of the system will be preserved as much as possible. Potential mechanisms underlying the changes are listed in the boxes under each graph. **(a)** In diaschisis, we see a reduction in the activity of structurally intact areas that are connected to a lesioned site. **(b)** In transneuronal degeneration, regions that are connected to a lesioned site show a structural deterioration over time (arrow). **(c)** Dedifferentiation occurs when the function of a system that is normally associated with a particular behavior is reduced, while the activity of other systems is increased. **(d)** Compensation occurs when other nodes of the affected neural system, or other neural systems, increase their activity to preserve behavior (the former is shown here). **(e)** If sufficient neural reserve is available following a lesion, there will be no behavioral impairment and no change in the activity of unaffected neural elements. **(f)** Degeneracy is evident when a second system assumes the function of the affected neural network, without any change in the activity of that second system (see Noppeney et al. (2004), for a discussion of other forms of degeneracy). Complex combinations of these responses may occur, and one may evolve into another over time. *Figure reproduced from Fornito et al. (2015) with permission.*

In this context, α/J is the *corrected* significance level that ensures the probability of making even *one* false rejection across the entire family of J tests is no more than α. The Bonferonni procedure follows from Boole's inequality, which asserts that the probability of rejecting the null hypothesis for at least one test in no more than the sum of the probabilities of rejecting each test.

The Sidak procedure (Sidak, 1967) is an alternative that offers a small gain in statistical power compared to the Bonferonni procedure, but requires that all the J tests are independent. With the Sidak procedure, we reject the null hypothesis for all p-values satisfying,

$$p_j \le 1 - (1-\alpha)^{1/J}, \quad j = 1, \dots, J, \tag{11.3}$$

where the right-hand side of this inequality comes from solving $\alpha = 1 - (1-\alpha_1)^J$ for α_1, the uncorrected significance level. Note that $(1-\alpha_1)^J$ is the probability that the null hypothesis is accepted for all J tests, assuming independence between the tests, and thus $1 - (1-\alpha_1)^J$ is the probability that the null hypothesis is rejected for at least one test.

The Holm-Bonferonni step-down procedure (Holm, 1979) is yet another alternative to the classic Bonferonni correction. Unlike the Sidak procedure, it does not assume independence between tests, although if the tests show positive dependence, the procedure becomes conservative. To ensure the FWER is no larger than a prescribed α, we identify the *smallest j* such that,

$$p_{(j)} > \frac{\alpha}{J+1-j}, \quad j = 1, \dots, J, \tag{11.4}$$

where $p_{(j)}$ denotes the jth smallest p-value. We then reject the null hypothesis for the $j-1$ smallest p-values; namely, the p-values that are smaller than $p_{(j)}$. All null hypotheses are accepted if the smallest p-value is greater than α/J. The Holm-Bonferonni procedure is universally more powerful than the classic Bonferonni correction.

11.2.4 Multiple Comparisons Under Dependence

Network measures computed across a family of nodes or connections are unlikely to be independent of each other. They usually show a positive dependence, although this issue has not been thoroughly studied and the general nature of any dependence is not known reliably. Positive dependence is especially evident in correlation-based networks. For example, if the neural activity at one node is disrupted, the functional connectivity (as measured by the correlation coefficient) between that node's activity and the activity at many other nodes may simultaneously show a disturbance. In this case, a test performed on each correlation coefficient will yield a set of test statistics that are positively dependent. Under this kind of positive dependence, the probability that the null hypothesis is accepted for all J tests is greater than $(1-\alpha_1)^J$, and thus $1-(1-\alpha_1)^J$ becomes a conservative estimate of the true FWER. As a result, we lose sensitivity for declaring effects as significant when there are dependencies between the tests.

Accounting for correlated tests when correcting for multiple comparisons is a complicated issue. When the test statistics are independent, we might commit $x > 0$ false positives on average whenever a familywise error is committed. Under positive dependence, although familywise errors are rarer, due to the Bonferonni and Sidak procedures becoming conservative, they are each likely to encompass $x' > x$ false positives. In other words, familywise errors are rarer but their consequences are more severe under positive dependence. This is because a single disturbance at a given node or connection can spread to neighboring nodes and connections, causing multiple test statistics to simultaneously reach significance and thus giving rise to clusters of false positives. This effect can be quantified with the generalized FWER (Lehmann and Romano, 2005). Clarke and Hall (2009) show that when the test statistics have typically observed null distributions, clusters of false positives are no more common than in the independent case.

Multiple comparisons under dependence is an issue that has been tackled in statistical genetics. A common approach is to determine the effective number of independent tests, $J_{\text{eff}} \leq J$, and then perform the Sidak procedure, for example, such that the null hypothesis is rejected for all p-values satisfying,

$$p_j \leq 1 - (1-\alpha)^{1/J_{\text{eff}}}, \quad j = 1, \dots J. \tag{11.5}$$

The effective number of tests can be estimated based on the **eigenvalues** (see Box 5.1) of the correlation matrix, since the eigenvalues quantify the extent of dependence between all pairs of tests (Li and Ji, 2005). Also see Leek and Storey (2008) for a related approach. Varoquaux et al. (2010) tackle the dependence issue in a fundamentally different way by representing edge-wise connectivity differences between two populations of networks as a series of independent and normally distributed perturbations of a group level covariance matrix.

In the connectomics literature, the dependence issue is either ignored, or resampling and permutation-based approaches are used. Permutation tests (Box 11.1) inherently ensure that any dependencies between tests carry through to the test statistic null distribution. Although generating empirical null distributions with permutation can be computationally burdensome, subsampling algorithms have been developed that improve computation times (Hinrichs et al., 2013). These factors, coupled with the ubiquity of parallel computing, make the application of permutation tests a natural and practical choice for many applications in statistical connectomics.

11.2.5 The False Discovery Rate

Bonferonni correction and related methods provide strong control of family-wise errors; that is, they ensure that the probability of a type I error does not exceed a desired α. However, they are generally considered too conservative for most practical applications in connectomics, particularly when the family is large. For example, if we have an undirected network of $N = 100$ nodes, $J = N(N-1)/2 = 4950$ (assuming a statistical test at each possible edge), and thus a typical FWER of $\alpha = 0.05$ mandates an uncorrected p-value of $0.05/4950 \approx 0.00001$. This uncorrected p-value corresponds to an effect size exceeding a Z-statistic of 4. Effects sizes of this magnitude may be rare in practice, leading to high false negative rates, or type II errors, where the null hypothesis is not rejected for many tests, even when it is false. It is for this reason that classic approaches to control the FWER in the strong sense are generally considered too conservative for performing inference on connectomic data.

Statistical power can be substantially improved if we are content to settle for a less conservative approach known as *weak* control of the FWER. Weak and strong control are the same when all the null hypotheses are true. However, with weak control, as soon as we reject the null hypothesis for at least one connection or node, control of the FWER is no longer guaranteed. In this case, we can be certain that an effect is present *somewhere* in the data, but we cannot be certain that we have localized that effect to a set of nodes or connections in a way that controls the FWER across the entire family of nodes or connections. Procedures that control the FWER in the weak sense usually rely on less conservative strategies for localizing effects to sets of nodes or connections. Control of the FDR is the most well-known such strategy.

The FDR is defined as,

$$\text{FDR} = E\left[\frac{V}{\max(1, S+V)}\right], \tag{11.6}$$

where V and S are random variables denoting the total number of false positives, or *false discoveries*, and the total number of true positives, respectively. Therefore, $S+V$ is equal to the total number of discoveries, and thus $V/(S+V)$ is the proportion of false discoveries. The max in the denominator ensures that the FDR is zero when no discoveries are made at all. The FDR has been highly influential and was one the first alternatives to the FWER that gained broad acceptance across a range of scientific fields, including brain imaging (Genovese et al., 2002).

In Equation (11.6), E is used to denote the average value of a variable. The FDR is therefore the average proportion of false discoveries. What then do we mean by an FDR of 0.05, for example? Of all the null hypotheses we reject, on average, we can expect 5% of them to be false positives. For instance, we might reject 100 null hypotheses, of which 5 are false positives, or equivalently, we might reject 20 null hypotheses, of which only one is a false discovery. In both cases, the FDR is 0.05. In this way, the FDR is adaptive in that the number of false discoveries (V) is normalized by the total number of discoveries ($S + V$). In other words, making one false discovery among 20 discoveries is probably acceptable, but most likely unacceptable among only two discoveries.

Controlling the FDR seeks to ensure that the proportion of false discoveries is on average less than a prescribed threshold α. Since we are controlling an average quantity, for any given experiment, the actual FDR may be lower or higher than α. However, if we perform many experiments, each time controlling the FDR at level α, we can be certain that the actual FDR averaged across these independent experiments is no more than α. In contrast, controlling the FWER guarantees the probability of even one false discovery is less than α for each experiment individually. Therefore, controlling the FDR provides greater power at the cost of an increased false positive rate.

Benjamini and Hochberg (1995) provide a well-known controlling procedure that ensures FDR $\leq \alpha$. We begin by sorting the p-values from smallest to largest, denoting the jth smallest p-value with $p_{(j)}$. We then identify the *largest* j such that,

$$p_{(j)} \leq \frac{\alpha j}{J}, \quad j = 1, \dots, J, \tag{11.7}$$

and reject the null hypotheses corresponding to the p-values $p_{(1)}, \dots, p_{(j)}$. Note that no rejections are made when $p_{(j)} > \alpha j / J$ for all $j = 1, \dots, J$. Rearranging this expression as $J p_{(j)} / j \leq \alpha$, we can see that $J p_{(j)} / j$ is a bound on the expected number of false discoveries. This is known as the Benjamini and Hochberg (BH) procedure and was first described by Simes (1986) as a procedure to control the FWER in the weak sense.

Storey (2002) has proposed a modification of the BH procedure that is more powerful when the proportion of true null hypotheses, denoted with π_0, is small. In these cases, the BH procedure can be overly conservative, because it actually controls the FDR at level $\pi_0 \alpha$, which is smaller than α if π_0 is small. Interested readers are referred to Storey (2002) for more details.

11.2.6   The Network-Based Statistic

The **network-based statistic** (NBS) is a network-specific approach to control the FWER in the weak sense when performing mass univariate testing on all connections in a network (Zalesky et al., 2010a). In analogy to statistical genetics, such analyses can be called connectome-wide analyses because they involve performing a hypothesis test at each and every element of the connectivity matrix.

The NBS is used in settings where each connection is associated with a test statistic and corresponding p-value, and the goal is to identify groups of connections showing a significant effect while controlling the FWER. The approach is nonparametric and can be applied to both functional and structural brain networks.

The NBS seeks to exploit the topological characteristics of an effect to boost the power with which that effect can be detected. In particular, the NBS utilizes the fact that typical effects of interest in brain networks, such as the pathological consequences of disease or the changes in brain activity evoked by a cognitive task, are seldom confined to a single locus. Rather, they are distributed throughout the brain in a manner that is ultimately constrained by the anatomical network topology (Fornito et al., 2015). Therefore, any such effects on brain networks are likely to encompass multiple connections and nodes, which form interconnected subnetworks. This is akin to the cascade of failures seen in engineered interdependent networks, such as power grids and the Internet, where a failure or component breakdown in one node rapidly spreads to affect other connected elements (Buldyrev et al., 2010). A corollary of this observation is that interesting variations in network connectivity are more likely to span multiple edges, rather than be confined to individual connections. The NBS defines these interconnected subnetworks as connected components and ascribes a familywise error corrected p-value to each subnetwork using permutation testing. Recall from Chapter 6 that a connected component is a set of nodes for which a **path** can be found linking any pair of nodes in that set.

The NBS is somewhat analogous to cluster-based approaches developed for performing inference on statistical parametric maps in human neuroimaging (Bullmore et al., 1999; Nichols and Holmes, 2001). Instead of identifying clusters of voxels in physical space, the NBS identifies connected subnetworks in topological space. The size of a subnetwork is most typically measured by the number of edges that it comprises.

Figure 11.4 provides an overview of the main steps of the NBS. First, a univariate test statistic is independently computed for each and every connection. In this case, we depict a t-statistic that tests the null hypothesis of equality in the mean connectivity weight between two populations: a "control" population and a "patient" population (Figure 11.4a). The result is a matrix of test statistic values with the same dimensions as the connectivity matrix (Figure 11.4b, left). Note that for an undirected network, the adjacency matrix is symmetric about the matrix diagonal so we only need to compute statistics for the upper triangular part of the matrix. Next, we apply a primary, component-forming threshold to the matrix of test statistic values (Figure 11.4b, right). This thresholded

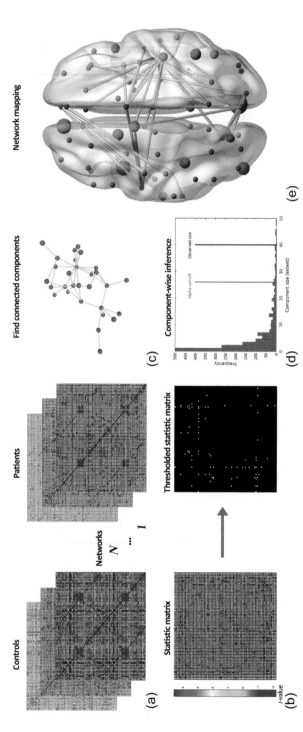

FIGURE 11.4 Key steps of the network-based statistic (NBS) analysis. The NBS methodology is illustrated with a comparison of functional connectivity networks in 15 healthy participants and 12 patients with chronic schizophrenia, as measured with resting-state functional MRI. Functional connectivity was measured for each pair of 74 anatomically defined regions using Pearson correlation between **wavelet** coefficients in the frequency interval $0.03 < f < 0.06$Hz. Details of the dataset are provided by Zalesky et al. (2010a). **(a)** We begin with two populations of connectivity matrices, one of controls (left) and one of patients (right). **(b)** A test statistic (in this case, a *t*-test) is computed at each and every matrix element, resulting in a matrix of statistic values (left). This matrix is then threshold using a primary, component-forming threshold to yield a thresholded and binarized statistic matrix (right). **(c)** The connected components of this thresholded, binarized statistic matrix are identified and the size of each (in terms of the number of links) is computed. Shown here is a topological projection of the largest connected component of the thresholded, binarized matrix depicted in the right panel of **(b)**. This component comprises 40 edges. **(d)** Data labels are permuted. In this case, the labels "control" and "patient" are randomly shuffled and reassigned to the connectivity matrices depicted in panel **(a)** and steps **(b–d)** are repeated. At each iteration, the size of the largest component is stored to generate an empirical null distribution of maximal component sizes, shown here. The red line indicates the cut-off for declaring a component size as statistically significant ($\alpha = 0.05$). The black line shows the observed size of the component illustrated in **(c)**. In this case, $p = 0.037$. **(e)** Projecting the network into anatomical space, we see that functional connectivity differences between patients and controls involve a distributed network of connections (yellow edges), largely centered on frontal and temporal areas. In this plot, node sizes correspond to the number of edges attached to each node in this network. Repeating the analysis using the FDR instead of the NBS found only one significant connection showing a difference between patients and controls (Zalesky et al., 2010a). The NBS thus offers a considerable gain in power, but cannot reject the null hypothesis at the level of individual edges.

matrix is then treated as a pseudo-network—a collection of edges showing an effect of interest at a nominal level of statistical confidence. This pseudo-network of test statistic values is then subjected to a breadth-first search (Box 6.3) to identify the number and size of connected components. These components can be interpreted as subnetworks of edges showing a common statistical effect of interest (Figure 11.4c). The size of each component is stored. The size of a connected component can be measured as the number of edges it comprises or the sum of the test statistic values associated with those edges. The latter approach is referred to as *cluster mass* in traditional image-based cluster statistics and can account for variation in effect sizes (Bullmore et al., 1999). Just counting the number of edges comprising a component ignores the effect size associated with each edge. If we sum the test statistic values, two components comprising the same number of edges can now have different sizes depending on the magnitude of the effect at each connection. In this way, the size of a cluster reflects a combination of both its topological extent and its effect size.

The observed component sizes are used by the NBS for statistical inference, so selecting an appropriate component-forming threshold is of critical importance. In principle, the threshold specifies the minimum test statistic value for which the null hypothesis can be potentially rejected and should be set according to the type of effects that are expected in the network. If we expect a strong effect restricted to a small subset of edges, the threshold should be conservative (e.g., convert the test statistics to p-values and remove all connections with $p < 0.001$, uncorrected), as it will result in smaller components comprising edges that respond strongly to the experimental manipulation. On the other hand, we might expect an effect to be weak at any individual edge, but to have a broad distribution across a large number of edges. In this case, we should use a less conservative threshold (e.g., $p < 0.05$, uncorrected). It is often difficult to know what type of effects should be expected, so it is useful to repeat analyses across a range of component-forming thresholds. Such an analysis allows us to understand whether effects are indeed strong but limited to a small number of edges, or weaker with a broader distribution across the network. In any case, it is important to remember that the FWER is controlled, at the level of connected components, irrespective of the primary threshold.

After the sizes of the observed components are computed, permutation testing (Box 11.1) is used to estimate a corrected p-value for the size of each observed component. In the example depicted in Figure 11.4, this involves randomly shuffling the labels assigned to each network so that the new "patient" and "control" groups comprise a mixture of actual patients and controls. The analysis is then repeated and the size of the *largest* component is stored. We repeat this procedure many times to generate an empirical null distribution of maximal component size. The corrected p-value for a component of size m is then given by the proportion of permutations for which the largest component is equal to or greater in size than m (Figure 11.4d).

Using the null distribution of maximal component size ensures control of the FWER. To understand why, recall that the FWER is the probability of one or more false positives at any connection. Therefore, as shown by Nichols and Hayasaka (2003), if we use s_i to denote the size of the ith component and α the desired FWER, it follows that,

$$
\begin{aligned}
FWER &= P\,(\geq 1 \text{ component declared significant}\,|\,H_0) \\
&= 1 - P\,(0 \text{ components declared significant}\,|\,H_0) \\
&= 1 - P\left(\bigcup_i \{s_i \leq t_\alpha\}\,|\,H_0\right) \\
&= 1 - P\,(\max\{s_i\} \leq t_\alpha) = \alpha,
\end{aligned}
\tag{11.8}
$$

where H_0 denotes the global null hypothesis. By storing the size of the largest component, the null distribution $F(t_\alpha) = P(\max\{s_i\} \leq t_\alpha)$ is estimated empirically. Each permutation contributes a new data point to this sample of the null distribution. The component size threshold, t_α, for an α-level significance is then determined such that $t_\alpha = F^{-1}(1 - \alpha)$. The crucial detail here is that ensuring the largest component is less than t_α guarantees that all components are less than this critical threshold. In particular, the above derivation shows that for all of the components to be of size less than t_α, we require that the largest component, $\max\{s_i\}$, is less than t_α.

With the NBS, the null hypothesis is always rejected at the level of components, not at the level of individual connections. Strictly speaking, it is therefore invalid to make inference about a specific connection embedded within a substantially larger component for which the null hypothesis is rejected. In other words, it is only valid to make inference about the component (subnetwork) as a whole. This is a consequence of weak control of the FWER. Thus, while the NBS can offer a considerable gain in statistical power (Zalesky et al., 2010a), this gain comes at the expense of not being able to localize effects to individual edges (e.g., Figure 11.4e).

11.2.7 Spatial Pairwise Clustering

Spatial pairwise clustering (SPC) is closely related to the NBS (Zalesky et al., 2012a). It is also a permutation-based approach that controls the FWER in the weak sense when performing mass univariate testing on all connections in a network. SPC differs from the NBS in the way that "clusters" are defined among connections in a network. Whereas the NBS defines clusters in terms of connected components, SPC uses a more stringent pairwise clustering approach that takes into account the network's spatial embedding.

To understand the difference between these clustering approaches, consider two connections (u_1, v_1) and (u_2, v_2). With the NBS, these two connections form

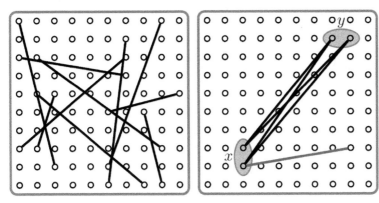

FIGURE 11.5 Connections forming a spatial pairwise cluster provide localised evidence for an effect. Connections distributed randomly (left) versus connections forming a spatial pairwise cluster (right). Each open circle is a node, while lines represent connections exceeding a predefined test statistic threshold. The pairwise cluster comprises four connections. The red connection is not part of the cluster because only one of its nodes is a neighbor. The presence of a pairwise cluster provides evidence of an effect between the two groups of nodes shaded gray. *Figure reproduced from Zalesky et al. (2012a) with permission.*

a cluster if $u_1 = u_2$ or $v_1 = v_2$, or equivalently, $u_1 = v_2$ or $v_1 = u_2$. A cluster defined as such is a connected component of size two. SPC defines clusters by taking into account the network's spatial embedding. In particular, (u_1, v_1) and (u_2, v_2) form a spatial pairwise cluster if the pair of nodes u_1 and u_2 as well as the pair of nodes v_1 and v_2 are spatial neighbors, or the same node. The concept of neighbors here is flexible. A pair of nodes can be considered spatial neighbors if they share a common border, or if the Euclidean distance between their centers does not exceed a given threshold. Figure 11.5 shows an example of a spatial pairwise cluster in a network that is embedded on a two-dimensional grid. Nodes are considered neighbors in this example if they are spatially **adjacent** relative to the grid.

The main advantage of SPC compared to the NBS is that effects can in some cases be localized with greater specificity to individual pairs of regions. The null hypothesis can therefore be rejected individually for each pair of regions. In contrast, the NBS is more likely to assimilate connectivity effects evident across distinct regions into a single, all-encompassing network. Figure 11.6 provides an example of this phenomenon in a study aiming to localize changes in human brain functional connectivity measured with **EEG** during a working memory task performed at different memory loads. SPC identified three separate spatial pairwise clusters, each corresponding to a different group of brain regions between which connectivity changed as a function of the working memory load. The null hypothesis can therefore be evaluated for each cluster individually. In contrast, the NBS identified a single subnetwork that

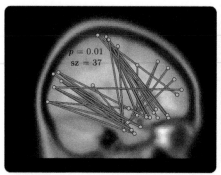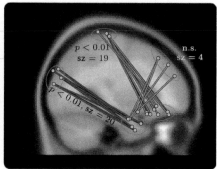

FIGURE 11.6 Comparison of connections identified by the NBS and by spatial pairwise clustering (SPC). Changes in functional brain connectivity across different working memory loads in healthy individuals were localized with the NBS (left) and SPC (right). Functional connectivity was measured with source analysis of EEG data. With the NBS, a single subnetwork can be declared significant, whereas SPC enables finer localization of this subnetwork, identifying three distinct spatial pairwise clusters, of which two can be declared significant ($p < 0.05$). The connections comprising a common cluster are colored identically. sz, size of the cluster; n.s., not significant. *Figure reproduced from Zalesky et al. (2012a) with permission.*

assimilated all three of these distinct spatial pairwise clusters identified with SPC. With the NBS, the null hypothesis was therefore only rejected at the level of the entire subnetwork.

SPC has a higher computational burden than the NBS. Moreover, the finer localization provided by SPC may be to the detriment of statistical power. In particular, a spatial pairwise cluster in itself may be too small in size to reach statistical significance; however, if it is assimilated into a connected component comprising other pairwise clusters, the connected component may be large enough for the null hypothesis to be rejected with the NBS.

SPC is most advantageous when performing inference on networks with high spatial resolution, such as voxel-based networks constructed with MRI or high-density source analysis of M/EEG. For coarser networks, it is more likely that a difference in connectivity between two regions is only sampled by a single pair of nodes, and thus a pairwise cluster cannot be formed. In these cases, the NBS may provide greater power. Variants of the SPC have also been developed for the analysis of dynamic functional connectivity networks (Hipp et al., 2011).

Clusters among connections in a network can in principle be defined in many different ways. Connected components and spatial pairwise clusters are perhaps the most obvious and intuitive definitions, hence their use in the NBS and SPC, respectively. Since there is no need to derive a theoretical null distribution when using permutation tests, the NBS and SPC methodologies can be trivially adapted to suit any sensible definition of a cluster. For example, in their analysis of functional connectivity in hypercapnic states, Ing and Schwarzbauer (2014) define clusters as spatially contiguous groups of nodes/voxels, where each node/voxel is associated with at least one suprathreshold connection. In this way, we can identify groups of voxels that are associated with a connectivity effect. Clusters in the context of the NBS could also be defined in terms of network modules (Chapter 9).

11.2.8 The Screening-Filtering Method

Meskaldji and colleagues (2014) propose a screening-filtering method to control the FWER across a family of J tests. A closely related method was originally called subnetwork-based analysis (Meskaldji et al., 2011). The approach is generic and not specific to connectomic analysis.

The J tests in a family are first partitioned into m subsets, where each subset comprises $s = J/m$ tests. A summary statistic is determined for each subset by summing the s test statistics associated with each test and normalizing this sum by $\sqrt{s}$, thus yielding a normally distributed summary statistic with zero mean and unity variance under the null hypothesis. The subsets are then divided into two classes according to whether or not their summary statistic exceeds a prescribed threshold. As part of this screening step, the null hypothesis is accepted for all tests comprising subsets with a summary statistic below the prescribed threshold. The remaining *positive* subsets are given further consideration in a filtering step. The rationale here is that if the prescribed threshold is chosen to control the FWER across the m subsets, then weak control is guaranteed across the family of J tests.

The filtering step involves dividing the original p-values corresponding to all the tests comprising the positive subsets with a relaxation coefficient, $r > 0$. The classic Bonferonni correction is then applied to the relaxed p-values. The computation of r is achieved with an approximation algorithm. For a given r, the expected number of false positives can be computed if the proportion of true null hypotheses per each positive subset is known. These proportions are unknown in practice, and thus the algorithm loops over all feasible proportions, returning the worst-case expected number of false positives. The algorithm continues incrementing r by a small quantity until the worst-case expected number of false positives exceeds a desired α.

The success of the screening-filtering method is contingent on a judicious partitioning of the J tests into m subsets. In particular, a partitioning is required in which the tests corresponding to true positives are all allocated to as few of the m subsets as possible. Such a partitioning requires prior knowledge about the location of true positives, which is of course unknown. Meskaldji and colleagues (2014) evaluate the power of the screening-filtering method using simulated datasets in which all the true positives are assigned to $m_1 = 2$, 5, or 10 *affected* subsets out of a total of $m = 20$ or 50 subsets. Each of the affected subsets comprises a proportion $\pi = 0.25$, 0.5, or 0.75 of true positives. Therefore, the total number of true positives is given by $J\pi m_1/m$. These parameter choices may provide a somewhat unrealistic representation of the partition quality that can be achieved in practice when we have no prior knowledge about the location of true positives. For example, when $J = 2000$, $m_1 = 5$, $m = 20$, and $\pi = 0.25$, in the absence of any prior knowledge, the probability of actually finding a given subset with $\pi \geq 0.25$ is in the order of 10^{-9}.

11.2.9   Sum of Powered Score

Kim et al. (2014) propose a sum of powered score (SPU) test, which was originally devised to identify rare variants in genome-wide association studies. Strictly speaking, the approach provides only a global test, and thus between-group differences cannot be localized to specific connections or nodes. If the null hypothesis is rejected, we can only claim that an effect is present somewhere in the network.

In the absence of any nuisance covariates, the simplest SPU test boils down to computing a score, U_j, for each hypothesis $j = 1, \ldots, J$, according to,

$$U_j = \sum_{i=1}^{n} X_{ij}(Y_i - \bar{Y}), \tag{11.9}$$

where X_{ij} is the value of the observation for the jth hypothesis of the ith individual, $\mathbf{Y} = (Y_1, \ldots, Y_n)$ is a binary vector partitioning the sample into two groups and $\bar{Y}$ is its mean. For example, X_{ij} might contain the edge weight for the jth connection of the ith individual. Intuitively, U_j measures the covariance between the observed data values $(X_{1j}, \ldots, X_{nj})$ and the demeaned grouping vector. The rationale here is that evidence against the null hypothesis grows as a function of the covariance between the observed data values and the grouping vector.

An omnibus test statistic is then computed according to $T_{\text{SPU}} = \sum_{j=1}^{J} U_j^{\gamma}$, where $\gamma \geq 1$ is a parameter that can be chosen to give more weight to larger elements of U. Care must be taken when setting γ, since odd values may render the test statistic insensitive to bidirectional effects. To perform statistical inference, a null distribution is computed for T_{SPU} using a permutation test (Box 11.1), thereby yielding a corrected p-value. Kim and colleagues (2014) also consider an adaptive test statistic $T_{\text{aSPU}} = \min_{\gamma \in \Gamma} P_{\text{SPU}}(\gamma)$, where $\Gamma = \{1, 2, \ldots, \infty\}$ and $P_{\text{SPU}}(\gamma)$ is the p-value computed for T_{SPU} with γ. This accounts for the arbitrariness in selecting γ. Once again, a permutation test is used to compute a p-value for T_{aSPU}.

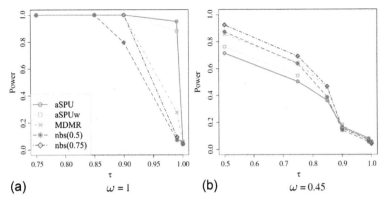

FIGURE 11.7 Comparison of the statistical power of the sum of powered score (SPU) and the network-based statistic (NBS). Between-group differences were simulated in a network comprising 2701 connections. The vertical axis (power) shows the probability of rejecting the global null hypothesis of no between-group difference anywhere in the network. The horizontal axis shows the proportion of connections at which the null hypothesis is true. The null hypothesis is true everywhere when the value of the horizontal axis is one, and thus this point represents the type 1 error rate. The parameter ω controls the effect size at each connection. In terms of the global null hypothesis, the NBS outperforms SPU when the effect size is small **(b)**, but SPU may be advantageous when the effect size is large and confined to a limited number of connections **(a)**. Note however that SPU does not localize rejection of the null hypothesis to particular connections or subnetworks. MDMR (multivariate matrix distance regression) is an alternative approach evaluated by Kim et al. (2014; see Section 11.3.2). Note that nbs(x) denotes the NBS with a primary threshold of x, and aSPU and aSPUw denote adaptive and weighted variants of the SPU, respectively. *Figure reproduced from Kim et al. (2014) with permission.*

To localize any effects to specific nodes or connections, Kim and colleagues suggest ranking the J tests based on their U score and plotting the true positive rate for the top-m ranked connections as a function of m. However, for any given value of m, we have no estimate of the FWER. Kim and colleagues used simulated data to compare the statistical power of the SPU test relative to the NBS. We can see in Figure 11.7 that the NBS outperforms SPU when the effect size is small, but SPU is advantageous when the effect size is large and confined to a limited number of connections.

11.3 MULTIVARIATE APPROACHES

As we have already seen, mass univariate hypothesis testing allows inference at the level of individual connectome elements and subnetworks. Although this offers tremendous localizing power, there may be fundamental limits to what we can learn about a connectome by studying its elements in isolation. This viewpoint follows from the notion of emergent phenomena—a fundamental tenet of systems biology—which contends that complex biological functions emerge from complex interactions that cannot be trivially reduced to isolated elements and processes (Bassett and Gazzaniga, 2011).

Multivariate approaches seek to recognize and learn complex patterns among multiple connectome elements and utilize these patterns for inferential classification or prediction. This is in contrast to the methods described earlier in the chapter, which focus on fitting a model to minimize the sum of squared errors. We may wish to classify a group of patients into putative disease subtypes, or predict a patient's diagnosis based on connectivity patterns. Unlike mass univariate hypothesis testing, multivariate approaches usually avoid the multiple comparisons problem because the significance of an entire connectivity pattern is evaluated using a single test (Varoquaux and Craddock, 2013).

Multivariate approaches encompass a broad range of methods, including pattern recognition algorithms, principal component and discriminant analysis, and machine and deep learning algorithms. Many of these methods have been around for decades, but their application to the field of connectomics is in its early stages. Some readers may be familiar with the application of these methods to functional MRI data, under the guise of multivoxel pattern analysis (MVPA; Norman et al., 2006). MVPA utilizes pattern classification and machine learning methods to discover patterns of activity across multiple voxels in functional MRI data that can be used to distinguish between distinct cognitive states. In this section, we briefly overview some of the main approaches that have been used in the analysis of brain connectivity data.

11.3.1 Support Vector Machines

Support vector machines (SVMs) are the most common multivariate approach used in the connectomics literature. SVMs in their simplest form are used to classify an individual into one of two groups based on the presence or absence of complex combinations of connectome features. The most discriminatory feature combinations are learned during a training phase in which the SVM is presented with individuals from both groups and told the particular group to which each individual belongs. An optimization problem is then solved to generate a model that optimally maps each individual to a high-dimensional feature space in such a way that the two groups are maximally separated. To classify new individuals, they are mapped to the high-dimensional feature space and assigned to one of the two groups based on which side of the gap they fall. The reader is referred to the classic tutorial by Burges (1998) for an in depth mathematical treatment of SVMs. The topic is also covered in numerous books on machine learning and artificial intelligence (e.g., Mohri et al., 2012).

To evaluate the classification accuracy of an SVM, we usually apportion the data into training and validation subsets using methods such as k-fold cross-validation. The training subset is used to train the SVM, after which the SVM is then asked to classify each individual comprising the validation subset. Ideally, the training and validation subsets should comprise datasets acquired independently using a range of different techniques to ensure generalizability across different settings, although this is not always possible. In any case, it is crucial that the individuals allocated to the validation subset are kept hidden during the training phase. Feature selection is used to identify network elements that provide the greatest power in predicting an individual's membership of a particular group. The t-statistic is an example of a simple feature selector, in the sense that networks elements with a high t-statistic are likely to be more predictive. While feature selection is an important step to prevent model over-fitting, it should never be performed on the validation subset, since doing so can substantially inflate the estimated classification accuracies.

SVMs can also be extended to analyze continuous outcome variables, using a variant called support vector regression. Following the work of Dosenbach et al. (2010) in which an individual's age was predicted on the basis of his/her resting-state functional connectivity (Figure 11.8a), a number of SVM-based studies have reported the ability to classify various outcomes, such as cognitive or disease state, based on connectivity measures (e.g., Cole et al., 2013; Ekman et al., 2012; Zeng et al., 2012; Figure 11.8b). A particular advantage of these methods is that they allow inferences about individual cases. For example, it is in principle possible with SVMs to measure some aspect of brain structure or function in a new patient, compare the measure to an existing data set (the training set) and determine the probability that the new person belongs in one group or the other (i.e., is this person a patient or control?).

In view of this potential, SVMs and machine learning more generally have been proposed as methods that can potentially assist in providing a more objective, personalized and biologically based diagnosis of brain disorders. The ultimate aim of this work is to be able to determine the probability that a given individual carries a diagnosis of a particular disease given their pattern of brain connectivity. However, much of the work to date has focused on simply classifying patients and controls based on measures of brain structure or function. These distinctions are often trivial (and less expensive) for most clinicians to make without recourse to multivariate analysis of network parameters. Predicting illness characteristics that are more difficult to discern based on clinical examination, such as treatment response and illness course, is a more challenging task. SVMs trained to detect connectome pathology that is predictive of a particular response to a drug, for example, will most likely offer the greatest potential to influence clinical decision-making.

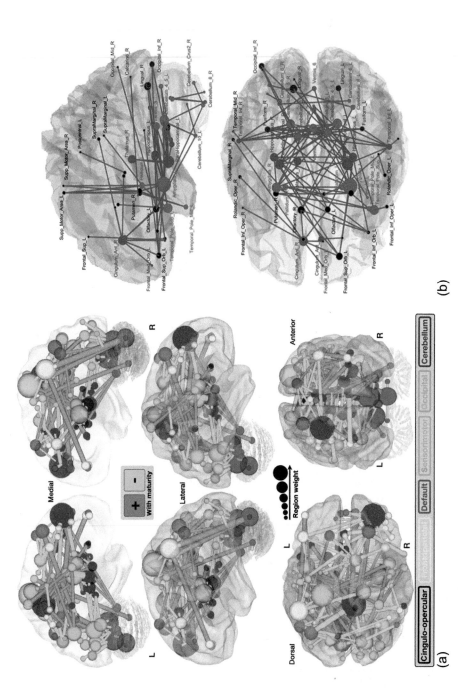

FIGURE 11.8 Combining connectomics and machine learning to investigate the demographic and clinical variability of human brain networks.
(a) Functional links, as measured with human resting-state functional MRI, that contribute to predicting participant age in a support vector machine. Edge thickness scales with weight; node size with regional strength. Links positively correlated with age are in orange, links negatively correlated with age are in green. Nodes are colored according to the broad functional system to which they belong (see key at bottom of figure). **(b)** Functional links, measured with human resting-state functional MRI, that contribute to the discrimination between patients with major depression and healthy controls. In this analysis, patients were classified with 94% accuracy based on functional connectivity measures alone. Nodes are colored according to the functional system to which they belong (blue is default mode network; purple is an affective system; green is visual cortex; red is cerebellum; and gray denotes other areas). **(a)** Reproduced from Dosenbach et al. (2010) and **(b)** from Zeng et al. (2012) with permission.

11.3.2 Other Multivariate Approaches

Shehzad et al. (2014) develop a novel multivariate approach to analyze brain connectivity data. Unlike SVMs, their method does not allow for inference at the level of individuals. It is rather a multivariate statistic that combines evidence across multiple connections to maximize statistical power, although this can be to the detriment of highly focal effects. First used in the analysis of functional MRI data, the method performs inference on connections, but is implemented in such a way that the null hypothesis is rejected at the level of spatially contiguous clusters of voxels, not connections, yielding outputs that are similar in presentation to classic functional MRI activation studies.

In the analysis of functional MRI data, the method proceeds as follows. For each individual, we compute the functional connectivity between a given gray-matter voxel and all other gray-matter voxels, and store all these values as a single vector, with one vector for each of n individuals. Using these vectors, we then form a symmetric $n \times n$ matrix expressing the "distance" between each pair of individuals in terms of their connectivity patterns. In particular, element (i,j) of the distance matrix is populated with the distance measure $\sqrt{2(1 - r_{ij})}$, where r_{ij} denotes the Pearson correlation coefficient between the vector of connectivity values for the ith and jth individual. This distance measure is zero for individuals with perfectly correlated connectivity patterns, one for uncorrelated patterns and two for perfectly negatively correlated patterns. Next, we use an established approach called multivariate distance matrix regression (MDMR) to test if the distances between individuals within the same group are shorter than the distances between individuals across different groups. MDMR is a multivariate test yielding a pseudo-F-statistic that assesses between-group differences or some other desired effect. This entire procedure is then repeated for all gray-matter voxels. To control the FWER across all voxels, a cluster-based permutation test is performed (Nichols and Holmes, 2001), which yields clusters of voxels for which the null hypothesis can be rejected. It is important to note that this approach does not test specific connections for an effect, but rather tests whether all the connections originating from a single node can cumulatively provide sufficient evidence to reject the null hypothesis at that node.

Deep learning and artificial neural networks are another group of promising multivariate approaches that are only beginning to be used to identify informative connectivity patterns in the connectome and other neuroimaging data (Plis et al., 2014). Partial least squares (PLS) is a well-established multivariate approach that can also be used to perform multivariate inference on brain networks (Krishnan et al., 2011; McIntosh et al., 1996). The basic idea of PLS is to identify latent variables that express the largest amount of covariation between a set of response variables (e.g., clinical diagnosis or symptom scores), and a set of predictor variables (e.g., brain network connectivity or topological measures). Permutation testing can be used to assess the significance of each latent variable. Canonical correlation analysis (CCA) is related to PLS and has been used to identify multivariate relations linking demographic and psychometric measures with patterns of human brain functional connectivity (Smith et al, 2015). CCA is essentially a doubly multivariate approach that involves analyzing "correlations of correlations", and which does not readily allow for the statistical identification of specific connections, subnetworks and regions that underpin these multivariate relations. While these multivariate approaches have not gained substantial traction in the connectomics literature, it is likely that they will play a greater role in future studies using multivariate inference to link components of network topology to clinical, cognitive and neurobiological data.

11.4 SUMMARY

We began this chapter with a look at methods for thresholding connectivity matrices. Thresholding is intended to suppress spurious connections that may arise from measurement noise and imperfect connectome reconstruction techniques. While thresholding is not a mandatory step before performing statistical inference, it can potentially improve statistical power and interpretability. Density-based thresholding treats differences in connection density across populations as a confound, whereas weight-based methods assume that differences in density reflect meaningful variation in the distribution of connectivity weights. Methods for local thresholding can additionally be used to address the problems posed by network fragmentation that can occur with global thresholding. However, the process of local thresholding can in and of itself introduce nontrivial topological structure. Integration over a range of thresholds avoids arbitrariness in the choice of threshold.

The second part of the chapter focused on methods for performing statistical inference on brain networks. Inference can be performed at the level of network-wide measures, which is referred to as omnibus or global testing. We can also test hypotheses at the level of individual network elements. This allows us to identify the specific network elements responsible for an effect. In other words, we can reject the null hypothesis at the level of individual nodes, connections or subnetworks. For example, we may want to pinpoint specific connections that substantially differ in their connectivity weight between two populations of individuals. This is referred to as mass univariate hypothesis testing. We saw that the multiple comparisons problem is particularly sizeable when mass univariate testing is performed on all connections.

Network-specific methods for dealing with the multiple comparisons problem have been developed that provide greater power than generic multiple comparisons such as the FDR. These network-specific methods exploit the fact that effects in a brain network, as with many other biological and engineered networks, are more likely to form interconnected clusters, rather than appearing across many disconnected network elements.

Finally, we briefly discussed multivariate approaches for statistical inference. The advantage of methods such as support vector machines is that inference can be made at the level of individuals. Multivariate methods are likely to play a greater role in the future of statistical connectomics, as datasets continue to increase in size and complexity.

Glossary

A

Adjacency matrix A matrix representation of a network in which each element encodes the presence of a weighted or binary connection (edge) between a unique pair of nodes.

Adjacent Two nodes can be adjacent topologically or spatially. Topological adjacency means that two nodes are directly connected. Spatial adjacency means that two nodes are close together in physical space. The two are not equivalent — for example, two nodes at the frontal and occipital poles of a human brain network are not physically adjacent, but they may be topologically adjacent if they share a direct connection.

Assortativity The tendency of nodes with similar properties to connect to each other. Most commonly investigated with respect to degree, such that assortativity measures the tendency of high-degree nodes to connect to other high-degree nodes, and low-degree nodes to connect to other low-degree nodes.

Autocorrelation The cross-correlation of a temporally evolving signal with itself at different points in time, which is estimated across a range of time lags of the signal. White noise has zero autocorrelation (except at a time lag of zero) whereas autocorrelated signals will show a dependence between time points and are sometimes called colored noise. A positively autocorrelated process is said to have memory, or persistence. In the frequency domain, positive autocorrelation is often reflected by $1/f$ scaling of spectral power. Neurophysiological signals generally show positive autocorrelation.

B

Betweenness A measure of topological centrality based on the fraction of shortest paths that pass through a given node or edge. A node or edge with high betweenness can be described as a mediator or broker of network integration.

Binary graph A graph in which edges are marked simply as either present or absent. Binary graphs can be either directed or undirected. They are also sometimes referred to as unweighted graphs, and can be contrasted with a weighted graph, in which edges are assigned values that index the strength of connectivity between nodes.

Block model A model of how subsets of nodes interact with each other. Each subset is treated as a single block and the model is used to specify relations between every pair of blocks. In connectomics, blocks have been used to represent core and peripheral nodes, or nodes in different modules. Stochastic block models are generalizations of the block model that allow for probabilistic relationships between nodes.

Blood-oxygenation-level-dependent (BOLD) signal A contrast signal, measured with functional magnetic resonance imaging, which is sensitive to changes in the magnetic properties of oxygenated and deoxygenated blood. Active neurons require oxygen to support their metabolic needs, prompting an influx of oxygenated blood into the area. This influx changes the magnetic properties of the local tissue, causing a signal contrast that is detected by the scanner. The BOLD signal is thus a hemodynamic marker of neuronal activity. Measurements can be localized at millimeter resolution throughout the brain. Temporal resolution is on the order of several seconds, which is slower than most fluctuations in neuronal activity.

Broad scale An intermediate class of degree distributions that is not unimodal, like the Poisson distribution, but is not entirely scale-free, like a power-law distribution. Broad-scale distributions are typically characterized by a power-law regime followed by a cut-off after which the tail of the distribution decays exponentially or according to a Gaussian function. An example is the exponentially truncated power law, which has often been found to provide a good fit to the degree distributions of brain networks.

C

Centrality The influence or importance of a node or edge in network function. This influence is defined in relation to the connection topology of the node or edge. There are many different centrality metrics, each of which emphasizes distinct aspects of topological importance.

Characteristic path length The mean of the shortest path lengths between all possible pairs of nodes in a graph. The shortest path in a binary graph is the sequence of edges between two distinct nodes that comprise the fewest possible edges (or minimum hop count). In weighted graphs, the shortest path is a sequence of edges whose sum of weights is minimum. Random graphs are associated with a short characteristic path length. Brain graphs typically have a characteristic path length that is comparable to a random graph. Regular, lattice-like networks have a long characteristic path length.

Closeness Two nodes are topologically close if they are connected by a short path. More generally, a node has high closeness centrality if it is connected, on average, by short paths to a large number of other nodes in the network. Nodal efficiency and closeness are equivalent measures.

Clustering coefficient A topological measure of the tendency for any two neighbors of a node to be directly connected to each other. If the neighbors are connected, the three nodes form a closed triangular motif. The clustering coefficient is typically high in regular, lattice-like networks and low in random graphs.

Communicability The weighted sum of the total number of walks between a pair of nodes. Walks comprising fewer hops are weighted to have a greater influence on the communicability between nodes. It is a measure of the efficiency with which information can be communicated in a network according to a diffusion process.

Community A subset of nodes that are more densely connected to each other than to other network nodes. The strong interconnectivity of nodes within a community implies some commonality or specialization of function. Communities are also often referred to as modules. Brain networks typically comprise a hierarchy of communities across different measurement scales.

Configuration model A generalization of the Erdös-Rényi random graph model that allows for an arbitrary degree sequence. It is often used as a null model in the assessment of network modularity.

Connected component A subset of nodes in which all node pairs can be linked by a path. Most networks consist of a single connected component that encompasses most, if not all, nodes. A network is fragmented if it comprises more than one connected component. In this case, the components represent subsets of nodes that cannot be linked by a path.

Connection density The number of edges in a graph, divided by the number of edges in a fully connected graph with the same number of nodes. Often this quantity is expressed as a percentage. For example, a network with 30% connection density contains links between 30% of all possible pairs of nodes.

Connectome A complete description of the anatomical connections between all elements of a nervous system. This description can be detailed at different resolution scales, from the level of individual neurons and synapses to macroscopic connectivity between large-scale neuronal populations. The term connectome is also often used to describe the organization of functional connectivity networks.

Connectomics The field of neuroscience concerned with the mapping and analysis of connectomes, at all resolution scales and across all species.

Cost Spatially embedded networks, like the brain, are associated with physical costs. Wiring cost is a key example of the cost of a connectome: every millimeter of axonal or dendritic wiring costs energy, biological material, and space within the limited intracranial volume. A pressure to minimize wiring costs can explain many, although not all, aspects of brain network organization.

Cycle A walk on a graph in which the first and last nodes are the same.

D

Degeneracy The capacity of different elements of a system to assume the same function. The elements of a degenerate system will generally mediate separate, yet overlapping, processes. Neural degeneracy can be measured by perturbing different elements of a nervous system and measuring the effects of the perturbation on a particular output of the system.

Degree The number of edges that connect a node to the rest of the network. Nodes with high degree are often described as network hubs. The minimum degree of a node is zero, in which case the node is isolated. The maximum degree is one less than the total number of nodes in the network.

Degree distribution The distribution of degree across the nodes of a graph. Brain networks typically have broad-scale or scale-free degree distributions, meaning that the probability of high-degree nodes (hubs) is greater than expected in a random graph.

Degree sequence The set of degree values across all nodes of a network. For example, a network comprising three nodes with a degree of 4, three nodes with a degree of 3, and one node with a degree of 1, has the degree sequence $\{4, 4, 4, 3, 3, 3, 1\}$. The sum of the elements of a degree sequence is always an even number, due to the fact that each edge connects two vertices and is thus counted twice.

Diaschisis The interruption of function in brain areas remote to a lesioned site, which is thought to arise from a deafferentation of the remote region.

Diffusion magnetic resonance imaging (MRI) A noninvasive imaging technique that measures the diffusion of water molecules in biological tissue. Axons act as barriers that constrain water diffusion in the brain. These constraints can be used to derive measures of axonal microstructure and to track the trajectories of macroscopic fiber bundles in living organisms. Commonly used to measure anatomical connectivity in human connectomics.

Directed graph A graph-based representation of a network in which each edge has a point of origin and a point of termination. Directionality is commonly represented as an arrowhead attached at the point of termination of each edge. Directed networks can be used to model putative directions of information flow in the network and to understand causal interactions between nodes. In connectomics, directed networks allow the distinction between afferent and efferent connections.

E

Economy Network economy generally refers to the trade-off between finite resources (costs) and the delivery of robust, adaptive, and efficient performance (value). In connectomics, cost is commonly quantified by the total physical length of all connections in a nervous system, or by the connection density of the network. Both quantities are proxy measures of wiring cost. Network function usually improves with network cost. For example, long-range axonal projections incur a significant cost, but greatly reduce the characteristic path length, and as a consequence, increase the efficiency with which information can be routed between nodes.

Edge A connection between a pair of nodes in a graph. Also sometimes referred to as a link or arc. Edges can be directed or undirected and weighted or binary.

Effective connectivity The causal influence that one neuronal system exerts over another. Typically inferred from a model of the neuronal interactions that cause fluctuations in measured neurophysiological signals. A model of effective connectivity results in a directed graph.

Efficiency Broadly, the efficiency of a network is defined in terms of how easy it is for pairs of nodes to communicate with each other. In connectomics, efficiency is most commonly measured as the inverse shortest path length, based on the intuition that a network with a low average shortest path length will facilitate more rapid and efficient communication. Other measures of efficiency that do not rely on routing along shortest paths, such as those based on diffusion and navigation models of communication, have also been proposed.

Eigenvalue A scalar, λ, that satisfies the equation $Ax = \lambda x$, where A is a matrix and x is the corresponding eigenvector. The magnitude of the eigenvalue describes whether the vector x, when multiplied by the matrix A, is stretched or shrunk. The polarity of the eigenvalue describes the direction of the vector. The eigenvalues are sometimes thought of as corresponding to the natural modes of a system. For example, the eigenvalues of an oscillating string correspond to its resonant frequencies. In the statistical technique known as principal component analysis, the eigenvalues correspond to the proportion of variance explained by each principal component.

Eigenvector A nonzero vector, x, that satisfies the equation $Ax = \lambda x$, where A is a matrix and λ is the corresponding eigenvalue. The eigenvectors of the system represented by the matrix A are those that change only in magnitude but not direction if the system is subjected to some linear, geometric transformation. For example, the eigenvectors of an oscillating string correspond to the shape of the oscillations. In principal component analysis, the eigenvectors correspond to the orthogonal dimensions of variance.

Electrocorticography (ECoG) An invasive technique for recording neuronal activity from electrodes placed directly on the cortical surface. It is also known as intracranial electroencephalography. The electrodes must be implanted surgically. The recordings reflect local field potential fluctuations from neural tissue located underneath and in the nearby vicinity of the electrode.

Electroencephalography (EEG) A noninvasive technique for recording neuronal activity from electrodes placed on the scalp surface. It records the weak electrical signals generated from underlying neural tissue. The technique has high temporal resolution but poor spatial resolution.

Electron microscopy A technique for imaging the passage of electrons through a tissue specimen. Different structures scatter electrons in different ways, providing a contrast for the resolution of anatomical features. The wavelength of electrons is smaller than the wavelength of photons, meaning that electron microscopy can be used to resolve very small structures on the nanometer scale.

Erdös-Rényi graph A probabilistic model for a random graph in which connections are randomly placed between pairs of nodes. The graph is characterized by two parameters: the number of nodes, N, and the probability of finding a connection between an arbitrary pair of nodes, p.

Exchangeability A statistical property that is broadly similar to independence of observations and therefore fundamental to the validity of permutation tests. Groups of observations are exchangeable if their joint probability distribution is symmetric across the groups. Exchangeability is violated when, for example, the size and variance of observations is different between groups.

Exponential random graph model A flexible class of models for generating networks displaying a diverse range of topological and/or spatial properties. Networks are sampled from an exponential probability distribution that is specified across the set of all binary graphs with a given number of nodes.

F

False discovery rate (FDR) An established method for dealing with multiple comparisons that is less stringent than control of the familywise error rate. The FDR is defined as the ratio of the number of false positives to the number of false and true positives. Most FDR controlling procedures control the *average* of this ratio at a prescribed level, meaning that the actual proportion of false discoveries in any single experiment may be higher or lower than the prescribed level.

Familywise error rate The probability of falsely rejecting at least one null hypothesis in a family of multiple hypotheses that are tested independently. The term is typically used in relation to the rate of type I errors — errors made when falsely rejecting the null hypothesis. A null hypothesis is often independently tested at every node or edge in a brain network, requiring some form of control of type I errors across the large number of tests.

Fourier analysis The decomposition of signals as a weighted sum of sinusoids of differing frequencies. Fourier analysis is fundamental to understanding the frequency and phase characteristics of neurophysiological signals and in generating phase-randomized surrogate time series.

Fractal A pattern that repeats across a range of temporal and/or spatial scales. Also referred to as self-similarity or scale invariance. Zooming in or out on a fractal does not reveal any new details; the same patterns are seen across all resolutions. A classic example is the fractal-like properties of a coastline, which explains the paradox of why the coastline of a landmass does not have a well-defined length. Fractals abound in nervous systems. Axonal and dendritic branching patterns are fractals in anatomical space. Neuronal avalanches — cascading bursts of activity in nervous systems — have a size distribution that follows a power law, which is a signature of their fractal organization in time.

Fragmentation Occurs when the nodes of a network do not comprise a single, connected component. A network is fragmented when there is at least one node that cannot be linked to any other node by a path.

Functional connectivity A statistical dependence between neurophysiological signals. In functional MRI, this dependency is most commonly measured using the correlation coefficient, such that two brain regions with correlated activity are said to be functionally connected. Partial correlation is a popular alternative and other measures such as mutual information can also be used to measure functional connectivity. Frequency-resolved and phase-based measures of dependency (for example, coherence and the phase locking value, respectively) are more commonly used in the analysis of electrophysiological signals, due to the higher temporal resolution afforded by these modalities. Functional connectivity between a pair of nodes does not necessarily imply that one node is causing activity in the other, or that the nodes share a direct anatomical connection.

Functional magnetic resonance imaging (functional MRI) A noninvasive method for measuring brain activity indirectly based on hemodynamic signals, most commonly the BOLD contrast. It is the only technique available for measuring the activity of all human brain regions simultaneously in vivo. It has a spatial resolution on the order of a few millimeters and a temporal resolution on the order of a few seconds.

G

Generative model A model of the processes that generated an observed property or set of properties. Generative models for networks include the Erdös-Rényi, Watts-Strogatz, and Barabási-Albert models, which specify simple rules for connecting the nodes of a network to generate properties such as random connectivity, small worldness, and scale-free degree distributions, respectively.

Geodesic In a binary graph, the sequence of edges (or path) between two distinct nodes that comprises the fewest possible edges. In a weighted graph, it is the sequence of edges whose sum of weights is minimum. A geodesic is also referred to as a shortest path.

Graph theory A branch of mathematics concerned with the analysis of systems of interacting elements, as represented by a graph of nodes connected by edges. Theoretical developments were originally focused on random graphs but in the last few decades the tools of graph theory have been extensively applied to the empirical analysis of real-life complex networks. Connectomics is one such application.

H

Hierarchical clustering A statistical method for clustering observations to reveal nested relationships, such that pairs of observations are grouped into larger and larger subsets until all observations are in the same cluster. The clustering is performed according to some measure of pairwise similarity between observations. In networks, the observations correspond to nodes and the similarity metric could be edge weight or some other measure of pairwise node similarity.

Hierarchical modularity Networks can often be decomposed into strongly interconnected subsets of nodes, called modules. In some networks, these modules can be further divided into submodules and so on across several scales of topological resolution, much like a Russian doll. This multiscale modular organization is referred to as hierarchical modularity.

Hierarchy Hierarchical relationships between the nodes of a network have traditionally been analyzed in one of two ways: (1) by characterizing the hierarchical modular organization of the network, in which modules contain nested submodules across several scales of topological resolution; or (2) by examining the relationship between node degree and node clustering. In the second sense, hierarchical organization is evident when high-degree nodes have low clustering, which implies that these nodes act as topological bridges between otherwise unconnected, low-degree nodes. A classic example is a star-like motif in which a single hub node connects other peripheral nodes that are not linked to each other. All communication must therefore pass through the hub node.

Hop count The number of edges comprising a path or some other sequence of edges in a network.

Hub An important or topologically central node in a network. Most commonly defined as a node with a relatively large number of connections; ie, high degree. Other measures of centrality can be used to define hubs and to distinguish between different types of hubs.

I

In-degree The number of incoming edges attached to a node. Can only be defined in directed graphs.

Independent component analysis A statistical method for decomposing a multivariate or mixed signal into independent, additive subcomponents. These subcomponents represent putative sources of the signal. ICA is commonly used in studies of human resting-state functional connectivity to identify independent spatial components with temporally coherent activity — so-called resting-state networks.

L

Laplacian matrix A transformation of a network's adjacency matrix defined by subtracting the adjacency matrix from a diagonal matrix of node degrees. The eigenvalues of the Laplacian matrix are informative with respect to the network's connected components, community structure, motifs, and bipartite organization. For example, the number of zero eigenvalues is the number of connected components and the smallest eigenvalue delineates the best partition of the network into two modules. The normalized form of the Laplacian specifies the transition probabilities governing a random walk.

Lattice A network embedded in physical space in which each node is connected to a fixed number of nearest neighbors, according to a recurring pattern, much like a grid. Lattices often have a mesh-like appearance and are the opposite of a random network with respect to many topological properties. Complex networks are defined as such because they are not completely regular like a lattice; nor are they completely random like an Erdös-Rényi graph.

Light microscopy An optical imaging technique that uses visible light to image small structures. Its resolution is thus limited by the visible wavelength of light.

Line graph A graph-based representation of a network in which the nodes of the network are represented as the edges of the line graph, and the edges of the network are represented as the nodes of the line graph. Line graphs are useful for understanding adjacencies between edges rather than nodes, and can be used to characterize the edge-centric community structure of a network.

Link A connection or edge between two nodes in a network, which can be directed or undirected and binary or weighted.

Local field potential (LFP) An electrical field potential caused by the superposition of electric currents generated by various neuronal processes within the vicinity of an electrode. Synchronized synaptic activity makes a major contribution to the signal, but the signal can also be affected by spiking output, fluctuations in glial function, and other physiological processes. LFPs are recorded using implanted electrodes or electrocorticography.

M

Magnetoencephalography (MEG) A noninvasive technique for measuring brain function based on the magnetic fields generated by the electrical signals caused by neuronal activity. It has excellent temporal resolution but poorer spatial resolution than MRI. MEG is insensitive to radially oriented neural sources whereas EEG is sensitive to both radially and tangentially oriented sources.

Markov chain Monte Carlo (MCMC) A well-established numerical method for generating samples from an intractable probability distribution, such as those that involve integrals that cannot be evaluated analytically. MCMC involves constructing a Markov chain with a stationary distribution that estimates the intractable probability distribution. After simulating the Markov chain for an initial number of steps, the state of the chain at each step represents a sample from the desired distribution. Exponential random graph models utilize the method to generate samples of networks.

Matrix A convention for representing data as a rectangular array of rows and columns. A graph comprising N nodes is defined by a $N \times N$ adjacency or connectivity matrix, where each row and column corresponds to a unique node and each element encodes the absence or presence of a connection between a unique pair of nodes. For weighted networks, each matrix element encodes a connection weight.

Maximal clique A clique is a subset of nodes that are fully connected. A maximal clique is a fully connected subset of nodes that is not a subset of any other clique.

Minimum spanning tree (MST) A subgraph that connects all nodes using exactly $N - 1$ edges, where N is the number of nodes in the graph. In weighted networks, it is a subgraph connecting all nodes with the minimum sum of edge weights.

Modularity The division of the nodes of a network into strongly interconnected subsets called modules or communities. Networks that can be decomposed into modules have, by definition, high connectivity between nodes belonging to the same module and weaker connectivity between nodes that belong to different modules.

Module A subset of nodes that are strongly interconnected with each other and sparsely interconnected with nodes in other modules. Also called a community.

Motif A subgraph of a network with a particular topological configuration. A node motif is a connected subgraph of M nodes linked by at least $M - 1$ edges. A path motif comprises a specific sequence of edges categorized into different types, such as the distinction between rich, feeder, and peripheral edges defined in relation to the rich-club organization of a network. Some motifs recur within a network with a frequency that is significantly greater than chance, suggesting that they represent topological building blocks for the network.

Multielectrode array An array of electrodes, often arranged as a grid, that can be used to record or deliver electrical signals to neural tissue either in vivo or in vitro.

Multiple comparisons problem A null hypothesis is often independently tested at every node or edge in a brain network, requiring some form of control of type I errors across the family of all network elements considered.

N

Neighbor Refers to a node that is directly connected to an index node. Thus, two nodes sharing a direct connection are said to be topological neighbors.

Network-based statistic (NBS) A nonparametric statistical technique for connectome-wide analyses conducted at the level of each and every edge of a network. The method retains control over the familywise error rate by testing the null hypothesis at the level of connected components of edges, rather than individual connections.

Node A fundamental element of a graph, also called a vertex, that is connected to other nodes by edges. In brain networks, nodes can correspond to individual neurons, neuronal populations, or macroscopic brain regions.

Node connected A property of a graph in which all nodes form a single connected component.

Nonstationarity Property of a signal — more specifically a stochastic process — whose statistical properties vary as a function of time. A nonstationary signal's mean, variance, and/or other higher order moments vary according to a given timescale. White noise is an example of a stationary process, whereas many physiological signals are nonstationary. Examples include heart rate, blood pressure, neural activity, and functional connectivity. These signals change over periods of minutes, hours, and days based on levels of physical activity, mental state, and so on.

Null model A graph that is matched to some basic properties of an observed graph, but is random in all other aspects. Null models are often used as a benchmark or reference point to determine whether a network displays a topological feature to a greater extent than expected under the null hypothesis. Matching is typically performed with respect to the number of nodes, degree distribution, and connection density, because these basic properties can trivially influence higher order network properties. To perform statistical inference, an ensemble of matched random graphs can be generated by independently sampling from the null model. Null models can also refer to methods for generating surrogate time series data when testing hypotheses about functional connectivity dynamics.

O

Out-degree The number of outgoing edges attached to a node. Can only be defined in directed graphs.

P

Path An ordered sequence of edges or nodes in a graph. A path cannot include multiple occurrences of the same node or edge, although this condition is sometimes reserved for the term simple path. In contrast, trails include sequences of edges that can repeatedly visit the same node, while walks are sequences of edges that can include multiple occurrences of the same edges and nodes.

Path length In binary or unweighted graphs, the number of edges comprising the ordered sequence that defines a path; also called the hop count. In weighted graphs, path length is defined as the sum of weights across the edges comprising a path. The shortest path in a binary graph is the sequence of edges between two distinct nodes that comprises the fewest possible edges (or minimum hop count). In weighted graphs, the shortest path or geodesic is a sequence of edges whose sum of weights is minimum.

Percolation Refers to processes that occur on a network as nodes and/or edges are added or deleted. Percolation is typically studied in relation to the size of the largest connected component of a graph. A classic example is the rapid emergence of a single giant component that encompasses a large fraction of nodes in an Erdös-Rényi graph at a given probability of connection. The probability at which this large component emerges is called the percolation threshold.

Permutation test A powerful, intuitive, and versatile method of nonparametric inference that is part of a more general class of data resampling techniques for statistical significance testing. The empirical distribution of a test statistic under the null hypothesis is estimated by repeatedly rearranging the labels of the observed data samples (on the assumption of exchangeability). A one-sided p-value is then given by the proportion of test statistics in the permuted data that are greater (or less) than the observed test statistic. The main limitations are computational tractability and applicability to complicated statistical designs.

Positron emission tomography (PET) An in vivo brain imaging technique that measures the cerebral uptake of a radioactive ligand that binds specifically to a target molecule of interest, such as a receptor or marker of metabolic activity. PET is most commonly used to measure glucose metabolism and regional cerebral blood flow. PET can provide a specific molecular signal, but has a low spatial and temporal resolution.

Preferential attachment A process that underlies a generative model for scale-free networks, in which the probability of a new node being connected to an existing node is greater for existing nodes with high degree, ie, hubs. This "rich-get-richer" growth process leads to the emergence of hubs and a power-law degree distribution. It is central to the Barabási-Albert model for generating scale-free networks.

R

Random walker An agent that traverses a graph by walking along its edges in a sequential but random manner. Random walks provide a useful metaphor in the analysis of diffusion processes on a graph. The next hop in a random walk is equally likely to be any of the connections outgoing from a node. In a weighted network, the probability that a random walk will progress along any given edge can be biased by the weight of that edge.

Redundancy The repetition of identical elements of a system.

Rich club A subset of high-degree nodes that are more densely connected to each other than expected by chance. A weighted rich club is defined as a subset of rich nodes (where richness is usually defined according to degree) that are connected by a larger fraction of the most highly weighted edges in the network than expected by chance.

S

Scale free A term often used to describe networks with power-law degree distributions, which have no clear modal value defining a characteristic scale of the system and in which the probability of finding highly connected nodes is greater than in a random network. Scale-free or power-law scaling is a signature of fractal or self-similar organization.

Simulated annealing A heuristic algorithm for numerically approximating the global optimum of a nontrivial objective function over a large search space when exhaustive enumeration of the search space is intractable. Has been used in connectomics for component placement optimization and fitting some generative models.

Single scale Single-scale networks are associated with a degree distribution that has a clear peak or modal value, which corresponds to the characteristic scale of the network. The probability of finding nodes with a degree that is much higher or lower than this modal value is very low.

Sliding window analysis A technique for mapping the dynamics of functional connectivity across time. Each window defines a continuous time interval and windows can be overlapping or mutually exclusive. Sliding a window across time defines consecutive time points. Functional connectivity is computed independently using the data in each window, thereby yielding a time-resolved measure. This is in contrast to the classic approach of using the data across all time points to compute a time-averaged measure of functional connectivity. The choice of window length is crucial and should be dictated by the frequency spectrum of the neurophysiological signal.

Small world A class of networks that show high clustering, much like a lattice, but with a short characteristic path length, much like a random graph. The classic, Watts-Strogatz generative model for a small-world network involves randomly rewiring an arbitrary proportion of edges in a lattice graph.

Source localization A class of techniques for inferring the neural sources that generate signals measured outside the skull using techniques such as MEG and EEG. The sources are reconstructed using biophysically informed mathematical models that account for the way in which the signals are conducted and attenuated as they pass through the brain, skull, and scalp.

Sparse matrix A computationally efficient structure for representing a matrix that predominantly comprises zeros, in which only nonzero edges are stored in the form of a list.

Strength The strength of a node is the sum of its edge weights. Node strength is the analog of node degree for a weighted network.

Structural connectivity The anatomical connections between neural elements. Equivalent to individual axons and synaptic contacts at the micro scale. At coarser scales, structural connectivity refers to the axonal tracts or white matter fiber bundles between different brain regions.

Structural covariance A term used mainly in analysis of structural MRI to describe the covariation of cortical thickness or grey matter volume between brain regions measured in multiple individuals. Many pairs of brain regions demonstrate positive structural covariance, which is often interpreted as a proxy measure of anatomical connectivity.

Structural equivalence A property defined at the level of pairs of nodes. Two nodes are structurally equivalent if they connect to exactly the same set of nodes; that is, they have identical neighborhoods. Also referred to as topological overlap.

Subgraph A subset of nodes and edges within a larger graph.

Support vector machine A machine learning technique that can be used for classification or regression. In its simplest form, a support vector machine recognizes patterns in a high-dimensional feature space that maximally separates two distinct categories of data. Currently used in connectomics mainly to predict a patient's diagnostic status (or other relevant outcome) based on brain network properties.

T

Tensor A geometric structure describing relations between vectors. Used in connectomics with respect to the diffusion tensor model, one of the classic models for analyzing diffusion weighted MRI data. While there are many representations of a tensor, the diffusion tensor used in MRI corresponds to the special case of a 3×3 symmetric matrix, which can be visualized as an ellipsoid. The shape and orientation of the ellipsoid are used to infer microstructural properties of underlying white matter tissue and local tangents of axonal bundles. A diffusion tensor is independently fitted to every voxel.

Topology A branch of mathematics concerned with the analysis of spaces whose properties are invariant to continuous spatial deformations. With respect to graphs, it refers to the pattern of interconnectivity between nodes. A graph's topology is invariant to the geographic layout of nodes in physical space. For example, we could enlarge, shrink, or twist a brain, but its connection topology will remain unchanged if we do not add or remove any connections.

Tract tracing An invasive method for reconstructing axonal fibers via the introduction (usually by injection) of a tracer substance (usually a fluorescent dye or virus) directly to neural tissue. The tracer is permeable to cellular membranes and is transferred between the soma and synaptic terminals via active axonal transport. Anterograde tracers are transported from the soma to the synapse and map the efferent projections from the injection site. Retrograde tracers are transported from the synapse and axon to the soma and map the afferent projections to the injection site.

Tractography A class of methods typically applied to diffusion MRI data to trace the trajectories of axonal fiber bundles based on preferred directions of water diffusion in the brain. In deterministic tractography, a single trajectory, called a streamline, is propagated from an initial seed location in a step-by-step manner according to the locally maximal direction of water diffusion, which is assumed to correspond with the orientation of the underlying axonal fiber bundle. In probabilistic tractography, the step-by-step propagation of streamlines is no longer deterministic but is rather sampled from a probability distribution which captures uncertainty in the estimated direction of water diffusion.

Trail A path that is allowed to visit the same nodes on multiple occasions. The same edge cannot be visited more than once on a trail.

Transcranial magnetic stimulation A technique for noninvasive manipulation of neuronal excitability. A magnetic coil is placed on the scalp surface. An induced current passes through the skull and modulates the excitability of the underlying neural tissue.

U

Undirected graph A graph-based representation of a network that indicates which pairs of nodes are connected, but does not indicate the direction (ie, the source and target) of a connection.

V

Vector autoregressive model A generalization of the autoregressive model that enables the linear interdependencies between multiple time series to be modeled. Used in connectomics to generate surrogate time series data to test hypotheses about the dynamics of functional connectivity.

Vertex Another name for a node in a graph.

W

Walk A path in a network that is allowed to visit the same nodes and edges on multiple occasions.

Watts-Strogatz model A generative model for a small-world network which starts with a lattice configuration and randomly rewires an arbitrary proportion of edges. Once a certain fraction of edges have been rewired, the network enters a regime in which the high clustering of the lattice is maintained but the average path length drops dramatically and is comparable to a random graph.

Wavelet analysis A method for decomposing signals and spatial processes in terms of both scale (approximate frequency) and location (space or time). Thus, a discrete wavelet transform of a time series will partition the total energy of the signal across a hierarchical basis set of independent wavelet coefficients. The wavelet domain is appropriate for analysis of fractal processes and has been used for preprocessing and resampling of human neuroimaging data prior to connectomic analysis.

Weighted graph A graph-based representation of a network in which edges are weighted as a function of the strength of connectivity between node pairs. Edge weights are typically scalars that can be positive or negative. This contrasts with a binary graph, in which the absence or presence of a connection is all that matters. When the network is displayed as a graph, variations in edge weight are often represented as variations in the thickness of the edges.

References

Achard, S., Bullmore, E., 2007. Efficiency and cost of economical brain functional networks. PLoS Comput. Biol. 3, e17.

Achard, S., Salvador, R., Whitcher, B., Suckling, J., Bullmore, E., 2006. A resilient, low-frequency, small-world human brain functional network with highly connected association cortical hubs. J. Neurosci. 26, 63–72.

Aertsen, A., Erb, M., Palm, G., 1994. Dynamics of functional coupling in the cerebral cortex: an attempt at a model-based interpretation. Physica D 75, 103–128.

Aertsen, A.M., Gerstein, G.L., Habib, M.K., Palm, G., 1989. Dynamics of neuronal firing correlation: modulation of "effective connectivity". J. Neurophysiol. 61, 900–917.

Agar, A., Alderson, R.H., Chescoe, D., 1974. Principles and Practice of Electron Microscope Operation. Elsevier, New York, NY.

Ahlfors, S.P., Han, J., Belliveau, J.W., Hämäläinen, M.S., 2010. Sensitivity of MEG and EEG to source orientation. Brain Topogr. 23, 227–232.

Ahn, Y.-Y., Bagrow, J.P., Lehmann, S., 2010. Link communities reveal multiscale complexity in networks. Nature 466, 761–764.

Ahn, Y.-Y., Jeong, H., Kim, B.J., 2006. Wiring cost in the organization of a biological neuronal network. Physica A 367, 531–537.

Aho, A.V., Hopcraft, J.E., Ullman, J.D., 1983. Data Structures and Algorithms. Addison-Wesley, Boston, MA.

Ahrens, M.B., Li, J.M., Orger, M.B., Robson, D.N., Schier, A.F., Engert, F., Portugues, R., 2012. Brain-wide neuronal dynamics during motor adaptation in zebrafish. Nature 485, 471–477.

Ahrens, M.B., Orger, M.B., Robson, D.N., Li, J.M., Keller, P.J., 2013. Whole-brain functional imaging at cellular resolution using light-sheet microscopy. Nat. Methods 10, 413–420.

Albert, R., Jeong, H., Barabási, A.-L., 2000. Error and attack tolerance of complex networks. Nature 406, 378–382.

Albert, R., Barabási, A.-L., 2002. Statistical mechanics of complex networks. Rev. Mod. Phys. 74 (1), 47.

Albert, R., DasGupta, B., Hegde, R., Sivanathan, G.S., Gitter, A., Gürsoy, G., Paul, P., Sontag, E., 2011. Computationally efficient measure of topological redundancy of biological and social networks. Phys. Rev. E 84, 036117.

Aldecoa, R., Marín, I., 2011. Deciphering network community structure by surprise. PLoS One 6, e24195.

Aldecoa, R., Marín, I., 2013. Surprise maximization reveals the community structure of complex networks. Sci. Rep. 3, 1–9.

Alexander-Bloch, A.F., Reiss, P.T., Rapoport, J., McAdams, H., Giedd, J.N., Bullmore, E.T., Gogtay, N., 2014. Abnormal cortical growth in schizophrenia targets normative modules of synchronized development. Biol. Psychiatry 76 (6), 438–446.

Alexander-Bloch, A.F., Giedd, J.N., Bullmore, E., 2013a. Imaging structural co-variance between human brain regions. Nat. Rev. Neurosci. 14, 322–336.

Alexander-Bloch, A.F., Gogtay, N., Meunier, D., Birn, R., Clasen, L., Lalonde, F., Lenroot, R., Giedd, J., Bullmore, E.T., 2010. Disrupted modularity and local connectivity of brain functional networks in childhood-onset schizophrenia. Front. Syst. Neurosci. 4, 147.

Alexander-Bloch, A.F., Lambiotte, R., Roberts, B., Giedd, J., Gogtay, N., Bullmore, E., 2012. The discovery of population differences in network community structure: new methods and applications to brain functional networks in schizophrenia. NeuroImage 59, 3889–3900.

Alexander-Bloch, A.F., Raznahan, A., Bullmore, E., Giedd, J., 2013b. The convergence of maturational change and structural covariance in human cortical networks. J. Neurosci. 33, 2889–2899.

Alexander-Bloch, A.F., Vértes, P.E., Stidd, R., Lalonde, F., Clasen, L., Rapoport, J., Giedd, J., Bullmore, E.T., Gogtay, N., 2013c. The anatomical distance of functional connections predicts brain network topology in health and schizophrenia. Cereb. Cortex 23, 127–138.

Alexander, D.C., 2005. Multiple-fiber reconstruction algorithms for diffusion MRI. Ann. N. Y. Acad. Sci. 1064, 113–133.

Alexander, D.C., Hubbard, P.L., Hall, M.G., Moore, E.A., Ptito, M., Parker, G.J., Dyrby, T.B., 2010. Orientationally invariant indices of axon diameter and density from diffusion MRI. Neuro-Image 52, 1374–1389.

Alivisatos, A.P., Chun, M., Church, G.M., Greenspan, R.J., Roukes, M.L., Yuste, R., 2012. The brain activity map project and the challenge of functional connectomics. Neuron 74, 970–974.

Alon, U., 2007. Network motifs: theory and experimental approaches. Nat. Rev. Genet. 8, 450–461.

Alstott, J., Breakspear, M., Hagmann, P., Cammoun, L., Sporns, O., 2009. Modeling the impact of lesions in the human brain. PLoS Comput. Biol. 5, e1000408.

Alstott, J., Bullmore, E., Plenz, D., 2014a. Powerlaw: a Python package for analysis of heavy-tailed distributions. PLoS One 9, e85777.

Alstott, J., Panzarasa, P., Rubinov, M., Bullmore, E.T., Vértes, P.E., 2014b. A unifying framework for measuring weighted rich clubs. Sci. Rep. 4, 7258.

Alvarez-Hamelin, J.I., Dall'Asta, L., Barrat, A., Vespignani, A., 2008. k-core decomposition of Internet graphs: hierarchies, self-similarity and measurement biases. Netw. Heterog. Media 3, 371–393.

Amaral, L.A., Scala, A., Barthelemy, M., Stanley, H.E., 2000. Classes of small-world networks. Proc. Natl. Acad. Sci. U. S. A. 97, 11149–11152.

Amunts, K., Malikovic, A., Mohlberg, H., Schormann, T., Zilles, K., 2000. Brodmann's areas 17 and 18 brought into stereotaxic space—where and how variable? NeuroImage 11, 66–84.

Andreotti, J., Jann, K., Melie-Garcia, L., Giezendanner, S., Abela, E., Wiest, R., Dierks, T., Federspiel, A., 2014. Validation of network communicability metrics for the analysis of brain structural networks. PLoS One 9 (12), e115503.

Anthonisse, J.M., 1971. The rush in a directed graph. Technical Report BN 9/71. Stichting Mathematisch Centrum, Amsterdam.

Antonopoulos, C.G., Srivastava, S., Pinto, S.E., Baptista, M.S., 2015. Do brain networks evolve by maximizing their information flow capacity? PLoS Comput. Biol. 11 (8), e1004372.

Anwander, A., Tittgemeyer, M., von Cramon, D.Y., Friederici, A.D., Knosche, T.R., 2007. Connectivity-based parcellation of Broca's area. Cereb. Cortex 17, 816–825.

Arenas, A., Díaz-Guilera, A., Pérez-Vicente, C.J., 2006. Synchronization reveals topological scales in complex networks. Phys. Rev. Lett. 96, 114102–114104.

Aru, J., Aru, J., Priesemann, V., Wibral, M., Lana, L., Pipa, G., Singer, W., Vicente, R., 2015. Untangling cross-frequency coupling in neuroscience. Curr. Opin. Neurobiol. 31, 51–61.

Assaf, Y., Blumenfeld-Katzir, T., Yovel, Y., Basser, P.J., 2008. AxCaliber: a method for measuring axon diameter distribution from diffusion MRI. Magn. Reson. Med. 59, 1347–1354.

Attwell, D., Laughlin, S.B., 2001. An energy budget for signaling in the grey matter of the brain. J. Cereb. Blood Flow Metab. 21, 1133–1145.

Avena-Koenigsberger, A., Goñi, J., Betzel, R.F., van den Heuvel, M.P., Griffa, A., Hagmann, P., Thiran, J.-P., Sporns, O., 2014. Using Pareto optimality to explore the topology and dynamics of the human connectome. Philos. Trans. R. Soc. Lond. B Biol. Sci. 369, 20130530.

Avena-Koenigsberger, A., Goñi, J., Sole, R., Sporns, O., 2015. Network morphospace. J. R. Soc. Interface 12, 20140881.

Averbeck, B.B., Latham, P.E., Pouget, A., 2006. Neural correlations, population coding and computation. Nat. Rev. Neurosci. 7, 358–366.

Axer, M., Amunts, K., Grassel, D., Palm, C., Dammers, J., Axer, H., Pietrzyk, U., Zilles, K., 2011. A novel approach to the human connectome: ultra-high resolution mapping of fiber tracts in the brain. NeuroImage 54, 1091–1101.

Azouz, R., 2005. Dynamic spatiotemporal synaptic integration in cortical neurons: neuronal gain, revisited. J. Neurophysiol. 94, 2785–2796.

Baars, B.J., 1989. A Cognitive Theory of Consciousness. Cambridge University Press, Cambridge, MA.

Baddeley, A., 1996. The fractionation of working memory. Proc. Natl. Acad. Sci. U. S. A. 93, 13468–13472.

Bak, P., 1996. How Nature Works: The Science of Self-Organized Criticality. Springer, New York, NY.

Bak, P., Tang, C., Wiesenfeld, K., 1987. Self-organized criticality: an explanation of the 1/f noise. Phys. Rev. Lett. 59 (4), 381.

Baker, S.T.E., Lubman, D.I., Yücel, M., Allen, N.B., Whittle, S., Fulcher, B.D., Zalesky, A., Fornito, A., 2015. Developmental changes in brain network hub connectivity in late adolescence. J. Neurosci. 35, 9078–9087.

Bakker, R., Wachtler, T., Diesmann, M., 2012. CoCoMac 2.0 and the future of tract-tracing databases. Front. Neuroinform. 6, 30.

Ball, G., Srinivasan, L., Aljabar, P., Counsell, S.J., Durighel, G., Hajnal, J.V., Rutherford, M.A., Edwards, A.D., 2013. Development of cortical microstructure in the preterm human brain. Proc. Natl. Acad. Sci. U. S. A. 110, 9541–9546.

Banerjee, A., Jost, J., 2009. Graph spectra as a systematic tool in computational biology. Discrete Appl. Math. 157, 2425–2431.

Banerjee, A., Jost, J., 2008. On the spectrum of the normalized graph Laplacian. Linear Algebra Appl. 428, 3015–3022.

Barabási, A.-L., 2002. Linked. Basic Books, Philadelphia, PA.

Barabási, A.L., Albert, R., 1999. Emergence of scaling in random networks. Science 286 (5439), 509–512.

Barahona, M., Pecora, L.M., 2002. Synchronization in small-world systems. Phys. Rev. Lett. 89, 054101.

Bargmann, C.I., Marder, E., 2013. From the connectome to brain function. Nat. Methods 10, 483–490.

Barnes, A., Bullmore, E.T., Suckling, J., 2009. Endogenous human brain dynamics recover slowly following cognitive effort. PLoS One 4, e6626.

Barnett, L., Buckley, C.L., Bullock, S., 2009. Neural complexity and structural connectivity. Phys. Rev. E 79, 051914.

Barnett, L., Buckley, C.L., Bullock, S., 2011. Neural complexity: a graph theoretic interpretation. Phys. Rev. E 83, 041906.

Barrat, A., Barthelemy, M., Pastor-Satorras, R., Vespignani, A., 2004. The architecture of complex weighted networks. Proc. Natl. Acad. Sci. U. S. A. 101, 3747–3752.

Bartels, A., Logothetis, N.K., Moutoussis, K., 2008. fMRI and its interpretations: an illustration on directional selectivity in area V5/MT. Trends Neurosci. 31, 444–453.

Barthélemy, M., 2011. Spatial networks. Phys. Rep. 499 (1), 1–101.

Bassett, D.S., Nelson, B.G., Mueller, B.A., Camchong, J., Lim, K.O., 2012. Altered resting state complexity in schizophrenia. NeuroImage 59 (3), 2196–2207.

Bassett, D.S., Gazzaniga, M.S., 2011. Understanding complexity in the human brain. Trends Cogn. Sci. 15 (5), 200–209.

Bassett, D.S., Bullmore, E., 2006. Small-world brain networks. Neuroscientist 12 (6), 512–523.

Bassett, D.S., Bullmore, E., Verchinski, B.A., Mattay, V.S., Weinberger, D.R., Meyer-Lindenberg, A., 2008. Hierarchical organization of human cortical networks in health and schizophrenia. J. Neurosci. 28, 9239–9248.

Bassett, D.S., Bullmore, E.T., Meyer-Lindenberg, A., Apud, J.A., Weinberger, D.R., Coppola, R., 2009. Cognitive fitness of cost-efficient brain functional networks. Proc. Natl. Acad. Sci. U. S. A. 106 (28), 11747–11752.

Bassett, D.S., Greenfield, D.L., Meyer-Lindenberg, A., Weinberger, D.R., Moore, S.W., Bullmore, E.T., 2010. Efficient physical embedding of topologically complex information processing networks in brains and computer circuits. PLoS Comput. Biol. 6, e1000748.

Bassett, D.S., Meyer-Lindenberg, A., Achard, S., Duke, T., Bullmore, E., 2006. Adaptive reconfiguration of fractal small-world human brain functional networks. Proc. Natl. Acad. Sci. U. S. A. 103, 19518–19523.

Bassett, D.S., Porter, M.A., Wymbs, N.F., Grafton, S.T., Carlson, J.M., Mucha, P.J., 2013a. Robust detection of dynamic community structure in networks. Chaos 23, 013142. 16.

Bassett, D.S., Wymbs, N.F., Porter, M.A., Mucha, P.J., Carlson, J.M., Grafton, S.T., 2011. Dynamic reconfiguration of human brain networks during learning. Proc. Natl. Acad. Sci. U. S. A. 108, 7641–7646.

Bassett, D.S., Wymbs, N.F., Rombach, M.P., Porter, M.A., Mucha, P.J., Grafton, S.T., 2013b. Task-based core-periphery organization of human brain dynamics. PLoS Comput. Biol. 9, e1003171.

Bassett, D.S., Yang, M., Wymbs, N.F., Grafton, S.T., 2015. Learning-induced autonomy of sensorimotor systems. Nat. Neurosci. 18, 744–751.

Bastos, A.M., Vezoli, J., Fries, P., 2015a. Communication through coherence with inter-areal delays. Curr. Opin. Neurobiol. 31, 173–180.

Bastos, A.M., Vezoli, J., Bosman, C.A., Schoffelen, J.-M., Oostenveld, R., Dowdall, J.R., De Weerd, P., Kennedy, H., Fries, P., 2015b. Visual areas exert feedforward and feedback influences through distinct frequency channels. Neuron 85, 390–401.

Batagelj, V., Zaveršnik, M., 2002. Generalized cores. arXiv cs/0202039v1.

Bauer, F., Jost, J., 2012. Bipartite and neighborhood graphs and the spectrum of the normalized graph Laplacian. Commun. Anal. Geom. 21, 787–845.

Bavelas, A., 1948. A mathematical model for group structures. Hum. Organ. 7, 16–30.

Bavelas, A., 1950. Communication patterns in task-oriented groups. J. Acoust. Soc. Am. 22, 271–282.

Bazzi, M., Porter, M.A., Williams, S., McDonald, M., 2014. Community detection in temporal multilayer networks, and its application to correlation networks. arXiv 1501.00040v2.

Beauchamp, M.A., 1965. An improved index of centrality. Behav. Sci. 10, 161–163.

Beckmann, C.F., DeLuca, M., Devlin, J.T., Smith, S.M., 2005. Investigations into resting-state connectivity using independent component analysis. Philos. Trans. R. Soc. Lond. B Biol. Sci. 360, 1001–1013.

Beggs, J.M., 2008. The criticality hypothesis: how local cortical networks might optimize information processing. Phil. Trans. R. Soc. A 366, 329–343.

Beggs, J.M., Plenz, D., 2003. Neuronal avalanches in neocortical circuits. J. Neurosci. 23 (35), 11167–11177.

Behrens, T.E., Woolrich, M.W., Jenkinson, M., Johansen-Berg, H., Nunes, R.G., Clare, S., Matthews, P.M., Brady, J.M., Smith, S.M., 2003. Characterization and propagation of uncertainty in diffusion-weighted MR imaging. Magn. Reson. Med. 50, 1077–1088.

Behrens, T.E.J., Berg, H.J., Jbabdi, S., Rushworth, M.F.S., Woolrich, M.W., 2007. Probabilistic diffusion tractography with multiple fibre orientations: what can we gain? NeuroImage 34, 144–155.

Behzadi, Y., Restom, K., Liau, J., Liu, T.T., 2007. A component based noise correction method (CompCor) for BOLD and perfusion based fMRI. NeuroImage 37, 90–101.

Bellman, R., 1958. On a routing problem. Q. Appl. Math. 16, 87–90.

Bendat, J.S., Piersol, A.G., 1986. Random Data: Analysis and Measurement Procedures. John Wiley & Sons, New York, NY.

Bender, E.A., Canfield, E.R., 1978. The asymptotic number of labeled graphs with given degree sequences. J. Comb. Theory 24, 296–307.

Benjamini, Y., Hochberg, Y., 1995. Controlling the false discovery rate: a practical and powerful approach to multiple testing. J. R. Stat. Soc. Ser. B 57 (1), 289–300.

Betzel, R.F., Griffa, A., Avena-Koenigsberger, A., Goñi, J., Thiran, J.-P., Hagmann, P., Sporns, O., 2014. Multi-scale community organization of the human structural connectome and its relationship with resting-state functional connectivity. Netw. Sci. 1, 353–373.

Betzel, R.F., Avena-Koenigsberger, A., Goñi, J., He, Y., de Reus, M.A., Griffa, A., Vértes, P.E., Mišic, B., Thiran, J.P., Hagmann, P., van den Heuvel, M., 2016. Generative models of the human connectome. Neuroimage 124, 1054–1064.

Betzig, E., Patterson, G.H., Sougrat, R., Lindwasser, O.W., 2006. Imaging intracellular fluorescent proteins at nanometer resolution. Science 313, 1642–1645.

Bialonski, S., Lehnertz, K., 2013. Assortative mixing in functional brain networks during epileptic seizures. Chaos 23, 033139. 10.

Birn, R.M., Diamond, J.B., Smith, M.A., Bandettini, P.A., 2006. Separating respiratory-variation-related fluctuations from neuronal-activity-related fluctuations in fMRI. NeuroImage 31, 1536–1548.

Biswal, B., Yetkin, F.Z., Haughton, V.M., Hyde, J.S., 1995. Functional connectivity in the motor cortex of resting human brain using echo-planar MRI. Magn. Reson. Med. 34, 537–541.

Blondel, V.D., Guillaume, J.L., Lambiotte, R., 2008. Fast unfolding of communities in large networks. J. Stat. Mech. Theory Exp. E10, P10008.

Blumenfeld, R.S., Bliss, D.P., Pérez, F., D'Esposito, M., 2014. CoCoTools: open-source software for building connectomes using the CoCoMac anatomical database. J. Cogn. Neurosci. 26, 722–745.

Boccaletti, S., Ivanchenko, M., Latora, V., Pluchino, A., Rapisarda, A., 2007. Detecting complex network modularity by dynamical clustering. Phys. Rev. E 75, 045102–045104.

Boccaletti, S., Latora, V., Moreno, Y., Chavez, M., Hwang, D.U., 2006. Complex networks: structure and dynamics. Phys. Rep. 424, 175–308.

Bock, D.D., Lee, W.-C.A., Kerlin, A.M., Andermann, M.L., Hood, G., Wetzel, A.W., Yurgenson, S., Soucy, E.R., Kim, H.S., Reid, R.C., 2011. Network anatomy and in vivo physiology of visual cortical neurons. Nature 471, 177–182.

Boguñá, M., Krioukov, D., 2012. Navigating ultrasmall worlds in ultrashort time. Phys. Rev. Lett. 102, 058701.

Boguñá, M., Krioukov, D., Claffy, K.C., 2009. Navigability of complex networks. Nat. Phys. 5, 74–80.

Bohland, J.W., Wu, C., Barbas, H., Bokil, H., Bota, M., Breiter, H.C., Cline, H.T., Doyle, J.C., Freed, P.J., Greenspan, R.J., Haber, S.N., Hawrylycz, M., Herrera, D.G., Hilgetag, C.C., Huang, Z.J., Jones, A., Jones, E.G., Karten, H.J., Kleinfeld, D., Kötter, R., Lester, H.A., Lin, J.M., Mensh, B.D., Mikula, S., Panksepp, J., Price, J.L., Safdieh, J., Saper, C.B., Schiff, N.D., Schmahmann, J.D., Stillman, B.W., Svoboda, K., Swanson, L.W., Toga, A.W., Van Essen, D.C., Watson, J.D., Mitra, P.P., 2009. A proposal for a coordinated effort for the determination of brainwide neuroanatomical connectivity in model organisms at a mesoscopic scale. PLoS Comput. Biol. 5 (3), e1000334.

Bohland, J.W., Bokil, H., Pathak, S.D., Lee, C.-K., Ng, L., Lau, C., Kuan, C., Hawrylycz, M., Mitra, P.P., 2010. Clustering of spatial gene expression patterns in the mouse brain and comparison with classical neuroanatomy. Methods 50, 105–112.

Bollobás, B., 1998. Random Graphs. Springer, New York, NY, pp. 215–252.

Bonacich, P., 1987. Power and centrality: a family of measures. Am. J. Sociol. 92, 1170–1182.

Bonacich, P., Lloyd, P., 2001. Eigenvector-like measures of centrality for asymmetric relations. Soc. Networks 23, 191–201.

Borgatti, S.P., 2005. Centrality and network flow. Soc. Networks 27, 55–71.

Borgatti, S.P., Everett, M.G., 2000. Models of core/periphery structures. Soc. Networks 21, 375–395.

Borgatti, S.P., Everett, M.G., 2006. A graph-theoretic perspective on centrality. Soc. Networks 28, 466–484.

Bota, M., Arbib, M.A., 2004. Integrating databases and expert systems for the analysis of brain structures: connections, similarities, and homologies. Neuroinformatics 2, 19–58.

Bota, M., Dong, H.-W., Swanson, L.W., 2003. From gene networks to brain networks. Nat. Neurosci. 6, 795–799.

Bota, M., Dong, H.-W., Swanson, L.W., 2005. Brain architecture management system. Neuroinformatics 3, 15–48.

Bota, M., Dong, H.W., Swanson, L.W., 2012. Combining collation and annotation efforts toward completion of the rat and mouse connectomes in BAMS. Front. Neuroinform. 6, 2.

Bota, M., Sporns, O., Swanson, L.W., 2015. Architecture of the cerebral cortical association connectome underlying cognition. Proc. Natl. Acad. Sci. U. S. A. 112, E2093–E2101.

Böttger, J., Schurade, R., Jakobsen, E., 2014. Connexel visualization: a software implementation of glyphs and edge-bundling for dense connectivity data using brainGL. Front. Neurosci. 8, 15.

Branco, T., Häusser, M., 2010. The single dendritic branch as a fundamental functional unit in the nervous system. Curr. Opin. Neurobiol. 20, 494–502.

Brandes, U., Delling, D., Gaertler, M., Görke, R., 2006. Maximizing modularity is hard. arXiv 0608255v2.

Breakspear, M., Brammer, M.J., Bullmore, E.T., Das, P., Williams, L.M., 2004. Spatiotemporal wavelet resampling for functional neuroimaging data. Hum. Brain Mapp. 23 (1), 1–25.

Brenner, N., Bialek, W., de Ruyter van Steveninck, R., 2000. Adaptive rescaling maximizes information transmission. Neuron 26, 695–702.

Brier, M.R., Mitra, A., McCarthy, J.E., Ances, B.M., Snyder, A.Z., 2015. Partial covariance based functional connectivity computation using Ledoit-Wolf covariance regularization. NeuroImage 121, 29–38.

Briggman, K.L., Bock, D.D., 2012. Volume electron microscopy for neuronal circuit reconstruction. Curr. Opin. Neurobiol. 22, 154–161.

Briggman, K.L., Denk, W., 2006. Towards neural circuit reconstruction with volume electron microscopy techniques. Curr. Opin. Neurobiol. 16, 562–570.

Brin, S., Page, L., 1998. The anatomy of a large-scale hypertextual web search engine. Comput. Netw. ISDN Syst. 30, 107–117.

Broca, P., 1861. Perte de la parole: ramollissement chronique et destruction partielle du lobe antérieur gauche du cerveau. Bull. Soc. Anthropol. 2, 235–238.

Brouwer, A.E., Haemers, W.H., 2012. Spectra of Graphs. Springer, New York, NY.

Brovelli, A., Ding, M., Ledberg, A., Chen, Y., Nakamura, R., Bressler, S.L., 2004. Beta oscillations in a large-scale sensorimotor cortical network: directional influences revealed by Granger causality. Proc. Natl. Acad. Sci. U. S. A. 101, 9849–9854.

Brummitt, C.D., D'Souza, R.M., 2012. Suppressing cascades of load in interdependent networks. Proc. Natl. Acad. Sci. U. S. A. 109, E680–E689.

Buckner, R.L., Krienen, F.M., 2013. The evolution of distributed association networks in the human brain. Trends Cogn. Sci. 17, 648–665.

Buckner, R.L., Andrews-Hanna, J.R., Schacter, D.L., 2008. The brain's default network: anatomy, function, and relevance to disease. Ann. N. Y. Acad. Sci. 1124, 1–38.

Buckner, R.L., Sepulcre, J., Talukdar, T., Krienen, F.M., Liu, H., Hedden, T., Andrews-Hanna, J.R., Sperling, R.A., Johnson, K.A., 2009. Cortical hubs revealed by intrinsic functional connectivity: mapping, assessment of stability, and relation to Alzheimer's disease. J. Neurosci. 29, 1860–1873.

Buckner, R.L., Snyder, A.Z., Shannon, B.J., LaRossa, G., Sachs, R., Fotenos, A.F., Sheline, Y.I., Klunk, W.E., Mathis, C.A., Morris, J.C., et al., 2005. Molecular, structural, and functional characterization of Alzheimer's disease: evidence for a relationship between default activity, amyloid, and memory. J. Neurosci. 25, 7709–7717.

Budd, J.M., Kisvárday, Z.F., 2012. Communication and wiring in the cortical connectome. Front. Neuroanat. 6, 42.

Buldyrev, S.V., Parshani, R., Paul, G., Stanley, H.E., Havlin, S., 2010. Catastrophic cascade of failures in interdependent networks. Nature 464, 1025–1028.

Bullmore, E., Sporns, O., 2009. Complex brain networks: graph theoretical analysis of structural and functional systems. Nat. Rev. Neurosci. 10, 186–198.

Bullmore, E., Sporns, O., 2012. The economy of brain network organization. Nat. Rev. Neurosci. 13, 336–349.

Bullmore, E., Brammer, M., Williams, S.C., Rabe-Hesketh, S., Janot, N., David, A., Mellers, J., Howard, R., Sham, P., 1996a. Statistical methods of estimation and inference for functional MR image analysis. Magn. Reson. Med. 35, 261–277.

Bullmore, E., Fadili, J., Maxim, V., Sendur, L., Whitcher, B., Suckling, J., Brammer, M., Breakspear, M., 2004. Wavelets and functional magnetic resonance imaging of the human brain. NeuroImage 23 (Suppl. 1), S234–S249.

Bullmore, E., Long, C., Suckling, J., Fadili, J., Calvert, G., Zelaya, F., Carpenter, T.A., Brammer, M., 2001. Colored noise and computational inference in neurophysiological (fMRI) time series analysis: resampling methods in time and wavelet domains. Hum. Brain Mapp. 12, 61–78.

Bullmore, E.T., Rabe-Hesketh, S., Morris, R.G., Williams, S.C., Gregory, L., Gray, J.A., Brammer, M.J., 1996b. Functional magnetic resonance image analysis of a large-scale neurocognitive network. NeuroImage 4, 16–33.

Bullmore, E.T., Suckling, J., Overmeyer, S., Rabe-Hesketh, S., Taylor, E., Brammer, M.J., 1999. Global, voxel, and cluster tests, by theory and permutation, for a difference between two groups of structural MR images of the brain. IEEE Trans. Med. Imaging 18 (1), 32–42.

Burges, C.J.C., 1998. A tutorial on support vector machines for pattern recognition. Data Min. Knowl. Disc. 2 (2), 121–167.

Buschman, T.J., Miller, E.K., 2007. Top-down versus bottom-up control of attention in the prefrontal and posterior parietal cortices. Science 315, 1860–1862.

Bush, E.C., Allman, J.M., 2003. The scaling of white matter to gray matter in cerebellum and neocortex. Brain Behav. Evol. 61, 1–5.

Butts, C.T., 2009. Revisiting the foundations of network analysis. Science 325, 414–416.

Buzsáki, G., 2006. Rhythms of the Brain. Oxford University Press, New York, NY.

Buzsáki, G., Anastassiou, C.A., Koch, C., 2012. The origin of extracellular fields and currents—EEG, ECoG, LFP and spikes. Nat. Rev. Neurosci. 13, 407–420.

Buzsáki, G., Draguhn, A., 2004. Neuronal oscillations in cortical networks. Science 304, 1926–1929.

Buzsáki, G., Mizuseki, K., 2014. The log-dynamic brain: how skewed distributions affect network operations. Nat. Rev. Neurosci. 15, 264–278.

Buzsáki, G., Logothetis, N., Singer, W., 2013. Scaling brain size, keeping timing: evolutionary preservation of brain rhythms. Neuron 80, 751–764.

Cabral, J., Hugues, E., Kringelbach, M.L., Deco, G., 2012. Modeling the outcome of structural disconnection on resting-state functional connectivity. NeuroImage 62, 1342–1353.

Calabrese, E., Badea, A., Cofer, G., Qi, Y., Johnson, G.A., 2015. A diffusion MRI tractography connectome of the mouse brain and comparison with neuronal tracer data. Cereb. Cortex 25 (11), 4628–4637.

Calamante, F., Smith, R.E., Tournier, J.D., Raffelt, D., Connelly, A., 2015. Quantification of voxel-wise total fibre density: investigating the problems associated with track-count mapping. NeuroImage 117, 284–293.

Callaway, D.S., Newman, M.E.J., Strogatz, S.H., Watts, D.J., 2000. Network robustness and fragility: percolation on random graphs. Phys. Rev. Lett. 85, 5468–5471.

Canolty, R.T., Knight, R.T., 2010. The functional role of cross-frequency coupling. Trends Cogn. Sci. 14, 506–515.

Chang, C., Glover, G.H., 2009. Relationship between respiration, end-tidal CO_2, and BOLD signals in resting-state functional MRI. NeuroImage 47, 1381–1393.

Chang, C., Glover, G.H., 2010. Time-frequency dynamics of resting-state brain connectivity measured with fMRI. NeuroImage 50, 81–98.

Chatterjee, N., Sinha, S., 2008. Understanding the mind of a worm: hierarchical network structure underlying nervous system function in C. elegans. Prog. Brain Res. 168, 145–153.

Chen, B.L., Hall, D.H., Chklovskii, D.B., 2006. Wiring optimization can relate neuronal structure and function. Proc. Natl. Acad. Sci. U. S. A. 103, 4723–4728.

Chen, J., Jann, K., Wang, D.J., 2015. Characterizing resting-state brain function using arterial-spin labeling. Brain Connect. 5, 527–542.

Chen, Y., Paul, G., Cohen, R., Havlin, S., Borgatti, S.P., Liljeros, F., Stanley, H.E., 2007. Percolation theory applied to measures of fragmentation in social networks. Phys. Rev. E 75, 046107.

Chen, Y., Wang, S., Hilgetag, C.C., Zhou, C., 2013. Trade-off between multiple constraints enables simultaneous formation of modules and hubs in neural systems. PLoS Comput. Biol. 9, e1002937.

Chen, Z.J., He, Y., Rosa-Neto, P., Germann, J., Evans, A.C., 2008. Revealing modular architecture of human brain structural networks by using cortical thickness from MRI. Cereb. Cortex 18 (10), 2374–2381.

Cherniak, C., 1990. The bounded brain: toward quantitative neuroanatomy. J. Cogn. Neurosci. 2, 58–68.

Cherniak, C., 1994. Component placement optimization in the brain. J. Neurosci. 14, 2418–2427.

Cherniak, C., Changizi, M., Won Kang, D., 1999. Large-scale optimization of neuron arbors. Phys. Rev. E 59, 6001–6009.

Chialvo, D.R., 2010. Emergent complex neural dynamics. Nat. Phys. 6 (10), 744–750.

Chiang, A.S., Lin, C.Y., Chuang, C.C., Chang, H.M., Hsieh, C.H., Yeh, C.W., Shih, C.T., Wu, J.J., Wang, G.T., Chen, Y.C., Wu, C.C., Chen, G.Y., Ching, Y.T., Lee, P.C., Lin, C.Y., Lin, H.H., Wu, C.C., Hsu, H.W., Huang, Y.A., Chen, J.Y., Chiang, H.J., Lu, C.F., Ni, R.F., Yeh, C.Y., Hwang, J.K., 2011. Three-dimensional reconstruction of brain-wide wiring networks in Drosophila at single-cell resolution. Curr. Biol. 21 (1), 1–11.

Chklovskii, D.B., 2004. Synaptic connectivity and neuronal morphology: two sides of the same coin. Neuron 43, 609–617.

Chklovskii, D.B., Koulakov, A.A., 2004. Maps in the brain: what can we learn from them? Annu. Rev. Neurosci. 27, 369–392.

Chklovskii, D.B., Schikorski, T., Stevens, C.F., 2002. Wiring optimization in cortical circuits. Neuron 34, 341–347.

Christie, P., Stroobandt, D., 2000. The interpretation and application of Rent's rule. IEEE Trans. VLSI Syst. 8, 639–648.

Chung, F.R.K., 1997. Spectral Graph Theory. American Mathematical Society, Providence, RI.

Chung, K., Deisseroth, K., 2013. CLARITY for mapping the nervous system. Nat. Methods 10, 508–513.

Clark, D.D., Sokoloff, L., 1999. Circulation and energy metabolism of the brain. In: Siegel, G.J., Agranoff, B.W., Albers, R.W., Fisher, S.K., Uhler, M.D. (Eds.), Basic Neurochemistry Molecular, Cellular and Medical Aspects. Lippincott-Raven, Philadelphia, PA, pp. 637–670.

Clarke, S., Hall, P., 2009. Robustness of multiple testing procedures against dependence. Ann. Stat. 37 (1), 332–358.

Clauset, A., Shalizi, C.R., Newman, M.E.J., 2009. Power-law distributions in empirical data. SIAM Rev. 51, 661–703.

Coan, A.C., Campos, B.M., Yasuda, C.L., Kubota, B.Y., Bergo, F.P., Guerreiro, C.A., Cendes, F., 2014. Frequent seizures are associated with a network of gray matter atrophy in temporal lobe epilepsy with or without hippocampal sclerosis. PLoS One 9, e85843.

Cohen, A.L., Fair, D.A., Dosenbach, N.U., Miezin, F.M., Dierker, D., Van Essen, D.C., Schlaggar, B.L., Petersen, S.E., 2008. Defining functional areas in individual human brains using resting functional connectivity MRI. NeuroImage 41, 45–57.

Cohen, M.X., 2014. Analyzing Neural Time Series Data: Theory and Practice. The MIT Press, Cambridge, MA.

Cohen, M.X., 2015. Effects of time lag and frequency matching on phase-based connectivity. J. Neurosci. Methods 250, 137–146.

Cohen, R., Erez, K., Ben-Avraham, D., Havlin, S., 2000. Resilience of the Internet to random breakdowns. Phys. Rev. Lett. 85, 4626–4628.

Cole, M.W., Reynolds, J.R., Power, J.D., Repovs, G., Anticevic, A., Braver, T.S., 2013. Multi-task connectivity reveals flexible hubs for adaptive task control. Nat. Neurosci. 16, 1348–1355.

Colizza, V., Flammini, A., Serrano, M.A., Vespignani, A., 2006. Detecting rich-club ordering in complex networks. Nat. Phys. 2, 110–115.

Collin, G., Sporns, O., Mandl, R.C.W., van den Heuvel, M.P., 2014. Structural and functional aspects relating to cost and benefit of rich club organization in the human cerebral cortex. Cereb. Cortex 24, 2258–2267.

Costa, L.D.F., Kaiser, M., Hilgetag, C.C., 2007. Predicting the connectivity of primate cortical networks from topological and spatial node properties. BMC Syst. Biol. 1, 16.

Craddock, R.C., James, G.A., Holtzheimer 3rd., P.E., Hu, X.P., Mayberg, H.S., 2012. A whole brain fMRI atlas generated via spatially constrained spectral clustering. Hum. Brain Mapp. 33, 1914–1928.

Cribben, I., Haraldsdottir, R., Atlas, L.Y., Wager, T.D., Lindquist, M.A., 2012. Dynamic connectivity regression: determining state-related changes in brain connectivity. NeuroImage 61 (4), 907–920.

Crofts, J.J., Higham, D.J., 2009. A weighted communicability measure applied to complex brain networks. J. R. Soc. Interface 6 (33), 411–414.

Crofts, J.J., Higham, D.J., Bosnell, R., Jbabdi, S., Matthews, P.M., Behrens, T.E.J., Johansen-Berg, H., 2011. Network analysis detects changes in the contralesional hemisphere following stroke. NeuroImage 54 (1), 161–169.

Crossley, N.A., Mechelli, A., Vértes, P.E., Winton-Brown, T.T., Patel, A.X., Ginestet, C.E., McGuire, P.K., Bullmore, E.T., 2013. Cognitive relevance of the community structure of the human brain functional coactivation network. Proc. Natl. Acad. Sci. U. S. A. 110, 11583–11588.

Crossley, N.A., Mechelli, A., Scott, J., Carletti, F., Fox, P.T., McGuire, P., Bullmore, E.T., 2014. The hubs of the human connectome are generally implicated in the anatomy of brain disorders. Brain 137, 2382–2395.

Csermely, P., London, A., Wu, L.Y., Uzzi, B., 2013. Structure and dynamics of core/periphery networks. J. Complex Netw. 1, 93–123.

Da Mota, B., Fritsch, V., Varoquaux, G., Banaschewski, T., Barker, G.J., Bokde, A.L.W., Bromberg, U., Conrod, P., Gallinat, J., Garavan, H., et al., 2014. Randomized parcellation based inference. NeuroImage 89, 203–215.

da Silva, F.L., 2004. Functional localization of brain sources using EEG and/or MEG data: volume conductor and source models. Magn. Reson. Imaging 22, 1533–1538.

da Silva, M.R., Ma, H., Zeng, A.-P., 2008. Centrality, network capacity, and modularity as parameters to analyze the core-periphery structure in metabolic networks. Proc. IEEE 96, 1411–1420.

Damoiseaux, J.S., Rombouts, S.A., Barkhof, F., Scheltens, P., Stam, C.J., Smith, S.M., Beckmann, C.F., 2006. Consistent resting-state networks across healthy subjects. Proc. Natl. Acad. Sci. U. S. A. 103, 13848–13853.

Danon, L., Díaz-Guilera, A., Duch, J., Arenas, A., 2005. Comparing community structure identification. J. Stat. Mech. Theory Exp. 2005, P09008.

Darvas, F., Pantazis, D., Kucukaltun-Yildirim, E., Leahy, R.M., 2004. Mapping human brain function with MEG and EEG: methods and validation. NeuroImage 23, S289–S299.

Deco, G., Jirsa, V.K., Robinson, P.A., Breakspear, M., Friston, K., 2008. The dynamic brain: from spiking neurons to neural masses and cortical fields. PLoS Comput. Biol. 4, e1000092.

de Haan, W., Mott, K., van Straaten, E.C.W., Scheltens, P., Stam, C.J., 2012. Activity dependent degeneration explains hub vulnerability in Alzheimer's disease. PLoS Comput. Biol. 8, e1002582.

de Lange, S.C., de Reus, M.A., van den Heuvel, M.P., 2014. The Laplacian spectrum of neural networks. Front. Comput. Neurosci. 7, 189.

de Luca, M., Beckmann, C.F., De Stefano, N., Matthews, P.M., Smith, S.M., 2006. fMRI resting state networks define distinct modes of long-distance interactions in the human brain. NeuroImage 29, 1359–1367.

de Pasquale, F., Della Penna, S., Snyder, A.Z., Marzetti, L., 2012. A cortical core for dynamic integration of functional networks in the resting human brain. Neuron 74, 753–764.

de Reus, M.A., van den Heuvel, M.P., 2013a. Estimating false positives and negatives in brain networks. NeuroImage 70, 402–409.

de Reus, M.A., van den Heuvel, M.P., 2013b. Rich club organization and intermodule communication in the cat connectome. J. Neurosci. 33, 12929–12939.

de Reus, M.A., van den Heuvel, M.P., 2014. Simulated rich club lesioning in brain networks: a scaffold for communication and integration? Front. Hum. Neurosci. 8, 647.

de Reus, M.A., Saenger, V.M., Kahn, R.S., van den Heuvel, M.P., 2014. An edge-centric perspective on the human connectome: link communities in the brain. Philos. Trans. R. Soc. Lond. B Biol. Sci. 369, 20130527.

Debanne, D., Bialowas, A., Rama, S., 2013. What are the mechanisms for analogue and digital signalling in the brain? Nat. Rev. Neurosci. 14 (1), 63–69.

Deco, G., Jirsa, V.K., 2012. Ongoing cortical activity at rest: criticality, multistability, and ghost attractors. J. Neurosci. 32, 3366–3375.

Deco, G., Kringelbach, M.L., 2014. Great expectations: using whole-brain computational connectomics for understanding neuropsychiatric disorders. Neuron 84, 892–905.

Deco, G., Jirsa, V.K., Robinson, P.A., Breakspear, M., Friston, K., 2008. The dynamic brain: from spiking neurons to neural masses and cortical fields. PLoS Comput. Biol. 4, e1000092.

Dehaene, S., Kerszberg, M., Changeux, J.P., 1998. A neuronal model of a global workspace in effortful cognitive tasks. Proc. Natl. Acad. Sci. U. S. A. 95, 14529–14534.

Delvenne, J.C., Yaliraki, S.N., 2010. Stability of graph communities across time scales. Proc. Natl. Acad. Sci. U. S. A. 107, 12755–12760.

Denk, W., Horstmann, H., 2004. Serial block-face scanning electron microscopy to reconstruct three-dimensional tissue nanostructure. PLoS Biol. 2. e329.

DeRobertis, E.D.P., Bennett, H.S., 1955. Some features of the submicroscopic morphology of synapses in frog and earthworm. J. Biophys. Biochem. Cytol. 1, 47–58.

Desikan, R.S., Segonne, F., Fischl, B., Quinn, B.T., Dickerson, B.C., Blacker, D., Buckner, R.L., Dale, A.M., Maguire, R.P., Hyman, B.T., Albert, M.S., Killiant, R.J., 2006. An automated labeling system for subdividing the human cerebral cortex on MRI scans into gyral based regions of interest. NeuroImage 31, 968–980.

Diamond, J.S., 2002. A broad view of glutamate spillover. Nat. Neurosci. 5, 291–292.

Dice, L.R., 1945. Measures of the amount of ecologic association between species. Ecology 26, 297–302.

Dijkstra, E.W., 1959. A note on two problems in connexion with graphs. Numer. Math. 1, 269–271.

Dong, H.W., 2008. The Allen Reference Atlas: A Digital Color Brain Atlas of C57BL/6J Male Mouse. John Wiley & Sons, Hoboken, NJ.

Doreian, P., Batagelj, V., Ferligoj, A., 2005. Generalized Blockmodeling. Cambridge University Press, New York, NY.

Dorogovtsev, S.N., Goltsev, A.V., Mendes, J.F.F., 2006a. k-core organization of complex networks. Phys. Rev. Lett. 96, 040601–040604.

Dorogovtsev, S.N., Goltsev, A.V., Mendes, J.F.F., 2006b. k-core architecture and k-core percolation on complex networks. Physica D 224, 7–19.

Dorogovtsev, S.N., Goltsev, A.V., Mendes, J.F.F., 2008. Critical phenomena in complex networks. Rev. Mod. Phys. 80, 1275–1335.

Dosenbach, N.U., Nardos, B., Cohen, A.L., Fair, D.A., Power, J.D., Church, J.A., Nelson, S.M., Wig, G.S., Vogel, A.C., Lessov-Schlaggar, C.N., et al., 2010. Prediction of individual brain maturity using fMRI. Science 329, 1358–1361.

Downes, J.H., Hammond, M.W., Xydas, D., Spencer, M.C., Becerra, V.M., Warwick, K., Whalley, B.J., Nasuto, S.J., 2012. Emergence of a small-world functional network in cultured neurons. PLoS Comput. Biol. 8 (5), e1002522.

Drakesmith, M., Caeyenberghs, K., Dutt, A., Lewis, G., David, A.S., Jones, D.K., 2015. Overcoming the effects of false positives and threshold bias in graph theoretical analyses of neuroimaging data. NeuroImage 118, 313–333.

Drossel, B., Schwabl, F., 1992. Self-organized critical forest-fire model. Phys. Rev. Lett. 69, 1629–1632.

Dulmage, A.L., Mendelsohn, N.S., 1958. Coverings of bipartite graphs. Can. J. Math. 10, 517–534.

Dunlop, J., Smith, D.G., 1994. Telecommunications Engineering. CRC Press, Boca Raton, FL.

Dwyer, D.B., Harrison, B.J., Yücel, M., Whittle, S., Zalesky, A., Pantelis, C., Allen, N.B., Fornito, A., 2014. Large-scale brain network dynamics supporting adolescent cognitive control. J. Neurosci. 34, 14096–14107.

Dwyer, T., Koren, Y., Marriott, K., 2006. IPSEP-COLA: an incremental procedure for separation constraint layout of graphs. IEEE Trans. Vis. Comput. Graph. 12, 821–828.

Eades, P., 1984. A heuristic for graph drawing. Congr. Numer. 42, 149–160.

Easley, D., Kleinberg, J., 2010. Networks, Crowds, and Markets: Reasoning About a Highly Connected World. Cambridge University Press, New York, NY.

Eguíluz, V.M., Chialvo, D.R., Cecchi, G.A., Baliki, M., Apkarian, A.V., 2005. Scale-free brain functional networks. Phys. Rev. Lett. 94, 018102.

Eickhoff, S.B., Stephan, K.E., Mohlberg, H., Grefkes, C., Fink, G.R., Amunts, K., Zilles, K., 2005. A new SPM toolbox for combining probabilistic cytoarchitectonic maps and functional imaging data. NeuroImage 25, 1325–1335.

Eidsaa, M., Almaas, E., 2013. s-core network decomposition: a generalization of k-core analysis to weighted networks. Phys. Rev. E 88, 062819.

Einevoll, G.T., Kayser, C., Logothetis, N.K., Panzeri, S., 2013. Modelling and analysis of local field potentials for studying the function of cortical circuits. Nat. Rev. Neurosci. 14, 770–785.

Ekman, M., Derrfuss, J., Tittgemeyer, M., Fiebach, C.J., 2012. Predicting errors from reconfiguration patterns in human brain networks. Proc. Natl. Acad. Sci. U. S. A. 109 (41), 16714–16719.

Engel, A.K., Gerloff, C., Hilgetag, C.C., Nolte, G., 2013. Intrinsic coupling modes: multiscale interactions in ongoing brain activity. Neuron 80, 867–886.

Ercsey-Ravasz, M., Markov, N.T., Lamy, C., Van Essen, D.C., Knoblauch, K., Toroczkai, Z., Kennedy, H., 2013. A predictive network model of cerebral cortical connectivity based on a distance rule. Neuron 80, 184–197.

Erdős, P., Rényi, A., 1959. On random graphs. Publ. Math. Debr. 6, 290–297.

Erdős, P., Rényi, A., 1960. On the evolution of random graphs. Publ. Math. Inst. Hung. Acad. Sci. 5, 17–61.

Esquivel, A.V., Rosvall, M., 2011. Compression of flow can reveal overlapping-module organization in networks. Phys. Rev. X 1, 021025. 11.

Estrada, E., Hatano, N., 2008. Communicability in complex networks. Phys. Rev. E 77, 036111.

Expert, P., Evans, T.S., Blondel, V.D., Lambiotte, R., 2011. Uncovering space-independent communities in spatial networks. Proc. Natl. Acad. Sci. U. S. A. 108, 7663–7668.

Fagiolo, G., 2007. Clustering in complex directed networks. Phys. Rev. E 76, 026107–026108.

Fair, D.A., Cohen, A.L., Power, J.D., Dosenbach, N.U.F., Church, J.A., Miezin, F.M., Schlaggar, B.L., Petersen, S.E., 2009. Functional brain networks develop from a "local to distributed" organization. PLoS Comput. Biol. 5, e1000381.

Feinberg, D.A., Moeller, S., Smith, S.M., Auerbach, E., Ramanna, S., Gunther, M., Glasser, M.F., Miller, K.L., Ugurbil, K., Yacoub, E., 2010. Multiplexed echo planar imaging for sub-second whole brain FMRI and fast diffusion imaging. PLoS One 5, e15710.

Felleman, D.J., Van Essen, D.C., 1991. Distributed hierarchical processing in the primate cerebral cortex. Cereb. Cortex 1, 1–47.

Fenno, L., Yizhar, O., Deisseroth, K., 2011. The development and application of optogenetics. Annu. Rev. Neurosci. 34, 389–412.

Ferizi, U., Schneider, T., Panagiotaki, E., Nedjati-Gilani, G., Zhang, H., Wheeler-Kingshott, C.A.M., Alexander, D.C., 2014. A ranking of diffusion MRI compartment models with in vivo human brain data. Magn. Reson. Med. 72, 1785–1792.

Fillard, P., Descoteaux, M., Goh, A., Gouttard, S., Jeurissen, B., Malcolm, J., Ramirez-Manzanares, A., Reisert, M., Sakaie, K., Tensaouti, F., Yo, T., Mangin, J.F., Poupon, C., 2011. Quantitative evaluation of 10 tractography algorithms on a realistic diffusion MR phantom. NeuroImage 56, 220–234.

Finger, S., Koehler, P.J., Jagella, C., 2004. The von Monakow concept of diaschisis: origins and perspectives. Arch. Neurol. 61, 283–288.

Fischl, B., van der Kouwe, A., Destrieux, C., Halgren, E., Ségonme, F., Salat, D.H., Busa, E., Seidman, L.J., Goldstein, J.M., Kennedy, D., Caviness, V., Makris, N., Rosen, B., Dale, A.M., 2004. Anatomically parcellating the human cereral cortex. Cereb. Cortex 14, 11–22.

Fisher, R.A., 1935. The Design of Experiments. Hafner, New York, NY.

Fodor, J.A., 1983. Modularity of Mind: An Essay on Faculty Psychology. MIT Press, Cambridge, MA.

Fornito, A., Bullmore, E.T., 2010. What can spontaneous fluctuations of the blood oxygenation-level-dependent signal tell us about psychiatric disorders? Curr. Opin. Psychiatry 23, 239–249.

Fornito, A., Bullmore, E.T., 2014. Reconciling abnormalities of brain network structure and function in schizophrenia. Curr. Opin. Neurobiol. 30C, 44–50.

Fornito, A., Harrison, B.J., Zalesky, A., Simons, J.S., 2012a. Competitive and cooperative dynamics of large-scale brain functional networks supporting recollection. Proc. Natl. Acad. Sci. U. S. A. 109 (31), 12788–12793.

Fornito, A., Yoon, J., Zalesky, A., Bullmore, E.T., Carter, C.S., 2011a. General and specific functional connectivity disturbances in first-episode schizophrenia during cognitive control performance. Biol. Psychiatry 70, 64–72.

Fornito, A., Yucel, M., Wood, S., Stuart, G.W., Buchanan, J.A., Proffitt, T., Anderson, V., Velakoulis, D., Pantelis, C., 2004. Individual differences in anterior cingulate/paracingulate morphology are related to executive functions in healthy males. Cereb. Cortex 14, 424–431.

Fornito, A., Zalesky, A., Breakspear, M., 2013. Graph analysis of the human connectome: promise, progress, and pitfalls. NeuroImage 80, 426–444.

Fornito, A., Zalesky, A., Breakspear, M., 2015. The connectomics of brain disorders. Nat. Rev. Neurosci. 16, 159–172.

Fornito, A., Zalesky, A., Bullmore, E.T., 2010. Network scaling effects in graph analytic studies of human resting-state FMRI data. Front. Syst. Neurosci. 4, 22.

Fornito, A., Zalesky, A., Bassett, D.S., Meunier, D., Ellison-Wright, I., Yücel, M., Wood, S.J., Shaw, K., O'Connor, J., Nertney, D., Mowry, B.J., Pantelis, C., Bullmore, E.T., 2011b. Genetic influences on cost-efficient organization of human cortical functional networks. J. Neurosci. 31 (9), 3261–3270.

Fornito, A., Zalesky, A., Pantelis, C., Bullmore, E.T., 2012b. Schizophrenia, neuroimaging and connectomics. NeuroImage 62, 2296–2314.

Fortunato, S., 2010. Community detection in graphs. Phys. Rep. 486, 75–174.

Fortunato, S., Barthelemy, M., 2007. Resolution limit in community detection. Proc. Natl. Acad. Sci. U. S. A. 104, 36–41.

Foster, J.G., Foster, D.V., Grassberger, P., Paczuski, M., 2010. Edge direction and the structure of networks. Proc. Natl. Acad. Sci. U. S. A. 107, 10815–10820.

Foti, N.J., Hughes, J.M., Rockmore, D.N., 2011. Nonparametric sparsification of complex multiscale networks. PLoS One 6 (2), e16431.

Fox, M.D., Raichle, M.E., 2007. Spontaneous fluctuations in brain activity observed with functional magnetic resonance imaging. Nat. Rev. Neurosci. 8, 700–711.

Fox, M.D., Buckner, R.L., Liu, H., Chakravarty, M.M., Lozano, A.M., Pascual-Leone, A., 2014. Resting-state networks link invasive and noninvasive brain stimulation across diverse psychiatric and neurological diseases. Proc. Natl. Acad. Sci. U. S. A. 111, E4367–E4375.

Fox, M.D., Snyder, A.Z., Vincent, J.L., Raichle, M.E., 2007. Intrinsic fluctuations within cortical systems account for intertrial variability in human behavior. Neuron 56, 171–184.

Fox, M.D., Snyder, A.Z., Vincent, J.L., Corbetta, M., Van Essen, D.C., Raichle, M.E., 2005. The human brain is intrinsically organized into dynamic, anticorrelated functional networks. Proc. Natl. Acad. Sci. U. S. A. 102, 9673–9678.

Fox, M.D., Snyder, A.Z., Zacks, J.M., Raichle, M.E., 2006. Coherent spontaneous activity accounts for trial-to-trial variability in human evoked brain responses. Nat. Neurosci. 9, 23–25.

Fox, M.D., Zhang, D., Snyder, A.Z., Raichle, M.E., 2009. The global signal and observed anticorrelated resting state brain networks. J. Neurophysiol. 101, 3270–3283.

Frank, S.A., 2009. The common patterns of nature. J. Evol. Biol. 22, 1563–1585.

Fred, A.L., Jain, A.K., 2003. Robust data clustering. IEEE Comput. Soc. Conf. Comput. Vis. Pattern Recogn. 2, 128–136.

Freeman, L.C., 1977. A set of measures of centrality based on betweenness. Sociometry 40, 35–41.

Freeman, L.C., 1979. Centrality in social networks. Conceptual clarification. Soc. Networks 1, 215–239.

French, L., Pavlidis, P., 2011. Relationships between gene expression and brain wiring in the adult rodent brain. PLoS Comput. Biol. 7 (1), e1001049.

Freud, S., 1891. On Aphasia: A Critical Study (E. Stengel, Trans.). Imago Publishing Company Limited.

Freud, S., 1895. Project for a scientific psychology. In: Strachey, J. (Ed.), The Standard Edition of the Complete Psychological Works of Sigmund Freud, vol. I. Hogarth, London.

Friedman, J., Hastie, T., Tibshirani, R., 2008. Sparse inverse covariance estimation with the graphical lasso. Biostatistics 9 (3), 432–441.

Fries, P., 2005. A mechanism for cognitive dynamics: neuronal communication through neuronal coherence. Trends Cogn. Sci. 9, 474–480.

Fries, P., Neuenschwander, S., Engel, A.K., Goebel, R., Singer, W., 2001. Rapid feature selective neuronal synchronization through correlated latency shifting. Nat. Neurosci. 4, 194–200.

Fries, P., Nikolic, D., Singer, W., 2007. The gamma cycle. Trends Neurosci. 30, 309–316.

Friston, K., 2009. Causal modelling and brain connectivity in functional magnetic resonance imaging. PLoS Biol. 7, e33.

Friston, K., Moran, R., Seth, A.K., 2013. Analysing connectivity with Granger causality and dynamic causal modelling. Curr. Opin. Neurobiol. 23, 172–178.

Friston, K.J., 1994. Functional and effective connectivity in neuroimaging: a synthesis. Hum. Brain Mapp. 2, 56–78.

Friston, K.J., 2011. Functional and effective connectivity: a review. Brain Connect. 1, 13–36.

Friston, K.J., Price, C.J., 2011. Modules and brain mapping. Cogn. Neuropsychol. 28, 241–250.

Friston, K.J., Buechel, C., Fink, G.R., Morris, J., Rolls, E., Dolan, R.J., 1997. Psychophysiological and modulatory interactions in neuroimaging. NeuroImage 6, 218–229.

Friston, K.J., Harrison, L., Penny, W., 2003. Dynamic causal modelling. NeuroImage 19, 1273–1302.

Friston, K.J., Kahan, J., Razi, A., Stephan, K.E., Sporns, O., 2014. On nodes and modes in resting state fMRI. NeuroImage 99, 533–547.

Friston, K.J., Li, B., Daunizeau, J., Stephan, K.E., 2011. Network discovery with DCM. NeuroImage 56, 1202–1221.

Frost, B., Diamond, M.I., 2009. Prion-like mechanisms in neurodegenerative diseases. Nat. Rev. Neurosci. 11, 155–159.

Fruchterman, T., Reingold, E.M., 1991. Graph drawing by force-directed placement. Softw. Pract. Exp. 21, 1129–1164.

Fukushima, M., Chao, Z.C., Fujii, N., 2015. Studying brain functions with mesoscopic measurements: advances in electrocorticography for non-human primates. Curr. Opin. Neurobiol. 32, 124–131.

Fulcher, B.D., Fornito, A., 2016. A transcriptional signature of hub connectivity in the mouse connectome. Proc. Natl. Acad. Sci. U. S. A. 113, 1435–1440.

Galen, 1976. De Locis Affectis. In: Siegel, R. (Ed.), Galen on the Affected Parts. S. Karger AG, Basel, Switzerland.

Gan, G., Ma, C., Wu, J., 2007. Data Clustering: Theory, Algorithms, and Applications. SIAM/ASA, Philadelphia, PA/Alexandria, VA. ASA-SIAM Series on Statistics and Applied Probability.

Garas, A., Schweitzer, F., Havlin, S., 2012. A k-shell decomposition method for weighted networks. New J. Phys. 14, 083030. 15.

Garcia-Lopez, P., Garcia-Marin, V., Freire, M., 2010. The histological slides and drawings of Cajal. Front. Neuroanat. 4, 9.

Gardner, W.A., 1992. A unifying view of coherence in signal processing. Signal Process. 29, 113–140.

Genovese, C.R., Lazar, N.A., Nichols, T., 2002. Thresholding of statistical maps in functional neuroimaging using the false discovery rate. NeuroImage 15 (4), 870–878.

Gerstein, G.L., Perkel, D.H., 1969. Simultaneously recorded trains of action potentials: analysis and functional interpretation. Science 164, 828–830.

Gerstein, G.L., Perkel, D.H., Subramanian, K.N., 1978. Identification of functionally related neural assemblies. Brain Res. 140, 43–62.

Geschwind, N., 1965a. Disconnexion syndromes in animals and man. Part I. Brain 88, 237–294.

Geschwind, N., 1965b. Disconnexion syndromes in animals and man. Part II. Brain 88, 585–644.

Geyer, C.J., Thompson, E.A., 1992. Constrained Monte Carlo maximum likelihood for dependent data. J. R. Stat. Soc. Ser. B 54, 657–699.

Gibson, H., Faith, J., Vickers, P., 2013. A survey of two-dimensional graph layout techniques for information visualisation. Inf. Vis. 12, 324–357.

Gibson, W.C., 1962. Pioneers in localization of function in the brain. J. Am. Med. Assoc. 180, 944–951.

Gilbert, E.N., 1959. Random graphs. Ann. Math. Stat. 30 (4), 1141–1144.

Ginestet, C.E., Simmons, A., 2011. Statistical parametric network analysis of functional connectivity dynamics during a working memory task. NeuroImage 55 (2), 688–704.

Ginestet, C.E., Nichols, T.E., Bullmore, E.T., Simmons, A., 2011. Brain network analysis: separating cost from topology using cost-integration. PLoS One 6, e21570.

Girvan, M., Newman, M.E.J., 2002. Community structure in social and biological networks. Proc. Natl. Acad. Sci. U. S. A. 99, 7821–7826.

Glahn, D.C., Winkler, A.M., Kochunov, P., Almasy, L., Duggirala, R., Carless, M.A., Curran, J.C., Olvera, R.L., Laird, A.R., Smith, S.M., Beckmann, C.F., Fox, P.T., Blangero, J., 2010. Genetic control over the resting brain. Proc. Natl. Acad. Sci. U. S. A. 107, 1223–1228.

Glasser, M.F., Van Essen, D.C., 2011. Mapping human cortical areas in vivo based on myelin content as revealed by T1- and T2-weighted MRI. J. Neurosci. 31, 11597–11616.

Gleeson, J.P., 2009. Bond percolation on a class of clustered random networks. Phys. Rev. E 80, 036107.

Godwin, D., Barry, R.L., Marois, R., 2015. Breakdown of the brain's functional network modularity with awareness. Proc. Natl. Acad. Sci. 112 (12), 3799–3804.

Gold, L., Lauritzen, M., 2002. Neuronal deactivation explains decreased cerebellar blood flow in response to focal cerebral ischemia or suppressed neocortical function. Proc. Natl. Acad. Sci. U. S. A. 99, 7699–7704.

Gollo, L.L., Mirasso, C., Sporns, O., Breakspear, M., 2014. Mechanisms of zero-lag synchronization in cortical motifs. PLoS Comput. Biol. 10, e1003548.

Gollo, L.L., Zalesky, A., Hutchison, R.M., van den Heuvel, M., Breakspear, M., 2015. Dwelling quietly in the rich club: brain network determinants of slow cortical fluctuations. Philos. Trans. R. Soc. Lond. B Biol. Sci. 370, 20140165.

Goltsev, A.V., Dorogovtsev, S.N., Mendes, J., 2006. k-core (bootstrap) percolation on complex networks: critical phenomena and nonlocal effects. Phys. Rev. E 73, 056101.

Gómez-Gardenes, J., Zamora-López, G., Moreno, Y., Arenas, A., 2010. From modular to centralized organization of synchronization in functional areas of the cat cerebral cortex. PLoS One 5, e12313.

Gómez, S., Jensen, P., Arenas, A., 2009. Analysis of community structure in networks of correlated data. Phys. Rev. E 80, 016114.

Gong, G., He, Y., Concha, L., Lebel, C., Gross, D.W., Evans, A.C., Beaulieu, C., 2009. Mapping anatomical connectivity patterns of human cerebral cortex using in vivo diffusion tensor imaging tractography. Cereb. Cortex 19, 524–536.

Goñi, J., Avena-Koenigsberger, A., Velez de Mendizabal, N., van den Heuvel, M.P., Betzel, R.F., Sporns, O., 2013. Exploring the morphospace of communication efficiency in complex networks. PLoS One 8, e58070.

Goñi, J., van den Heuvel, M.P., Avena-Koenigsberger, A., Velez de Mendizabal, N., Betzel, R.F., Griffa, A., Hagmann, P., Corominas-Murtra, B., Thiran, J.P., Sporns, O., 2014. Resting-brain functional connectivity predicted by analytic measures of network communication. Proc. Natl. Acad. Sci. U. S. A. 111 (2), 833–838.

Good, B.H., de Montjoye, Y.A., Clauset, A., 2010. Performance of modularity maximization in practical contexts. Phys. Rev. E 81, 046106.

Good, P., 1994. Permutation Tests. A Practical Guide to Resampling Methods for Testing Hypotheses. Springer Verlag, New York, NY.

Grady, D., Thiemann, C., Brockmann, D., 2012. Robust classification of salient links in complex networks. Nat. Commun. 3, 864.

Graham, D., 2014. Routing in the brain. Front. Comput. Neurosci. 8, 44.

Graham, D., Rockmore, D., 2011. The packet switching brain. J. Cogn. Neurosci. 23 (2), 267–276.

Granovetter, M., 1978. Threshold models of collective behavior. Am. J. Sociol. 83, 1420–1443.

Gray, C.M., Singer, W., 1989. Stimulus-specific neuronal oscillations in orientation columns of cat visual cortex. Proc. Natl. Acad. Sci. U. S. A. 86, 1698–1702.

Gray, C.M., König, P., Engel, A.K., Singer, W., 1989. Oscillatory responses in cat visual cortex exhibit inter-columnar synchronization which reflects global stimulus properties. Nature 338, 334–337.

Greenblatt, R.E., Pflieger, M.E., Ossadtchi, A.E., 2012. Connectivity measures applied to human brain electrophysiological data. J. Neurosci. Methods 207, 1–16.

Gregoriou, G.G., Gotts, S.J., Zhou, H., Desimone, R., 2009. High-frequency, long-range coupling between prefrontal and visual cortex during attention. Science 324, 1207–1210.

Gregory, S., 2007. An algorithm to find overlapping community structure in networks. In: Kok, J.N., Koronacki, J., Lopez de Mantaras, R., Matwin, S., Mladenič, D., Skowron, A. (Eds.), Knowledge Discovery in Databases: PKDD 2007. Springer, Berlin, pp. 91–102.

Gregory, S., 2010. Finding overlapping communities in networks by label propagation. New J. Phys. 12, 103018.

Grienberger, C., Konnerth, A., 2012. Imaging calcium in neurons. Neuron 73, 862–885.

Guevara, M.A., Corsi-Cabrera, M., 1996. EEG coherence or EEG correlation? Int. J. Psychophysiol. 23, 145–153.

Guimerà, R., Amaral, L.N., 2005. Functional cartography of complex metabolic networks. Nature 433, 895–900.

Guimerà, R., Mossa, S., Turtschi, A., Amaral, L.A.N., 2005. The worldwide air transportation network: anomalous centrality, community structure, and cities' global roles. Proc. Natl. Acad. Sci. U. S. A. 102, 7794–7799.

Guimerà, R., Sales-Pardo, M., Amaral, L., 2004. Modularity from fluctuations in random graphs and complex networks. Phys. Rev. E 70, 025101.

Gulyás, A., Bíró, J.J., Kőrösi, A., Rétvári, G., Krioukov, D., 2015. Navigable networks as Nash equilibria of navigation games. Nat. Commun. 6, 7651.

Gururangan, S.S., Sadovsky, A.J., MacLean, J.N., 2014. Analysis of graph invariants in functional neocortical circuitry reveals generalized features common to three areas of sensory cortex. PLoS Comput. Biol. 10, e1003710–e1003712.

Hagmann, P., 2005. From diffusion MRI to brain connectomics. PhD Thesis, Ecole Polytechnique Fédérale de Lausanne, Lausanne.

Hagmann, P., Cammoun, L., Gigandet, X., Meuli, R., Honey, C.J., Wedeen, V.J., Sporns, O., 2008. Mapping the structural core of human cerebral cortex. PLoS Biol. 6, e159.

Hagmann, P., Kurant, M., Gigandet, X., Thiran, P., Wedeen, V.J., Meuli, R., Thiran, J.P., 2007. Mapping human whole-brain structural networks with diffusion MRI. PLoS One 2, e597.

Haimovici, A., Tagliazucchi, E., Balenzuela, P., Chialvo, D.R., 2013. Brain organization into resting state networks emerges at criticality on a model of the human connectome. Phys. Rev. Lett. 110, 178101.

Hämäläinen, M., Hari, R., Ilmoniemi, R.J., Knuutila, J., 1993. Magnetoencephalography—theory, instrumentation, and applications to noninvasive studies of the working human brain. Rev. Mod. Phys. 65, 413–497.

Harary, F., Palmer, E.M., 1973. Graphical Enumeration. Academic Press, New York, NY.

Hari, R., Parkkonen, L., Nangini, C., 2010. The brain in time: insights from neuromagnetic recordings. Ann. N. Y. Acad. Sci. 1191, 89–109.

Harlow, J.M., 1848. Passage of an iron rod through the head. Boston Med. Surg. J. 39, 389–393.

Harriger, L., van den Heuvel, M.P., Sporns, O., 2012. Rich club organization of macaque cerebral cortex and its role in network communication. PLoS One 7, e46497.

Harrison, B.J., Pujol, J., Ortiz, H., Fornito, A., Pantelis, C., Yucel, M., 2008. Modulation of brain resting-state networks by sad mood induction. PLoS One 3, e794.

Hawellek, D.J., Hipp, J.F., Lewis, C.M., Corbetta, M., Engel, A.K., 2011. Increased functional connectivity indicates the severity of cognitive impairment in multiple sclerosis. Proc. Natl. Acad. Sci. U. S. A. 108, 19066–19071.

Hayasaka, S., Laurienti, P.J., 2010. Comparison of characteristics between region-and voxel-based network analyses in resting-state fMRI data. NeuroImage 50, 499–508.

Hayworth, K.J., Kasthuri, N., Schalek, R., 2006. Automating the collection of ultrathin serial sections for large volume TEM reconstructions. Microsc. Microanal. 12, 86–87.

Hazy, T.E., Frank, M.J., O'reilly, R.C., 2007. Towards an executive without a homunculus: computational models of the prefrontal cortex/basal ganglia system. Philos. Trans. R. Soc. Lond. B Biol. Sci. 362, 1601–1613.

He, B., Yang, L., Wilke, C., Yuan, H., 2011. Electrophysiological imaging of brain activity and connectivity—challenges and opportunities. IEEE Trans. Biomed. Eng. 58, 1918–1931.

He, B.J., 2011. Scale-free properties of the functional magnetic resonance imaging signal during rest and task. J. Neurosci. 31, 13786–13795.

He, B.J., Raichle, M.E., 2009. The fMRI signal, slow cortical potential and consciousness. Trends Cogn. Sci. 13, 302–309.

He, B.J., Snyder, A.Z., Zempel, J.M., Smyth, M.D., Raichle, M.E., 2008. Electrophysiological correlates of the brain's intrinsic large-scale functional architecture. Proc. Natl. Acad. Sci. U. S. A. 105, 16039–16044.

He, B.J., Zempel, J.M., Snyder, A.Z., Raichle, M.E., 2010. The temporal structures and functional significance of scale-free brain activity. Neuron 66, 353–369.

He, Y., Chen, Z.J., Evans, A.C., 2007. Small-world anatomical networks in the human brain revealed by cortical thickness from MRI. Cereb. Cortex 17, 2407–2419.

Hebb, D.O., 1949. The Organization of Behavior. A Neuropsychological Theory. Routledge, New York, NY.

Helmstaedter, M., 2013. Cellular-resolution connectomics: challenges of dense neural circuit reconstruction. Nat. Methods 10, 501–507.

Helmstaedter, M., Briggman, K.L., Turaga, S.C., Jain, V., Seung, H.S., Denk, W., 2013. Connectomic reconstruction of the inner plexiform layer in the mouse retina. Nature 500, 168–174.

Henderson, J.A., Robinson, P.A., 2011. Geometric effects on complex network structure in the cortex. Phys. Rev. Lett. 107, 018102.

Henderson, J.A., Robinson, P.A., 2013. Using geometry to uncover relationships between isotropy, homogeneity, and modularity in cortical connectivity. Brain Connect. 3, 423–437.

Henderson, J.A., Robinson, P.A., 2014. Relations between the geometry of cortical gyrification and white-matter network architecture. Brain Connect. 4, 112–130.

Herculano-Houzel, S., 2012. The remarkable, yet not extraordinary, human brain as a scaled-up primate brain and its associated cost. Proc. Natl. Acad. Sci. U. S. A. 109, 10661–10668.

Herculano-Houzel, S., Mota, B., Wong, P., Kaas, J.H., 2010. Connectivity-driven white matter scaling and folding in primate cerebral cortex. Proc. Natl. Acad. Sci. U. S. A. 107, 19008–19013.

Hesselmann, G., Kell, C.A., Eger, E., Kleinschmidt, A., 2008. Spontaneous local variations in ongoing neural activity bias perceptual decisions. Proc. Natl. Acad. Sci. U. S. A. 105, 10984–10989.

Hilgetag, C.C., Goulas, A., 2016. Is the brain really a small-world network? Brain Struct. Funct. 1–6. http://dx.doi.org/10.1007/s00429-015-1035-6.

Hilgetag, C.C., Kaiser, M., 2004. Clustered organization of cortical connectivity. Neuroinformatics 2, 353–360.

Hilgetag, C.C., Burns, G.A., O'Neill, M.A., Scannell, J.W., Young, M.P., 2000. Anatomical connectivity defines the organization of clusters of cortical areas in the macaque monkey and the cat. Philos. Trans. R. Soc. Lond. B Biol. Sci. 355, 91–110.

Hinrichs, C., Ithapu, V.K., Sun, Q., Johnson, S.C., Singh, V., 2013. Speeding up permutation testing in neuroimaging. Adv. Neural Inf. Process. Syst. 2013, 890–898.

Hipp, J.F., Engel, A.K., Siegel, M., 2011. Oscillatory synchronization in large-scale cortical networks predicts perception. Neuron 69, 387–396.

Hirokawa, N., Niwa, S., Tanaka, Y., 2010. Molecular motors in neurons: transport mechanisms and roles in brain function, development, and disease. Neuron 68, 610–638.

Hirschberger, M., Qi, Y., Steuer, R.E., 2007. Randomly generating portfolio-selection covariance matrices with specified distributional characteristics. Eur. J. Oper. Res. 177 (3), 1610–1625.

Holm, S., 1979. A simple sequentially rejective multiple test procedure. Scand. J. Stat. 6 (2), 65–70.

Holme, P., 2005. Core-periphery organization of complex networks. Phys. Rev. E 72, 046111–046114.

Honey, C.J., Sporns, O., 2008. Dynamical consequences of lesions in cortical networks. Hum. Brain Mapp. 29, 802–809.

Honey, C.J., Kötter, R., Breakspear, M., Sporns, O., 2007. Network structure of cerebral cortex shapes functional connectivity on multiple time scales. Proc. Natl. Acad. Sci. U. S. A. 104, 10240–10245.

Honey, C.J., Sporns, O., Cammoun, L., Gigandet, X., Thiran, J.P., Meuli, R., Hagmann, P., 2009. Predicting human resting-state functional connectivity from structural connectivity. Proc. Natl. Acad. Sci. U. S. A. 106, 2035–2040.

Hong, H., Choi, M.Y., Kim, B.J., 2002. Synchronization on small-world networks. Phys. Rev. E 65, 026139.

Horwitz, B., 2003. The elusive concept of brain connectivity. NeuroImage 19 (2), 466–470.

Hubel, D.H., Wiesel, T.N., 1959. Receptive fields of single neurones in the cat's striate cortex. J. Physiol. 148, 574–591.

Hubert, L., Arabie, P., 1985. Comparing partitions. J. Classif. 2, 193–218.

Huettel, S.A., Song, A.W., McCarthy, G., 2014. Functional Magnetic Resonance Imaging, third ed. Sinauer Associates, Sunderland, MA.

Humphries, M.D., Gurney, K., 2008. Network "small-world-ness": a quantitative method for determining canonical network equivalence. PLoS One 3 (4), e0002051.

Humphries, M.D., Gurney, K., Prescott, T.J., 2006. The brainstem reticular formation is a small-world, not scale-free, network. Proc. Biol. Sci. 273, 503–511.

Hunter, D.R., Handcock, M.S., Butts, C.T., Goodreau, S.M., Morris, M., 2008. ergm: a package to fit, simulate and diagnose exponential-family models for networks. J. Stat. Softw. 24 (3), 1–29. nihpa54860.

Hutchison, R.M., Womelsdorf, T., Allen, E.A., Bandettini, P.A., Calhoun, V.D., Corbetta, M., Della Penna, S., Duyn, J.H., Glover, G.H., Gonzalez-Castillo, J., et al., 2013. Dynamic functional connectivity: promise, issues, and interpretations. NeuroImage 80, 360–378.

Ing, A., Schwarzbauer, C., 2014. Cluster size statistic and cluster mass statistic: two novel methods for identifying changes in functional connectivity between groups or conditions. PLoS One 9 (6), e98697.

Irimia, A., Van Horn, J.D., 2014. Systematic network lesioning reveals the core white matter scaffold of the human brain. Front. Hum. Neurosci. 8, 51.

Irimia, A., Chambers, M.C., Torgerson, C.M., Van Horn, J.D., 2012. Circular representation of human cortical networks for subject and population-level connectomic visualization. NeuroImage 60, 1340–1351.

Isenberg, T., 2015. A survey of illustrative visualization techniques for DTI-based fiber tracking. In: Schulz, T., Hotz, I. (Eds.), Visualization and Processing of Higher Order Descriptors for Multi-Valued Data. Springer, London, pp. 235–256.

Iturria-Medina, Y., Sotero, R.C., Canales-Rodriguez, E.J., Aleman-Gomez, Y., Melie-Garcia, L., 2008. Studying the human brain anatomical network via diffusion-weighted MRI and graph theory. NeuroImage 40, 1064–1076.

Jaccard, P., 1912. The distribution of the flora in the alpine zone. New Phytol. 11, 37–50.

Jain, A.K., Murty, M.N., Flynn, P.J., 1999. Data clustering: a review. ACM Comput. Surv. 31, 264–323.

Jarrell, T.A., Wang, Y., Bloniarz, A.E., Brittin, C.A., Xu, M., Thomson, J.N., Albertson, D.G., Hall, D.H., Emmons, S.W., 2012. The connectome of a decision-making neural network. Science 337, 437–444.

Jennings, J.H., Stuber, G.D., 2014. Tools for resolving functional activity and review connectivity within intact neural circuits. Curr. Biol. 24, R41–R50.

Jiang, Z.-Q., Zhou, W.-X., 2008. Statistical significance of the rich-club phenomenon in complex networks. New J. Phys. 10, 043002–043010.

Johansen-Berg, H., Rushworth, M.F.S., 2009. Using diffusion imaging to study human connectional anatomy. Annu. Rev. Neurosci. 32, 75–94.

Johansen-Berg, H., Rushworth, M.F., Bogdanovic, M.D., Kischka, U., Wimalaratna, S., Matthews, P.M., 2002. The role of ipsilateral premotor cortex in hand movement after stroke. Proc. Natl. Acad. Sci. U. S. A. 99, 14518–14523.

Jones, D.K., Knösche, T.R., Turner, R., 2013. White matter integrity, fiber count, and other fallacies: the do's and dont's of diffusion MRI. NeuroImage 73, 239–254.

Jung, J.C., 2004. In vivo mammalian brain imaging using one- and two-photon fluorescence micro-endoscopy. J. Neurophysiol. 92, 3121–3133.

Juran, J., 2005. Critical Evaluations in Business and Management. Routledge, New York, NY.

Jutla, I., Jeub, L., Mucha, P.J., 2011. A generalized Louvain method for community detection implemented in MATLAB. http://netwiki.amath.unc.edu/GenLouvain.

Kaiser, M., 2008. Mean clustering coefficients: the role of isolated nodes and leafs on clustering measures for small-world networks. New J. Phys. 10, 083042.

Kaiser, M., Hilgetag, C.C., 2004. Modelling the development of cortical systems networks. Neurocomputing 58–60, 297–302.

Kaiser, M., Hilgetag, C.C., 2006. Nonoptimal component placement, but short processing paths, due to long-distance projections in neural systems. PLoS Comput. Biol. 2, e95.

Kaiser, M., Goerner, M., Hilgetag, C.C., 2007a. Criticality of spreading dynamics in hierarchical cluster networks without inhibition. New J. Phys. 9, 1–13.

Kaiser, M., Martin, R., Andras, P., Young, M.P., 2007b. Simulation of robustness against lesions of cortical networks. Eur. J. Neurosci. 25, 3185–3192.

Kamada, T., Kawai, S., 1989. An algorithm for drawing general undirected graphs. Inf. Process. Lett. 31, 7–15.

Kanai, R., Rees, G., 2011. The structural basis of inter-individual differences in human behaviour and cognition. Nat. Rev. Neurosci. 12, 231–242.

Kandel, E.R., Markram, H., Matthews, P.M., Yuste, R., Koch, C., 2013. Neuroscience thinks big (and collaboratively). Nat. Rev. Neurosci. 14, 659–664.

Kaplan, T.D., Forrest, S., 2008. A dual assortative measure of community structure. arXiv 0801.3290v1.

Karrer, B., Levina, E., Newman, M.E.J., 2008. Robustness of community structure in networks. Phys. Rev. E 77, 046119.

Kashtan, N., Alon, U., 2005. Spontaneous evolution of modularity and network motifs. Proc. Natl. Acad. Sci. U. S. A. 102, 13773–13778.

Katz, L., 1953. A new status index derived from sociometric analysis. Psychometrika 18, 39–43.

Kelly, A.M., Uddin, L.Q., Biswal, B.B., Castellanos, F.X., Milham, M.P., 2008. Competition between functional brain networks mediates behavioral variability. NeuroImage 39, 527–537.

Kenet, T., Bibitchkov, D., Tsodyks, M., Grinvald, A., Arieli, A., 2003. Spontaneously emerging cortical representations of visual attributes. Nature 425, 954–956.

Kennedy, H., Knoblauch, K., Toroczkai, Z., 2013. Why data coherence and quality is critical for understanding interareal cortical networks. NeuroImage 80, 37–45.

Kerr, J.N.D., Greenberg, D., Helmchen, F., 2005. Imaging input and output of neocortical networks in vivo. Proc. Natl. Acad. Sci. U. S. A. 102, 14063–14068.

Khundrakpam, B.S., Reid, A., Brauer, J., Carbonell, F., Lewis, J., Ameis, S., Karama, S., Lee, J., Chen, Z., Das, S., et al., 2013. Developmental changes in organization of structural brain networks. Cereb. Cortex 23, 2072–2085.

Killworth, P., Bernard, H., 1978. The reversal small world experiment. Soc. Networks 1, 159–192.

Kim, J., Wozniak, J.R., Mueller, B.A., Shen, X., Pan, W., 2014. Comparison of statistical tests for group differences in brain functional networks. NeuroImage 101, 681–694.

Kim, Y., Son, S.-W., Jeong, H., 2010. Finding communities in directed networks. Phys. Rev. E 81, 016103–016109.

Kitano, H., 2004. Biological robustness. Nat. Rev. Genet. 5, 826–837.

Kitzbichler, M.G., Henson, R.N.A., Smith, M.L., Nathan, P.J., Bullmore, E.T., 2011. Cognitive effort drives workspace configuration of human brain functional networks. J. Neurosci. 31 (22), 8259–8270.

Kitzbichler, M.G., Khan, S., Ganesan, S., Vangel, M.G., Herbert, M.R., Hämäläinen, M.S., Kenet, T., 2015. Altered development and multifaceted band-specific abnormalities of resting state networks in autism. Biol. Psychiatry 77, 794–804.

Kitzbichler, M.G., Smith, M.L., Christensen, S.R., Bullmore, E., 2009. Broadband criticality of human brain network synchronization. PLoS Comput. Biol. 5, e1000314.

Kivelä, M., Arenas, A., Barthelemy, M., Gleeson, J.P., Moreno, Y., Porter, M.A., 2014. Multilayer networks. J. Complex Netw. 2, 203–271.

Kiviniemi, V., Starck, T., Remes, J., Long, X., Nikkinen, J., Haapea, M., Veijola, J., Moilanen, I., Isohanni, M., Zang, Y.F., et al., 2009. Functional segmentation of the brain cortex using high model order group PICA. Hum. Brain Mapp. 30, 3865–3886.

Klar, T.A., Jakobs, S., Dyba, M., Egner, A., Hell, S.W., 2000. Fluorescence microscopy with diffraction resolution barrier broken by stimulated emission. Proc. Natl. Acad. Sci. U. S. A. 97, 8206–8210.

Kleinberg, J.M., 2000. Navigation in a small world. Nature 406, 845.

Kleinfeld, D., Bharioke, A., Blinder, P., Bock, D.D., Briggman, K.L., Chklovskii, D.B., Denk, W., Helmstaedter, M., Kaufhold, J.P., Lee, W.C.A., et al., 2011. Large-scale automated histology in the pursuit of connectomes. J. Neurosci. 31, 16125–16138.

Klemm, K., Eguíluz, V.M., 2002. Highly clustered scale-free networks. Phys. Rev. E 65, 036123.

Klimm, F., Bassett, D.S., Carlson, J.M., Mucha, P.J., 2014. Resolving structural variability in network models and the brain. PLoS Comput. Biol. 10, e1003491. 22.

Klyachko, V.A., Stevens, C.F., 2003. Connectivity optimization and the positioning of cortical areas. Proc. Natl. Acad. Sci. U. S. A. 100, 7937–7941.

Knösche, T.R., Anwander, A., Liptrot, M., Dyrby, T.B., 2015. Validation of tractography: comparison with manganese tracing. Hum. Brain Mapp. 36 (10), 4116–4134.

Kobourov, S.G., 2013. Force-directed drawing algorithms. In: Tamassia, R. (Ed.), Handbook of Graph Drawing and Visualization. CRC Press, Boca Raton, FL, pp. 383–408.

Kopell, N., Ermentrout, G.B., Whittington, M.A., Traub, R.D., 2000. Gamma rhythms and beta rhythms have different synchronization properties. Proc. Natl. Acad. Sci. U. S. A. 97, 1867–1872.

Kraitchik, M., 1942. Mathematical Recreations. W.W. Norton, New York, NY, pp. 209–211, §8.4.1.

Krishnan, A., Williams, L.J., McIntosh, A.R., Abdi, H., 2011. Partial least squares (PLS) methods for neuroimaging: a tutorial and review. NeuroImage 56 (2), 455–475.

Kruskal, J., 1964. Multidimensional scaling by optimizing goodness of fit to nonmetric hypothesis. Psychometrika 29, 1–27.

Krzywinski, M., Birol, I., Jones, S.J., Marra, M.A., 2012. Hive plots—rational approach to visualizing networks. Brief. Bioinform. 13, 627–644.

Kuncheva, L., Hadjitodorov, S.T., 2004. Using diversity in cluster ensembles. IEEE Int. Conf. Syst. Man Cybern. 2, 1214–1219.

Kundu, P., Brenowitz, N.D., Voon, V., Worbe, Y., Vértes, P.E., Inati, S.J., Saad, Z.S., Bandettini, P.A., Bullmore, E.T., 2013. Integrated strategy for improving functional connectivity mapping using multiecho fMRI. Proc. Natl. Acad. Sci. U. S. A. 110, 16187–16192.

Lachaux, J.P., Rodriguez, E., Martinerie, J., Varela, F.J., 1999. Measuring phase synchrony in brain signals. Hum. Brain Mapp. 8, 194–208.

Lago-Fernández, L.F., Huerta, R., Corbacho, F., Sigüenza, J.A., 2000. Fast response and temporal coherent oscillations in small-world networks. Phys. Rev. Lett. 84, 2758–2761.

Lancichinetti, A., Fortunato, S., 2009. Community detection algorithms: a comparative analysis. Phys. Rev. E 80, 056117.

Lancichinetti, A., Fortunato, S., 2011. Limits of modularity maximization in community detection. Phys. Rev. E 84, 066122–066128.

Lancichinetti, A., Fortunato, S., 2012. Consensus clustering in complex networks. Sci. Rep. 2, 336.

Lancichinetti, A., Radicchi, F., Ramasco, J.J., 2010. Statistical significance of communities in networks. Phys. Rev. E 81, 046110–046119.

Lanciego, J.L., Wouterlood, F.G., 2011. A half century of experimental neuroanatomical tracing. J. Chem. Neuroanat. 42, 157–183.

Landman, B.S., Russo, R.L., 1971. On a pin versus block relationship for partitions of logic graphs. IEEE Trans. Comput. C-20, 1469–1479.

Latora, V., Marchiori, M., 2001. Efficient behavior of small-world networks. Phys. Rev. Lett. 87, 198701.

Latora, V., Marchiori, M., 2003. Economic small-world behavior in weighted networks. Eur. Phys. J. B 32, 249–263.

Latora, V., Marchiori, M., 2007. A measure of centrality based on network efficiency. New J. Phys. 9, 188.

Laughlin, S.B., Sejnowski, T.J., 2003. Communication in neuronal networks. Science 301, 1870–1874.

Leek, J.T., Storey, J.D., 2008. A general framework for multiple testing dependence. Proc. Natl. Acad. Sci. U. S. A. 105 (48), 18718–18723.

Lehmann, E.L., Romano, J.P., 2005. Generalizations of the familywise error rate. Ann. Stat. 33 (3), 1138–1154.

Leicht, E.A., Newman, M.E.J., 2008. Community structure in directed networks. Phys. Rev. Lett. 100, 118703–118704.

Lennie, P., 2003. The cost of cortical computation. Curr. Biol. 13, 493–497.

Leonardi, N., Van De Ville, D., 2015. On spurious and real fluctuations of dynamic functional connectivity during rest. NeuroImage 104, 430–436.

Lerch, J.P., Worsley, K., Shaw, W.P., Greenstein, D.K., Lenroot, R.K., Giedd, J., Evans, A.C., 2006. Mapping anatomical correlations across cerebral cortex (MACACC) using cortical thickness from MRI. NeuroImage 31, 993–1003.

Lewis, C.M., Baldassarre, A., Committeri, G., Romani, G.L., Corbetta, M., 2009. Learning sculpts the spontaneous activity of the resting human brain. Proc. Natl. Acad. Sci. U. S. A. 106, 17558–17563.

Leyzorek, M., Gray, R.S., Johnson, A.A., Ladew, W.C., Meaker Jr., S.R., Petry, R.M., Seitz, R.N., 1957. Investigation of Model Techniques. First Annual Report, 6 June 1956–1 July 1957. A Study of Model Techniques for Communication Systems. Case Institute of Technology, Cleveland, OH.

Li, J., Ji, L., 2005. Adjusting multiple testing in multilocus analyses using the eigenvalues of a correlation matrix. Heredity (Edinb.) 95 (3), 221–227.

Li, L., Rilling, J.K., Preuss, T.M., Glasser, M.F., 2012. The effects of connection reconstruction method on the interregional connectivity of brain networks via diffusion tractography. Hum. Brain Mapp. 33, 1894–1913.

Li, S.C., Lindenberger, U., Sikström, S., 2001. Aging cognition: from neuromodulation to representation. Trends Cogn. Sci. 5, 479–486.

Li, Y., Jewells, V., Kim, M., Chen, Y., Moon, A., Armao, D., Troiani, L., Markovic-Plese, S., Lin, W., Shen, D., 2013. Diffusion tensor imaging based network analysis detects alterations of neuroconnectivity in patients with clinically early relapsing-remitting multiple sclerosis. Hum. Brain Mapp. 34 (12), 3376–3391.

Li, Y., Liu, Y., Li, J., Qin, W., Li, K., Yu, C., Jiang, T., 2009. Brain anatomical network and intelligence. PLoS Comput. Biol. 5, e1000395.

Liang, X., Zou, Q., He, Y., Yang, Y., 2013. Coupling of functional connectivity and regional cerebral blood flow reveals a physiological basis for network hubs of the human brain. Proc. Natl. Acad. Sci. U. S. A. 110, 1929–1934.

Lichtheim, L., 1885. On aphasia. Brain 7 (4), 433–484.

Lichtman, J.W., Livet, J., Sanes, J.R., 2008. A technicolour approach to the connectome. Nat. Rev. Neurosci. 9 (6), 417–422.

Lichtman, J.W., Denk, W., 2011. The big and the small: challenges of imaging the brain's circuits. Science 334, 618–623.

Lichtman, J.W., Pfister, H., Shavit, N., 2014. The big data challenges of connectomics. Nat. Neurosci. 17, 1448–1454.

Limpert, E., Stahel, W.A., Abbt, M., 2001. Log-normal distributions across the sciences: keys and clues. Bioscience 51, 341–352.

Lindquist, M.A., Xu, Y., Nebel, M.B., Caffo, B.S., 2014. Evaluating dynamic bivariate correlations in resting-state functional MRI: a comparison study and a new approach. NeuroImage 101, 531–546.

Linkenkaer-Hansen, K., Nikouline, V.V., Palva, J.M., Ilmoniemi, R.J., 2001. Long-range temporal correlations and scaling behavior in human brain oscillations. J. Neurosci. 21, 1370–1377.

Lo, C.-Y.Z., Su, T.-W., Huang, C.-C., Hung, C.-C., Chen, W.-L., Lan, T.-H., Lin, C.-P., Bullmore, E.T., 2015. Randomization and resilience of brain functional networks as systems-level endophenotypes of schizophrenia. Proc. Natl. Acad. Sci. U. S. A. 112, 9123–9128.

Logothetis, N.K., 2002. The neural basis of the blood-oxygen-level-dependent functional magnetic resonance imaging signal. Philos. Trans. R. Soc. Lond. B Biol. Sci. 357, 1003–1037.

Logothetis, N.K., 2008. What we can do and what we cannot do with fMRI. Nature 453, 869–878.

Logothetis, N.K., Pauls, J., Augath, M., Trinath, T., Oeltermann, A., 2001. Neurophysiological investigation of the basis of the fMRI signal. Nature 412, 150–157.

Lohmann, G., Margulies, D.S., Horstmann, A., Pleger, B., Lepsien, J., Goldhahn, D., Schloegl, H., Stumvoll, M., Villringer, A., Turner, R., 2010. Eigenvector centrality mapping for analyzing connectivity patterns in fMRI data of the human brain. PLoS One 5, e10232.

Lohse, C., Bassett, D.S., Lim, K.O., Carlson, J.M., 2014. Resolving anatomical and functional structure in human brain organization: identifying mesoscale organization in weighted network representations. PLoS Comput. Biol. 10 (10), e1003712.

Lopes da Silva, F., 2013. EEG and MEG: relevance to neuroscience. Neuron 80, 1112–1128.

López-Muñoz, F., Boya, J., Alamo, C., 2006. Neuron theory, the cornerstone of neuroscience, on the centenary of the Nobel Prize award to Santiago Ramón y Cajal. Brain Res. Bull. 70, 391–405.

Luce, R.D., 1950. Connectivity and generalized cliques in sociometric group structure. Psychometrika 15, 169–190.

Luce, R.D., Perry, A.D., 1949. A method of matrix analysis of group structure. Psychometrika 14, 95–116.

Lütcke, H., Gerhard, F., Zenke, F., Gerstner, W., 2013. Inference of neuronal network spike dynamics and topology from calcium imaging data. Front. Neural Circuits 7, 201.

Lynall, M.E., Bassett, D.S., Kerwin, R., McKenna, P.J., Kitzbichler, M., Muller, U., Bullmore, E., 2010. Functional connectivity and brain networks in schizophrenia. J. Neurosci. 30, 9477–9487.

MacMahon, M., Garlaschelli, D., 2015. Community detection for correlation matrices. Phys. Rev. X 5, 021006–021034.

MacQueen, J., 1967. Some methods for classification and analysis of multivariate observations. In: Proceedings of the Fifth Berkeley Symposium on Mathematical Statistics and Probability, vol. 1, pp. 281–297.

Mantini, D., Perrucci, M.G., Del Gratta, C., Romani, G.L., Corbetta, M., 2007. Electrophysiological signatures of resting state networks in the human brain. Proc. Natl. Acad. Sci. U. S. A. 104, 13170–13175.

Marder, E., 2015. Understanding brains: details, intuition, and big data. PLoS Biol. 13, e1002147. 6.

Marder, E., Goaillard, J.-M., 2006. Variability, compensation and homeostasis in neuron and network function. Nat. Rev. Neurosci. 7, 563–574.

Margulies, D.S., Böttger, J., Watanabe, A., Gorgolewski, K.J., 2013. Visualizing the human connectome. NeuroImage 80, 445–461.

Markov, N.T., Ercsey-Ravasz, M., Lamy, C., Ribeiro Gomes, A.R., Magrou, L., Misery, P., Giroud, P., Barone, P., Dehay, C., Toroczkai, Z., et al., 2013a. The role of long-range connections on the specificity of the macaque interareal cortical network. Proc. Natl. Acad. Sci. U. S. A. 110, 5187–5192.

Markov, N.T., Ercsey-Ravasz, M., Van Essen, D.C., Knoblauch, K., Toroczkai, Z., Kennedy, H., 2013b. Cortical high-density counterstream architectures. Science 342, 1238406.

Markov, N.T., Ercsey-Ravasz, M.M., Ribeiro Gomes, A.R., Lamy, C., Magrou, L., Vezoli, J., Misery, P., Falchier, A., Quilodran, R., Gariel, M.A., Sallet, J., Gamanut, R., Huissoud, C., Clavagnier, S., Giroud, P., Sappey-Marinier, D., Barone, P., Dehay, C., Toroczkai, Z., Knoblauch, K., Van Essen, D.C., Kennedy, H., 2014. A weighted and directed interareal connectivity matrix for macaque cerebral cortex. Cereb. Cortex 24 (1), 17–36.

Markov, N.T., Vezoli, J., Chameau, P., Falchier, A., Quilodran, R., Huissoud, C., Lamy, C., Misery, P., Giroud, P., Ullman, S., et al., 2013c. Anatomy of hierarchy: feedforward and feedback pathways in macaque visual cortex. J. Comp. Neurol. 522, 225–259.

Marrelec, G., Kim, J., Doyon, J., Horwitz, B., 2009. Large-scale neural model validation of partial correlation analysis for effective connectivity investigation in functional MRI. Hum. Brain Mapp. 30 (3), 941–950.

Marrelec, G., Krainik, A., Duffau, H., Pélégrini-Issac, M., Lehéricy, S., Doyon, J., Benali, H., 2006. Partial correlation for functional brain interactivity investigation in functional MRI. NeuroImage 32, 228–237.

Maslov, S., Sneppen, K., 2002. Specificity and stability in topology of protein networks. Science 296, 910–913.

Mason, M.F., Norton, M.I., Van Horn, J.D., Wegner, D.M., Grafton, S.T., Macrae, C.N., 2007. Wandering minds: the default network and stimulus-independent thought. Science 315, 393–395.

Maxim, V., Sendur, L., Fadili, J., Suckling, J., Gould, R., Howard, R., Bullmore, E., 2005. Fractional Gaussian noise, functional MRI and Alzheimer's disease. NeuroImage 25, 141–158.

McIntosh, A.R., Bookstein, F.L., Haxby, J.V., Grady, C.L., 1996. Spatial pattern analysis of functional brain images using partial least squares. NeuroImage 3 (3 Pt 1), 143–157.

Mechelli, A., Friston, K.J., Frackowiak, R.S., Price, C.J., 2005. Structural covariance in the human cortex. J. Neurosci. 25, 8303–8310.

Meilă, M., 2007. Comparing clusterings—an information based distance. J. Multivar. Anal. 98, 873–895.

Meskaldji, D.-E., Ottet, M.-C., Cammoun, L., Hagmann, P., Meuli, R., Eliez, S., Thiran, J.-P., Morgenthaler, S., 2011. Adaptive strategy for the statistical analysis of connectomes. PLoS One 6 (8), e23009.

Meskaldji, D.-E., Vasung, L., Romascano, D., Thiran, J.P., Hagmann, P., Morgenthaler, S., Van De Ville, D., 2014. Improved statistical evaluation of group differences in connectomes by screening-filtering strategy with application to study maturation of brain connections between childhood and adolescence. NeuroImage 108, 251–264.

Mesulam, M.M., 1990. Large-scale neurocognitive networks and distributed processing for attention, language, and memory. Ann. Neurol. 28, 597–613.

Mesulam, M.M., 1998. From sensation to cognition. Brain 121, 1013–1052.

Meunier, D., Achard, S., Morcom, A., Bullmore, E., 2009a. Age-related changes in modular organization of human brain functional networks. NeuroImage 44, 715–723.

Meunier, D., Lambiotte, R., Bullmore, E.T., 2011. Modular and hierarchically modular organization of brain networks. Front. Neurosci. 4, 200.

Meunier, D., Lambiotte, R., Fornito, A., Ersche, K.D., Bullmore, E.T., 2009b. Hierarchical modularity in human brain functional networks. Front. Neuroinform. 3, 37.

Michel, C.M., Murray, M.M., Lantz, G., Gonzalez, S., 2004. EEG source imaging. Clin. Neurophysiol. 115, 195–222.

Micheloyannis, S., Pachou, E., Stam, C.J., Breakspear, M., Bitsios, P., Vourkas, M., Erimaki, S., Zervakis, M., 2006a. Small-world networks and disturbed functional connectivity in schizophrenia. Schizophr. Res. 87, 60–66.

Micheloyannis, S., Pachou, E., Stam, C.J., Vourkas, M., Erimaki, S., Tsirka, V., 2006b. Using graph theoretical analysis of multi channel EEG to evaluate the neural efficiency hypothesis. Neurosci. Lett. 402, 273–277.

Milgram, S., 1967. The small world problem. Psychol. Today 2, 60–67.

Miller, K.J., Sorensen, L.B., Ojemann, J.G., den Nijs, M., 2009. Power-law scaling in the brain surface electric potential. PLoS Comput. Biol. 5, e1000609.

Milo, R., Kashtan, N., Itzkovitz, S., Newman, M.E.J., Alon, U., 2004. On the uniform generation of random graphs with prescribed degree sequences. arXiv, cond-mat/0312028v2.

Milo, R., Shen-Orr, S., Itzkovitz, S., Kashtan, N., Chklovskii, D., Alon, U., 2002. Network motifs: simple building blocks of complex networks. Science 298, 824–827.

Minzenberg, M.J., Laird, A.R., Thelen, S., Carter, C.S., Glahn, D.C., 2009. Meta-analysis of 41 functional neuroimaging studies of executive function in schizophrenia. Arch. Gen. Psychiatry 66, 811–822.

Mišić, B., Betzel, R.F., Nematzadeh, A., Goñi, J., Griffa, A., Hagmann, P., Flammini, A., Ahn, Y.Y., Sporns, O., 2015. Cooperative and competitive spreading dynamics on the human connectome. Neuron 86 (6), 1518–1529.

Mišić, B., Sporns, O., McIntosh, A.R., 2014. Communication efficiency and congestion of signal traffic in large-scale brain networks. PLoS Comput. Biol. 10 (1), e1003427.

Mitchison, G., 1991. Neuronal branching patterns and the economy of cortical wiring. Proc. Biol. Sci. 245, 151–158.

Mitra, P.P., 2014. The circuit architecture of whole brains at the mesoscopic scale. Neuron 83, 1273–1283.

Mitzenmacher, M., 2004. A brief history of generative models for power law and lognormal distributions. Internet Math. 1, 226–251.

Modha, D.S., Singh, R., 2010. Network architecture of the long-distance pathways in the macaque brain. Proc. Natl. Acad. Sci. U. S. A. 107, 13485–13490.

Mohri, M., Rostamizadeh, A., Talwalkar, A., 2012. Foundations of Machine Learning. MIT Press, Cambridge, MA.

Molloy, M., Reed, B., 1995. A critical point for random graphs with a given degree sequence. Random Struct. Algoritm. 6, 161–179.

Moody, J., 2001. Race, school integration, and friendship segregation in America. Am. J. Sociol. 107, 679–716.

Moreno-Dominguez, D., Anwander, A., Knösche, T.R., 2014. A hierarchical method for whole-brain connectivity-based parcellation. Hum. Brain Mapp. 35, 5000–5025.

Morgan, J.L., Lichtman, J.W., 2013. Why not connectomics? Nat. Rev. Neurosci. 10, 494–500.

Mori, S., Zhang, J., 2006. Principles of diffusion tensor imaging and its applications to basic neuroscience research. Neuron 51, 527–539.

Mossa, S., Barthélemy, M., Eugene Stanley, H., Nunes Amaral, L.A., 2002. Truncation of power law behavior in "scale-free" network models due to information filtering. Phys. Rev. Lett. 88, 138701.

Motter, A.E., de Moura, A.P., Lai, Y.C., Dasgupta, P., 2002. Topology of the conceptual network of language. Phys. Rev. E 65 (6), 065102.

Mucha, P.J., Richardson, T., Macon, K., Porter, M.A., Onnela, J.-P., 2010. Community structure in time-dependent, multiscale, and multiplex networks. Science 328, 876–878.

Muff, S., Rao, F., Caflisch, A., 2005. Local modularity measure for network clusterizations. Phys. Rev. E 72, 056107. 4.

Nagasaka, Y., Shimoda, K., Fujii, N., 2011. Multidimensional recording (MDR) and data sharing: an ecological open research and educational platform for neuroscience. PLoS One 6, e22561–e22567.

Nelson, S.M., Cohen, A.L., Power, J.D., Wig, G.S., Miezin, F.M., Wheeler, M.E., Velanova, K., Donaldson, D.I., Phillips, J.S., Schlaggar, B.L., et al., 2010. A parcellation scheme for human left lateral parietal cortex. Neuron 67, 156–170.

Nepusz, T., Négyessy, L., Bazsó, F., 2008. Fuzzy communities and the concept of bridgeness in complex networks. Phys. Rev. E 77, 016107.

Newman, M.E.J., 2002. Assortative mixing in networks. Phys. Rev. Lett. 89, 208701–208704.

Newman, M.E.J., 2003a. The structure and function of complex networks. SIAM Rev. 45 (2), 167–256.

Newman, M.E.J., 2003b. Mixing patterns in networks. Phys. Rev. E 67, 026126. 13.

Newman, M.E.J., 2004a. Analysis of weighted networks. Phys. Rev. E 70, 056131.

Newman, M.E.J., 2004b. Coauthorship networks and patterns of scientific collaboration. Proc. Natl. Acad. Sci. U. S. A. 101 (Suppl. 1), 5200–5205.

Newman, M.E.J., 2004c. Fast algorithm for detecting community structure in networks. Phys. Rev. E 69, 066133.

Newman, M.E.J., 2005a. Power laws, Pareto distributions and Zipf's law. Contemp. Phys. 46, 323–351.

Newman, M.E.J., 2005b. A measure of betweenness centrality based on random walks. Soc. Networks 27, 39–54.

Newman, M.E.J., 2006. Modularity and community structure in networks. Proc. Natl. Acad. Sci. U. S. A. 103, 8577–8582.

Newman, M.E.J., 2009. Random graphs with clustering. Phys. Rev. Lett. 103, 05870.

Newman, M.E.J., 2010. Networks. An Introduction. Oxford University Press, Oxford.

Newman, M.E.J., 2013. Spectral methods for community detection and graph partitioning. Phys. Rev. E 88, 042822.

Newman, M.E.J., Girvan, M., 2004. Finding and evaluating community structure in networks. Phys. Rev. E 69, 026113.

Newman, M.E.J., Strogatz, S.H., Watts, D.J., 2001. Random graphs with arbitrary degree distributions and their applications. Phys. Rev. E 64, 026118.

Nichols, T.E., Holmes, A.P., 2001. Nonparametric permutation tests for functional neuroimaging: a primer with examples. Hum. Brain Mapp. 15, 1–25.

Nichols, T., Hayasaka, S., 2003. Controlling the familywise error rate in functional neuroimaging: a comparative review. Stat. Methods Med. Res. 12 (5), 419–446.

Nicosia, V., Mangioni, G., Carchiolo, V., 2009. Extending the definition of modularity to directed graphs with overlapping communities. J. Stat. Mech. Theory Exp. P03024.

Nicosia, V., Vértes, P.E., Schafer, W.R., Latora, V., Bullmore, E.T., 2013. Phase transition in the economically modeled growth of a cellular nervous system. Proc. Natl. Acad. Sci. 110 (19), 7880–7885.

Nieuwenhuys, R., 2013. The myeloarchitectonic studies on the human cerebral cortex of the Vogt-Vogt School, and their significance for the interpretation of functional neuroimaging data. Brain Struct. Funct. 218, 303–352.

Niven, J.E., Laughlin, S.B., 2008. Energy limitation as a selective pressure on the evolution of sensory systems. J. Exp. Biol. 211 (11), 1792–1804.

Noppeney, U., Friston, K.J., Price, C.J., 2004. Degenerate neuronal systems sustaining cognitive functions. J. Anat. 205, 433–442.

Norman, K.A., Polyn, S.M., Detre, G.J., Haxby, J.V., 2006. Beyond mind-reading: multi-voxel pattern analysis of fMRI data. Trends Cogn. Sci. 10 (9), 424–430.

Nunez, P.L., Srinivasan, R., 2006. Electrical Fields of the Brain. Oxford University Press, New York, NY.

O'Shea, J., Johansen-Berg, H., Trief, D., Gobel, S., Rushworth, M.F., 2007. Functionally specific reorganization in human premotor cortex. Neuron 54, 479–490.

Oh, S.W., Harris, J.A., Ng, L., Winslow, B., Cain, N., Mihalas, S., Wang, Q., Lau, C., Kuan, L., Henry, A.M., Mortrud, M.T., Ouellette, B., Nguyen, T.N., Sorensen, S.A., Slaughterbeck, C.R., Wakeman, W., Li, Y., Feng, D., Ho, A., Nicholas, E., Hirokawa, K.E., Bohn, P., Joines, K.M., Peng, H., Hawrylycz, M.J., Phillips, J.W., Hohmann, J.G., Wohnoutka, P., Gerfen, C.R., Koch, C., Bernard, A., Dang, C., Jones, A.R., Zeng, H., 2014. A mesoscale connectome of the mouse brain. Nature 508 (7495), 207–214.

Oláh, S., Füle, M., Komlósi, G., Varga, C., Báldi, R., Barzó, P., Tamás, G., 2009. Regulation of cortical microcircuits by unitary GABA-mediated volume transmission. Nature 461, 1278–1281.

Oldham, M.C., Konopka, G., Iwamoto, K., Langfelder, P., Kato, T., Horvath, S., Geschwind, D.H., 2008. Functional organization of the transcriptome in human brain. Nat. Neurosci. 11 (11), 1271–1282.

Olshausen, B.A., Field, D.J., 2004. Sparse coding of sensory inputs. Curr. Opin. Neurobiol. 14 (4), 481–487.

Onnela, J.-P., Saramäki, J., Kertész, J., Kaski, K., 2005. Intensity and coherence of motifs in weighted complex networks. Phys. Rev. E 71, 065103.

Opsahl, T., Agneessens, F., Skvoretz, J., 2010. Node centrality in weighted networks: generalizing degree and shortest paths. Soc. Networks 32, 245–251.

Opsahl, T., Colizza, V., Panzarasa, P., Ramasco, J., 2008. Prominence and control: the weighted rich-club effect. Phys. Rev. Lett. 101, 168702.

Orden, A., 1956. The transshipment problem. Manag. Sci. 2 (3), 276–285.

Osten, P., Margrie, T.W., 2013. Mapping brain circuitry with a light microscope. Nat. Methods 10, 515–523.

Ostojic, S., Brunel, N., Hakim, V., 2009. How connectivity, background activity, and synaptic properties shape the cross-correlation between spike trains. J. Neurosci. 29, 10234–10253.

Ozaktas, H.M., 1992. Paradigms of connectivity for computer circuits and networks. Opt. Eng. 31, 1563–1567.

O'Reilly, R.C., 2006. Biologically based computational models of high-level cognition. Science 314, 91–94.

Palla, G., Barabási, A.-L., Vicsek, T., 2007. Quantifying social group evolution. Nature 446, 664–667.

Palla, G., Derenyi, I., Farkas, I., Vicsek, T., 2005. Uncovering the overlapping community structure of complex networks in nature and society. Nature 435, 814–818.

Palva, J.M., Monto, S., Kulashekhar, S., Palva, S., 2010. Neuronal synchrony reveals working memory networks and predicts individual memory capacity. Proc. Natl. Acad. Sci. 107 (16), 7580–7585.

Pannese, E., 1999. The Golgi stain: invention, diffusion and impact on neurosciences. J. Hist. Neurosci. 8, 132–140.

Passingham, R.E., Stephan, K.E., Kötter, R., 2002. The anatomical basis of functional localization in the cortex. Nat. Rev. Neurosci. 3, 606–616.

Patel, A.X., Bullmore, E.T., 2016. A wavelet-based estimator of the degrees of freedom in denoised fMRI time series for probabilistic testing of functional connectivity and brain graphs. NeuroImage. http://dx.doi.org/10.1016/j.neuroimage.2015.04.052.

Patel, A.X., Kundu, P., Rubinov, M., Jones, P.S., Vértes, P.E., Ersche, K.D., Suckling, J., Bullmore, E.T., 2014. A wavelet method for modeling and despiking motion artifacts from resting-state fMRI time series. NeuroImage 95, 287–304.

Paus, T., Keshavan, M., Giedd, J.N., 2008. Why do many psychiatric disorders emerge during adolescence? Nat. Rev. Neurosci. 9, 947–957.

Pavlovic, D.M., Vértes, P.E., Bullmore, E.T., Schafer, W.R., Nichols, T.E., 2014. Stochastic block-modeling of the modules and core of the *Caenorhabditis elegans* connectome. PLoS One 9, e97584. 16.

Peasarin, F., 2002. Multivariate Permutation Tests: With Applications in Biostatistics. Wiley, New York, NY.

Peixoto, T.P., Bornholdt, S., 2012. Evolution of robust network topologies: emergence of central backbones. Phys. Rev. Lett. 109, 118703.

Penfield, W., Jasper, H., 1954. Epilepsy and the Functional Anatomy of the Human Brain. Little, Brown and Company, Boston, MA.

Petermann, T., Thiagarajan, T.C., Lebedev, M.A., Nicolelis, M.A.L., Chialvo, D.R., Plenz, D., 2009. Spontaneous cortical activity in awake monkeys composed of neuronal avalanches. Proc. Natl. Acad. Sci. U. S. A. 106, 15921–15926.

Pitman, E.J.G., 1937. Significance tests which may be applied to samples from any population. Suppl. J. R. Stat. Soc. 4, 119–130. 225–232 (parts I & II).

Plaza, S.M., Scheffer, L.K., Chklovskii, D.B., 2014. Toward large-scale connectome reconstructions. Curr. Opin. Neurobiol. 25, 201–210.

Plis, S.M., Hjelm, D.R., Salakhutdinov, R., Allen, E.A., Bockholt, H.J., Long, J.D., Johnson, H.J., Paulsen, J.S., Turner, J.A., Calhoun, V.D., 2014. Deep learning for neuroimaging: a validation study. Front. Neurosci. 8, 229.

Power, J.D., Barnes, K.A., Snyder, A.Z., Schlaggar, B.L., Petersen, S.E., 2012. Spurious but systematic correlations in functional connectivity MRI networks arise from subject motion. NeuroImage 59, 2142–2154.

Power, J.D., Cohen, A.L., Nelson, S.M., Wig, G.S., Barnes, K.A., Church, J.A., Vogel, A.C., Laumann, T.O., Miezin, F.M., Schlaggar, B.L., et al., 2011. Functional network organization of the human brain. Neuron 72, 665–678.

Power, J.D., Schlaggar, B.L., Lessov-Schlaggar, C.N., Petersen, S.E., 2013. Evidence for hubs in human functional brain networks. Neuron 79, 798–813.

Prichard, D., Theiler, J., 1994. Generating surrogate data for time series with several simultaneously measured variables. Phys. Rev. Lett. 73, 951–954.

Priester, C., Schmitt, S., Peixoto, T.P., 2014. Limits and trade-offs of topological network robustness. PLoS One 9, e108215–e108219.

Rademacher, J., Bürgel, U., Zilles, K., 2002. Stereotaxic localization, intersubject variability, and interhemispheric differences of the human auditory thalamocortical system. NeuroImage 17, 142–160.

Rademacher, J., Caviness Jr., V.S., Steinmetz, H., Galaburda, A.M., 1993. Topographical variation of the human primary cortices: implications for neuroimaging, brain mapping, and neurobiology. Cereb. Cortex 3, 313–329.

Raichle, M.E., MacLeod, A.M., Snyder, A.Z., Powers, W.J., Gusnard, D.A., Shulman, G.L., 2001. A default mode of brain function. Proc. Natl. Acad. Sci. U. S. A. 98, 676–682.

Raj, A., Chen, Y.H., 2011. The wiring economy principle: connectivity determines anatomy in the human brain. PLoS One 6, e14832.

Raj, A., Kuceyeski, A., Weiner, M., 2012. A network diffusion model of disease progression in dementia. Neuron 73, 1204–1215.

Rajah, M.N., 2005. Region-specific changes in prefrontal function with age: a review of PET and fMRI studies on working and episodic memory. Brain 128, 1964–1983.

Ramón y Cajal, S., 1995. Histology of the Nervous System (N. Swanson, L.W. Swanson, Trans.). Oxford University Press, New York, NY

Rand, W.M., 1971. Objective criteria for the evaluation of clustering methods. J. Am. Stat. Assoc. 66, 846–850.

Ravasz, E., Barabási, A.L., 2003. Hierarchical organization in complex networks. Phys. Rev. E 67, 026112.

Rehme, A.K., Eickhoff, S.B., Rottschy, C., Fink, G.R., Grefkes, C., 2012. Activation likelihood estimation meta-analysis of motor-related neural activity after stroke. NeuroImage 59, 2771–2782.

Reichardt, J., Bornholdt, S., 2006. Statistical mechanics of community detection. Phys. Rev. E 74, 016110.

Reveley, C., Seth, A.K., Pierpaoli, C., Silva, A.C., Yu, D., Saunders, R.C., Leopold, D.A., Ye, F.Q., 2015. Superficial white matter fiber systems impede detection of long-range cortical connections in diffusion MR tractography. Proc. Natl. Acad. Sci. U. S. A. 112, E2820–E2828.

Ringo, J.L., 1991. Neuronal interconnection as a function of brain size. Brain Behav. Evol. 38, 1–6.

Riva-Posse, P., Choi, K.S., Holtzheimer, P.E., McIntyre, C.C., Gross, R.E., Chaturvedi, A., Crowell, A.L., Garlow, S.J., Rajendra, J.K., Mayberg, H.S., 2014. Defining critical white matter pathways mediating successful subcallosal cingulate deep brain stimulation for treatment-resistant depression. Biol. Psychiatry 76, 963–969.

Rivera-Alba, M., Vitaladevuni, S.N., Mishchenko, Y., Lu, Z., Takemura, S.-Y., Scheffer, L., Meinertzhagen, I.A., Chklovskii, D.B., de Polavieja, G.G., 2011. Wiring economy and volume exclusion determine neuronal placement in the *Drosophila* brain. Curr. Biol. 21, 2000–2005.

Roberts, J.A., Perry, A., Lord, A.R., Roberts, G., Mitchell, P.B., Smith, R.E., Calamante, F., Breakspear, M., 2016. The contribution of geometry to the human connectome. NeuroImage 124, 379–393.

Robins, G., Pattison, P., Kalish, Y., Lusher, D., 2007. An introduction to exponential random graph (p^*) models for social networks. Soc. Networks 29 (2), 173–191.

Robinson, P., Henderson, J., Matar, E., Riley, P., Gray, R., 2009. Dynamical reconnection and stability constraints on cortical network architecture. Phys. Rev. Lett. 103, 108104.

Rocca, J., 1997. Galen and the ventricular system. J. Hist. Neurosci. 6, 227–239.

Rochat, Y., 2009. Closeness centrality extended to unconnected graphs: the harmonic centrality index. In: Applications of Social Network Analysis.

Roelfsema, P.R., Engel, A.K., König, P., Singer, W., 1997. Visuomotor integration is associated with zero time-lag synchronization among cortical areas. Nature 385, 157–161.

Rombach, M.P., Porter, M.A., Fowler, J.H., Mucha, P.J., 2014. Core-periphery structure in networks. SIAM J. Appl. Math. 74, 167–190.

Rose, F.C., 2009. Cerebral localization in antiquity. J. Hist. Neurosci. 18, 239–247.

Ross, D.T., Ebner, F.F., 1990. Thalamic retrograde degeneration following cortical injury: an excitotoxic process? Neuroscience 35, 525–550.

Rossa, F.D., Dercole, F., Piccardi, C., 2013. Profiling core-periphery network structure by random walkers. Sci. Rep. 3, 1–8.

Rosvall, M., Bergstrom, C.T., 2008. Maps of random walks on complex networks reveal community structure. Proc. Natl. Acad. Sci. U. S. A. 105, 1118–1123.

Rubinov, M., Sporns, O., 2011. Weight-conserving characterization of complex functional brain networks. NeuroImage 56, 2068–2079.

Rubinov, M., Sporns, O., 2010. Complex network measures of brain connectivity: uses and interpretations. NeuroImage 80, 426–444.

Rubinov, M., Knock, S.A., Stam, C.J., Micheloyannis, S., Harris, A.W., Williams, L.M., Breakspear, M., 2009. Small-world properties of nonlinear brain activity in schizophrenia. Hum. Brain Mapp. 30 (2), 403–416.

Rubinov, M., Sporns, O., Thivierge, J.P., Breakspear, M., 2011. Neurobiologically realistic determinants of self-organized criticality in networks of spiking neurons. PLoS Comput. Biol. 7, e1002038.

Rubinov, M., Ypma, R.J.F., Watson, C., Bullmore, E.T., 2015. Wiring cost and topological participation of the mouse brain connectome. Proc. Natl. Acad. Sci. U. S. A. 112, 10032–10037.

Rumelhart, D.E., McLelland, J.L., The PDP Research Group, 1986. Parallel Distributed Processing: Explorations in the Microstructure of Cognition. Foundations, vol. 1. MIT Press, Cambridge, MA.

Sadovsky, A.J., MacLean, J.N., 2013. Scaling of topologically similar functional modules defines mouse primary auditory and somatosensory microcircuitry. J. Neurosci. 33, 14048–14060.

Sadovsky, A.J., MacLean, J.N., 2014. Mouse visual neocortex supports multiple stereotyped patterns of microcircuit activity. J. Neurosci. 34, 7769–7777.

Sales-Pardo, M., Guimerà, R., Moreira, A.A., Amaral, L.A.N., 2007. Extracting the hierarchical organization of complex systems. Proc. Natl. Acad. Sci. U. S. A. 104, 15224–15229.

Salton, G., 1989. Automatic Text Processing: The Transformation, Analysis and Retrieval of Information by Computer. Addison-Wesley, Reading, MA.

Salvador, R., Martinez, A., Pomarol-Clotet, E., Sarro, S., Suckling, J., Bullmore, E., 2007. Frequency based mutual information measures between clusters of brain regions in functional magnetic resonance imaging. NeuroImage 35, 83–88.

Salvador, R., Suckling, J., Coleman, M.R., Pickard, J.D., Menon, D., Bullmore, E., 2005. Neurophysiological architecture of functional magnetic resonance images of human brain. Cereb. Cortex 15, 1332–1342.

Samu, D., Seth, A.K., Nowotny, T., 2014. Influence of wiring cost on the large-scale architecture of human cortical connectivity. PLoS Comput. Biol. 10, e1003557. 24.

Saramäki, J., Kivelä, M., Onnela, J.-P., Kaski, K., Kertész, J., 2007. Generalizations of the clustering coefficient to weighted complex networks. Phys. Rev. E 75, 027105. 4.

Satterthwaite, T.D., Elliott, M.A., Gerraty, R.T., Ruparel, K., Loughead, J., Calkins, M.E., Eickhoff, S.B., Hakonarson, H., Gur, R.C., Gur, R.E., et al., 2013. An improved framework for confound regression and filtering for control of motion artifact in the preprocessing of resting-state functional connectivity data. NeuroImage 64, 240–256.

Scannell, J.W., 1997. Determining cortical landscapes. Nature 386, 452.

Schaeffer, S.E., 2007. Graph clustering. Comput. Sci. Rev. 1, 27–64.

Schleicher, A., Amunts, K., Geyer, S., Morosan, P., Zilles, K., 1999. Observer-independent method for microstructural parcellation of cerebral cortex: a quantitative approach to cytoarchitectonics. NeuroImage 9, 165–177.

Scholkmann, F., Kleiser, S., Metz, A.J., Zimmermann, R., Mata Pavia, J., Wolf, U., Wolf, M., 2014. A review on continuous wave functional near-infrared spectroscopy and imaging instrumentation and methodology. NeuroImage 85 (Pt 1), 6–27.

Scholtens, L.H., Schmidt, R., de Reus, M.A., van den Heuvel, M.P., 2014. Linking macroscale graph analytical organization to microscale neuroarchitectonics in the macaque connectome. J. Neurosci. 34 (36), 12192–12205.

Scholvinck, M.L., Friston, K.J., Rees, G., 2011. The influence of spontaneous activity on stimulus processing in primary visual cortex. NeuroImage 80, 297–306.

Schomer, D.L., da Silva, F.H., 2005. Niedermeyer's Electroencephalography: Basic Principles, Clinical Applications, and Related Fields. Lippincott Williams & Wilkins, Philadelphia, PA.

Schreiber, T., Schmitz, A., 2000. Surrogate time series. Physica D 142, 346–382.

Schrijver, A., 2012. On the history of the shortest path problem. Doc. Math. Optimization Stories 155–167.

Schrödel, T., Prevedel, R., Aumayr, K., Zimmer, M., Vaziri, A., 2013. Brain-wide 3D imaging of neuronal activity in Caenorhabditis elegans with sculpted light. Nat. Methods 10, 1013–1020.

Schroeter, M.S., Charlesworth, P., Kitzbichler, M.G., Paulsen, O., Bullmore, E.T., 2015. Emergence of rich-club topology and coordinated dynamics in development of hippocampal functional networks in vitro. J. Neurosci. 35, 5459–5470.

Scoville, W.B., Milner, B., 1957. Loss of recent memory after bilateral hippocampal lesions. J. Neurol. Neurosurg. Psychiatry 20, 11–21.

Seehaus, A.K., Roebroeck, A., Chiry, O., Kim, D.S., Ronen, I., Bratzke, H., Goebel, R., Galuske, R.A.W., 2013. Histological validation of DW-MRI tractography in human postmortem tissue. Cereb. Cortex 23, 442–450.

Seeley, W.W., Crawford, R.K., Zhou, J., Miller, B.L., Greicius, M.D., 2009. Neurodegenerative diseases target large-scale human brain networks. Neuron 62, 42–52.

Seghier, M.L., Friston, K.J., 2013. Network discovery with large DCMs. NeuroImage 68, 181–191.

Seghier, M.L., Lee, H.L., Schofield, T., Ellis, C.L., Price, C.J., 2008. Inter-subject variability in the use of two different neuronal networks for reading aloud familiar words. NeuroImage 42, 1226–1236.

Seidman, S.B., 1983. Network structure and minimum degree. Soc. Networks 5, 269–287.

Sejnowski, T.J., Churchland, P.S., Movshon, J.A., 2014. Putting big data to good use in neuroscience. Nat. Neurosci. 17, 1440–1441.

Serrano, M.A., Boguñá, M., Vespignani, A., 2009. Extracting the multiscale backbone of complex weighted networks. Proc. Natl. Acad. Sci. U. S. A. 106 (16), 6483–6488.

Serrano, M.A., Boguñá, M., Pastor-Satorras, R., 2006. Correlations in weighted networks. Phys. Rev. E 74, 055101.

Shallice, T., 1988. From Neuropsychology to Mental Structure. Cambridge University Press, New York, NY.

Shanahan, M., 2010. Metastable chimera states in community-structured oscillator networks. Chaos 20, 013108.

Shanahan, M., Wildie, M., 2012. Knotty-centrality: finding the connective core of a complex network. PLoS One 7, e36579.

Shanahan, M., Bingman, V.P., Shimizu, T., Wild, M., Guntukun, O., 2013. Large-scale network organization in the avian forebrain: a connectivity matrix and theoretical analysis. Front. Comput. Neurosci. 7, 1–17.

Sharir, M., 1981. A strong connectivity algorithm and its applications to data flow analysis. Comp. Math. Appl. 7, 67–72.

Shehzad, Z., Kelly, C., Reiss, P.T., Craddock, C.R., Emerson, J.W., McMahon, K., Copland, D.A., Castellanos, F.X., Milham, M.P., 2014. A multivariate distance-based analytic framework for connectome-wide association studies. NeuroImage 93 (Pt 1), 74–94.

Shen-Orr, S.S., Milo, R., Mangan, S., Alon, U., 2002. Network motifs in the transcriptional regulation network of *Escherichia coli*. Nat. Genet. 31, 64–68.

Shen, H.W., Cheng, X.Q., Guo, J.F., 2009. Quantifying and identifying the overlapping community structure in networks. J. Stat. Mech. Theory Exp. 2009, P07042.

Sherbondy, A.J., Rowe, M., Alexander, D., 2010. MicroTrack: an algorithm for concurrent projectome and microstructure estimation. MICCAI 13, 183–190.

Shi, J., Malik, J., 2000. Normalized cuts and image segmentation. IEEE Trans. Pattern Anal. Mach. Intell. 22 (8), 888–905.

Shih, C.-T., Sporns, O., Yuan, S.-L., Su, T.-S., Lin, Y.-J., Chuang, C.-C., Wang, T.-Y., Lo, C.-C., Greenspan, R.J., Chiang, A.-S., 2015. Connectomics-based analysis of information flow in the *Drosophila* brain. Curr. Biol. 25, 1249–1258.

Shimbel, A., 1953. Structural parameters of communication networks. Bull. Math. Biophys. 15, 501–507.

Shoval, O., Alon, U., 2010. SnapShot: network motifs. Cell 143, 326–326. e1.

Shulman, G.L., Fiez, J.A., Corbetta, M., Buckner, R.L., Miezin, F.M., Raichle, M.E., Petersen, S.E., 1997. Common blood flow changes across visual tasks: II. Decreases in cerebral cortex. J. Cogn. Neurosci. 9, 648–663.

Sidak, Z.K., 1967. Rectangular confidence regions for the means of multivariate normal distributions. J. Am. Stat. Assoc. 62 (318), 626–633.

Siegel, M., Donner, T.H., Engel, A.K., 2012. Spectral fingerprints of large-scale neuronal interactions. Nat. Rev. Neurosci. 13, 121–134.

Silver, R.A., 2010. Neuronal arithmetic. Nat. Rev. Neurosci. 11, 474–489.

Simes, R.J., 1986. An improved Bonferroni procedure for multiple tests of significance. Biometrika 73 (3), 751–754.

Simon, H.A., 1955. On a class of skew distribution functions. Biometrika 42, 425.

Simon, H.A., 1962. The architecture of complexity. Proc. Am. Philos. Soc. 106, 467–482.

Simons, J.S., Henson, R.N., Gilbert, S.J., Fletcher, P.C., 2008. Separable forms of reality monitoring supported by anterior prefrontal cortex. J. Cogn. Neurosci. 20, 447–457.

Simpson, S.L., Bowman, F.D., Laurienti, P.J., 2013. Analyzing complex functional brain networks: fusing statistics and network science to understand the brain. Stat. Surv. 7, 1–36.

Simpson, S.L., Moussa, M.N., Laurienti, P.J., 2012. An exponential random graph modeling approach to creating group-based representative whole-brain connectivity networks. Neuro-Image 60 (2), 1117–1126.

Singer, W., 1999. Neuronal synchrony: a versatile code for the definition of relations? Neuron 24, 49–65. 111–125.

Singer, W., 2013. Cortical dynamics revisited. Trends Cogn. Sci. 17, 616–626.

Singer, W., Gray, C.M., 1995. Visual feature integration and the temporal correlation hypothesis. Annu. Rev. Neurosci. 18, 555–586.

Skudlarski, P., Jagannathan, K., Anderson, K., Stevens, M.C., Calhoun, V.D., Skudlarska, B.A., Pearlson, G., 2010. Brain connectivity is not only lower but different in schizophrenia: a combined anatomical and functional approach. Biol. Psychiatry 68, 61–69.

Skudlarski, P., Jagannathan, K., Calhoun, V.D., Hampson, M., Skudlarska, B.A., Pearlson, G., 2008. Measuring brain connectivity: diffusion tensor imaging validates resting state temporal correlations. NeuroImage 43, 554–561.

Smilkov, D., Kocarev, L., 2010. Rich-club and page-club coefficients for directed graphs. Physica A 389, 2290–2299.

Smit, D.J., Stam, C.J., Posthuma, D., Boomsma, D.I., de Geus, E.J., 2008. Heritability of "small-world" networks in the brain: a graph theoretical analysis of resting-state EEG functional connectivity. Hum. Brain Mapp. 29, 1368–1378.

Smith, S.M., Fox, P.T., Miller, K.L., Glahn, D.C., Fox, P.M., Mackay, C.E., Filippini, N., Watkins, K.E., Toro, R., Laird, A.R., et al., 2009. Correspondence of the brain's functional architecture during activation and rest. Proc. Natl. Acad. Sci. U. S. A. 106, 13040–13045.

Smith, R.E., Tournier, J.-D., Calamante, F., Connelly, A., 2013. SIFT: spherical-deconvolution informed filtering of tractograms. NeuroImage 67, 298–312.

Smith, S.M., Miller, K.L., Moeller, S., Xu, J., Auerbach, E.J., Woolrich, M.W., Beckmann, C.F., Jenkinson, M., Andersson, J., et al., 2012. Temporally-independent functional modes of spontaneous brain activity. Proc. Natl. Acad. Sci. U. S. A. 109, 3131–3136.

Smith, S.M., Miller, K.L., Salimi-Khorshidi, G., Webster, M., Beckmann, C.F., Nichols, T.E., Ramsey, J.D., Woolrich, M.W., 2011. Network modelling methods for FMRI. NeuroImage 54, 875–891.

Smith, R.E., Tournier, J.D., Calamante, F., Connelly, A., 2015. Enabling dense quantitative assessment of brain white matter connectivity using streamlines tractography. NeuroImage 119, 338–351.

Smith, S.M., Nichols, T.E., Vidaurre, D., Winkler, A.M., Behrens, T.E., Glasser, M.F., Ugurbil, K., Barch, D.M., Van Essen, D.C., Miller, K.L., 2015. A positive-negative mode of population covariation links brain connectivity, demographics and behavior. Nat Neurosci. 18 (11), 1565–1567.

Sneppen, K., Trusina, A., Rosvall, M., 2005. Hide-and-seek on complex networks. Europhys. Lett. 69 (5), 853–859.

Snijders, T.A.B., Nowicki, K., 1997. Estimation and prediction for stochastic blockmodels for graphs with latent block structure. J. Classif. 14, 75–100.

Song, H.F., Kennedy, H., Wang, X.-J., 2014. Spatial embedding of structural similarity in the cerebral cortex. Proc. Natl. Acad. Sci. U. S. A. 111, 16580–16585.

Song, S., Abbott, L.F., 2001. Cortical development and remapping through spike timing-dependent plasticity. Neuron 32, 339–350.

Sonuga-Barke, E.J.S., Castellanos, F.X., 2007. Spontaneous attentional fluctuations in impaired states and pathological conditions: a neurobiological hypothesis. Neurosci. Biobehav. Rev. 31, 977–986.

Sporns, O., 2006. Small-world connectivity, motif composition, and complexity of fractal neuronal connections. BioSystems 85, 55–64.

Sporns, O., 2011a. Networks of the Brain. MIT Press, Cambridge, MA.

Sporns, O., 2011b. The non-random brain: efficiency, economy, and complex dynamics. Front. Comput. Neurosci. 5, 5.

Sporns, O., 2012. Discovering the Human Connectome. MIT Press, Cambridge, MA.

Sporns, O., Kötter, R., 2004. Motifs in brain networks. PLoS Biol. 2, e369.

Sporns, O., Zwi, J.D., 2004. The small world of the cerebral cortex. Neuroinformatics 2, 145–162.

Sporns, O., Chialvo, D.R., Kaiser, M., Hilgetag, C.C., 2004. Organization, development and function of complex brain networks. Trends Cogn. Sci. 8 (9), 418–425.

Sporns, O., Honey, C.J., Kötter, R., 2007. Identification and classification of hubs in brain networks. PLoS One 2, e1049.

Sporns, O., Tononi, G., Edelman, G.M., 2000. Theoretical neuroanatomy: relating anatomical and functional connectivity in graphs and cortical connection matrices. Cereb. Cortex 10, 127–141.

Sporns, O., Tononi, G., Kötter, R., 2005. The human connectome: a structural description of the human brain. PLoS Comput. Biol. 1, e42.

Stam, C.J., 2014. Modern network science of neurological disorders. Nat. Rev. Neurosci. 15 (10), 683–695.

Stam, C.J., 2004. Functional connectivity patterns of human magnetoencephalographic recordings: a "small-world" network? Neurosci. Lett. 355, 25–28.

Stam, C.J., van Dijk, B.W., 2002. Synchronization likelihood: an unbiased measure of generalized synchronization in multivariate data sets. Physica D 163, 236–251.

Stam, C.J., Jones, B.F., Nolte, G., Breakspear, M., Scheltens, P., 2007. Small-world networks and functional connectivity in Alzheimer's disease. Cereb. Cortex 17, 92–99.

Stam, C.J., Tewarie, P., van Dellen, E., van Straaten, E.C.W., Hillebrand, A., van Mieghem, P., 2014. The trees and the forest: characterization of complex brain networks with minimum spanning trees. Int. J. Psychophysiol. 92, 129–138.

Stephan, K.E., Kamper, L., Bozkurt, A., Burns, G.A., Young, M.P., Kötter, R., 2001. Advanced database methodology for the collation of connectivity data on the macaque brain (CoCoMac). Philos. Trans. R. Soc. Lond. B 356 (1412), 1159–1186.

Stephan, K.E., 2013. The history of CoCoMac. NeuroImage 80, 46–52.

Stephan, K.E., Kötter, R., 1999. One cortex—many maps: an introduction to coordinate-independent mapping by objective relational transformation (ORT). Neurocomputing 26–27, 1049–1054.

Stephan, K.E., Kasper, L., Harrison, L.M., Daunizeau, J., den Ouden, H.E., Breakspear, M., Friston, K.J., 2008. Nonlinear dynamic causal models for fMRI. NeuroImage 42, 649–662.

Stephan, K.E., Zilles, K., Kötter, R., 2000. Coordinate-independent mapping of structural and functional data by objective relational transformation (ORT). Philos. Trans. R. Soc. Lond. B Biol. Sci. 355, 37–54.

Stettler, D.D., Yamahachi, H., Li, W., Denk, W., Gilbert, C.D., 2006. Axons and synaptic boutons are highly dynamic in adult visual cortex. Neuron 49, 877–887.

Stojmenovic, I., 2002. Position based routing in ad hoc networks. IEEE Commun. Mag. 40 (7), 128–134.

Storey, J.D., 2002. A direct approach to false discovery rates. J. R. Stat. Soc. Ser. B 64 (3), 479–498.

Strong, S.P., Koberle, R., de Ruyter van Steveninck, R.R., Bialek, W., 1998. Entropy and neural spike trains. Phys. Rev. Lett. 80, 197.

Supper, J., Spangenberg, L., Planatscher, H., Dräger, A., Schröder, A., Zell, A., 2009. BowTieBuilder: modeling signal transduction pathways. BMC Syst. Biol. 3, 67.

Takemura, S.-Y., Bharioke, A., Lu, Z., Nern, A., Vitaladevuni, S., Rivlin, P.K., Katz, W.T., Olbris, D.J., Plaza, S.M., Winston, P., et al., 2013. A visual motion detection circuit suggested by *Drosophila* connectomics. Nature 500, 175–181.

Telesford, Q., Simpson, S.L., Burdette, J.H., Hayasaka, S., Laurienti, P.J., 2011a. The brain as a complex system: using network science as a tool for understanding the brain. Brain Connect. 1, 295–308.

Telesford, Q.K., Joyce, K.E., Hayasaka, S., Burdette, J.H., Laurienti, P.J., 2011b. The ubiquity of small-world networks. Brain Connect. 1, 367–375.

Terry, J.R., Benjamin, O., Richardson, M.P., 2012. Seizure generation: the role of nodes and networks. Epilepsia 53, e166–e169.

Thomas, C., Ye, F.Q., Irfanoglu, M.O., Modi, P., Saleem, K.S., Leopold, D.A., Pierpaoli, C., 2014. Anatomical accuracy of brain connections derived from diffusion MRI tractography is inherently limited. Proc. Natl. Acad. Sci. U. S. A. 111, 16574–16579.

Tomasi, D., Wang, G.J., Volkow, N.D., 2013. Energetic cost of brain functional connectivity. Proc. Natl. Acad. Sci. U. S. A. 110, 13642–13647.

Tononi, G., Cirelli, C., 2014. Sleep and the price of plasticity: from synaptic and cellular homeostasis to memory consolidation and integration. Neuron 81, 12–34.

Tononi, G., Edelman, G.M., Sporns, O., 1998. Complexity and coherency: integrating information in the brain. Trends Cogn. Sci. 2, 474–484.

Tononi, G., Sporns, O., Edelman, G.M., 1994. A measure for brain complexity: relating functional segregation and integration in the nervous system. Proc. Natl. Acad. Sci. U. S. A. 91, 5033–5037.

Tononi, G., Sporns, O., Edelman, G.M., 1999. Measures of degeneracy and redundancy in biological networks. Proc. Natl. Acad. Sci. U. S. A. 96, 3257–3262.

Torricelli, A., Contini, D., Pifferi, A., Caffini, M., Re, R., Zucchelli, L., Spinelli, L., 2014. Time domain functional NIRS imaging for human brain mapping. NeuroImage 85 (Pt 1), 28–50.

Tournier, J.-D., Mori, S., Leemans, A., 2011. Diffusion tensor imaging and beyond. Magn. Reson. Med. 65, 1532–1556.

Tournier, J.D., Calamante, F., Gadian, D.G., Connelly, A., 2004. Direct estimation of the fiber orientation density function from diffusion-weighted MRI data using spherical deconvolution. NeuroImage 23, 1176–1185.

Towlson, E.K., Vértes, P.E., Ahnert, S.E., Schafer, W.R., Bullmore, E.T., 2013. The rich club of the C. *elegans* neuronal connectome. J. Neurosci. 33, 6380–6387.

Traag, V.A., Bruggeman, J., 2009. Community detection in networks with positive and negative links. Phys. Rev. E 80, 036115.

Traag, V.A., Van Dooren, P., Nesterov, Y., 2011. Narrow scope for resolution-limit-free community detection. Phys. Rev. E 84, 016114.

Trusina, A., Rosvall, M., Sneppen, K., 2005. Communication boundaries in networks. Phys. Rev. Lett. 94 (23), 238701.

Tsodyks, M., Kenet, T., Grinvald, A., Arieli, A., 1999. Linking spontaneous activity of single cortical neurons and the underlying functional architecture. Science 286, 1943–1946.

Tuch, D.S., Reese, T.G., Wiegell, M.R., Wedeen, V.J., 2003. Diffusion MRI of complex neural architecture. Neuron 40, 885–895.

Tufte, E.R., 1990. Envisioning Information. Graphics Press, Cheshire, CT.

Tutte, W.T., 1963. How to draw a graph. Proc. Lond. Math. Soc. 13, 743–768.

Tzourio-Mazoyer, N., Landeau, B., Papathanassiou, D., Crivello, F., Etard, O., Delcroix, N., Mazoyer, B., Joliot, M., 2002. Automated anatomical labeling of activations in SPM using a macroscopic anatomical parcellation of the MNI MRI single-subject brain. NeuroImage 15, 273–289.

Vaishnavi, S.N., Vlassenko, A.G., Rundle, M., Snyder, A.Z., Mintun, M.A., Raichle, M.E., 2010. Regional aerobic glycolysis in the human brain. Proc. Natl. Acad. Sci. U. S. A. 107, 17757–17762.

Van De Ville, D., Seghier, M.L., Lazeyras, F., Blu, T., Unser, M., 2007. WSPM: wavelet-based statistical parametric mapping. NeuroImage 37 (4), 1205–1217.

van den Heuvel, M., Mandl, R., Hulshoff Pol, H., 2008a. Normalized cut group clustering of resting-state FMRI data. PLoS One 3, e2001.

van den Heuvel, M.P., Sporns, O., 2013a. Network hubs in the human brain. Trends Cogn. Sci. 17 (12), 683–696.

van den Heuvel, M.P., Sporns, O., 2011. Rich-club organization of the human connectome. J. Neurosci. 31, 15775–15786.

van den Heuvel, M.P., Sporns, O., 2013b. An anatomical substrate for integration among functional networks in human cortex. J. Neurosci. 33, 14489–14500.

van den Heuvel, M.P., Kahn, R.S., Goñi, J., Sporns, O., 2012. High-cost, high-capacity backbone for global brain communication. Proc. Natl. Acad. Sci. U. S. A. 109, 11372–11377.

van den Heuvel, M.P., Mandl, R.C.W., Stam, C.J., Kahn, R.S., Hulshoff Pol, H.E., 2010. Aberrant frontal and temporal complex network structure in schizophrenia: a graph theoretical analysis. J. Neurosci. 30, 15915–15926.

van den Heuvel, M.P., Scholtens, L.H., de Reus, M.A., 2016. Topological organization of connectivity strength in the rat connectome. Brain Struct. Funct. http://dx.doi.org/10.1007/s00429-015-0999-6.

van den Heuvel, M.P., Sporns, O., Collin, G., Scheewe, T., Mandl, R.C.W., Cahn, W., Goñi, J., Hulshoff Pol, H.E., Kahn, R.S., 2013a. Abnormal rich club organization and functional brain dynamics in schizophrenia. JAMA Psychiatry 70, 783–792.

van den Heuvel, M.P., Stam, C.J., Boersma, M., Pol, H.E.H., 2008b. Small-world and scale-free organization of voxel-based resting-state functional connectivity in the human brain. NeuroImage 43, 528–539.

van den Heuvel, M.P., Stam, C.J., Kahn, R.S., Hulshoff Pol, H.E., 2009. Efficiency of functional brain networks and intellectual performance. J. Neurosci. 29, 7619–7624.

van den Heuvel, M.P., van Soelen, I.L., Stam, C.J., Kahn, R.S., Boomsma, D.I., Hulshoff Pol, H.E., 2013b. Genetic control of functional brain network efficiency in children. Eur. Neuropsychopharmacol. 23, 19–23.

Van Essen, D.C., Ugurbil, K., 2012. The future of the human connectome. NeuroImage 62 (2), 1299–1310.

Van Essen, D.C., 1997. A tension-based theory of morphogenesis and compact wiring in the central nervous system. Nature 385, 313–318.

Van Essen, D.C., 2005. A population-average, landmark- and surface-based (PALS) atlas of human cerebral cortex. NeuroImage 28, 635–662.

Van Essen, D.C., 2013. Cartography and connectomes. Neuron 80, 775–790.

Van Essen, D.C., Smith, S.M., Barch, D.M., Behrens, T.E., Yacoub, E., Ugurbil, K., WU Minn HCP Consortium, 2013. The WU-Minn Human Connectome Project: an overview. NeuroImage 80, 62–79.

van Grootheest, D.S., Cath, D.C., Beekman, A.T., Boomsma, D.I., 2005. Twin studies on obsessive-compulsive disorder: a review. Twin Res. Hum. Genet. 8, 450–458.

Van Horn, J.D., Irimia, A., Torgerson, C.M., Chambers, M.C., Kikinis, R., Toga, A.W., 2012. Mapping connectivity damage in the case of Phineas Gage. PLoS One 7, e37454.

van Mieghem, P., 2012. Graph Spectra for Complex Networks. Cambridge University Press, Cambridge, MA.

van Wijk, B.C.M., Stam, C.J., Daffertshofer, A., 2010. Comparing brain networks of different size and connectivity density using graph theory. PLoS One 5, e13701.

Varela, F., Lachaux, J.P., Rodriguez, E., Martinerie, J., 2001. The brainweb: phase synchronization and large-scale integration. Nat. Rev. Neurosci. 2, 229–239.

Variano, E.A., Lipson, H., 2004. Networks, dynamics, and modularity. Phys. Rev. Lett. 92, 188701.

Varoquaux, G., Baronnet, F., Kleinschmidt, A., Fillard, P., Thirion, B., 2010. Detection of brain functional-connectivity difference in post-stroke patients using group-level covariance modeling. Med. Image Comput. Comput. Assist. Interv. 13 (Pt 1), 200–208.

Varoquaux, G., Craddock, R.C., 2013. Learning and comparing functional connectomes across subjects. NeuroImage 80, 405–415.

Varshney, L.R., Chen, B.L., Paniagua, E., Hall, D.H., Chklovskii, D.B., 2011. Structural properties of the *Caenorhabditis elegans* neuronal network. PLoS Comput. Biol. 7 (2), e1001066.

Ventura-Antunes, L., Mota, B., Herculano-Houzel, S., 2013. Different scaling of white matter volume, cortical connectivity, and gyrification across rodent and primate brains. Front. Neuroanat. 7, 3.

Verplaetse, P., Dambre, J., Stroobandt, D., Campenhout, J.V., 2001. On partitioning vs. placement rent properties. In: International Workshop on System-Level Interconnect Prediction. pp. 33–40.

Vértes, P.E., Alexander-Bloch, A.F., Gogtay, N., Giedd, J.N., Rapoport, J.L., Bullmore, E.T., 2012. Simple models of human brain functional networks. Proc. Natl. Acad. Sci. 109 (15), 5868–5873.

Vértes, P.E., Bullmore, E.T., 2015. Growth connectomics—the organization and reorganization of brain networks during normal and abnormal development. J. Child Psychol. Psychiatry 56 (3), 299–320.

Vicente, R., Gollo, L.L., Mirasso, C.R., Fischer, I., Pipa, G., 2008. Dynamical relaying can yield zero time lag neuronal synchrony despite long conduction delays. Proc. Natl. Acad. Sci. U. S. A. 105, 17157–17162.

Viger, F., Latapy, M., 2005. Random generation of large connected simple graphs with prescribed degree distribution. Comput. Comb. 3595, 440–449.

Vincent, J.L., Patel, G.H., Fox, M.D., Snyder, A.Z., Baker, J.T., Van Essen, D.C., Zempel, J.M., Snyder, L.H., Corbetta, M., Raichle, M.E., 2007. Intrinsic functional architecture in the anaesthetized monkey brain. Nature 447, 83–86.

Vinh, N.X., Epps, J., Bailey, J., 2010. Information theoretic measures for clusterings comparison: variants, properties, normalization and correction for chance. J. Mach. Learn. Res. 11, 2837–2854.

Vogelstein, J.T., Packer, A.M., Machado, T.A., Sippy, T., Babadi, B., Yuste, R., Paninski, L., 2010. Fast nonnegative deconvolution for spike train inference from population calcium imaging. J. Neurophysiol. 104, 3691–3704.

von Economo, C., 1929. Cytoarchitectonics of the Human Cerebral Cortex. Oxford University Press, London.

von Monakow, C., 1969. Diaschisis. In: Pribram, K.H. (Ed.), Mood, States and Mind. Penguin, Baltimore, MD, pp. 27–36.

von Stein, A., Sarnthein, J., 2000. Different frequencies for different scales of cortical integration: from local gamma to long range alpha/theta synchronization. Int. J. Psychophysiol. 38, 301–313.

Wang, C., Lizardo, O., Hachen, D., 2014. Algorithms for generating large-scale clustered random graphs. Netw. Sci. 2 (3), 403–415.

Wang, J., Wang, L., Zang, Y., Yang, H., Tang, H., Gong, Q., Chen, Z., Zhu, C., He, Y., 2009. Parcellation-dependent small-world brain functional networks: a resting-state fMRI study. Hum. Brain Mapp. 30, 1511–1523.

Wang, P., Lü, J., Yu, X., 2014. Identification of important nodes in directed biological networks: a network motif approach. PLoS One 9, e106132. 15.

Wang, Q., Sporns, O., Burkhalter, A., 2012. Network analysis of corticocortical connections reveals ventral and dorsal processing streams in mouse visual cortex. J. Neurosci. 32, 4386–4399.

Wang, S.P., Pei, W.J., 2008. First passage time of multiple Brownian particles on networks with applications. Physica A 387 (18), 4699–4708.

Warren, D.E., Power, J.D., Bruss, J., Denburg, N.L., Waldron, E.J., Sun, H., Petersen, S.E., Tranel, D., 2014. Network measures predict neuropsychological outcome after brain injury. Proc. Natl. Acad. Sci. U. S. A. 111, 14247–14252.

Watts, D.J., 2002. A simple model of global cascades on random networks. Proc. Natl. Acad. Sci. U. S. A. 99, 5766–5771.

Watts, D.J., Strogatz, S.H., 1998. Collective dynamics of "small-world" networks. Nature 393, 440–442.

Wedeen, V.J., Hagmann, P., Tseng, W.-Y.I., Reese, T.G., Weisskoff, R.M., 2005. Mapping complex tissue architecture with diffusion spectrum magnetic resonance imaging. Magn. Reson. Med. 54, 1377–1386.

Weimann, J.M., Marder, E., 1994. Switching neurons are integral members of multiple oscillatory networks. Curr. Biol. 4, 896–902.

Weiner, N., 1948. Cybernetics: Or Control and Communication in the Animal and the Machine. Hermann & Cie/MIT Press, Paris/Cambridge, MA, ISBN: 978-0-262-73009-9.

Welker, W., 1990. Why does cerebral cortex fissure and fold? A review of determinants of sulci an gyri. In: Jones, E.G., Peters, A. (Eds.), Cerebral Cortex. Plenum Press, New York, NY, pp. 3–136.

Wernicke, C., 1906. Grundriss der Psychiatrie in klinischen Vorlesungen. Thieme, Leipzig.

Wernicke, C., 1970. The aphasic symptom-complex: a psychological study on an anatomical basis. Arch. Neurol. 22 (3), 280.

White, J.G., Southgate, E., Thomson, J.N., Brenner, S., 1986. The structure of the nervous system of the nematode Caenorhabditis elegans. Philos. Trans. R. Soc. Lond. B Biol. Sci. 314, 1–340.

Wig, G.S., Laumann, T.O., Petersen, S.E., 2014. An approach for parcellating human cortical areas using resting-state correlations. NeuroImage 93, 276–291.

Wig, G.S., Schlaggar, B.L., Petersen, S.E., 2011. Concepts and principles in the analysis of brain networks. Ann. N. Y. Acad. Sci. 1224, 126–146.

Wilke, C., Worrell, G., He, B., 2010. Graph analysis of epileptogenic networks in human partial epilepsy. Epilepsia 52, 84–93.

Winterer, G., Weinberger, D.R., 2004. Genes, dopamine and cortical signal-to-noise ratio in schizophrenia. Trends Neurosci. 27, 683–690.

Wolf, L., Goldberg, C., Manor, N., Sharan, R., Ruppin, E., 2011. Gene expression in the rodent brain is associated with its regional connectivity. PLoS Comput. Biol. 7 (5), e1002040.

Womelsdorf, T., Valiante, T.A., Sahin, N.T., Miller, K.J., Tiesinga, P., 2014. Dynamic circuit motifs underlying rhythmic gain control, gating and integration. Nat. Rev. Neurosci. 17, 1031–1039.

Wong, E., Baur, B., Quader, S., Huang, C.H., 2012. Biological network motif detection: principles and practice. Brief. Bioinform. 13, 202–215.

Xia, M., He, Y., 2011. Magnetic resonance imaging and graph theoretical analysis of complex brain networks in neuropsychiatric disorders. Brain Connect. 1 (5), 349–365.

Xie, J., Kelley, S., Szymanski, B.K., 2013. Overlapping community detection in networks. ACM Comput. Surv. 45, 1–35.

Xie, J., Szymanski, B.K., Liu, X., 2011. SLPA: uncovering overlapping communities in social networks via a speaker-listener interaction dynamic process. In: IEEE 11th International Conference on Data Mining Workshops (ICDMW), pp. 344–349.

Yatsenko, D., Josić, K., Ecker, A.S., Froudarakis, E., Cotton, R.J., Tolias, A.S., 2015. Improved estimation and interpretation of correlations in neural circuits. PLoS Comput. Biol. 11, e1004083. 28.

Yeo, B.T., Krienen, F.M., Sepulcre, J., Sabuncu, M.R., Lashkari, D., Hollinshead, M., Roffman, J.L., Smoller, J.W., Zollei, L., Polimeni, J.R., et al., 2011. The organization of the human cerebral cortex estimated by intrinsic functional connectivity. J. Neurophysiol. 106, 1125–1165.

Yeterian, E.H., Pandya, D.N., 1991. Prefrontostriatal connections in relation to cortical architectonic organization in rhesus monkeys. J. Comp. Neurol. 312, 43–67.

Yip, A.M., Horvath, S., 2007. Gene network interconnectedness and the generalized topological overlap measure. BMC Bioinf. 8, 22.

Young, M.P., 1992. Objective analysis of the topological organization of the primate cortical visual system. Nature 358, 152–155.

Young, M.P., Hilgetag, C.C., Scannell, J.W., 2000. On imputing function to structure from the behavioural effects of brain lesions. Philos. Trans. R. Soc. Lond. B Biol. Sci. 355, 147–161.

Young, M.P., Scannell, J.W., O'Neill, M.A., Hilgetag, C.C., Burns, G., Blakemore, C., 1995. Nonmetric multidimensional scaling in the analysis of neuroanatomical connection data and the organization of the primate cortical visual system. Philos. Trans. R. Soc. Lond. B Biol. Sci. 348, 281–308.

Yu, S., Huang, D., Singer, W., Nikolic, D., 2008. A small world of neuronal synchrony. Cereb. Cortex 18, 2891–2901.

Yule, G.U., 1925. A mathematical theory of evolution, based on the conclusions of Dr. J. C. Willis, F. R.S. Philos. Trans. R. Soc. Lond. B Biol. Sci. 213, 21–87.

Yuste, R., 2015. From the neuron doctrine to neural networks. Nat. Rev. Neurosci. 16, 487–497.

Zalesky, A., 2008. DT-MRI fiber tracking: a shortest paths approach. IEEE Trans. Med. Imaging 27 (10), 1458–1471.

Zalesky, A., 2009. To burst or circuit switch? IEEE/ACM Trans. Networking 17 (1), 305–318.

Zalesky, A., Breakspear, M., 2015. Towards a statistical test for functional connectivity dynamics. NeuroImage 114, 466–470.

Zalesky, A., Fornito, A., 2009. A DTI-derived measure of cortico-cortical connectivity. IEEE Trans. Med. Imaging 28, 1023–1036.

Zalesky, A., Cocchi, L., Fornito, A., Murray, M.M., Bullmore, E., 2012a. Connectivity differences in brain networks. NeuroImage 60 (2), 1055–1062.

Zalesky, A., Fornito, A., Bullmore, E., 2012b. On the use of correlation as a measure of network connectivity. NeuroImage 60 (4), 2096–2106.

Zalesky, A., Fornito, A., Bullmore, E.T., 2010a. Network-based statistic: identifying differences in brain networks. NeuroImage 53, 1197–1207.

Zalesky, A., Fornito, A., Cocchi, L., Gollo, L.L., Breakspear, M., 2014. Time-resolved resting-state brain networks. Proc. Natl. Acad. Sci. U. S. A. 111, 10341–10346.

Zalesky, A., Fornito, A., Egan, G.F., Pantelis, C., Bullmore, E., 2012c. The relationship between regional and inter-regional functional connectivity deficits in schizophrenia. Hum. Brain Mapp. 33 (11), 2535–2549.

Zalesky, A., Fornito, A., Harding, I.H., Cocchi, L., Yucel, M., Pantelis, C., Bullmore, E.T., 2010b. Whole-brain anatomical networks: does the choice of nodes matter? NeuroImage 50, 970–983.

Zalesky, A., Fornito, A., Seal, M.L., Cocchi, L., Westin, C.F., Bullmore, E.T., Egan, G.F., Pantelis, C., 2011. Disrupted axonal fiber connectivity in schizophrenia. Biol. Psychiatry 69 (1), 80–89.

Zalesky, A., Vu, H.L., Rosberg, Z., Wong, E.W.M., Zukerman, M., 2007. OBS contention resolution performance. Perform. Eval. 64 (4), 357–373.

Zamfirescu, T., 1976. On longest paths and circuits in graphs. Math. Scand. 38, 211–239.

Zamora-López, G., Zhou, C., Kurths, J., 2010. Cortical hubs form a module for multisensory integration on top of the hierarchy of cortical networks. Front. Neuroinform. 4, 1.

Zatorre, R.J., Fields, R.D., Johansen-Berg, H., 2012. Plasticity in gray and white: neuroimaging changes in brain structure during learning. Nat. Neurosci. 15, 528–536.

Zeki, S., 2001. Localization and globalization in conscious vision. Annu. Rev. Neurosci. 24, 57–86.

Zeng, L.L., Shen, H., Liu, L., Wang, L., Li, B., Fang, P., Zhou, Z., Li, Y., Hu, D., 2012. Identifying major depression using whole-brain functional connectivity: a multivariate pattern analysis. Brain 135 (Pt 5), 1498–1507.

Zhang, X., Martin, T., Newman, M.E.J., 2015. Identification of core-periphery structure in networks. Phys. Rev. E 91, 032803.

Zhou, H., 2003. Network landscape from a Brownian particle's perspective. Phys. Rev. E 67, 041908.

Zhou, J., Gennatas, E.D., Kramer, J.H., Miller, B.L., Seeley, W.W., 2012. Predicting regional neurodegeneration from the healthy brain functional connectome. Neuron 73, 1216–1227.

Zhou, S., Mondragón, R.J., 2004. The rich-club phenomenon in the Internet topology. IEEE Commun. Lett. 8, 180–182.

Zilles, K., Palomero-Gallagher, N., Grefkes, C., Scheperjans, F., Boy, C., Amunts, K., Schleicher, A., 2002. Architectonics of the human cerebral cortex and transmitter receptor fingerprints: reconciling functional neuroanatomy and neurochemistry. Eur. Neuropsychopharmacol. 12, 587–599.

Zingg, B., Hintiryan, H., Gou, L., Song, M.Y., Bay, M., Bienkowski, M.S., Foster, N.N., Yamashita, S., Bowman, I., Toga, A.W., Dong, H.W., 2014. Neural networks of the mouse neocortex. Cell 156, 1096–1111.

Zipf, G., 1936. The Psychobiology of Language. Routledge, London.

Zipf, G., 1949. Human Behavior and the Principle of Least Effort. Addison-Wesley, New York, NY.

Zuo, X.N., Ehmke, R., Mennes, M., Imperati, D., Castellanos, F.X., Sporns, O., Milham, M.P., 2012. Network centrality in the human functional connectome. Cereb. Cortex 22, 1862–1875.

Index

Note: Page numbers followed by *b* indicate boxes, *f* indicate figures, and *t* indicate tables.